SOME CHARACTERISTIC OPTICAL SPECTRA

(a) CONTINUOUS SPECTRUM OF FILAMENT LAMP

(b) LINE SPECTRUM OF HYDROGEN ATOM

(c) LINE SPECTRUM OF NEON ATOM

(d) ABSORPTION LINE SPECTRUM OF SUN

(e) BAND SPECTRUM OF HYDROGEN MOLECULE

The spectra obtained when light is dispersed by prisms or diffraction gratings can carry information on the composition, physical state, distance and motion of the light-emitting body and the presence of gravitational, electric, or magnetic fields. This is possible even when the source is an otherwise inaccessible astronomical object.

Optical spectra are classified as emission spectra or absorption spectra, depending on whether the critical features are bright or dark. All spectra shown above are emission spectra except (d). All are line spectra except (a) and (e). Under high resolution (e) is found to consist of individual, sharp emission lines. Band spectra and continuous spectra are also obtainable in absorption.

For further details on spectra, review pages 532–535.

PHYSICS

Principles and Insights

About the Author

Ira M. Freeman received his doctor's degree at the
University of Chicago, after which he spent two years in
Frankfurt, Germany as a Fellow of the Institute of
International Education, working in experimental optical
spectroscopy. As a member of the Princeton expedition
in 1945, he took the first color photographs of a total
solar eclipse. In 1950–51, he was on leave from his staff
position at Rutgers University to serve as Program
Specialist in natural sciences at UNESCO in Paris.
Dr. Freeman was the first scientist to propose the idea of
dumping dangerous radioactive wastes in space. He is a
fellow of the American Association for the Advancement
of Science and a member of the Science Policy
Foundation of London. His principal professional interests
are in the fields of theoretical physics and public
education in science.

Second Edition

PHYSICS
Principles and Insights

IRA M. FREEMAN   *Professor of Physics*
Rutgers · The State University of New Jersey

McGRAW-HILL BOOK COMPANY

NEW YORK   ST. LOUIS   SAN FRANCISCO   DÜSSELDORF
JOHANNESBURG   KUALA LUMPUR   LONDON
MEXICO   MONTREAL   NEW DELHI   PANAMA   RIO DE JANEIRO
SINGAPORE   SYDNEY   TORONTO

PHYSICS: Principles and Insights

Library of Congress Cataloging in Publication Data

Freeman, Ira Maximilian, 1905–
    Physics.

    Bibliography: p.
    1. Physics. I. Title.
QC21.2.F72 1973   530   72-7329
ISBN 0-07-021938-9

This book was set in Vega Light by York Graphic Services,
Inc. The editors were Jack L. Farnsworth and Stuart D.
Boynton; the designer was J. E. O'Connor; and the
production supervisor was John A. Sabella. The drawings
were done by York Graphic Services, Inc.
The printer was The Murray Printing Company; the binder,
Rand McNally & Company.

CONTENTS

Part four   ELECTRICITY

Part five   THE FUNDAMENTAL STRUCTURE OF MATTER

APPENDIX

BIBLIOGRAPHY

ANSWERS to Selected Problems and Questions

INDEX

PREFACE

"Physics: Principles and Insights" is intended for a first course in physics for students who do not plan to specialize in science but wish to gain an understanding of the subject appropriate to well-informed members of society. One of the most distinguished exponents of this kind of understanding was Enrico Fermi. His ideas are aptly expressed in a review of his Collected Papers, vol. II, that appeared in the *Scientific American,* June, 1966:

> . . . his approach, as some of his students and co-workers have remarked, was intuitive and geometrical rather than analytical, which made his ideas accessible even to those who were not versed in the art of complex mathematical transactions. Fermi never forgot that he was dealing with the physical world, where there is a basic need not to overintellectualize, and that it is not enough, if one wishes to understand and control what is happening, merely to satisfy theoretical passions by devising symmetrical logical and solvable equations. Fermi wanted the books to balance, but he also wanted to keep within the budget of reality.

It is my hope that users of this book will recognize in it the spirit of Fermi's view. In applying such ideas, there must be reasonable limitation on the number of topics included as well as on the extent to which they are detailed.

The cultivation of physical intuition, a prime objective of this text, requires an effort to get to the heart of the main concepts of physics by making results seem plausible as well as logical—precisely what the word "insights" in the title of the book is intended to convey. At the same time, it is frequently necessary to point out the pitfalls of blind reliance on what is often called "common sense." Since no serious treatment of the subject can sidestep the quantitative and analytical aspect completely, there is a moderate amount of mathematical expression and development in this book, but it is not assumed that the student has had mathematics experience beyond simple trigonometry.

A noteworthy aim of physics is the devising of successful conceptual models of physical phenomena, and much that modern physics has accomplished toward an understanding of the universe can be credited to the great extent to which this has become feasible. In addition, the progress of physics has been advanced significantly by the possibility of making operational definitions, and the student is reminded of this throughout the text.

The place of history in introductory physics instruction continues to be widely debated. On this point, I have sought to convey a sense of the man-made nature of our science without detailing the false leads and blind alleys that sometimes harassed the pioneers of physics. Suggestions for pursuing the historical aspects at greater length are given by references for further reading.

Relativity and quantum theory are integral, established parts of physics, and I believe that even in an elementary course these concepts must permeate the entire discussion rather than be relegated to separate chapters. Some relativity ideas and relations are brought in without formal proof as early as the chapter on kinematics. Quantum notions cannot be introduced as promptly but can be anticipated in a general way by indicating certain instances of quantization occurring in classical physics. Here, as well as throughout the rest of the book, I have encouraged a progressive development of the student's ability to handle abstract ideas while keeping the intellectual demands within the realm of the possible.

Classical mechanics demands a rather sophisticated level of comprehension if it is to be properly understood and assimilated, and it would be a mistake to plunge abruptly into this topic, especially for a student of limited motivation or capacity for scientific thinking. For this reason, I have included an early chapter in which the physical and chemical behavior of matter is introduced mainly in a descriptive way. The remaining topics follow in a workable natural sequence, involving no great surprises.

At the end of the main part of each chapter there is a programmed review designed to test the student's knowledge of the key concepts, laws, and definitions appearing in the text, followed by a brief list of reading references and a set of questions and problems having rather a broad range of difficulty.

Rationalized mks units, in the form of the International System, are used throughout. English units appear to a limited extent at the beginning in order to provide an association with more common experience.

This revised edition takes into account suggestions and comments contributed by users. It was feasible to incorporate many of these points without significantly changing the general arrangement and treatment of the material. In response to a number of requests, a concise treatment of ray optics has been added in the new edition. The overall length of the text has not been increased.

In keeping with current trends and interests, brief expositions of some environmental questions related to physical principles and methods have been included as an adjunct to the main text. Examples are: Environmental Pollution and the Phases of Matter, Global Energy Needs and Resources, Noise Pollution, Radiation Hazards, etc.

A student who meets physics concepts for the first time should be helped by language that is clear, unambiguous, and as free of technical jargon as possible, so that he may concentrate on assimilating ideas and principles. To this end, the entire text was carefully reexamined and restyled for the new edition. In addition, redesign of the book has enabled the ground to be covered in fewer pages.

IRA M. FREEMAN

Part One
SPACE, MATTER, AND MOTION

Photograph facing page 1
Vibration patterns produced in a liquid film by means of sound waves are revealed by a shadow technique. A given pattern can be made to break suddenly into a new configuration by increasing the sound intensity. (Courtesy of Hans Jenny and J. C. Stuten, Dornach, Switzerland.)

Chapter 1
THE WAY OF SCIENCE

In today's world, science and its applications are dominant forces, changing our way of life and our institutions with unprecedented intensity and speed. This circumstance alone makes some knowledge of the nature of science indispensable to every thoughtful person. We must become aware not only of the potentialities, past accomplishments, and future possibilities of science but of its limitations and misuses as well.

This introductory chapter presents a brief description of the nature of the scientific venture, set against a compact sketch of the historical development of the subject that traces the eventual emergence of physics as a separate science.

THE SCIENTIFIC ENTERPRISE

1.1 SCIENCE IN TODAY'S WORLD

See Ref. 1.11 (end of this chapter).

The present pace of scientific activity has never before been experienced in the history of civilization. From its quiet and insignificant beginnings in the minds of a few solitary thinkers, science became the pursuit of dabblers, amateurs, and gentlemen with time on their hands and ultimately evolved into a fully developed profession. Today it often attains the status of big business—and big government—in projects involving tens of thousands of scientists and multibillion-dollar budgets. In the three centuries since the beginning of modern science, scientific activity measured by the number of its practitioners or their rate of publication has been doubling about every 15 years (Fig. 1.1). The upsweep has been so rapid that, according to reliable estimates, some 80 to 90 percent of all the scientists who ever lived are alive today, so that almost all the scientific research ever undertaken has been conducted within the lifetimes of present-day scientists.

The term scientist was coined only a little more than a century ago.

The public tends to confuse size with greatness, so that modern science has sometimes been given a venerated position in contemporary life. Occasionally the reverse is true when the advance of science is blamed for

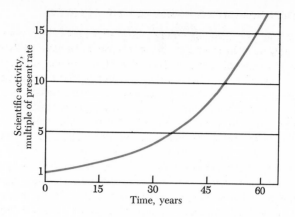

FIGURE 1.1 The ever-increasing rise of scientific activity is represented by a curve of the form shown here. If the present rate were to continue, there would be about a tenfold increase every half century. Some estimates make the rate even higher. A leveling off is, of course, inevitable but has not yet appeared.

some of the world's social and political difficulties. At least part of the trouble is a failure to distinguish between science itself and the technology based on it.

To many people science means color television, jet airplanes, polio vaccine, space travel, and nuclear power plants. All these have little to do directly with the true spirit of science, which is primarily a way of viewing the natural world rather than a way of shaping it to serve man's needs and desires. Warren Weaver makes this point in an effective statement:

Warren Weaver is a distinguished mathematician, administrator, and interpreter of science to the public. This quotation is from "Goals for Americans" (1960).

For science is not technology, it is not gadgetry, it is not some mysterious cult, it is not a great mechanical monster. Science is an adventure of the human spirit. It is essentially an artistic enterprise, stimulated largely by curiosity, served largely by disciplined imagination, and based largely on faith in the reasonableness, order, and beauty of the universe of which man is a part.

In discussing the "two cultures," the British novelist and scientist C. P. Snow (see Ref. 1.4) warns of the serious consequences of the split that exists between the modern scientific culture and the traditional literary culture. Snow contends that the hope for reduction of this momentous threat to the integrity of our society lies in education reforms for people on both sides of the intellectual chasm. He believes that the disciples of the literary culture are in the main more culpable than those on the scientific side, largely because of a tradition that science is essentially a difficult subject, reserved for the few. In support of his view he says:

Nearly all intelligent people can learn something about science and scientists if they are brought up against them properly. It is very stupid to attempt to make everybody into technologists; but it is essential that everybody, including the technologists themselves, should understand something of the intellectual and human meaning of what the technologists are about. I don't think that that task is beyond us, though it will need a drastic rethinking about education, both in the United States and Great Britain.

1.2 SOME DISTINGUISHING FEATURES OF SCIENCE

Science is not the only way of looking at the universe. The poet, the musician, the artist, and the theologian have different but equally valid points of view, each deserving of respect. One difference between science and other fields of endeavor is the characteristic striving of scientists to discover pattern and order in nature, leading to statements on which universal agreement can be obtained. This point, to be discussed in more detail below, is mentioned here in order to contrast it with the more personal, individual involvement connected with creative effort and appreciation in, say, the arts.

There is a second fundamental difference between science and other intellectual or aesthetic enterprises. The long history of the development of scientific thought is by no means an unbroken series of superb triumphs and brilliant achievements. Rather, like any other human undertaking, it is marked by frequent errors and many failures, by regressions as well as advances. A closeup view of the rapidly rising curve of scientific accomplishment would reveal that it is jagged indeed.

George Sarton (1884–1956), outstanding American historian of science, believed that "the acquisition and systemization of positive knowledge is the only human activity that is truly cumulative and progressive."

In spite of such fluctuations, the progress of science is unmistakable. The main reason for this is a feature that distinguishes science from almost all other endeavors: *Scientific knowledge is cumulative and sequential;* the bulk of verified truth is handed on from one generation of scientists to the next. This body of information may undergo much modification and revision in the process, but the important point is that it continues to evolve as it grows. It is not possible to say this in a broad sense for any other branch of knowledge. In such subjects as political history there is a continually increasing body of factual information; however, this type of growth differs importantly from the development of science in that the latter alone deserves to be called sequential. This characteristic is the true mark of science, quite apart from the material benefits it offers.

As an introduction to the procedures that science has developed for investigating and understanding nature, there follows a brief and compact sketch of the early history of scientific thought. Many additional details will be brought out later at appropriate places in this book.

ORIGINS AND GROWTH OF PHYSICAL SCIENCE

Activities that would now be called scientific arose when, in some remote era, people began to show curiosity concerning their environment and to record what they saw. In time, reflection on these observations led to the idea that nature is knowable and dependable, and that as a result predictions could be made which would give man a certain degree of control over his environment. This approach was a great advance over the belief in magic and supernatural forces with which primitive man tried to influence natural events. Thus the new way of viewing the universe was characterized by an implicit confidence in the idea that nature is uniform and orderly—that it operates according to "laws."

1.3 THE RISE OF SCIENCE

The actual birth of science took place in prehistoric times, probably in Egypt and Babylonia, more than 2,000 years before our era when practical needs induced people to construct calendars and devise methods of measuring land. In the process, they accumulated astronomical information that enabled them to measure time. They were also able to predict eclipses of the sun and determine the length of the year to within 2 or 3 minutes.

True progress in science did not begin until about the sixth century B.C. with the rise of the Greek civilization. The next 500 years was the age of the great philosophers of antiquity—Thales, Pythagoras, Plato, Aristotle, Archimedes, and others.

Aristotle (384–322 B.C.) laid the foundations for the scientific outlook on questions about nature. His influence in many fields—including politics, ethics, rhetoric, and metaphysics as well as science—was still evident 20 centuries after his time. One of Aristotle's greatest and more lasting contributions to knowledge was his promotion of inductive reasoning. The synthetic, or **inductive,** method of reasoning is the method of proceeding from

Aristotle. (*Courtesy of Historisches Bildarchiv Handke-Bad Berneck.*)

particular facts to generalizations concerning them. He stated his thesis in this way: "We must not accept a general principle from logic only, but we must prove its applicability to each fact; for it is in facts that we must seek general principles, and these must always accord with facts, from which induction is the pathway to general laws."

However, in much of his work Aristotle apparently failed to follow his own doctrine. Too often he showed himself willing to draw hasty inferences based on insufficient evidence and to cling to a previous conclusion in the face of new observations that contradicted it. This was in line with the prevailing idea that the value of an explanation was to be measured only by its universality and its ability to satisfy the mind.

Of all the prominent Greek thinkers, perhaps only Archimedes may be classed as scientific in the modern sense. He discovered some of the basic laws governing mechanisms and floating bodies and brought them to bear on a variety of practical problems. To Archimedes we owe the first applications of mathematics to the description of nature. He was very much in advance of his time.

Except for the work of Archimedes, Greek science, with its preference for "armchair philosophy" over experiment, had a certain sterility and left few specific results of lasting value. But whatever the shortcomings of their science, the Greeks made two outstanding contributions: They originated the concept of *pure science,* which is the quest after knowledge for its own sake, and they developed the use of *mathematical reasoning* as the first step toward exact science.

Roman civilization contributed little to the development of science, since the people were interested almost exclusively in practical applications. They built great libraries to preserve the results of Greek culture but added little that was new in scientific thought. By A.D. 500 the Roman Empire began to disintegrate, and the Arabian civilization that followed did little more than maintain the Greek body of learning.

In this brief survey only two names need be mentioned in the entire period from the Greeks to the Renaissance: Ptolemy and Alhazen. Claudius Ptolemy of Alexandria, who lived in the second century, made original investigations in optics, geography, and astronomy. His greatest undertaking was to collect the known astronomical knowledge of that time in the form of an encyclopedic work called the "*Almagest*" (A.D. 139). The Arabian Alhazen (965?–1038) contributed importantly to mathematics, astronomy, and especially to many phases of optics.

The prevailing intellectual influence during the Middle Ages was the Church. A fusion of its teachings with those of Aristotle, brought about largely by St. Thomas Aquinas in the thirteenth century, developed into the great movement known as Scholasticism, which dominated the important centers of learning in Europe. It established Aristotle as the great authority in science.

1.4   THE RENAISSANCE; WORK OF COPERNICUS

The turbulent transitional period between medieval and modern times—the Renaissance—was marked by intense intellectual activity. Such men as Copernicus, Kepler, Galileo, and Newton completely freed science from the

(a)   (b)

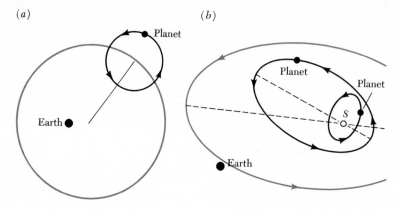

FIGURE 1.2 (a) In one scheme used by Ptolemy, a planet is assumed to move uniformly around a circle whose center, in turn, mover uniformly around a larger circle. The earth is located near, but not exactly at, the center of the larger circle. (b) In the copernican system the orbit of each planet, including the earth, is an ellipse with the sun occupying one focus. This diagram is not drawn to scale.

domination of aristotelian thought and introduced the modern scientific outlook based on observation and experiment.

First in importance among the scientific achievements of the Renaissance was the development of the view that the sun, rather than the earth, is the central body in our system of sun, moon, and planets. Up to the beginning of the sixteenth century, the prevailing idea was still that of an earth-centered universe, as described by Ptolemy. He pictured the earth at "the center of the heavens," with the stars located on a sphere that turns once a day around a fixed axis. To account for the observed motions of the planets, which appear to move relative to the background of the stars, he was forced to resort to a succession of devices (Fig. 1.2a), stubbornly holding to the traditional belief that only uniform motions along circular paths were admissible. Successive refinements enabled this scheme to account for more and more of the observed features of planetary motions, and it gained wide acceptance in spite of its complexity.

Copernicus' manuscript, "De orbium coelestium revolutionibus," was printed in 1543, after its author's death.

Ptolemy's earth-centered theory of the structure of the universe remained the accepted idea for about 13 centuries. Then the Polish scholar Nikolaus Copernicus (1473–1543) saw that a much simpler interpretation of observed planetary motions could be obtained by taking the sun as the central body, giving the earth the status of a mere planet revolving about the sun, together with the other planets (Fig. 1.2b). Resulting from this degradation of the earth's role, a great wave of opposition and harsh criticism confronted Copernicus' views when, near the end of his life, he consented

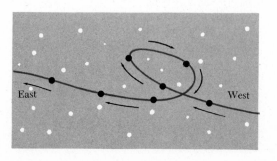

FIGURE 1.3 Part of the path of Venus as seen against the background of the stars. Positions of the planet are shown at one-month intervals. For some weeks, the planet is seen to be in retrograde (east-to-west) motion. This could be explained on the ptolemaic system but has a simpler interpretation on the copernican system as an effect of the relative motion of the earth and the observed planet as they go around the sun.

to their publication. In the present age of freedom of thought and expression it is hard to appreciate the courage that was required to advance an idea so contrary to current belief.

1.5 KEPLER, GALILEO, AND NEWTON

In addition to his work in astronomy, Kepler was obliged to practice astrology while holding a royal appointment in Prague. See Ref. 1.3.

Copernicus' sun-centered theory of what is now called the solar system was only a qualitative, pictorial description. The German astronomer Johannes Kepler (1571–1630) became interested in trying to find a general mathematical formulation of the copernican system. Using the very extensive and accurate observations of planetary positions made by his teacher, the Danish astronomer Tycho Brahe, he puzzled over the problem for nearly a quarter of a century, to emerge finally with a satisfactory solution. This solution was an empirical one, the result of trying various methods of combining the observed planetary positions rather than a deduction from a preconceived scheme. It agreed with the observations better than the ptolemaic formulation did.

Galileo. (*Yerkes Observatory.*)

The key to Kepler's success was his willingness to abandon the traditional idea of circular paths when he realized that the apparently complicated motions of the planets could be reduced to ordered simplicity by assuming that the orbit of each body is a curve called an ellipse. By 1619 he had stated the results of his work in the form of three laws (see Chap. 8). More than half a century later, Newton was able to make a momentous advance and actually deduce Kepler's laws by assuming the existence of a force of attraction between the sun and each planet. The name of this force is *gravitation* (Chap. 8).

For bio-graphical material on Galileo, see Refs. 1.2 and 1.7.

The true spirit of modern scientific inquiry had its first clear expression in the work of the gifted Italian scientist Galileo Galilei (1564–1642), who was born in the same year as Shakespeare. Galileo insisted on systematic observation and experimentation as the essential basis of scientific work, and this is the outstanding characteristic of the science of today. It is often said that he replaced the question "Why?", which Aristotle put to nature, with "How?". Although his own observations with the telescope confirmed

Newton. (*British Information Service.*)

Roger Bacon (1214?– 1294?), who urged the impor- tance of ex- perimenta- tion 400 years ear- lier, was twice impris- oned for his views.

many features of the copernican system, Galileo was forced to deny his support of these ideas when he came into serious disagreement with the authority of the Church (Ref. 1.7). Many aspects of Galileo's work in physics will be discussed in some detail in following chapters.

Newton was born in England in 1642, within a year of Galileo's death. Although men such as Copernicus and Kepler had launched the modern scientific method, it remained for Newton to bring it to full development. Galileo had gone a great part of the way, but Newton completed the process by confirming theory by additional experimentation. Moreover, Newton's use of mathematical reasoning in shaping theory was much more profound and extensive than that of his predecessors. With his keen insight and sound judgment he was able to lay the foundations of mechanics—and there- fore a great part of physical science—as well as make contributions of first importance in optics and pure mathematics. Newton's genius for synthesizing and generalizing experience has rarely been equaled in the history of human thought, and even a brief look at some of his original writings will show the breadth of his interests and his mastery of scientific exposition.

See, for example, Ref. 1.6.

1.6   THE SCOPE OF PHYSICS

A century ago there was no clear-cut division between the physical sciences as we know them now—physics, chemistry, astronomy, geology, meteor- ology, etc. Instead, all that was known in these fields was included in the single term *natural philosophy.* Before long, differentiation and splitting off began: Astronomy and chemistry developed methods and procedures of their own, and at the same time the invention of new means of motive power such as the steam engine and the electric motor gave rise to the profession of engineering. Of the original content of natural philosophy there remained a number of seemingly miscellaneous topics—mechanics, heat, sound, electricity, magnetism, light—and these came to be called, collectively, *physics.*

The term physics de- rives from the Greek word phys- ike, mean- ing the sci- ence or knowledge of nature.

The division of the subject into such branches is largely a matter of convenience because the boundaries between some sections are diffuse or almost nonexistent. There are many interrelations that will become apparent as we proceed.

Within the present century two great conceptual schemes, relativity and quantum theory, have together succeeded in permeating and fusing several large areas of physics. Various attempts have been made, notably by Einstein, to construct a **unified field theory** which would include both gravitational and electromagnetic forces. Thus far, these efforts have not succeeded.

WHAT SCIENTISTS DO

What is the general nature of the scientific approach to problems—the procedure anticipated by Aristotle and developed with particular success by Galileo and by Newton? It must be stated at the outset that there is no rigid "scientific method" for studying nature, no set procedure that is always followed in scientific investigation. The specific steps that lead to a scientific advance necessarily depend on the nature of the area or problem in question, and a single, invariable formula does not exist.

The fundamental ingredients of any science are *experiment* and *theory*. Their role and their relation to each other will now be discussed.

1.7   OBSERVATION AND EXPERIMENT

Every natural science must make contact with the world as we know it, and this means that observation of nature is an essential part of the procedure. This seemingly obvious fact was not recognized by Greek science, which was based largely on speculation rather than on experience with actual phenomena. It was Galileo who first established the importance of drawing this distinction. Einstein makes this point in an apt statement in "The World as I See It" (1934):

Pure logical thinking cannot yield us any knowledge of the empirical world; all knowledge of reality starts from experience and ends in it. Propositions arrived at by purely logical means are completely empty as regards reality. Because Galileo saw this, and particularly because he drummed it into the scientific world, he is the father of modern physics—indeed of modern science altogether.

In discussions on the methodology of science, the terms *observation* and *experiment* are frequently used interchangeably. There is a slight distinction: Observation implies merely the contemplation of phenomena, whereas experimentation presupposes the deliberate arrangement of new situations that the scientist can then observe and record. In this sense astronomy may be regarded primarily as an observational science, where we must watch passively while the phenomena unfold. On the other hand, in physics and chemistry, for instance, we can often "set the stage" for events to take place under conditions where we have some degree of control, and this constitutes experimentation. In either case, the essential point is the attempt to extract reliable, direct information from nature itself.

One aspect of making observations can be universally taken for granted: *An observation made on one occasion may be expected again on any other occasion when the same essential conditions prevail.* This proposition, known as the **uniformity** (or constancy) **of nature,** expresses the conviction that natural phenomena are independent of the particular times and places at which they occur. It has been the subject of much discussion on the part of philosophers. Some regard the principle as a mere article of faith, whereas others hold it to be an inherent property of nature. The Austrian scientist-philosopher Ernst Mach (1838–1916) said: "All our efforts to mirror the world in thought would be futile if we found nothing permanent in the varied changes of things."

It is often quite difficult to decide what constitutes essential conditions.

Whether the doctrine of the dependability and orderliness of nature is only a creation of the human mind or something more fundamental, there is no question about its usefulness as a rule of procedure. It is the scientist's denial of the ancient belief in a world ruled by capricious demons and deities.

1.8 SCIENTIFIC LAWS

If scientists were to stop with the collecting of data, they would be left with nothing more than a catalog of unrelated facts comparable in spirit and purpose to a telephone directory or a stock inventory. But some inherent urge of the human intellect makes them go further, trying to correlate and classify the available information in order to generalize their experience. This area of scientific activity complements experiment and deals with **theory, hypothesis,** and **scientific laws.** These terms are extremely difficult to define and interpret universally, but a brief description of their role in science follows.

A scientific law is a general statement expressing a correlation between observations. Often it can be given the form of a mathematical relation (formula). For example, Kepler was able to discover certain regularities among the observations of planetary positions (page 9). He expressed these correlations in the form of three scientific laws whose mathematical description will be discussed in Chap. 8. Similarly, other well-known laws of physics embrace connections between observational data, each in a limited field: Hooke's law of elasticity (page 44) is a statement of the simple relation that exists under certain conditions between the elastic deformations of a solid body and the forces causing them; Boyle's law (page 223) expresses, also by a simple formula, the general relation between the volume and the pressure of a gas under specified circumstances. There is also a law that gives the time of a complete swing of a pendulum in terms of its length and the strength of the earth's gravity.

A scientific law is arrived at by the logical process of synthesis. Synthesis, or **induction,** means proceeding from a collection of particular facts (observations in the case of science) to a general statement embracing them. Unlike a civil or moral law, there is no element of compulsion or obligation about a law of nature. A natural law is simply a concise statement summarizing what is observed in a certain group of phenomena.

A scientific law, once attained, serves as a shortcut to experience. For instance, suppose the generalization is "A piece of iron will sink when placed in water." Having arrived at this conclusion a long time ago through seeing innumerable pieces of iron sink to the bottom, we feel it unnecessary to test this proposition with each piece of iron we use. We are confident that, if placed in water, it will go to the bottom. And yet there is no absolute guarantee that the very next time the experiment is tried the iron will sink. We can never be completely sure. A little reflection shows that all generalizations, being necessarily based on a finite number of cases, can at best be said to have a given *probability* of being true.

This probability increases rapidly with the number of experiences. If a law has already been successful in encompassing a very large number of observations, it may for all practical purposes be assumed valid unless and until a contradictory observation is reported. In the example of the experiments on iron immersed in water, it is quite conceivable that a certain piece of iron, when placed under water, will slowly rise to the surface. Examination of this object may show it to have a large cavity inside, in which case harmony is again restored, provided that we already are familiar with a law dealing with flotation or buoyancy.

Once firmly established, a scientific law is capable of foretelling what may be expected to happen in a particular case. In this way it serves to make the repetition of certain experiments unnecessary. The course to be followed is that of logical deduction. **Deduction** means proceeding from the general to the particular. This scheme is *analytical* in nature; it is just the reverse of the process of *induction,* or synthesis (page 13), by which a physical law was obtained. An example is the case of flotation in the preceding paragraph. There is a general law to the effect that an object will float in water if its average density is less than that of water. Iron has a density considerably greater than that of water. The inescapable conclusion is that an object made of solid iron will not float.

1.9 HYPOTHESIS AND THEORY

Although a scientific law encompasses a multitude of facts, it remains inherently limited in scope and meaning. Science goes beyond these limitations by trying to formulate conceptual connections between such laws. The process is called the setting up of a **hypothesis.** A hypothesis is a working guess as to how the laws in a given area are connected in a logical system.

Details of Newton's gravitational hypothesis are given in Chap. 8.

An excellent example of an effective hypothesis is the one invented by Newton to describe the structure of the solar system. Kepler's laws had given an accurate and satisfactory correlation of the motions of the planets, but as mere statements of fact they remained empirical and detached. It was not evident why the observed motions should correspond to those particular descriptions rather than to some others that might be conceived. Newton, with superb insight, realized that Kepler's laws in turn could be accounted for by means of a hypothesis. His hypothesis was the assumption that there exists a force of attraction between the sun and each planet. Guided by the specific mathematical form of the Kepler laws, he was able to show how the magnitude of the assumed force must depend on the masses of the attracting bodies and the distances between them. He did this a full century before such a force was detected and measured by actual experiment.

Since the setting up of a hypothesis is a creative act of the imagination, it it understandable that no hypothesis can be considered unique or final. The history of science is rich in examples of the replacement of one conceptual scheme by another. On what grounds should a given hypothesis be abandoned for a different one? There is no simple, direct answer. Some authorities justly maintain that such attributes as simplicity, generality, and elegance of form are important criteria. Others are inclined to the view that *predictability* is the main issue: A good hypothesis must suggest new experiments. In this sense, theory is said to be a link between past and future observations.

Newton's assumption of a force of attraction between the sun and the planets accounted for more than the particular set of observations on which Kepler based his laws. It also *predicted* the motions of the satellites of the planets, of comets around the sun, and of double stars around each other.

A more recent theory, general relativity, seeks to explain the same phenomena but starts from an entirely different hypothesis. It says nothing of forces of attraction but bases the explanation of planetary motions on an assumed deformation of space caused by the presence of a massive body such as the sun. This starting point proves sufficient to account for all that Newton's theory covers, but it goes much further. The structure based on Einstein's hypothesis predicted new astronomical phenomena which were subsequently confirmed by observation. On this ground alone, the relativity theory of gravitation must be judged superior to Newton's.

It is impossible to draw a sharp distinction between the terms *hypothesis* and *theory*. A hypothesis of wide generality attains the status of a scientific theory if it confirms already known effects or suggests outcomes for new experiments which are then confirmed in the laboratory. For example, in Chap. 10 we shall examine the consequences of the hypothesis—originally nothing more than a vague conjecture—that matter consists of molecules in a continual state of motion. The results that followed from this assumption have become so numerous and so well substantiated that it is no longer referred to as the kinetic-molecular hypothesis but as the kinetic-molecular *theory*, thus expressing a high degree of confidence in the idea.

1.10 MAKING CONCEPTUAL MODELS

In constructing a theory, use is made of some kind of idealized approximation to some part of the world of experience. Such a mental image or idealized approximation is called a **conceptual model.** It is a most important idea in modern science. Unlike a literal model, a conceptual model is not a facsimile, a mock-up, or an image. In fact, it need not be visualizable at all. It may consist of nothing more tangible than a set of mathematical relationships or a computer program. For example, although it is possible to construct a literal model of the solar system—a mechanical planetarium made up of painted metal spheres, rods, gears, clockwork, etc.—there also is a conceptual model of the solar system in the form of Kepler's three laws.

In many respects the terms *theory* and *conceptual model* have the same meaning. Each is a description consistent with the salient facts observed in a given area of experience. Caution is required in guessing what the salient facts are; in most cases, if any progress is to be made, we must find a way to exclude enough factors of lesser importance to allow analysis of the situation and the calculation of consequences. Thus the setting up of a model requires that we *idealize* real situations.

Closely allied with the idea of a model is the concept of an isolated system. An **isolated system** is an assemblage or combination of things that is completely uninfluenced by anything outside it. Although such an entity does not exist in any strict sense, this condition is often closely approached in nature, and the concept proves to be an extremely workable one. As Warren Weaver aptly says:

The quotation is from Science and the Citizen, Science, Dec. 13, 1957.

The physical world happens to be put together in such a way (I consider this one of God's really bright ideas) that one can usefully take it apart and study an isolated bit of it at a time.

In attempting to understand the motion of a falling object, Galileo replaced the actual situation by a simplified model consisting of a particle falling a limited distance in a vacuum near the surface of the earth. This

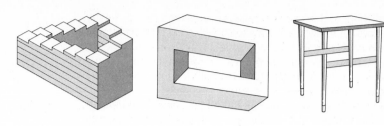

FIGURE 1.4 Not every physical situation can be interpreted pictorially. It is equally true that not every picture represents a possible physical situation. Our intuitions are frustrated when we attempt to interpret in three dimensions such drawings as the ones shown here.

led to a simple, workable result (Sec. 5.6, page 104). Aristotle, on the contrary, apparently tried to grasp the phenomenon in all its complexity. He failed.

Models, like theories, must be subject to revision in the light of new knowledge. We have already referred to the kinetic-molecular theory of the structure of matter. In the earlier form of the model on which the theory was based, matter was regarded as a collection of infinitesimally small particles, assumed to be continually in motion. This simple model was spectacularly successful in accounting for a great many observations. Later it was found possible to bring additional facts into line by altering the model, replacing the pointlike molecules by structures composed of atoms. Still later, the atoms themselves were found to have a structure, and the incor-

The word model, in the sense used in science, was first applied to the structure of matter.

poration of this fact into the model led to an explanation of certain observations on the radiations given off by atoms. At this stage the model remained a mechanical one of electrons orbiting around the central nucleus. It soon appeared that this in turn had great limitations, and eventually a more general atomic model was devised. But the last version, which is the best we have today, is no longer visualizable. Gone are the atoms pictured as billiard balls and the neat scheme of electrons whirling around the nucleus. In place of this, there is left a set of mathematical symbols, to be manipulated in a prescribed way. In other areas of physics, too, the trend has been in this direction. Giving up the luxury of pictorializing nature seems to be the price demanded for the greater power and increased generality of the more abstract models (see chap. 4 of Ref. 1.9).

In recognizing the fact that models and theories are subject to evolution and revision, we automatically admit that ultimate truth and absolute certainty are unattainable goals in science. The concern of science is the pursuit of understanding rather than the attainment of some nonexistent finality. The idea is well expressed by Armitage (Ref. 1.3):

We have first to understand what is meant by the ''truth'' of a scientific theory. In science we employ the word ''truth'' in a rather special sense, differing from its use in everyday life. In judging the truth of a theory we apply two main tests. On the one hand we expect a theory to be able to take hold of a mass of facts which seem isolated and meaningless and to bind them together into an intelligible system so that we can see the connections between them. On the other hand we expect a theory to be able to suggest lines of enquiry along which further investigations can be fruitfully made. These may lead to fresh discoveries resulting, perhaps, in the overthrow of the old theory and the setting up of a new one able to account for the new discoveries as well as for the facts previously known. For no theory in science is regarded as the final truth, only as a step, or as a useful tool, towards the attainment of a further measure of truth.

It seems proper, then, to think of observations (experiments), correlated and grouped into laws, as forming a foundation on which to build a structure that is not unique but which may be changed, extended, or even replaced by a new one. This view of the nature of science brings out the fundamental character of observation, but it should not be concluded that observations always must *precede* the devising of a theory. Scientists often set up conceptual schemes without seeking an experimental basis beforehand. General relativity is an excellent example. Einstein was not motivated by some new gravitational experiments. Instead, he began on what were essentially speculative grounds and arrived, by mathematical procedures, at equations from which certain predictions could be made. He had such conviction as to the validity of his ideas that he is said to have expressed no interest whatever in seeing them tested experimentally. However, while maintaining rightfully that no amount of experimentation could unequivocally prove his theories, Einstein recognized that it would require only *one* experiment to disprove them.

1.11   THOUGHT EXPERIMENTS

Placing speculation ahead of experiment would seem to be a reversion to the aristotelian view, were it not frequently and successfully applied by many modern thinkers. This becomes clear when we recognize that theorists sometimes avail themselves of a type of imagined experimental situation called a **thought experiment.**

The term thought experiment is a literal translation of the German Gedanken-experiment.

Galileo was able to arrive at numerous important conclusions about motion by first trying to conjecture what the results of certain thought experiments would be. Then he tested these ideas by actual experiment. Einstein also made use of numerous thought experiments in developing the ideas of relativity. Some of them described the sending of light signals from one place to another, the comparison of the readings of clocks in various situations, and so on. The design of thought experiments, regulated by sound scientific judgment, has been found to pay off handsomely, but the results must eventually be validated by direct experiment.

1.12   MATHEMATICS AND THE EXACT SCIENCES

The role of mathematics in science is a double one: It is a tool as well as a language. In physics, the framework of natural laws leading to the establishment of a theory is usually constructed with the help of mathematics used as a sort of formalized logic. It often happens that, once the theory is attained, the best way to describe it is by using mathematical symbols. As already mentioned, there are some instances where pictorial representation fails completely and mathematical description seems to be the only possible answer.

In certain situations mathematical formulation actually promotes intuitive understanding. The mathematical relations describing electric currents in a particular kind of circuit are identical in form with those representing the swinging of a pendulum suspended in oil. Another example is the correspondence between certain formulas for the strength of magnetic forces and the equations for the flow of liquids. In each instance, one phenomenon that is comparatively easy to visualize is linked with another that is more abstract. The correspondence is not between the phenomena as such but between the models which are found expedient to make for them. Nevertheless, when the laws of an abstract phenomenon can be expressed in the same form as those of a concrete one, we get a better understanding of the former.

Sometimes a mathematical formulation is not possible or feasible, so that ordinary language must take the place of mathematics in both roles. However, mathematics is far more rigorous and its concepts more precise than those of language. Any science that is able to make extensive use of mathematical symbolism and procedures is justly called an **exact** science.

This discussion of the philosophy and methodology of science has been far from complete. Several important topics have not been mentioned at all or have been touched upon only briefly: cause and effect, the statistical nature of knowledge, operationalism versus intuitionism, the process and role of measurement, the nature of explanation, etc. Some of these will be brought up later in this text, after more background material is available for illustrating the general principles. It will prove more profitable to consider some of the specific topics of the following chapters in the light of these principles than to extend the general discussion at this point.

Programmed review

Instructions: Cover the part of the page that follows this paragraph with a card before reading further. The numbered questions are based on the main points of the chapter. Slide the card down to uncover the first question. After you have formulated a complete response, preferably in writing, move the card down to reveal the answer. If your answer differs essentially from the one given, reread the part of the text that is concerned (relevant section numbers are given in brackets following each response.) Continue as before until all questions have been satisfactorily answered.

1.   What is meant by saying that scientific knowledge is cumulative?

Science deals with a growing body of established fact into which newly acquired knowledge is incorporated.   [1.2]

2.   Identify the main defect of Greek science.

The belief that appeal to the intellect rather than to observation was the necessary and sufficient basis for scientific understanding.   [1.3]

3.   Mention some positive, lasting contributions of the Greeks to science.

The promotion of inductive reasoning; the concept of pure science as the pursuit of knowledge for its own sake; the use of mathematical reasoning.   [1.3]

4.   Copernicus and Kepler may both be called theoretical astronomers. In what significant way did their contributions differ?

Copernicus proposed a *qualitative* scheme for the arrangement and motions of the astronomical bodies, Kepler discovered empirically a *quantitative* description of the solar system.   [1.4, 1.5]

5. In what way did Galileo contribute to the support of the copernican picture?

> By constructing an astronomical telescope and making observations which were in harmony with this idea. (Details will be found in chap. 10 of Ref. 1.1).
>
> [1.5]

6. What is meant by the uniformity of nature? Can this proposition be proved formally? If so, in what way?

> Natural phenomena repeat themselves whenever the same essential conditions prevail. The principle has no logical proof; it is a generalization from experience and seems to be an adequate basis for scientific activity.   [1.7]

7. Define *scientific law*. By what logical process is a law attained?

> A general statement expressing a correlation between observations. The process is *induction:* proceeding from the particular to the general.   [1.8]

8. Define *hypothesis, theory.*

> A hypothesis is a conjecture as to how a set of laws can be connected in a consistent scheme. After a hypothesis has become very general and well substantiated by observation, it may be called a theory.   [1.9]

9. What is generally considered the main function of a good hypothesis or theory?

> Predictability: the power to suggest new experiments.   [1.9]

10. Describe what is meant by a *conceptual model,* as the term is used in science.

> An idealized and simplified representation of some area of experience. It may have either descriptive or mathematical form and is sometimes hard to differentiate from the theory with which it is associated.   [1.10]

11. What is a *thought experiment*?

> An imagined experimental situation, usually constructed as an aid to arriving at a working hypothesis.   [1.11]

12. Describe the two principal uses of mathematics in an exact science such as physics.

> (*a*) As a form of logic that can be used to develop a theory and (*b*) as a language for expressing results quantitatively.   [1.12]

For further reading

For an annotated Bibliography of all books referred to in the text, see Appendix.

1.1. *Holton, G., and D. H. D. Roller* "Foundations of Modern Physical Science." Read chaps. 8, 14, and 15 on method in science. Chapter 10 gives a good account of Galileo's work in astronomy.

1.2. *Butterfield, H.* "The Origins of Modern Science." Read chap. 2 on the work of Copernicus.

1.3.   *Armitage, A.*   "The World of Copernicus."

1.4.   *Snow, C. P.*   "The Two Cultures; and a Second Look."

1.5.   *Andrade, E. N.*   "Sir Isaac Newton." A brief, readable biography.

1.6.   *Newton, I.   "Principia Mathematica."* "Opticks." Classical examples of Newton's scientific writings; for browsing.

1.7.   *De Santillana, G.*   "The Crime of Galileo."

1.8.   *Walker, M.*   "The Nature of Scientific Thought." A good introduction to scientific methodology. See particularly the discussion of models.

1.9.   *March, A.,* and *Ira M. Freeman*   "The New World of Physics."

1.10.   *Weaver, W.*   The Imperfections of Science, *Am. Scientist,* vol. 49, March, 1961, p. 99. An urbane statement about some of the limitations of scientific explanation.

1.11.   *Price, D. J.*   "Little Science, Big Science." A study of the way science and scientists operate today.

1.12.   *Harsanyi, Z.*   "The Star Gazer." An absorbing novelized account of Galileo's life, instructive for the picture it presents of the times in which he lived.

Questions and problems

1.1.   In Ref. 1.1, Galileo is quoted as saying (p. 53): "The knowledge of a single effect apprehended through its cause opens the mind to understand and ascertain other facts without need of recourse to experiment." Does this mean that Galileo denied the necessity and importance of experimentation? Consult Ref. 1.1.

1.2.   How many distinct principal definitions of the word *law* can you find in an unabridged dictionary? Compare the definition given for a physical law with the discussion in the text.

1.3.   After a 2 year investigation by a group of scientists supported by the United States Air Force there still remain a number of reported sightings of unidentified flying objects (UFOs, or "flying saucers") for which definite physical causes have not been established. How should these cases be regarded from a scientific standpoint?

1.4.   The British mathematician Karl Pearson (1857–1936) said a civil law is a *pre*scription, whereas a natural law is a *de*scription. Comment briefly on this apt statement.

1.5.   In the light of what you have read in this chapter, discuss the thesis: "Science *does* repeat itself, while history is only alleged to do so."

1.6.   (*a*) In logic, what is meant by a *categorical syllogism*? Look up this term if you are not sure. (*b*) To which part, if any, of such an argument does a law of nature correspond? (*c*) Is such a syllogism an example of deductive or of inductive reasoning? Justify your answer, giving one or two examples.

1.7.   Write down a law with which you are familiar from some field other than science (language, morality, economics, etc.). Compare its character with that of some physical laws, such as the ones mentioned on page 13.

1.8. Write a paragraph on what is meant by saying that a physical law is "correct."

1.9. Milton's "Paradise Lost," Book VIII, published in 1667, contains a number of references to the planets. Examine each of them and decide whether it assumes a sun-centered or an earth-centered system.

1.10. Explain how a ship model or a model of the DNA molecule made of colored balls differs from a conceptual model.

1.11. In building a model of an atom for a science-fair project a pupil used small, green rubber balls to represent electrons. Does this mean that electrons are round, green, and springy? In your opinion, is the model a valid one?

1.12. List Galileo's principal astronomical observations and state to what extent each one supports the copernican theory. Chapter 10 of Ref. 1.1 will be helpful.

1.13. Kepler and Newton were both theoretical scientists. In what important way did the nature of their contributions to knowledge of the solar system differ?

1.14. The classical divisions of physics are named on page 11. For which of these does man have senses capable of responding to the phenomena directly? Enumerate. If, for any of them, it has not yet been shown that a human sense exists, do you know of any evidence for such abilities on the part of other species?

1.15. You are assigned the task of determining experimentally the general description of the motion of a heavy object, such as a stone, when it is allowed to fall. For this purpose you are given a measuring stick and an electronic timing device so that you can observe the position of the stone at various times after it is released. Besides the recording of times and distances, what other observations in the following list would you judge to be relevant? With each selection, justify your reason for classifying it either as relevant or irrelevant, and try to think of all possible ways in which your argument may be challenged: (*a*) air temperature; (*b*) weight of the stone; (*c*) time of day when the experiment is performed; (*d*) shape of the stone; (*e*) air pressure; (*f*) position of the moon at the time the experiment is performed; (*g*) color of the stone; (*h*) altitude of the laboratory above sea level. By the time you finish this book you will have more definite factual information on which to base your decisions.

Chapter 2
SPACE, TIME, AND MEASUREMENT

Everyone has some general idea of what the words "space" and "time" signify, but in science the meanings of these and many other ordinary words must be made as clear and precise as possible.

It was pointed out in the preceding chapter that physics deserves to be called an exact science because it concerns itself as far as possible with things that can be measured, and particularly with measurements and relations between them that can be expressed through *numbers*. For this reason, it is necessary to know how physical measurements are made and what the measuring process involves. This chapter takes up these subjects as they relate to space and time.

2.1 EVENTS AND SPACE

Science deals with happenings in nature that may be called **events.** The falling of a stone is an event, and so is the breakdown of an atom or the birth of a star. Of the countless events constantly taking place in the universe, science selects a few of the simpler ones and tries to correlate them in order to arrive, if possible, at some understanding of what occurs.

One of the first things we realize in thinking about natural events is that they take place at different locations and at different times. Some occur quite nearby, whereas others happen far away and become known only after some time has passed. Lightning darts out of a distant storm cloud and a quarter of a minute later the faint sound of thunder is heard. A star flares up somewhere in the depths of the universe, but astronomers do not find out about this event until perhaps centuries later because it takes the flash of light that long to cross the intervening space.

By considering a variety of occurrences such as these, we realize that natural events are usually separated in space and time. The theory of relativity recognized the principle that space and time are not independent, unrelated concepts; on the contrary, they are merely two aspects of a single entity called **space-time.** The link is provided by the fact that, in order to become aware of a distant event, one must wait for the arrival of a light signal from it.

Ever since the launching of Sputnik I in 1957, our time in history is often called the "Space Age," and to most people the word "space" suggests the parts of the universe far away from the earth. But space is not something found only among the planets and the stars; it exists "right here" as well as "out there." All of us have an inborn feeling of being surrounded by space, which can be thought of as extending out in every direction. Although some of our most fundamental intuitive concepts such as space and time are very difficult to define precisely, it is possible to describe operational ways of measuring *intervals* of space and time, as will be shown below.

"Middle" is used here in a logarithmic or proportional sense: The ratio of star distance to man's size is about the same as the ratio of man's size to particle size (see page 36).

From a human point of view, there is an enormous range in size of the regions of space known to science. There are tiny particles inside a nucleus that are separated by distances a million billion times smaller than the height of a man, whereas the distance to even the nearest star is more than 10 million billion times this height. Somewhere near the middle of this range is the narrow span of familiar distances in which we live.

2.2 ORDINARY DISTANCES AND THEIR MEASUREMENT

Determining the length of an object or the distance between two points is probably the simplest type of measurement that can be made in scientific work or in everyday life. It is the familiar kind of measurement made when using a ruler or yardstick. However, quite different procedures for measurement must be used if the distance to be found is very small or very large on the human scale or if the end points are not accessible.

Choice of a unit.

Before it is possible to make a measurement of any kind, a suitable **unit** must be chosen. The size of this unit can be a matter of choice but for ease in stating the result, it should not differ much in size from the object to be measured. For example, the width of this book could well be expressed in inches, whereas the distance between two cities is better expressed in miles. It is quite proper to say that the book is $7\frac{1}{2}$ in. wide. To say that it is 0.00001183 mi wide would be just as correct but very awkward, and the result would be hard to visualize. In the same way, it probably means more to say that the road distance between Los Angeles and San Francisco is 431 mi rather than 27,300,000 in.

Making and using a measuring rod.

In order to measure the lengths of ordinary objects, a unit of suitable size is selected and transferred to a standard rod or stick by making two marks. For convenience several such units are marked, one after another, from one end of the rod and numbered consecutively as on a foot rule or yardstick. Then, to measure the length of an object directly, the "zero" (beginning point) of the scale is placed at one end of the object and the position of the other end is noted on the scale.

In general, the result will not be a whole number of units, and the end of the object to be measured will probably fall somewhere between two marks on the measuring rod. If each unit space has been subdivided into fractional parts, a closer reading can be made. But there are practical limits to how far such subdivision can be carried, and the final digit in the measurement will be the nearest tenth of the smallest marked division, estimated by eye (see Fig. 2.1).

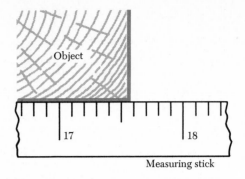

FIGURE 2.1 The position of the end of the object is read on the meter stick. The fraction of the smallest scale division is estimated here as 0.7, so that the complete position reading becomes 17.57, the final digit being uncertain.

2.3 STANDARDS OF LENGTH; THE METER

Long before the rise of science, a great many different and unrelated length units were in use in various places and even among various trades. In the **English system,** the inch, foot, yard, and mile are still in common use in English-speaking countries. This system of measure is not a very good choice for scientific work. One disadvantage is that the different-sized units have no simple relation to each other. For instance, there are 12 in. in a foot, 3 ft in a yard, 5,280 ft in a mile. Changing a length measurement from one of these units to another requires some arithmetic, using one or more of the above factors. The same is true when working with weight units and with other kinds of units in the English system.

Advantage of a decimal-based system of measure.

In France, in 1791 during the Revolution, the new government appointed a committee to work out an improved system of measure. This group recognized the great advantage of putting each type of unit on a *decimal* basis. Units of various sizes were made to differ from each other by *powers of 10,* and a measurement could be changed from one unit to another simply by moving the decimal point. The plan developed by the committee is called the **metric system,** and is the only system used for scientific work all over the world. It has also come into general and commercial use in almost all countries.

At an international conference on weights and measures in 1960, the length of the standard meter was officially fixed at 1,650,763.73 wavelengths of the orange light produced by atoms of krypton 86.

The basic standard of length in the metric system is called the **meter.** This unit is kept in the form of a special metal bar at the International Bureau of Weights and Measures in France. There are accurate copies in other countries; the U.S. Bureau of Standards has two. The standard meter has been carefully measured against the length of the waves of orange light coming from a lamp filled with krypton gas so that, even if all the meter bars were destroyed, the unit could be reproduced so accurately that a new bar would not differ from the old ones by as much as one ten-millionth of a meter.

2.4 METRIC UNITS OF LENGTH

Table 2.1 shows some of the most commonly used metric length units. Each is related to the basic unit, the meter, by some multiple of 10. The name of each unit is formed by attaching the proper prefix to the word "meter." The same plan is followed in naming the units used to measure other kinds of physical quantities (Table A.2, page 631).

TABLE 2.1 METRIC UNITS OF LENGTH

UNIT	ABBREVIATION	SIZE IN METERS	RELATION TO ENGLISH UNITS
Kilometer	km	1,000	1 km = 0.621 mi
			= 3,281 ft
Meter	m	1	1 m = 39.37 in.
Decimeter	dm	0.1	
Centimeter	cm	0.01	2.54 cm = 1 in.
Millimeter	mm	0.001	

For a more complete table of multiples of units and their standard prefixes and abbreviations see Table A.2, page 631.

Table 2.1 shows how simple it is to change a measurement from one length to another if the metric system is used.

Worked example 2.1   Express 0.0027 km in centimeters.

Solution   Since there are 1,000 m in a kilometer and 100 cm in a meter, there are $100 \times 1,000 = 100,000$ cm in a kilometer. Then 0.0027 km is $0.0027 \times 100,000$ cm; when the decimal point is moved five places to the right to multiply by 100,000, the result is 270 cm. Thus

$$0.0027 \text{ km} \times 1,000 \, \frac{\text{m}}{\text{km}} \times 100 \, \frac{\text{cm}}{\text{m}} = 270 \text{ cm}$$

2.5 REFERENCE FRAMES AND COORDINATES

The position of an object can be located by using distance measurements in connection with a reference frame, to be defined below.

A good way of locating things on a flat surface is to use a **rectangular coordinate system.** Any two straight lines are drawn, crossing each other at right angles (Fig. 2.2*a*). For convenience, one line is shown horizontal, the other vertical. The first is called the *x axis,* and the other the *y axis.* The exact location of any point on the paper can now be specified by stating its perpendicular distance from each axis. The first of these two distances locates the point side to side; the other locates it in the up-and-down

(*a*)   (*b*)

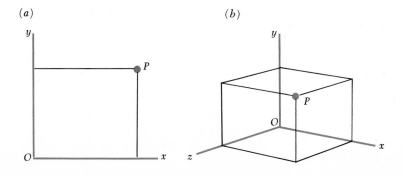

FIGURE 2.2 Coordinate systems in two and three dimensions. The position of a point in a plane can be located by means of two numbers, and in space by means of three numbers.

direction. Together, the numbers representing these two coordinates tell exactly where the point is on the paper.

The xy coordinate system can give the position of any point that is in the same plane as the paper, but it cannot be used for locating points anywhere else in space. To do this, a third axis is constructed, standing perpendicular to the other two (Fig. 2.2b). The three coordinates that locate a point in space are now the distances from three planes, each plane passing through a pair of the axes. It is comparable to locating an object in a room by stating its distance from the floor, from the west wall, and from the north wall.

To locate a point on a given straight line, only one measurement is needed—the distance of the point, measured either to one side or the other, from some agreed-upon starting place. A line has only one dimension. To locate a point on a plane, two measurements, such as the x and y coordinates, are needed. A plane has two dimensions. Finally, to locate a point in space, three measurements are needed: $x, y,$ and z coordinates. And that is the end of the story, for nobody is able to picture more than three dimensions for ordinary space: up and down, side to side, toward and away. According to our experience and intuitions, space is three-dimensional.

The rigid framework consisting of three coordinate axes is one example of a **reference frame.**

2.6 AREA AND VOLUME MEASUREMENT

It was pointed out above that (1) for measuring length or distance an agreed-upon basic unit of length is needed and that (2) it is convenient to have a set of length units of various sizes, all related by powers of 10.

Suppose it is required to measure the *area* of a flat surface. To do this, another unit is needed because a surface has two dimensions, requiring that all area units also have two dimensions. It is simplest and most convenient to choose a *square* unit, and the side of each square may be taken equal to one of the length units already set up. Some suitable area units are the square inch, square yard, square centimeter, square meter, etc. The abbreviations for these units are best written in.^{2}, yd^2, cm^2, m^2 but are read "square inches," "square yards," etc.

Direct measure-ment of area. In measuring a length, copies of the length unit are laid off, one after another, until the whole distance is covered. Similarly, in measuring an area, copies of the unit square are put down like floor tiles, until the whole area is covered. This procedure is applicable to any area, no matter how irregular its shape. If, for example, the area to be measured is that of a lake on a map, a transparent sheet ruled into unit squares can be laid over the map. Then simply counting the number of squares that lie inside the boundary curve gives the area of the lake directly. In order to reduce errors at the edge, where part of a unit square may fall outside the curve, smaller units may be used, as shown in Fig. 2.3.

This direct method of measuring area by laying down unit squares is not often used in actual work. In many cases, the surface to be measured has some simple geometric shape: a rectangle, circle, triangle, the surface of a sphere, etc. By using the proper formulas from geometry, the area can

(a)   (b)

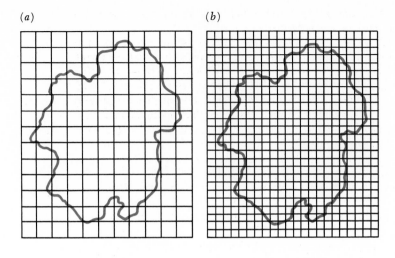

FIGURE 2.3 The area enclosed by any plane figure can be estimated by counting squares on a grid laid over the figure. The smaller the size of the squares, the closer the estimate can come to the true value of the area.

be computed from length measurements made on certain parts of the figure. For instance, the area of a circle is given by $A = \pi r^2$, where r is the radius, and the area of a triangle is equal to half the base times the altitude.

There are mechanical instruments, called planimeters, for finding the area of a plane figure of any shape, regular or not. Still other ways of measuring area can be devised. No matter which method is used, the final result of an area measurement is a *number* followed by a proper area *unit*: 247 in.^{2}, 78.3 km^2, etc.

Measure-ment of vol-ume.

The measurement of the volume of a three-dimensional region of space involves the methods used for one and two dimensions. The chosen unit may be a *cube* whose edge is equal to one of the length units already used. It may be a cubic inch (in.^{3}), cubic foot (ft^3), cubic meter (m^3), etc. If the volume to be measured is that of a solid object, it may be imagined cut into unit-sized cubes. The volume is found by counting the total number of unit cubes and making allowance for the partly filled cubes at the surface. The volume of a quantity of liquid can be measured directly by finding how many times a container of unit volume can be filled by it.

The formulas of geometry are often used to calculate the volumes of solid bodies that have simple shapes. For example, if an object is known to be a true sphere, its volume can be found by measuring the radius and using the formula $V = 4\pi r^3/3$. A liquid has no fixed shape, but it can be poured into a graduate, which is a cylindrical vessel with a scale marked along its side.

Worked example 2.2

How far apart must the 1-cm^3 divisions be engraved on the side of a graduated cylinder if the inside diameter of the cylinder is 26.0 mm?

Solution   From geometry, the volume of a circular cylinder is $V = \pi r^2 h$, where r is the radius of the base and h is the altitude. Since 1 cm = 10 mm, 1 cm^3 = (10)^3 mm^3. From the formula, (10 mm)^3 = (3.14)(13 mm)$^2 h$ mm, so that $h = (10)^3/(3.14)(13)^2$, or $h = 1.88$ mm.

Practice set*   The numerical part of each answer is given at the foot of the page. Be sure to check your results before reading on.

1.   Compute the total surface area of a cube whose edge is 4.0 cm.
2.   The two perpendicular sides of a right triangle are 3.0 and 5.0 cm long, respectively. What is the area of the triangle?
3.   A trough in the form of a half cylinder has semicircular bases of radius 1 ft and is 3 ft long. How many cubic feet of water can it hold?

2.7 TIME AND ITS MEANING

Events are observed to follow one after another, some happening earlier and some later. When we line them up mentally, we get an idea of the passing of time. In fact, it is hardly possible to conceive of time without the events that we use to mark its passage.

The concept of time is connected with some kind of motion or change: The moon, which was low in the eastern sky at twilight, is now high overhead and will be far down toward the west in the morning. A tree is several inches taller this year than it was last year. Over a period of thousands of years, rivers wear away their banks and cut deep canyons into solid rock. Time passes.

Physics deals with change and motion, and so it is important to find out how to make reliable measurements of time. It is usually necessary to know the duration of a time span—the interval between events—but it soon becomes apparent that estimates or judgment cannot be trusted. An hour at the dentist's seems much longer than an hour at the beach. It is very difficult to convey a satisfactory idea of time and virtually impossible to make a strict definition of it because *there is nothing more familiar or more basic to which the concept can be referred.* The same difficulty exists in connection with attempts to define space in anything but an intuitive way.

The remedy adopted by most scientists is a point of view proposed by the American physicist P. W. Bridgman (1882–1961). It is called **operationalism.** According to the idea of operationalism, the proper way to describe or define a physical quantity is by setting forth a prescription as to how the quantity is to be produced and measured rather than trying to convey its meaning by relating it to other concepts. Thus, on page 23, the measurement of length was prescribed by the procedure of laying off a standard rod in end-to-end fashion; this is apparently the best way to convey what is meant by the concept "length." In much the same way, an operational definition of time, based on the use of clocks, will be given below. Not all scientists agree completely with the strict operational point of view. Many believe that intuition must continue to play a prominent part in our understanding of natural phenomena.

Another difficulty connected with the measurement of time is the matter of comparing time at different places on earth. In the past, people regulated their activities by the position of the sun in the sky—that is, by the rotation of the earth. Someone living in Michigan sees the sun reach its highest point in the sky later than a person in New York but earlier than one in California.

*Ans.: 96; 7.5; 4.7.

Such differences in time caused great confusion when the railroads were built across the country. Each town on the line kept its own local time. Then standard time zones were set up across the United States and Canada, and eventually this plan was adopted all over the world. All clocks in one zone show the same time and differ from the clocks in the zone on either side by 1 hr. Travelers going from one time zone to another must reset their watches, and anyone who makes a long east-west jet flight knows how troublesome and disturbing it is to adjust to the time change.

The theory of relativity was responsible for some fundamental changes in our ideas of time. It pointed out that the only operational method of comparing two clocks at different locations is to send a timing signal from one to the other. But even the fastest signal that can be used—a flash of light—takes some time for the trip—a delay that depends on how far apart the two stations are. One consequence is that there is no meaning to such questions as: What time is it now on Mars?

Some other results of relativity, to be described later, show that a clock in a rocket that is moving rapidly past an observer would seem to him to keep slightly slower time than the same clock mounted on his wall, and that a clock would slow down if taken from the earth to a more massive body such as Jupiter.

Although people ordinarily think of space and time as completely different and distinct, the theory of relativity shows that they are really connected, forming the space-time continuum (page 22). In this sense, time is often referred to as the fourth dimension, and an event can be represented as a point in four-dimensional *space-time.*

Our observations of events may be thought of as snapshots which can be strung together like the frames of a motion-picture film. The running film gives the impression of things happening one after another, in their proper order, as time passes. Changing the speed of the projector can make time appear to pass more quickly or more slowly than normal. However, the film is never observed to run backward in the sense that weathered rocks become whole again or omelettes revert to raw eggs! Time appears to flow in one direction, at least in the world as we perceive it.

2.8 MEASUREMENT OF TIME

Time, as a background for events, seems to flow at a uniform rate when compared with, for instance, the pulse rate or other recurrent physiological sensations. This brings up the problem of how a uniform flow of this kind can be marked or measured objectively. Apparently what is needed is something that repeats regularly or that runs at a steady rate. The rotation of the earth on its axis seems to be suitable for time measurement simply because there appears to be no imaginable cause for any drastic change in its rate. The basic time unit, the **second,** was originally defined in terms of this rotation, according to the following scheme:

For an observer at any given place on earth, the sun reaches its highest point in the sky at a time called local noon. The interval from one noon to the next is called one **solar day.** The noon-to-noon interval is not perfectly constant because (1) the earth's axis is inclined and (2) the earth does not move along its orbit with constant speed. To allow for these

*Original defi-
nition of the
second.*

variations, the *average* length of all solar days through the year is used as a standard. For historical reasons, the average solar day was divided into 24 hr, the hour into 60 min, and the minute into 60 sec. This fixed the second as 1/24 of 1/60 of 1/60, or 1/86,400 of an average solar day.

The second itself is the most commonly used time unit for scientific work. It is perfectly proper to construct other units based on the second by using any of the prefixes shown in the tables on pages 25 or 631. In scientific work, only a very few units are commonly used: the second, millisecond, and microsecond for short time intervals and the day and the year for long ones.

Time measurement usually means the scaling off of a time *interval,* just as distance measurement means finding the length of a distance interval. If a time-measuring device (clock) is to be useful, it must in some way signal or mark off equal intervals of time. A swinging pendulum goes

A NEW DEFINITION OF THE SECOND; ATOMIC CLOCKS

Although by ordinary standards the rotation of the earth on its axis is quite constant, there are factors that can produce minute variations. Very slight irregular changes are caused by the melting or accumulation of polar ice and by the winds. There is also a gradual slowing up of the rate of turning caused by tides in the ocean, making the day longer by an estimated 0.0018 sec per century.

Careful study shows that the year (the time the earth takes to go once around the sun) is more nearly constant than the day as measured by the earth's rotation on its axis. In 1955, the International Astronomical Union decided to base the second directly on the year, making it officially "1/31,556,925.97474 of the length of the tropical year 1900. . . ."

The number of seconds in a year was written out to thirteen digits because now there are clocks that can measure time with this degree of exactness. It has been found possible to use the vibration of atoms to regulate clocks with an accuracy of 1 part in 10 billion. The steady rate of vibration of a molecule of ammonia gas can be stepped down by electric circuits until it is suitable for controlling an electric clock. And now there are atomic clocks, regulated by the spinning motion of an electron in an atom.

The modern trend in the science of measurement is to seek ways of basing the fundamental units on reproducible natural phenomena rather than on particular objects. In 1960, the International Conference on Weights and Measures adopted a definition of the second based on the behavior of the cesium atom, making 1 sec officially equal to the time of 9,192,631,770 vibrations of the outermost electron of such an atom.

The above figures imply an accuracy of one part in several hundred billion for time measurements. An accuracy of one part in a million billion is now a possibility, which would correspond in exactness to the measurement of the sun's distance from the earth to within the diameter of a human hair.

Atomic clocks are being used in atomic research, in making exact measurements of the variations in the earth's motions, as well as in improving the accuracy of radar navigation. In 1972, the American scientists J. C. Hafele and R. Keating flew atomic clocks around the world in opposite directions and obtained results consistent with the slowing down of moving clocks predicted by the theory of relativity.

through the lowest point of its path at equal time intervals, and it was used as the first reliable regulator for mechanical clocks. The balance wheel and hairspring of a watch serve the same purpose. A slice cut from a quartz crystal can be made to vibrate at a very steady rate by applying an electric voltage. The best crystal-controlled clocks are not in error by more than a few hundredths of a second in a year. The most accurate clocks so far constructed are guided by the vibrations of atoms or parts of atoms (see photograph below).

2.9 DENARY NOTATION

At the beginning of this chapter it was pointed out that physics, as an exact science, makes use of measured quantities. Measurements by themselves do not make up a science; they are only the raw results of observations or experiments. Before such data can be useful in describing nature, they

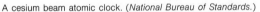

A cesium beam atomic clock. (*National Bureau of Standards.*)

TABLE 2.2 SOME TIME UNITS COMMONLY USED IN PHYSICS

Year	yr	31,570,000 sec
Day	day	86,400
Second	sec	1
Millisecond	msec	0.001
Microsecond	μsec	0.000 001

For a more complete table of multiples of units and their standard prefixes and abbreviations see page 631.

must be expressed in numbers in some consistent way, and then the numbers must be examined for trends or relationships. In most cases, several items of data must be used in an arithmetical calculation to get a desired result.

Science often deals with numbers that are extremely large or extremely small because this is sometimes more convenient than using a variety of units of different size. Writing such large numbers and computing with them in the ordinary decimal form are troublesome. Many zeros usually are needed, either before or after the decimal point, and there is a great chance of making a mistake in counting or recording them.

There is a compact and convenient method of writing numbers by using powers (exponents) of 10, called the **denary** notation, from a Latin word meaning "containing ten." For instance, the number 100, expressed as an integer power of 10, is written 10^2 (read: "ten squared"), and this means that 10 is used as a factor twice. In the same way, 1,000 is 10^3 ("ten cubed"), meaning that this number is given by using 10 as a factor three times; 10,000 becomes 10^4 ("ten to the fourth power"), and so on.

For a review of the laws of exponents, see the Appendix, page 627.

It is possible to write *any* number in the denary notation—not only those which are multiples of 10—and this notation is particularly useful for very large or very small numbers. As an example, the distance across the earth's orbit in miles, is 186 million, ordinarily written 186,000,000. This number is not merely the digit 1 followed by a certain number of zeros, and so it cannot be written simply as a power of 10 standing alone. The digits 1, 8, and 6 have to be included in some way.

Rewriting an ordinary number in denary form.

To do this, the size of the given number is reduced by dividing it by 10 as many times as necessary until it becomes a number between 1 and 10. In the example, this means dividing by 10 eight times, that is, dividing by 10^8, or moving the decimal point eight places to the left. The result is 1.86. To leave the value of the number the same as before, the 1.86 must then be multiplied by 10^8. This appears to be undoing what went before. However, the multiplication by 10^8 is now merely *indicated* by writing the final result as 1.86×10^8.

Now, in place of a long string of zeros, there is a compact expression showing at a glance that the number in question is of the general order of size of 10^8 (a hundred million). The exponent gives this information and makes the counting of zeros unnecessary.

The advantages are just as great when working with very small numbers.

Worked example 2.3

The diameter of the nucleus of an atom of uranium is about 0.000 000 000 0017 cm. Write this in denary form.

Solution Multiplying the number as it stands by 10^{12} would make it 1.7. To restore the value, multiplication by 10^{-12} is required. Therefore the original number may be written 1.7×10^{-12} cm, which is of the order of size of a trillionth of a centimeter.

Practice set* Change each of the following numbers to the denary form:

1. 4,803,000,000
2. 0.00012
3. 2.81/10,000

For numbers of moderate size, perhaps between 10^{-3} and 10^3, there is comparatively little advantage in using the denary expressions unless other numbers in the problem are already given in this form. The next section shows how computations can be simplified by keeping all the numbers in denary form.

2.10 CALCULATIONS

Many problems in physics involve the substitution of numbers, obtained from laboratory measurements or in some other way, into a general formula or mathematical relationship. After substitution, the indicated arithmetic must be carried out to get a single number that represents the value of the result that is sought. For instance, suppose the problem is to compute the volume of a metal cylinder from measurements of its diameter and height. From geometry, the formula for the volume of a circular cylinder is

$$V = \frac{\pi d^2 h}{4}$$

where d is the diameter of the base and h is the altitude. It is now necessary to substitute the measured values of d and h, insert the numerical value of $\pi (=3.14)$, and do the necessary arithmetic to get the value of V, the volume of the cylinder.

In another example, if the calculation to be made is

$$\frac{(71,200,000)(0.0000963)}{(0.00145)^2}$$

a good procedure is the following:

1. Write all the numbers in denary form, keeping the numerical factors between 1 and 10:

    $$\frac{(7.12 \times 10^7)(9.63 \times 10^{-5})}{(1.45 \times 10^{-3})^2}$$

2. The answer, too, will consist of a numerical factor and a power of 10. The exponent of the 10 is found first by combining the given exponents. In the numerator, $7 - 5 = 2$. In the denominator, $(2)(-3) = -6$. The final exponent will be given by $2 - (-6)$, or 8.
3. Now the *numerical factor* of the result is determined by multiplying 7.12 by 9.63 and dividing by $(1.45)^2$. This can be done by paper-and-pencil arithmetic, but it is much quicker and easier to learn the simple opera-

* *Ans.:* 4.803×10^9; 1.2×10^{-4}; 2.81×10^{-4}.

tions of multiplication and division on a slide rule, using it for all such calculations. A 10-in. slide rule is inexpensive and will save much time and effort in the classroom and in the laboratory.

Locating the decimal point in slide-rule calculations.

The slide-rule answer to the above computation is 338. The instrument gives these three digits but does not show where to put the decimal point. Is the number 3.38, 0.338, or something else? One of the best ways to find out is to make a rough estimate by rounding off the factors given in the problem: $(7.12)(9.63)/(1.45)^2$ is roughly $(7)(10)/2 = 35$, so that the true value must be 33.8, since this is closer to 35 in magnitude than, say, 3.38 or 338.

4. Finally, the two parts of the calculation can be combined to give the result 33.8×10^8 or, since the numerical factor is always kept between 1 and 10, this is better written 3.38×10^9.

Worked example 2.4

What fraction of the light sent out by the sun hits the earth?

Solution   It is reasonable to assume that the sun radiates equally strongly in all directions. At the earth's distance, all the radiation flows through the surface of a huge sphere whose radius is the distance of the earth from the sun, 9.26×10^7 mi. The surface area of a sphere of radius R is given by $A = 4\pi R^2$, and so the area in this problem is $4\pi(9.26 \times 10^7)^2$ mi². Instead of computing the numerical value of this quantity, it is better to leave it in this form and proceed with the next step in the calculation.

The amount of radiation that hits the earth depends on its *cross-section* area, which is πr^2, where r is the earth's radius, 3,960 mi. Then the cross-sectional area, in square miles, is $\pi(3.96 \times 10^3)^2$. The fraction of the radiation hitting the earth will be the ratio of these two areas:

$$F = \frac{\pi(3.96 \times 10^3)^2}{4\pi(9.26 \times 10^7)^2}$$

After both terms of the fraction have been divided by the factor π we have

$$F = \frac{(3.96)^2}{4(9.26)^2} \times 10^{-8} = 0.0458 \times 10^{-8} = 4.58 \times 10^{-10}$$

This shows that less than half a billionth of the sun's total radiation strikes the earth.

Practice set*

Carry out the indicated calculations:

1. $(4.2 \times 10^6)(2.0 \times 10^{31})$
2. $(7 \times 10^{21})(5 \times 10^{-13})$
3. $\dfrac{(4 \times 10^{-5})(3 \times 10^{11})}{6 \times 10^4}$

2.11   ORDER OF MAGNITUDE; ESTIMATES

Although there is a continuing search for more and more precise values of physical quantities, there are many instances where it is enough to know an approximate figure. Sometimes even a rough estimate will convey adequate useful information.

*Ans.: 8.4×10^{37}; 3.5×10^9; 200.

Long ago, people believed that the sun and the moon were of equal size and at the same distance from the earth, as they appear to be. The first crude distance measurements, made by the parallax method, showed that the moon is a few hundred thousand miles away and the sun about a hundred million. Later, refined measurements gave the two distances as 239,000 and 92,900,000. But it was the early estimate (that the sun is roughly a thousand times as distant as the moon) that represented the big step forward in man's knowledge of the solar system.

Expressing any quantity to the nearest power of 10 gives the *order of magnitude* of the quantity. In the above instance, the distance of the sun may be written in denary form as 9.29×10^7 mi, and that of the moon as 2.39×10^5 mi. Since 9.29 is closer to 10 than to 1, whereas 2.39 is closer to 1 than to 10, these two numbers have the respective orders of magnitude of 10^8 and 10^5. Further, the ratio is $10^8/10^5 = 10^3 = 1,000$, as mentioned in the preceding paragraph. If the more exact values of the distances are used, the result is close to 400. This is still well within a factor of 10 of the estimated 1,000. The two numbers 400 and 1,000 are roughly of the same order of magnitude.

The making of order-of-magnitude estimates is interesting and can even be enjoyable. No time-consuming computations are required, and usually many simplifying assumptions can be made. For instance, in estimating volume, a sphere can be replaced by a cube whose edge is the same as the diameter of the sphere. The two volumes have the same order of magnitude, as a simple calculation will show.

Practical hints on making estimates.

For making an estimate, each quantity must be inserted to the nearest power of 10. It soon becomes apparent that an overestimate of one factor in a product can be balanced by making a slight underestimate of another factor. In division, the effect of a too large factor in the numerator can be reduced by overestimating one or more factors in the denominator. For example, in making an estimate of the area of a rectangle whose sides measure 62 and 550 m, it may seem that 10^2 and 10^3 should be used as the factors in the estimate, since 62 is closer to 10^2 than to 10, and 550 is closer to 10^3 than to 10^2. But both figures happen to be near the half-way point, and so it is a good idea to estimate one on the lower side and the other on the upper. If 10 and 10^3 are taken, the product is 10^4, which compares well with the exact result of 3.41×10^4. Notice, too, that both dimensions of the rectangle had to be expressed in the *same* length unit, meters. The resulting area, whether estimated or computed exactly, is expressed in square meters.

Worked example 2.5

Estimate (order of magnitude) how many grains of rice would be required to fill the Empire State Building.

Solution The required number may be taken equal to the volume of the building divided by the volume of a single grain. Think of both the actual building and the rice grain replaced by rectangular solids. The order of size of each dimension of the base of the building is perhaps 10^2 ft and the height is of the order of 10^3 ft. This makes the volume of the building $10^2 \times 10^2 \times 10^3$, or 10^7 ft^3. The dimensions of the rice grain will also have to be expressed in feet. The cross-section of a grain will be of the order of 10^{-2} ft in each direction. This is somewhat of an overestimate. The length of a grain will also be of the order of 10^{-2} ft, and this is a slight underestimate, which will tend to compensate for the other estimate being too large. The volume of a grain will then be of the order of $(10^{-2}$ ft$)^3 = 10^{-6}$ ft^3. Dividing the estimated volume of the building by that of a single grain, we

TABLE 2.3   ORDER OF MAGNITUDE OF DISTANCE AND TIME

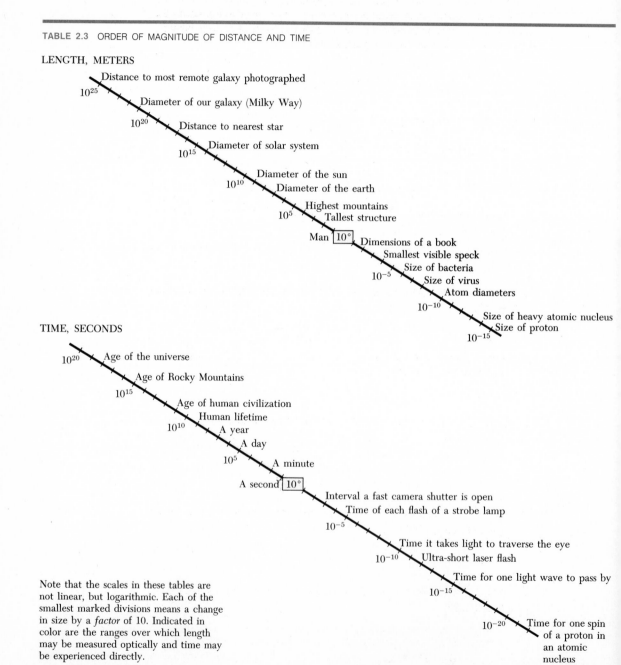

LENGTH, METERS

Distance to most remote galaxy photographed
10^{25}
Diameter of our galaxy (Milky Way)
10^{20}
Distance to nearest star
Diameter of solar system
10^{15}
Diameter of the sun
10^{10}
Diameter of the earth
Highest mountains
10^5 Tallest structure
Man 10^0 Dimensions of a book
Smallest visible speck
Size of bacteria
10^{-5} Size of virus
Atom diameters
10^{-10}
Size of heavy atomic nucleus
Size of proton
10^{-15}

TIME, SECONDS

10^{20} Age of the universe
Age of Rocky Mountains
10^{15}
Age of human civilization
Human lifetime
10^{10} A year
A day
10^5 A minute
A second 10^0
Interval a fast camera shutter is open
Time of each flash of a strobe lamp
10^{-5}
Time it takes light to traverse the eye
10^{-10} Ultra-short laser flash
Time for one light wave to pass by
10^{-15}
10^{-20} Time for one spin of a proton in an atomic nucleus

Note that the scales in these tables are not linear, but logarithmic. Each of the smallest marked divisions means a change in size by a *factor* of 10. Indicated in color are the ranges over which length may be measured optically and time may be experienced directly.

get 10^7 ft³/10^{-6} ft³ $= 10^{13}$. Thus the building is estimated to hold 10^{13} (10 million million) grains; this estimate is probably within a factor of 10 of the correct number.

Several other matters of practical importance in connection with the processes of measurement and computation are discussed in the Appendix, pages 621–629. These sections will serve as a reminder of certain essential procedures and should be consulted at this point.

Programmed review

Instructions: See page 18.

1. What is meant by a *unit* of measure?

    A quantity of specified size, such that any other quantity of the *same kind* can be measured by comparing it with the unit.   [2.2]

2. In principle, what constitutes the operation of measuring a physical quantity?

    Finding how many times a selected unit is contained in the quantity to be measured.   [2.2]

3. Has the meter been measured in terms of some more fundamental length?

    Yes, by comparing it with the wavelength of a certain kind of light.   [2.3]

4. If the meter is selected as the unit of length, name the corresponding unit of area; of volume.

    Square meter (m²); cubic meter (m³).   [2.6]

5. Describe a three-dimensional rectangular coordinate system (reference frame).

    A set of three mutually perpendicular lines (axes) drawn through a point. The position of any other point in space may be specified by giving its distances from the three planes determined by the axes.   [2.5]

6. Name the fundamental unit of time, and mention the basis on which it is defined.

    The second, originally defined in terms of the motion of the earth but now specified on the basis of atomic vibrations.   [2.8]

7. What is meant by the denary notation for expressing numbers used in science?

    Writing the number as a power of 10. A number, when so expressed, will consist of a decimal with indicated multiplication by 10 raised to a suitable power.   [2.9]

8. Explain the meaning of the term "order of magnitude" of a number.

    The value to the nearest power of 10.   [2.11]

9. How can the order of magnitude be obtained at sight?

    With the decimal part of the number adjusted to be between 1 and 10, the
    order of magnitude is merely the exponent of 10 itself if the decimal is less
    than 5, or the next higher integer if the decimal part is 5 or greater.   [2.11]

For further reading

2.1.   *Astin, A. V.* Standards of Measurement, *Sci. Am.,* June, 1968.
2.2.   *Campbell, N.* "What Is Science?" Chapter on measurement. Re-
       produced in J. R. Newman (ed.), "The World of Mathematics," pp.
       1797–1813.
2.3.   *Lyons, H.* Atomic Clocks, *Sci. Am.,* February, 1957.
2.4.   *Whitrow, G. J.* "The Natural Philosophy of Time."
2.5.   *McKie, D.* The Origins of the Metric System, *Endeavour,* vol. 32,
       1963, p. 24. Official definitions of the meter and the second: Report
       of the 11th General Conference of Weights and Measures. See *Am.
       J. Phys.,* February, 1961, p. 94.
2.6.   *Sawyer, W. W.* "Mathematician's Delight," chap. 8.

Questions and problems

Wherever convenient, make use of the Conversion Factors given in the Appendix,
pages 629–630.

2.1.   The x and y coordinates of a point are 12 and 5 cm, respectively.
       How far is this point from the origin?

2.2.   Would it be feasible to use two coordinate axes that are *not* mutually
       perpendicular in order to locate points in a plane? Could Question
       2.1 be answered for such an oblique coordinate system? What piece
       of information, in addition to the two coordinates of the point, would
       be needed to answer the question?

2.3.   Look up the description of (*a*) polar coordinates and (*b*) spherical
       coordinates in a book on geometry.

2.4.   Measure the thickness of a dime by finding the total height of a
       stack of 10 or 20 of these coins with a metric ruler. Why is this
       more accurate than measuring one of the coins with the same ruler?

2.5.   Measure the two dimensions of the cover of this book in inches
       and again in centimeters. Compute the *ratio* of height to width in
       each kind of unit. Are the resulting numbers essentially the same?
       Explain. Save the measurements for use in Prob. 2.9.

2.6.   Under normal conditions, the speed of sound in air is about 330
       m/sec. What is this in feet per second? In miles per hour? Use
       information from Table 2.1.

2.7.   The speedometer of a European automobile is calibrated in kilome-
       ters per hour. Could such a car be driven at 75 km/hr and still
       be within the law in a state where the speed limit is 50 mi/hr?

2.8.   What is the cost of 4,000 m of wire if the price is quoted at 15 cents per 100.ft?

2.9.   Use the measurements made in Prob. 2.5 to compute the number of square centimeters equivalent to one square inch.

2.10. An American-made 9-by-12-ft rug is placed in a European living room whose floor measures 4.5 by 5.2 m. Express the area of the uncovered part of the floor in square meters.

2.11. A circle 5 cm in diameter is cut out of a sheet of insect screening having 18 wires to the inch. Approximately how many squares will the piece contain?

2.12   The plastic spheres used for size reference in the electron microscope have a diameter of about 1/14,000 in. Express this in denary notation, rounding off to three digits, and calculate how many such spheres, placed in contact, in square array, would cover a square area 0.1 mm on a side.

2.13. From information given in Table 2.1, find the number of cubic centimeters in a cubic inch.

2.14. A volume of 0.123 m^3 of water is added to 3.52 ft^3 of water. Express the total volume in cubic feet.

2.15. The sides of a triangle measure 11.5, 14.5, and 20.0 cm. Find the area. Consult a trigonometry book for a suitable relation.

2.16. Make a scale drawing of the triangle in Prob. 2.15. Measure the altitude, considering the 20-cm side to be the base, and use the result to compute the area. Compare with the value calculated previously.

2.17. A conical wine glass is 8 cm deep, and the rim has a diameter of 8 cm. How many cubic centimeters of liquid will it contain when filled to a point that is 1 cm vertically below the rim?

2.18. In the metric system the name *liter* is given to a volume of 1,000 cm^3. How many liters of oil are needed to fill a spherical tank whose inside radius is 7.5 m?

2.19. Express each of the following verbal statements in denary notation:
   a. Number of animal species on earth: 2 million.
   b. Number of insect species: 80 percent of the above.
   c. Number of galaxies in the universe: 100 billion.
   d. Number of cells in the human body: 1 million billion.

2.20. Solve the following proportion for x, using all numbers in denary form:

$$\frac{x}{0.000023} = \frac{7.8 \times 10^7}{4.1 \times 10^4}$$

2.21. Estimate, to the nearest power of 10, how many cars there are in a bumper-to-bumper traffic jam 2 mi long on a three-lane highway. Assume the average length of a car to be 16 ft.

Chapter 3
MATTER: ITS PHYSICAL AND CHEMICAL BEHAVIOR

In Chap. 2 the concept of space was introduced in an intuitive way as something extending out in all directions—an indefinite expanse in which events of various kinds are observed to happen. This difficult concept takes on physical meaning only when we realize that space is populated by objects composed of **matter.**

All methods of measuring space are necessarily bound up with material bodies of some kind, an obvious example of which is a measuring stick. What about surveying methods, using triangulation? There must be a distant object upon which to sight. Even space measurements by means of radar require something ''out there'' to reflect the waves back to the observer. It becomes clear that what is determined whenever space is measured is the distance to or between material objects. Space takes on an aspect of reality only because of an order or arrangement of the things in it.

A universe entirely devoid of matter would be of comparatively minor scientific interest. There would be little for an observer to do and nothing with which an experimenter could work. In fact, there could be no scientist in the first place!

This chapter describes some of the general characteristics of matter and traces the development of the concepts and theories that have been devised to account for its behavior.

3.1 ATOMS

Not only do material objects punctuate space but they themselves occupy regions of space. It is natural to regard a piece of matter as the tangible, continuous, and permanent object it appears to be, but this view proves to be much too restricted. From earliest times, people wondered about the constitution of matter and tried to find a rational explanation of what they observed when substances were subjected to a variety of conditions. They found it hard to imagine that there was no limit to the extent an object could be mechanically broken down, and so they made the assumption that all matter must be composed of ultimate, *indivisible* particles which

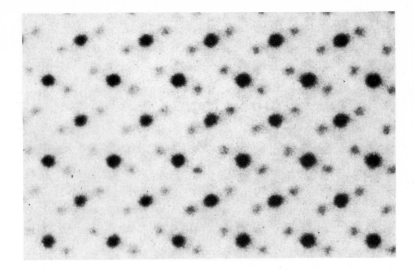

Arrangement of atoms in a crystal of marcasite, FeS_2, as shown by an x-ray shadowgraph. The large dark spots are iron atoms, the smaller ones sulfur atoms. Magnification is about 40 million diameters. (*Courtesy of M. J. Buerger.*)

The Greek word atomos *means, essentially, "indivisible."*

were named **atoms.** The introduction of this idea is ascribed to the Greek philosophers Democritus and Leucippus, about 400 B.C. Democritus believed that "nothing exists except atoms and empty space, all else is mere conjecture."

The atomic concept as a model for the structure of matter has been greatly extended and verified so that it now forms a firm basis for modern physics and chemistry. This concept pictures matter as composed of simple and, in some sense, fundamental particles. Compared with ordinary objects, these submicroscopic bits of matter are largely outside the range of first-hand experience. They cannot be handled, inspected, or calipered directly, yet there is overwhelming evidence for their existence, and countless indirect measurements of their characteristics have been made.

In attempting to make a conceptual model for the structure of matter it is often allowable, for some purposes, to think of an atom as an object that occupies a definite portion of space, but it will be shown that such a model is adequate only for certain purposes.

3.2 PHASES OF MATTER

Experience shows that the many different kinds of matter found in our surroundings can be sorted into three broad categories called physical **states** or **phases.** One category includes **solids,** such as stone or steel. The second comprises **liquids,** such as water or oil. The third includes **gases,** such as air or steam. Liquids and gases are known collectively as **fluids,** the phases of matter that can flow. It is possible to produce phase changes in all kinds of matter, and such transitions are often brought about by changes in the conditions of the environment, such as temperature, pressure, and so on.

The three phases of matter can usually be distinguished by observing how a sample behaves with respect to changes in its form or the amount of space it occupies (Fig. 3.1). A solid is characterized by its strong tendency to preserve both its original shape and volume, except when very strong forces are brought to bear on it.

A liquid lacks rigidity of form and will flow even under the influence of its own weight. However, like a solid, it has a definite volume and resists strongly any attempt to change that volume. Some materials, such as glass and certain plastics, appear to be solid at ordinary temperatures but are not true solids according to a more rigorous specification given below.

Gases have neither rigidity of shape nor definite volume. Any sample of a gas, no matter how small, will fill completely any vessel to which it is admitted.

Another method of classification, more precise than the above, is based on the atomic model of the internal structure of matter. This is a logical extension of Democritus' original "hunch" that matter is made up of aggregates of microscopic particles (atoms). According to the atomic concept, all materials are either **crystalline** or **amorphous** in structure.

"Amorphous" is from a Greek word meaning "without form."

In a crystalline solid the atoms are closely packed and held close to certain regularly spaced fixed locations by electric forces acting between the atoms. The spacing and arrangement of all these positions form an imaginary framework called the **space lattice** of the crystal. Each kind of crystal has its characteristic lattice whose form and dimensions can be investigated by means of x rays (Ref. 3.3). Two relatively simple examples of lattice structure are represented in Fig. 3.2.

In actual mineral crystals the atoms are not all arranged in the ideal way just described. There are many places where the lattice has **dislocations,** regions where the lattice is locally distorted and where the arrangement of the atoms departs from regularity (see Fig. 3.3a). Crystal **defects** may also be present: Some lattice points are unoccupied, or an occasional

Solid

Liquid

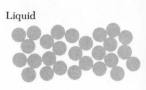

Gas

FIGURE 3.1 Schematic drawing showing the difference in molecular aggregation in a solid, a liquid, and a gas. The molecules must be thought of as constantly in motion.

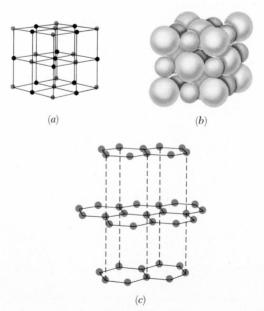

(a)   (b)

(c)

FIGURE 3.2 (a) The space lattice of common salt (NaCl) has a cubic structure. Black dots represent positions of Na atoms; colored dots, positions of Cl atoms. (b) A ball model of this structure. (c) Lattice structure of graphite. Dots represent the positions of the carbon atoms.

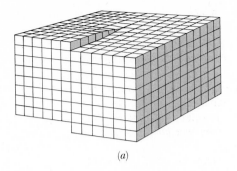

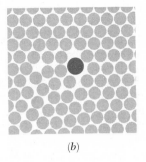

(a)

(b)

FIGURE 3.3 Typical imperfections in crystals: (a) a step dislocation; (b) an impurity defect. Imperfections such as these can move through the crystal.

foreign atom may be present among the ones that belong there (Fig. 3.3b). Such impurities can produce great changes in the mechanical, electrical, and other properties of some crystals, sometimes leading to important technical applications (page 427).

Large single crystals can often be grown from a solution or from a melted sample of the material. The external shape of such a crystal reveals features of the invisible lattice structure, but in most cases of crystallization the solid that forms is made up of a rigid mass of small individual crystals, stuck together in random positions. A single crystal of the same material is about a hundred times as strong as this aggregate.

3.3 ELASTICITY AND HOOKE'S LAW

A perfectly rigid, unyielding solid body does not exist. Under the influence of applied forces, objects made of even such strong materials as steel can be stretched, compressed, twisted, or bent. Any change in shape of a solid can be looked upon as a combination of two kinds of strain: 1) tension or compression and 2) shear. Think of a rectangular, solid block, indicated in Fig. 3.4. Pure stretch results when two oppositely directed outward tension forces are applied to the ends of the specimen. Linear compression is the result of a pair of opposing inward-directed forces. Shear is produced when the two forces act *along* rather than perpendicular to two opposite

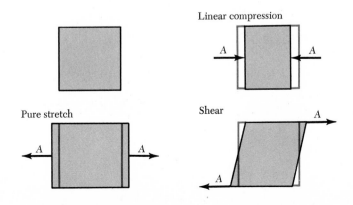

FIGURE 3.4 Basic deformations of a solid body. A rectangular block compressed, stretched, sheared by pairs of equal, opposite forces (shown as arrows).

ends of the specimen. Both stretching and compression are accompanied by slight changes in the cross-section of the sample.

The twisting of a shaft is a special example of shear (Fig. 3.5a), and the bending of a beam involves stretching on one side together with compression on the other (Fig. 3.5b).

The extent to which a body deforms under an applied force depends on such factors as the material of which it is made, its shape, and the temperature at which the experiment is carried out. Experience shows that, for deformations that are not very large, the extent of *the displacement is proportional to the applied force*. This generalization, called **Hooke's law,** is a typical example of an empirical rule, one derived from direct observation rather than from theory.

Robert Hooke (1635–1703) was a gifted English experimenter, a contemporary and scientific rival of Newton. Other important work of his will be described later, especially in connection with light.

The simple stretching of a steel wire is an example of the applicability of Hooke's law. If one end of a wire is clamped firmly in a support and increasing amounts of weight are hung from the lower end of the wire, the resulting increases in length will be found to be strictly proportional to the weight of the added load. Moreover, if kept within a certain limit, the process is reversible; the wire recovers its original length exactly after the load is removed (Fig. 3.6).

If the applied tension force is called F and the corresponding elongation of the wire s, Hooke's law may be written

$$F = ks \qquad\qquad 3.1$$

The value of the **force constant** k is found to depend on the particular wire specimen used and on the temperature.

The limiting point referred to above is called the **proportional limit** of the specimen, and the tension-elongation graph (Fig. 3.7) is a straight line up to this point. With further loading, this linearity no longer holds. As the tension is increased, a metal specimen is found to yield more readily than before and the curve bends over. During this part of the test, the metal actually flows. If the load is now removed, the sample does not return perfectly to its original length; it has acquired a permanent "set."

With continued loading, the specimen begins to pinch in at one place and rapidly reaches the breaking point. For brittle substances such as cast

FIGURE 3.5   Two examples of elastic deformation: (a) twisting of a rod; (b) bending of a beam.

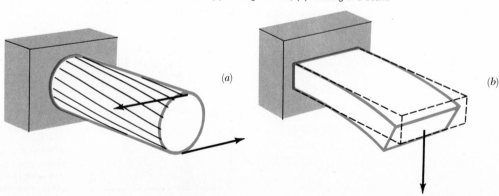

(a) (b)

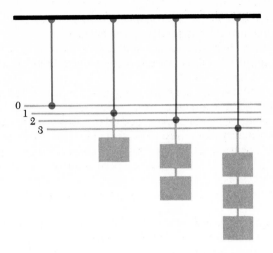

FIGURE 3.6 When a wire is stretched, its increase in length is proportional to the stretching force. This is strictly true, however, only over a limited range.

iron or glass, the breaking point follows closely upon the elastic limit, whereas more ductile materials such as copper or mild steel have a much longer region of plastic flow.

Hooke's law is valid, within the proper limit, for any kind of deformation of any phase of matter. The constant of proportionality k is always the ratio of the applied stress to the resulting strain. The k appropriate to each of the three principal kinds of stress—tension, compression, and shear—may be expressed as a combination of certain dimensions of the sample and a constant that is characteristic of the material of which it is made.

3.4 ATOMS AND THE PHYSICAL STRUCTURE OF MATTER

The preceding observations about the elastic behavior of materials can be explained by the atomic model described in Sec. 3.2. The elastic properties are connected with internal, that is, *interatomic,* forces in the material. These forces tend to resist changes in the relative positions of the atoms.

In a crystal, the internal forces are often pictured crudely as imaginary springs connecting the atoms. Since the atoms of elastic solids resist attempts to force them closer together or farther apart, these springs must be thought of as resisting compression as well as extension. A better

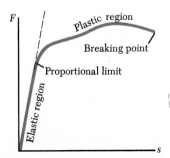

FIGURE 3.7 Tension-elongation curve for a metal. The straight portion represents the elastic region, where Hooke's law is valid.

mechanical model of a solid would be an aggregate of very sticky rubber balls in contact (Fig. 3.8). In devising models such as these, we use familiar large-scale entities such as balls and springs. But these objects themselves owe their behavior to precisely the forces that the model was devised to explain. This points up the inadequacy of mechanical models. Nevertheless, they continue to be used at times to satisfy the scientist's natural desire to visualize subjects he investigates.

The force exerted by one atom on another is found to be represented by a curve like the one in Fig. 3.9. The equilibrium distance, where the forces of attraction and repulsion balance, is usually about 3×10^{-10} m. If the atoms are forced closer together, there is a rapidly increasing repulsion. If the atoms are pulled farther apart, there is a more moderate force of attraction which becomes gradually weaker with increasing separation of atoms. Notice that Hooke's law is merely another way of saying that the force curve can be considered to be fairly straight in the immediate neighborhood of the equilibrium point. This makes the atom displacements very nearly proportional to the applied forces.

The crystal structures of many metals and certain other substances are determined by the tendency of the atoms to get as close together as possible. This kind of "close packing" can be attained in more than one way, as can be demonstrated by placing layers of marbles in a box (see Ref. 3.8).

Hooke simulated crystal shapes by stacking musket balls.

Most crystals are composed of two or more kinds of atoms (see Fig. 3.2), and there are crystals in which the lattice points are occupied by entire molecules, resulting in great complexity of form.

A true solid has the orderly internal structure characteristic of crystals. Many common materials that *seem* solid are not true solids. They are

Amorphous substances.

amorphous substances of which rubber, wool, glass, and plastics are examples. In many such substances the atoms tend to form chainlike molecules which may be thousands of atoms long; this fact is responsible for the characteristic behavior of these materials.

When a crystalline solid is heated, the atoms or molecules vibrate more and more vigorously around their normal lattice positions (Chap. 10). They

FIGURE 3-8 Crude mechanical representation of the forces between atoms by means of elastic spheres that adhere to each other.

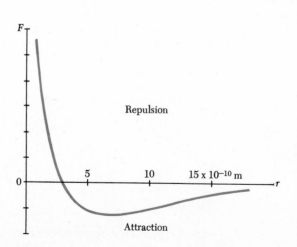

FIGURE 3-9 Graph of the force between two atoms of a solid. At a critical distance the force changes from one of repulsion to one of attraction.

X-ray examination shows that even in liquids some slight order exists, but it extends over only a few atoms or molecules.

are still held by the attraction of their neighbors, but with continued heating they finally break away altogether as the solid melts. The resulting mass of disordered molecules, virtually in contact but free to slide about, constitutes the liquid phase. The remaining force between molecules shows up in the resistance to flow of one layer over another. It is called **viscosity** and is exhibited especially by "thick" liquids such as molasses, paraffin, and even glass. A piece of glass or sealing wax will shatter under a sudden, sharp blow, acting like a brittle solid, but under a moderate, prolonged force it will flow like a very viscous liquid. Many features of the behavior of amorphous solids and liquids are still not completely understood, and so it is not always possible to make a sharp distinction between these two phases.

Liquid-surface effects.

There is another way in which the remaining attractive forces between molecules of a liquid become evident: They hold the molecules close enough together to give the liquid a definite surface. A molecule deep inside the body of liquid has no net force acting on it, apart from its weight, since the nearby molecules are uniformly distributed around it. However, a molecule at or very near the surface has fewer neighbors above it than in other directions, making the upward pull on such a molecule less than the downward pull (Fig. 3.10). The net downward pull, acting at all points, makes the free surface behave like a stretched membrane that tends to become as small as possible. This phenomenon is called **surface tension.** It manifests itself in the tendency of a free, unsupported body of liquid (a falling raindrop, for example) to become spherical because a sphere has a smaller surface area for a given volume of material than any other shape (see Prob. 3.8).

Surface tension produces notable effects where the surface of the liquid meets the sides of its container. Water molecules are more strongly attracted by a clean, solid surface than by other water molecules. As a result, the liquid surface curves abruptly upward where it touches the side of the vessel. If one end of a glass tube of small diameter is dipped into water, the liquid inside the tube rises some distance above the level outside. This phenomenon is observed for many other liquids and various tube materials. It is called **capillarity** and has many interesting technical and biological applications (Ref. 3.4).

Because of the regularity of arrangement of the atoms in crystals, it might seem that the solid phase would be the best understood form of matter.

Surface of liquid

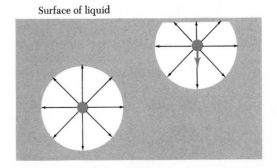

FIGURE 3-10 The net force exerted on a molecule that is near the surface is responsible for the surface tension of the liquid.

THE PHYSICS OF ENVIRONMENTAL POLLUTION

Many of the most common instances of air and water pollution can be described as unwanted mixtures of matter in various phases. One kind of mixture, called a **solution,** consists of individual atoms or molecules of one substance dispersed in another. Although solutions of solids in liquids (such as sugar in water) are most familiar, other phases also form solutions. Air itself may be considered a mutual solution of gases, mainly nitrogen and oxygen.

A frequent kind of air pollution occurs when certain gases are released into the atmosphere. The burning of fuels produces gases such as carbon monoxide, carbon dioxide, and sulfur dioxide. When these substances are present in the atmosphere even in minute quantities, they are dangerous to health and can cause great damage to property.

Another important class of phase mixtures has the name **colloids.** Unlike true solutions, the dispersed particles in a colloid usually consist of aggregates of atoms or molecules. These particles are not large enough to settle out appreciably nor are they visible in a microscope. They range in size from about 10^{-7} to 10^{-4} cm. Most of the human body is composed of colloids. Colloids, like solutions, are made up of various combinations of the three primary phases of matter. Table 3.1 classifies colloids as sols, gels, emulsions, and aerosols and gives examples of these types.

Most smoke consists of specks of ash or carbon, but some may be composed of metals or metal compounds. Liquid chemical wastes can poison streams and lakes to the extent that the water becomes unfit to drink. Fish and other forms of water life in whose bodies the contaminants become concentrated are killed off or rendered unsuitable for food. Even the oceans are in danger of becoming contaminated by oils and chemicals that are dumped into them.

Pollution of the environment is unquestionably one of the most serious problems of our time. Some experts believe that, before long, continued misplacement of materials, together with the physical, chemical, and biological processes that result, can lead to conditions where life on earth may become impossible.

TABLE 3.1   COLLOIDS

dispersed in →	SOLID	LIQUID	GAS
SOLID	Some alloys (others may be solutions)	Sols Starch "solution," milk of magnesia, mud, sludges	Aerosols Smoke
LIQUID	Gels (fine solid network containing liquid) Jellies, gelatin	Emulsions Milk (butterfat globules suspended in water)	Aerosols Fog, aerosol sprays
GAS	Foams Foam rubber, foamed plastics	Foams Shaving cream, fire-quenching foam	Not distinguished from true solution

However, in solids the closeness of the atoms and molecules greatly complicates the situation. This is also true for liquids, where the regularity of arrangement is almost entirely absent. The theory of liquids is still very incomplete.

The behavior of gases is simplest of all because the molecules are widely separated in space most of the time. This makes them independent of each other, and they are virtually unaffected by mutual attraction as long as the gas is not too highly compressed. The molecular model of a gas will be discussed in some detail in Chap. 10.

3.5 PHYSICAL AND CHEMICAL PROPERTIES; MASS

The three million or so different substances known to modern chemistry are distinguished from each other by their **intrinsic properties.** Each pure substance has a unique set of such characteristics: color, texture, melting point, as well as a large number of mechanical, thermal, optical, electrical, magnetic, and other properties.

Some of the elastic properties of materials have already been discussed.

In addition to such physical attributes, each substance has chemical properties that describe its behavior in chemical changes such as burning, rusting, decomposition, etc. A complete theory of matter would be one that could predict and account for the whole set of properties of any substance. This aim has been realized only partially up to the present time.

However, there is an extremely useful characteristic possessed by a given sample of any kind of matter regardless of its phase, temperature, chemical composition, etc. This characteristic of each material body is called its **rest mass.** Although the strict operational definition of rest mass is based (Chap. 6) on the extent to which applied forces affect the motion of a body, a somewhat intuitive but useful working definition can be given at this point: The rest mass of a body is a measure of the quantity of matter in it.

As an example of this notion, it seems reasonable to assume that there is more matter in a cubic foot of iron than in a cubic foot of cork. For any *pure* substance, whose atoms or molecules are all alike, the rest mass of a sample can be specified in principle by counting the atoms or molecules. A gold nugget containing 9×10^{23} gold atoms is three times as massive as one containing 3×10^{23} atoms. Twice a given mass of sugar would sweeten, to a given taste, just twice the number of gumdrops as the original amount. Pending a more exact formulation, the counting of atoms of a given kind best conveys the idea of the measurement of mass at this stage of our discussion.

The adjective "rest" applied to the term "mass" implies that the mass of a body is not a strictly invariable quantity. Experimentally, its value is found to depend to some extent on the *motion* of the body, especially at speeds approaching that of light. However, as long as the speed is low, the change is extremely small. The rest mass is the minimum value to which the mass of a body tends as the speed decreases to zero (rest).

In very violent physical or chemical changes, particularly those involving the nucleus of the atom, some of the rest mass of matter seems to disappear. The theory of relativity explains these changes in rest mass, but the operational definition must be put off to a later chapter because the expla-

The term energy comes from a Greek word for agency or force.

nation involves a physical entity called **energy.** Purely as a working description, energy may be said to be some sort of physical intermediary, different from matter as we ordinarily conceive it but capable of causing changes in matter. In somewhat the same sense in which the word is applied to certain people, energy might be described as an agency that "gets things done."

Since most physical situations do not involve extreme speeds or highly energetic processes, the mass of an object ordinarily does not depart measurably from its rest mass, and we often drop the qualifying objective "rest" and speak simply of the mass of the body.

3.6   STANDARDS OF MASS; THE KILOGRAM

The founders of the metric system provided a fundamental standard mass in the form of a cylinder of platinum alloy kept at the International Bureau of Weights and Measures, along with the standard meter bar. As in the case of length measurement, auxiliary units were chosen that differ from the primary unit by powers of 10. Table 3.2 lists some of the more frequently used metric mass units.

Originally it was intended to base the mass unit on the unit of length through the use of a standard substance, and for this purpose the kilogram was taken to be the mass of 1,000 cm^3 of water measured at the temperature at which it has its maximum density. Accurate measurement shows, however, that a standard kilogram of water occupies 1,000.027 cm^3 at the temperature referred to. This volume has been named the **liter,** and 0.001 liter is called a **milliliter** (ml). For all ordinary purposes, the milliliter and the cubic centimeter may be considered identical.

Density is defined on page 54.

One liter equals 1.057 U.S. liquid quarts.

How can the masses of ordinary objects be found in terms of the standard kilogram? It is a fact of experience that this can be done in a consistent and simple manner by *weighing* the body in question against standard masses, using an equal-arm balance. It should not be assumed that *weight* and *mass* are the same thing—a common misconception. They are fundamentally different kinds of physical quantities. Mass, according to the provisional definition on page 49, is a measure of *how much matter* a body

TABLE 3.2   METRIC UNITS OF MASS

UNIT	ABBREVIATION	RATIO TO PRIMARY UNIT	RELATION TO ENGLISH UNITS
Metric ton		1,000	
Kilogram	kg	1	Equivalent to mass of a 2.2046-lb object
Gram	g	10^{-3}	453.59 g equivalent to a 1-lb mass
Milligram	mg	10^{-6}	

Note that, in comparison with the metric length units in Table 2.1, the primary mass unit carries a prefix (kilo-). For a more complete table of multiples of units and their standard prefixes and abbreviations see page 631.

TABLE 3.3 ORDER OF MAGNITUDE OF MASSES

MASS, KILOGRAMS

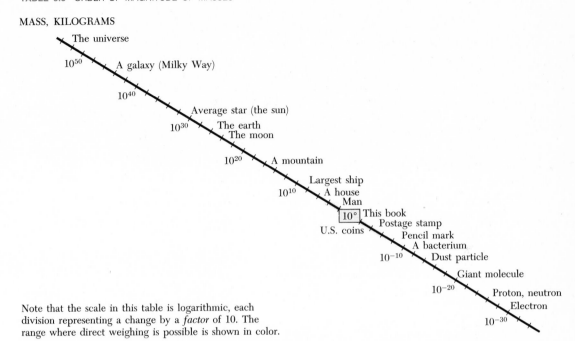

The universe
10^{50} — A galaxy (Milky Way)
10^{40}
Average star (the sun)
10^{30} — The earth
The moon
10^{20} — A mountain
Largest ship
10^{10} — A house
Man
10^{0} — This book
Postage stamp
U.S. coins — Pencil mark
A bacterium
10^{-10} — Dust particle
Giant molecule
10^{-20} — Proton, neutron
Electron
10^{-30}

Note that the scale in this table is logarithmic, each division representing a change by a *factor* of 10. The range where direct weighing is possible is shown in color.

contains; weight, on the other hand, is a *force*. Weight is merely a convenient name for the force of gravitational attraction between an object and the earth or other astronomical body near which the object happens to be located. A standard kilogram mass taken to the moon would still have a *mass* of 1 kg there, for the amount of matter in the body is the same as before. However, the *weight* of the object—the force of the moon's gravity on it—would be only about one-sixth of its weight on earth. Mass is a more fundamental quantity than weight, and the relation between the two will be explored more fully in Chap. 6.

3.7 CONSERVATION OF MASS

Comparison of masses, using an equal-arm balance, shows that this property is *transitive:* If a mass A is found to balance against any standard mass (such as the kilogram), and another mass B is also found to balance against the standard, then A will exactly balance against B. The mass of an object is found to be virtually unaffected by such operations as breaking the object into pieces, changing its phase, or even allowing it to undergo chemical changes. This fact, which is fundamental to all of chemistry, is sometimes called the law of **conservation of mass,** or conservation of matter. It is valid to a high order of accuracy for ordinary physical and chemical processes. However, as stated previously, this law is only approximately true and will be shown to be part of a broader conservation principle involving both mass and energy.

Direct weighing can check the conservation of mass to about 1 part in 10 million for samples of moderate size.

3.8   DIMENSIONALITY; SYSTEMS OF UNITS

Physics deals with measured physical quantities and their relations to each other. These two circumstances—that measurement is possible and that correlations between the resulting numbers can be found—characterize physics as the prime example of an exact science.

Physical quantities can be sorted into groups, all members of one group being basically of the same kind. For example, all lengths (distances) are of the same physical nature but do not belong in the same category as either masses or times. Although the operation of repeatedly laying down a foot rule to measure the distance across a room is not identical with the operations involved in measuring the distance to the moon, the two procedures are equivalent. All lengths or distances are said to have the same physical *dimension, L*, regardless of the particular *units* (inches, meters, miles, etc.) in which they may be measured.

Fortunately, many physical quantities are related dimensionally so that it is not necessary to employ as many different dimensions as there are quantities. A simple example is the relationship between length, area, and volume described in Sec. 2.6: An area is the product of two lengths, and so its physical dimension is L^2; a volume is given by the product of three lengths, and its dimension is L^3. In a later chapter it will be shown that both length and time are involved in specifying the speed of motion of an object. Also, a physical quantity called density is expressible dimensionally as a combination of length and mass. Many other examples will be given.

Funda-mental and derived quantities. As a result of the way in which they are defined, all physical quantities are found to be expressible as various combinations of a very small number of fundamental dimensions. The selection of the set of fundamental dimensions is a matter of taste and convenience. For mechanics, the branch of physics taken up in the next few chapters, the simplest choice is length L, mass M, and time T. All other mechanical quantities needed are derived quantities and can be expressed as various combinations of these three.

In discussing the phenomena of electricity and magnetism, a physical quantity called electric current will be introduced as an additional independent member of the set of fundamental physical dimensions. In 1960, an international conference on weights and measures officially added two more quantities, temperature and light source intensity, making six fundamental dimensions altogether. The complete set is the basis of the **International System.**

Dimension-less quanti-ties. Certain quantities used in mathematics and science are **dimensionless** because their nature cannot be specified in terms of fundamental units. Thus a biologist may express the rate of growth of a colony of bacteria in "organisms per hour," and a public health technician may describe the concentration of dust in the air in "particles per cubic inch." But "organisms" and "particles" are dimensionless because they are pure numbers, the result of counting individuals and not expressible by means of the set of dimensioned quantities. Consequently, the physical dimensions of the two quantities mentioned above are $1/T$ and $1/L^3$, respectively. Other pure numbers, such as 4 or π or $\sin \theta$ which may appear as factors of some of the terms in scientific formulas, do not contribute to the dimensions of these terms.

Some physical quantities are dimensionless because they express the

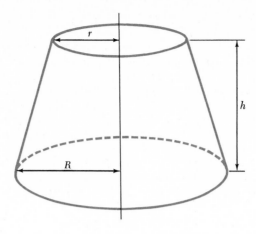

FIGURE 3-11 Computing the total surface area of a solid object having the shape of a lampshade.

ratio of two other quantities that have the *same* dimensions. An example, described in the next section, is called relative density. Other dimensionless quantities to be defined later deal with material properties such as heat capacity, the ability to bend light rays, electric and magnetic behavior, etc.

An important and useful aspect of the idea of dimensions is the matter of consistency: *In any valid physical equation, every term must have the same dimension.* This is merely a statement of the impossibility of adding apples to chairs or subtracting bicycles from books. Addition and subtraction are meaningful only if the things involved are all of the same kind. Dimensional consistency is often helpful in detecting mistakes in the derivation or recording of scientific formulas.

Worked example 3.1

The total surface of the frustum of a right circular cone (lampshade-shaped solid) is given by the geometric formula

$$A = \pi(R + r)\sqrt{h^2 + (R - r)^2} + \pi(R^2 + r^2)$$

where R is the radius of the larger base, r is that of the smaller, and h is the altitude (see Fig. 3.11). Show that the formula is dimensionally correct.

Solution Each of the quantities R, r, and h has the dimension L, and π is a pure number (dimensionless). Then the dimension of the first term on the right is $L\sqrt{L^2} = L^2$, and that of the second term is also L^2. Thus each term in the equation is dimensionally an area, as it must be.

Practice set*

Although no two of the quantities below are stated in the same units, arrange them in the following groups according to their physical dimensions: (a) L; (b) L^2; (c) L/M; (d) M/L^3; (e) all others. Use a dictionary to find the meaning of any unit with which you are not familiar.

1. 5,000 m²   5. 7.5 ft/sec   8. 1.3 g/cm³
2. 2.5 kg/m³   6. 3 rods   9. 15 knots
3. 12 acres   7. 27 cm²/kg-m   10. 3.1 hectares
4. 4 kg/m

*Ans.: 6; 1, 3, 10; 7; 2, 8; 4, 5, 9.

Two commonly used metric L, M, T systems.
The definition of mass now takes its place alongside those of length and time, making it possible to set up *systems of units* for the fundamental quantities L, M, and T. Based on metric units of various kinds, two systems have come into general scientific use. They are named (1) the *centimeter-gram-second* (cgs) *system* and (2) the *meter-kilogram-second* (mks) *system,* according to the fundamental units of L, M, and T adopted. The latter system proves to be more practical and convenient in most applications and will be used almost exclusively in this book.

3.9 DENSITY

One might hear a statement such as "Iron is heavier than wood." This statement cannot be literally true, since an iron nail is certainly not as heavy as a wooden log. Evidently some property of a *material* rather than of a particular object is involved, and its specification requires space units as well as weight. Thus the statement above should be changed to "A piece of iron is heavier than a piece of wood *of the same volume.*" The **weight density** of a material is defined as the weight of any sample divided by its volume. A cubic foot of iron is found to weigh about 490 lb, so that the weight density of iron may be stated as 490 lb/ft³.

It is often convenient to refer the weight density of a material to that of water as a standard. The ratio may be called the **relative density** of the material. Thus water has a weight density of 62.4 lb/ft³, and so the relative density of iron is 490/62.4 = 7.85. This is a *pure number,* since it is the ratio of two quantities having the same dimensions, lb/ft³.

Inasmuch as the weight of a body changes with its location in the universe (page 51), the value of the weight density will also change. For this reason it is useful to define a more fundamental quantity, called the **mass density:**

$$D = \frac{M}{V} \qquad\qquad 3.2$$

The mass density of a substance is measured by the mass per unit volume. The physical dimensions of this quantity are M/L^3. Values expressed in the mks unit, kg/m³, are tabulated in Table 3.4 for several ordinary substances. Values encountered elsewhere in the universe range from the estimated density of matter in "empty" space, of the order of 10^{-20} kg/m³ (or roughly one atom per cubic centimeter), to that of an atomic nucleus, 2×10^{17} kg/m³, or about a billion tons per cubic inch.

Practice set*
1. A cubical block of stone 0.50 m on an edge has a mass of 375 kg. What is its mass density?
2. What is its relative density?
3. What volume of hydrochloric acid (mass density 1,200 kg/m³) must be measured out for a chemical process requiring 240 kg of this acid?

From the definition, the mass density of a given material may be deter-

*Ans.: 3,000; 3; 0.2.

TABLE 3.4 DENSITIES OF FAMILIAR MATERIALS

SUBSTANCE	MASS DENSITY, kg/m³	RELATIVE DENSITY
Solids:		
Platinum	21,400	Since the mass density
Gold	19,300	of water is 1,000 kg/m³,
Lead	11,400	the relative density of
Iron	7,900	each substance is the
Aluminum	2,700	value in the preceding
Magnesium	1,750	column divided by 1,000.
Ice	917	
Wood (pine)	500	
Cork	240	
Liquids:		
Mercury	13,600	
Sea Water	1,030	
Pure water	1,000	
Petroleum	878	
Alcohol, ethyl	790	
Gases (at standard temperature and pressure):		
Air	1.29	
Helium	0.178	
Hydrogen	0.090	

mined by dividing the mass of a sample of *any* size by its volume. This is true for homogeneous substances, but not otherwise. If the substance is not homogeneous, the mass of a sample divided by its volume gives the *average* mass density of that sample. It is not always easy to decide whether a material is homogeneous. Substances that appear to be so on a large scale may show great inhomogeneities on a small scale. The value obtained for the density will then depend strongly on the size of the sample chosen for the determination.

Worked example 3.2 In redesigning a certain machine, a solid iron ball of mass 0.200 kg is to be replaced by an aluminum ball of the same size. What is the saving in mass?

Solution From Eq. 3.2 the volume of the ball is given by $V = M/D$. With the value of D for iron from Table 3.4, this becomes

$$V = \frac{0.200 \text{ kg}}{7,900 \text{ kg/m}^3} = 2.53 \times 10^{-5} \text{ m}^3$$

The aluminum ball will have the same volume. Its mass, given by $M = DV$, will be

$$M = (2,700 \text{ kg/m}^3)(2.53 \times 10^{-5} \text{ m}^3) = 0.068 \text{ kg}$$

The difference in mass amounts to 0.200 − 0.068, or 0.132 kg.

3.10   PHYSICAL EVIDENCE FOR ATOMS AND MOLECULES

It has already been pointed out that atoms and molecules, unlike ordinary objects, are beyond direct experience. However, in science less direct evidence than that offered to our immediate perceptions must often be accepted. In this sense, individual atoms and molecules can be counted, measured, and weighed.

There are physical means of measuring atoms, using the field-ion microscope, electron microscope, x rays, and other methods to be discussed later. All such methods agree in yielding molecular sizes of the order of a few times 10^{-10} m, while some "giant" molecules found in living matter are about a hundred times as large.

A simple way of estimating the size of a molecule is to place a very small but measured volume of oil on a clean water surface. The oil is observed to spread out over the surface to cover a definite area, and this area is found to be accurately proportional to the volume of oil used. This must mean that the film spreads out until it attains a certain minimum thickness.

A detailed study by x-ray and electron-microscope methods shows that such an oil layer consists of a single layer of rodlike molecules of oil standing endwise on the water surface, like blades of grass on a lawn. Since these oil molecules are virtually in contact, side by side, the order of size (length) of a molecule is found by dividing the volume of the original drop by the area over which it is observed to spread. In this way, such oil molecules are found to have sizes of the order of 10^{-9} m (see Prob. 3.16).

3.11   ELEMENTS AND COMPOUNDS; LAW OF DEFINITE PROPORTIONS

Robert Boyle (1627–1691) was a British alchemist and natural philosopher. In physics he is perhaps best known for the gas law named for him. Antoine Lavoisier (1743–1794) was a great French chemist who lost his life during the French Revolution.

The suggestion of Democritus regarding the atomicity of matter was kept alive through the ages, but it remained little more than a conjecture and did not produce any quantitative results until the eighteenth century. With the development of modern chemistry, a whole new line of evidence for atoms and molecules came into existence when purely chemical means of measuring these entities were devised. These ideas will now be traced briefly.

From the chemical point of view, the notion of the atomicity of matter is connected with the idea of **chemical elements.** Since earliest times, people had tried to explain the great variety of available materials in terms of a small number of fundamental substances. Aristotle believed that all materials on earth were made up of fire, air, water, and earth. In the Middle Ages mercury, salt, and sulfur were thought to be the only elements. Around the middle of the seventeenth century, an operational definition (page 28) of a chemical element was suggested by Boyle and later refined by Lavoisier. Their definition is as follows: An element is a substance that cannot be broken down into simpler substances by chemical processes.

With this criterion available, modern chemistry was able to get its start. By Lavoisier's time, about 30 elements had been identified. Of about 110 elements known at present, 92 are found in nature, and the rest can be

In the field-ion microscope, an electrified needle point repels gas atoms that spread out and strike a fluorescent screen to make patterns of the kind shown here. Since the gas atoms come predominantly from places where metallic atoms are located on the needle, the bright dots reveal the arrangement of the atoms of the metal. (*Central Scientific Company.*)

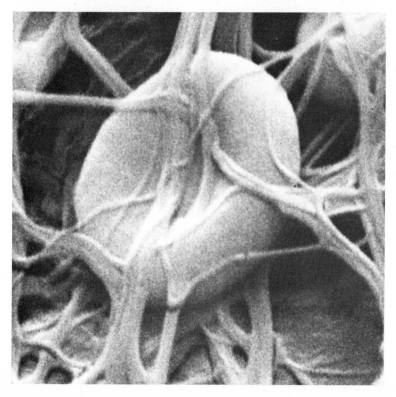

Electron micrograph of a red blood cell enmeshed in fibrin in a blood clot. The magnification attained here is about 10,000. The image was formed in a scanning electron microscope, in which an electron beam sweeps over the object. Reflected electrons form the picture, as in a TV picture tube. (*Courtesy of Emil Bernstein and Eila Kairinen, Gillette Company Research Institute.*)

prepared artificially (Chap. 22). All the rest of the more than 3 million pure substances known to modern chemistry are **compounds** formed by combining these relatively few elements in various ways. A chemical compound is a new substance having properties different from those of the elements composing it. It may be formed directly from its constituent elements or by interactions of other elements and compounds. Typical chemical actions are rusting, combustion, dissolution by acids, etc.

The majority of familiar materials are not single compounds but physical mixtures in which the ingredients keep most of their original properties. Mixtures can be separated out again by comparatively simple means, such as evaporation, solution, passage through filters, etc.

The true distinction between mixtures and compounds is described by an empirical law. This **law of definite proportions** states that each compound contains its constituent elements in a definite proportion by mass. For example, when any sample of pure water is decomposed into its constituent elements, hydrogen and oxygen, the masses of these gases are always in a constant ratio whose magnitude is about $1:8$. This ratio is a characteristic of the compound water. Conversely, when water forms by the chemical combination of hydrogen and oxygen, the two elements are needed in this same ratio. Any amount of one or the other that is present in excess is simply left behind after the reaction. On the other hand, solutions of salt in water can be made in arbitrary proportions. However, these solutions will differ from each other in certain properties ("saltiness," density, etc.) whereas all samples of pure water have identical properties.

3.12   MULTIPLE PROPORTIONS; THE WORK OF DALTON

Some pairs of elements are found to be capable of forming more than one compound, and this introduces a complication. For instance, in addition to the ability of hydrogen and oxygen to form the compound water, they can unite to produce a quite different compound called hydrogen peroxide. Analysis of this substance shows that the two elements are present in the mass ratio $1:16$, as compared with the ratio $1:8$ for water. Also, the two most abundant gases in the atmosphere, nitrogen and oxygen, can combine chemically in five distinct ways. Referred to a given mass of nitrogen, the masses of oxygen in these compounds are exactly 1, 2, 3, 4, and 5. Such simple relationships, which are found to hold also for compounds of more than two elements, are summarized in a generalization called the **law of multiple proportions.** The law of multiple proportions states that, if two elements are able to form several distinct compounds by combining in more than one proportion by mass, the various proportions are always in simple ratios to each other.

As a numerical illustration of multiple proportions, consider two compounds of iron and oxygen. Chemical analysis shows that one of these compounds, called ferrous oxide, contains oxygen and iron in the ratio of 0.286, by mass. In any sample of ferrous oxide, this is the invariable ratio of the two constituents. In another oxygen-iron compound, ferric oxide, the ratio is always found to be 0.429. Each of these compounds, by itself, illustrates the law of *definite* proportions. But the two numbers are not unrelated: The ratio 0.429 divided by the ratio 0.286 is exactly 3:2, an example of the law of *multiple* proportions.

John Dalton (1766–1844) was a self-educated scientist who taught mathematics and natural philosophy in Manchester. His work did much to make chemistry a separate branch of science.

The remarkable simplicity of the two laws of chemical combination was brilliantly explained by the English chemist Dalton at the beginning of the nineteenth century. His theory furnished the basis for modern atomic concepts.

Dalton made the assumption that each element is composed of small, indivisible particles (atoms) that are all alike but that differ in properties from one element to another. He further assumed that the atoms retain their identity during chemical reactions and that when elements combine to form a compound, certain numbers of atoms of each element join to form a group called a **molecule** of the compound. All molecules of a given compound are identical.

The proportion laws stated above follow readily and naturally from the following assumptions: Each kind of atom may be considered to have a characteristic mass, different from that of any other kind. Then, in the formation of a molecule of a compound, certain integral numbers of atoms of each of the constituent elements are brought together. This accounts for the constancy of the ratio of the masses of these elements in the particular compound under consideration.

For example, suppose that the atoms of element A have a mass of 2 units and those of B have a mass of 3 units, and that a molecule of the given compound of the two is made up of 3 atoms of A to 1 of B. Since the compound is made up of nothing more than such molecules, chemical analysis will show that the two elements are always present in the definite mass ratio of $(3 \times 2):(1 \times 3)$, or 2:1. If there exists another compound of A and B whose molecules contain, say, 2 atoms of A to 1 of B, the ratio will be $(2 \times 2):(1 \times 3)$, or 4:3—another simple ratio. The proportions for the two compounds will be in the ratio $(2/1):(4/3)$, or 3:2, which is an illustration of the law of multiple proportions.

Historically, the law of multiple proportions was obtained experimentally only after it suggested itself to Dalton in 1803, as a result of his hypothesis.

In summarizing the chemical approach to the atomic concept, it must be pointed out that the two proportion laws, based on quantitative experiments with gross matter, led compellingly to the idea of the existence of atoms. Dalton's great accomplishment was the realization that the *whole-number* aspects of these laws reflect the *atomicity* of matter.

Besides providing chemical evidence for the existence of atoms, Dalton's conceptual model pointed the way to the calculation of the relative masses of the atoms, culminating in the chemist's table of atomic weights. But before this could be achieved, certain assumptions made in Dalton's work had to be modified in the light of some developments from the physical side, to be discussed in Chap. 10.

Programmed review

Instructions: See page 18.

1.   What is the fundamental assumption of the atomic theory of matter?

    All matter is composed of indivisible particles called atoms.   [3.1]

2. Name the three principal phases of matter.

Solid, liquid, gas.   [3.2]

3. State Hooke's law of elasticity in its general form.

For sufficiently small deformations, the deformation is proportional to the applied force.   [3.3]

4. Describe the atomic picture of a crystalline solid.

Over limited regions, the atoms or other entities are arranged on an imaginary framework called the space lattice.   [3.2]

5. What is meant by the surface tension of a liquid?

The tendency for the free surface to act like a stretched membrane. This is a result of the interatomic forces of attraction.   [3.4]

6. Give a working definition of the property of a body known as its mass.

A quantity that measures the amount of matter or the numbers of atoms of various kinds in the body.   [3.5]

7. State the principle of the conservation of mass.

The mass of a sample of matter is unaffected by any physical or chemical changes it may undergo (very approximately true).   [3.7]

8. Compare the terms *dimensions* and *units* when applied to a physical quantity.

The dimensions indicate the physical nature of the quantity, e.g., a length, a density, etc. Each dimension may be expressed in a choice of units; length, for example, in centimeters, feet, miles, etc.   [3.8]

9. State the law of definite proportions.

In each compound, the elements are present in a definite proportion by mass.   [3.11]

10. State the law of multiple proportions.

If two elements form several compounds by combining in more than one proportion by mass, the various proportions bear simple ratios to each other.   [3.12]

For further reading

3.1. *Nash, L. K.* "The Atomic-Molecular Theory." An excellent historical reference.
3.2. *Müller, E. W.* Atoms Visualized, *Sci. Am.,* June, 1957. The field-ion microscope described by its inventor.
3.3. *Holden, A.,* and *P. Singer* "Crystals and Crystal Growing." Includes instructions for growing mineral crystals of several types.
3.4. *Boys, C. V.* "Soap Bubbles and the Forces Which Mould Them." A classic on bubbles and other surface-tension phenomena.

3.5. *Bernal, J.* The Structure of Liquids, *Sci. Am.,* August, 1960.

3.6. *Jaffe, B.* ''Crucibles: The Story of Chemistry.'' Great chemists and their work.

3.7. *Evans, J. C.* Units and Standards of Measurement in the Physical Sciences, in Penguin Science Survey 1964·A.

3.8. *Jones, G., J. Rotblat,* and *G. Whitrow* ''Atoms and the Universe.'' The making of models of solids by laying down marbles is described on pp. 125–128.

3.9. *Kitaigorodskiy, A.* ''Order and Disorder in the World of Atoms.'' A readable, nonmathematical discussion.

Questions and problems

3.1. When a 50-lb weight is hung by a wire 12 ft long, the wire stretches $\frac{1}{8}$ in. but still remains within the elastic limit. What is the value of the force constant k for this wire? (Express in appropriate units.)

3.2. Compute the value of the force constant of a steel wire of length 20 ft if it stretches 0.010 ft under a tension of 23 lb.

3.3. A spiral spring is 6 in. long and has a force constant of 10 lb/ft. Compute the force constant of a 3-ft length of the same kind of spring. HINT: Consider what happens to each 6-in. length of the longer spring as it is stretched.

3.4. The value of the force constant for the stretching of a certain steel wire is 2.3×10^3 lb/ft. A cable is made up of a parallel bundle of 20 such wires. How much does the cable stretch when it supports a load of 5,000 lb?

3.5. A cast-iron column in the form of a hollow cylinder is used to support the floor beams of a house. In tension tests on a 1-ft-long section of the column, a force of 2,130 lb is found to stretch the sample by 5×10^{-4} in. How much will a column 8 ft long compress under a load of 4,000 lb?

3.6. A steel measuring tape used in surveying is accurate only when stretched with a prescribed force. If such a tape is used with too great a stretching force, will measured distances be too large or too small? Will the departures in accuracy be greater for large distances or small ones? Does the use of the wrong stretching force affect the *precision* to which the readings can be made—that is, how well they agree among themselves?

3.7. In closest packing, how many spheres can be made to surround a given sphere, all in contact? Try it by sticking marbles together with plastic cement. Answer the same question for the two-dimensional case by seeing how many coins can be laid on a table, just touching a central coin and touching each other.

3.8. On page 47 it was stated that a given volume of material has a minimum surface area when it is in the form of a sphere. Show that this is plausible by computing the total surface area of 64 cm^3 of putty when in the shape of a cube and again when rolled into a ball.

3.9. A certain object, when weighed here on earth by means of an equal-arm balance, can be counterpoised by 26 standard kilograms. The weighing, by the same procedure, is repeated on the surface of the moon. How many standard kilograms are now required for balance?

3.10. Express the dimensions of the following quantities in terms of L, M, and T:
   a. Length of plastic fiber extruded each second in nylon manufacture.
   b. Mass per unit length of a suspension-bridge cable.
   c. Relative density of a substance.
   d. Rate at which the surface area of a balloon increases as it is inflated.
   e. Rate at which the volume of the balloon increases.
   f. Number of passengers carried by a suburban railroad each month.

3.11. The mass of a gold atom is 3.28×10^{-22} g. Calculate (a) the mass of the smaller nugget referred to on page 49 and (b) the area of the sheet of gold leaf that could be made from it if beaten out to a thickness of 10^{-5} cm.

3.12. Chemists sometimes find it convenient to speak of the **specific volume** of a substance, defined as the volume occupied by unit mass of the material. How is specific volume related to mass density?

3.13. A cubic foot of solid quartz weighs 162 lb, and a cubic foot of quartz sand weighs only 144 lb because of spaces between the grains. What fraction of the volume of this sand is occupied by voids?

3.14. One square meter of a special thin copper foil used in a physics research problem has a mass of 2.2×10^{-3} kg. The relative density of copper is 8.9. How many atoms thick is this foil (order of magnitude only)?

3.15. According to a law of flotation discovered by the Greek philosopher Archimedes in the second century B.C., a solid body will just float in a liquid if the average density of the solid is equal to that of the liquid. How thick a wall should an aluminum sphere have in order to do this? Express the result as a fraction of the radius.

3.16. A speck of oil of mass 1.1×10^{-6} kg is placed on a clean water surface and is observed to spread out over a roughly circular area of radius 0.18 m. If the relative density of the oil is 0.90, compute the thickness of the film and hence the size of an oil molecule. Estimate also the volume of such a molecule, assuming it to be a cylinder whose diameter is about one-fourth of its length.

3.17. What volume of oil of relative density 0.90 would be needed to cover the surface of a swimming pool of area 70 m² if the length of a molecule of this oil is 1×10^{-9} m? Assume the oil layer to be one molecule thick.

3.18.   A wholesaler sells tea and coffee in cartons, each of which contains 10 bags of tea and 4 bags of coffee. Each bag of tea weighs 0.75 lb and each bag of coffee 2.5 lb. After unpacking a shipment, a grocer finds that he has 30 lb of tea on hand. How many cartons did he order? How many pounds of coffee did he buy? To which of the laws of chemical combination is this situation analogous? Explain.

Chapter 4
VECTORS

4.1 MECHANICS AND EXPERIENCE

Earlier chapters of this book introduced certain ideas such as space, time, and matter, which are fundamental subjects of discourse in the study of the physical world. In particular, methods of measuring the position and extension of bodies in space were described, and the phases and properties of matter were discussed.

An important aspect of phenomena in space has not yet been touched upon, and that is the matter of *direction*. Since we are interested in physics rather than abstract geometry, our involvement with direction in space is bound up with the motion of material bodies and therefore with **force,** the agency that affects motion.

The part of physics concerned with the action of forces on the motion and condition of material bodies is called **mechanics.** The next few chapters will be concerned with this subject.

Mechanics attained a high degree of development early in the history of science, and people turned to it for the explanation of many diverse observations and experiences. Its evolution was comparatively rapid because the subject deals largely with simple devices that can be visualized and handled—weights, springs, levers, pulleys, etc. Our acquisition of mechanical judgment and intuition begins almost at birth: Standing, walking, lifting weights, and riding a bicycle are mechanical operations. It is interesting to note that all organisms, regardless of what other senses they may have, respond to mechanical stimuli.

This chapter shows how to deal with a class of physical quantities having direction in space and how to apply these methods to forces in particular.

4.2 FORCE AND ITS MEASUREMENT

Any mechanical operation that we perform involves a push or a pull, as exemplified by the direct use of our muscles. Tools, machines, or engines can be used to reduce the muscular effort needed for a given task or to replace it altogether. In any event, it may be said that we or the devices exert **force.** Forces are capable of deforming bodies, deflecting their motions, speeding them up, or slowing them down. In order to account for the varied physical phenomena with which we have experience, it has become necessary to assume the existence of a number of apparently unrelated types of force. Gravitational force is one kind. Other forces are related to electricity, to magnetism, and to the subatomic particles of matter, and these will be discussed later. For the present, a force may be taken to be anything capable of producing or affecting motion.

Gravitational force is familiar in the form of the *weight* of objects, and it is often convenient to use forces of this kind to prescribe the measurement of force in general. This can be done by taking the unit of force to be the attraction of the earth for a standard mass located at a standard place, say at sea level and 45° north latitude. In this way it is possible to define, operationally, forces of one pound, one ton, etc. Then, by means of hanging "weights" (really suspended masses), forces of desired amounts can be exerted. These forces can be made to act not only in the downward direction but—by using pulleys, levers, or other mechanical elements—in any other direction in space.

A convenient method of exerting measured forces is by means of a spring scale. First the divisions of the instrument are determined by hanging standard masses from the spring and marking the rest positions of the pointer accordingly. Thus the position of the pointer for a 5-lb suspended body is marked "5 lb." Strictly, it would be better to label this point "5-lb wt" or "5-lb force," indicating that the scale shows *force* rather than mass. Such units of force will be used only temporarily, pending a better definition in Chap. 6.

Many of the words used in ordinary affairs have been taken over in scientific work, where they are given special, technical meaning. Force is one of them. In everyday life one may hear such phrases as "the force of a body," or "the force working on a body," etc. These expressions are vague and should never be used in science. Instead, it should be said that a force *is applied to* a body, *acts on* a body, *is exerted by* one body on another. Such distinctions must be observed because in science it is essential to say exactly what is meant, insofar as the limitations of language permit.

4.3 SCALAR AND VECTOR QUANTITIES

Many of the quantities discussed in physics as well as in everyday affairs can be adequately specified by stating a *number,* along with its appropriate units. These are called **scalar** quantities. Examples are numerous: Age and height are scalars, and so is the volume of a stone, the temperature of a room, or the relative density of a new plastic.

However, there are some of the quantities which require, for their complete description, a statement of *direction in space* in addition to a numerical magnitude and its units. Force is an example. It is not sufficient to say, "A force of 10 lb acts on this body." If we want to investigate the effect of this force, its direction must be stated as well. The complete statement might be: "The body is acted upon by a force of 10 lb (how much) toward

"Downward" means simply the direction toward the center of the earth, as indicated by the rest position of a mass suspended by a cord (plumb line).

the northwest (direction)." Notice that, in describing a direction in space, the coordinate system used must be specified or readily inferred. In this example, it is the local geographical system, and the implication is that the force lies in a horizontal plane. *Weight* is a force, and it would seem necessary to say something about its direction, although in this case the vertically downward direction is understood.

An even simpler example of a quantity involving direction in space is called a **displacement.** A displacement is a description of the net change in location of an object. Regardless of the particular route it may take, the displacement of an airplane going from New York to Philadelphia could be described as "90 mi in a direction 37° west of south."

Any quantity, such as force or displacement, that requires a statement of direction in space as well as of magnitude is called a **vector** quantity. Other useful vectors such as velocity, acceleration, momentum, electric field intensity, and many more will be introduced from time to time.

Practice set* Which of the following quantities are vectors?

1. Capacity of an oil tank
2. Mass of a hydrogen atom
3. Total force of the wind on the side of a house
4. Interval between eclipses of the sun
5. Density of a gold alloy
6. Description of the result of a move in a game of chess
7. Area of a parking lot
8. Number of pages in this book

Vectors lend themselves naturally to diagrammatic representation. A vector is depicted by drawing a straight-line segment whose direction is that of the vector and whose length is made proportional to the magnitude of the vector. The segment starts at the point where the vector is assumed to act, and an arrowhead, pointing in the appropriate sense, is drawn at the other end of the segment. Figure 4.1 shows a force of 200 lb applied to the point A and acting in a direction 30° north of east. The vector was drawn to a scale such that 1 in. represents a 100-lb force.

*Ans.: Only 3 and 6 are vectors.

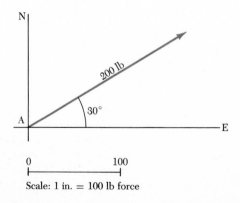

FIGURE 4.1 Vector representing a force. The accompanying scale gives the magnitude of the vector in terms of its length on the drawing.

4.4 COMBINING VECTORS

Can two or more forces act on an object at the same time? To answer this question one need only think of attaching several cords to a body and pulling on each in a chosen direction with a definite amount of force (Fig. 4.2). This situation suggests two further questions: Is there a single force that could take the place of (produce the same net effect as) the existing set of forces? If so, how could its direction and magnitude be found? Obviously, if such a force could be determined, the study of many mechanical problems would be considerably simplified.

To answer these questions about forces, consider first the case of several *displacements* applied to an object. The reason for doing this is that the effects of displacements are more directly visualized than those produced by forces. Moreover, experience shows that forces, which are also representable by vectors, can be combined in exactly the same way as displacements.

To keep the situation as simple as possible, think of two displacements applied, one after the other, to a **particle.** A particle—an important concept in mechanics—is any object that is small enough for the purpose at hand. It may also be described as a "point mass" and becomes a convenient model for a material body so small that any internal structure, rotation, or deformations that it may have are of no importance to the particular problem. Even an electron, small as it is, on a human scale, cannot be treated as a particle in some situations where it must be considered a

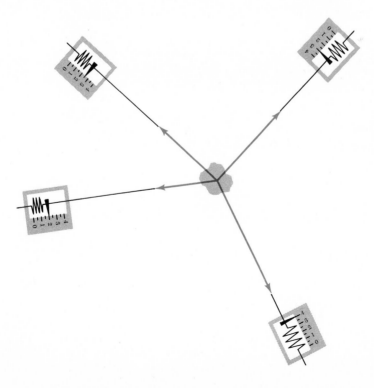

FIGURE 4.2 Body acted upon by several forces simultaneously, Can the net effect of all these forces be found?

spinning body. On the other hand, for many problems in astronomy, the whole earth may be considered a particle. On diagrams, a particle is indicated by a mathematical point.

Figure 4.3a shows a displacement **a** of 11 in. to the right applied to a particle originally located at A. The result is to bring the particle to location B. A second displacement **b** of 7 in. in the indicated direction leaves the particle at C. Evidently the same result could have been obtained by moving the particle directly from A to C as indicated by the single displacement vector **R**. This equivalence may be expressed as an equation:

(Displacement from A to B) + (displacement from B to C)
$$= (\text{displacement from } A \text{ to } C) \quad \textbf{4.1}$$

If it is agreed to represent vectors always by boldface letters and to represent their *magnitudes* by the corresponding letters in italics, the above relation can be written

$$\mathbf{a} + \mathbf{b} = \mathbf{R} \qquad\qquad \textbf{4.2}$$

Recall the geometric axiom about the shortest distance between two points.

It is essential to notice that this does *not* say that the sum of the *magnitudes* (lengths) of the first two vectors is equal to that of the third; Fig. 4.3a shows that $a + b$ must in general be greater than R.

The vector **R** is called the **resultant** of vectors **a** and **b**, and the process of finding the resultant is referred to as **combination, composition,** or **addition** of vectors. The word "addition" is used here in other than its usual arithmetic sense. This points up the operational meaning of the vector concept. Thus Eq. 4.2 states only that the operation of moving a particle from A to B, followed by that of moving it from B to C, is equivalent to the single operation of moving it from A to C.

Order in which vectors are added is immaterial.

Vector addition, like ordinary arithmetical addition, can be carried out in any desired order and the result will be the same. Figure 4.3b shows the addition of the same two vectors that appeared in Fig. 4.3a, but in reverse order. The resultant is the same as before. Vectors have an obvious property: They may be moved around in space without changing their essential characteristics. As long as direction and length are preserved, it is the same vector. This suggests that, in the above example, we can think of the two displacements as being applied *at the same time,* rather than one after the other. This can be realized by having two experimenters move a particle simultaneously, one moving it to the right a distance of 11 in. while the other, paying no attention to what his partner is doing, takes care to move it in the direction of the slanting vector for a total distance of 7 in. The net effect would be to land the particle at C, as before.

(a) (b)

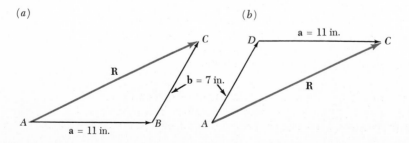

FIGURE 4.3 The order in which the vectors are taken is immaterial; the resultant is always the same. This figure illustrates this fact for two vectors **a** and **b**.

4.5 THE POLYGON RULE

The scheme just described for combining two vectors by placing them end to end can be generalized for any number of vectors. For example, three displacements **a, b,** and **c** may be applied to a particle (Fig. 4.4) by first laying off **a** followed by **b** and then drawing the resultant **r.** We are now entitled to consider **a** and **b** as having been replaced by their combined effect **r.** Then we can combine **r** in turn with the remaining vector **c** by laying off **c** from the end of **r.** Proceeding as before, we find that the resultant of these two is **R,** which is the actual resultant of the three original vectors **a, b,** and **c.**

Evidently we need not have troubled to draw the intermediate resultant **r** but could have proceeded directly by laying off the given vectors, tail to head, and then drawing **R** at once. The scheme is not restricted to three vectors but holds for any number. Furthermore, as already indicated, the same resultant will be obtained regardless of the order in which the original vectors are taken (see Prob. 4.2).

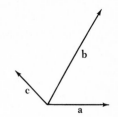

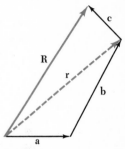

FIGURE 4.4 Resultant of three vectors. The vectors are laid off, end to end, in any order. The side that closes the polygon represents the resultant.

Newton intuitively understood this characteristic of vectors, for he stated a principle of the independence of forces which asserts that each of several acting forces produces its own effect, independent of the others.

The procedure may be summed up in the **polygon rule:** *To find the resultant of a number of vectors, lay off the vectors end to end in any order. The single vector drawn from the initial point of the first vector to the end point of the last will be the resultant of the set of given vectors.* The rule gets its name from the fact that the given vectors, together with their resultant, form a closed polygon. The shape of this polygon depends only on the order in which the vectors are drawn; it may be convex, or some sides may cross through others. The important requirement is that the directions be right; this is accomplished by placing the initial point (tail) of each given vector on the end point (arrowhead) of the one before it, and drawing the resultant from the tail of the *first vector* to the head of the last one of the set. Examine Fig. 4.5 with these points in mind.

It must be understood that only vectors of the *same* kind can be combined. For instance, it is meaningless to add a displacement vector to a force vector.

The most direct way of finding the resultant of a number of vectors by means of the polygon rule is to make a careful drawing, to some convenient scale, laying off the vectors and measuring the magnitude and direction of their resultant.

FIGURE 4.5 The order in which the vectors are taken determines the shape of the polygon but not the magnitude or direction of the resultant.

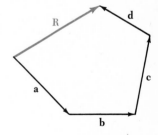

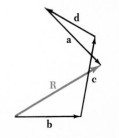

Practice set

Find graphically the resultant of two forces that act at a point: One is a force of 9 lb, acting toward the right; the other is one of 15 lb acting upward and to the right at an angle of 45° to each of these directions. It is convenient to use graph paper or quadrille paper, which is ruled off in squares. Choose a suitable scale, letting, say, one division represent one pound force. In any case, do not make the drawing too small because this will reduce the accuracy with which you can measure the resultant. A drawing of this kind should take up at least half a notebook page. Measure the direction of the resultant with respect to the smaller force by using a protractor. Compare your results with those of Prob. 4.8.

It is more accurate and more elegant to compute a resultant by trigonometry than to measure it from a scale drawing. Consider any two vectors **a** and **b,** the angle between them being θ (Fig. 4.6). Laying off these vectors in succession, we see the resultant is **R**, making an angle ϕ with the vector **a.** The problem is to determine the magnitude of **R** and the value of ϕ.

The former can be found immediately by applying the law of cosines to the triangle ABC:

$$R^2 = a^2 + b^2 + 2ab \cos \theta \qquad 4.3$$

In using this law, remember to take into account the algebraic sign of the $\cos \theta$ term; this will be negative if θ is between 90 and 180°.

To find the direction of **R,** use the sine law of trigonometry:

$$\frac{\sin \phi}{\sin (180° - \theta)} = \frac{b}{R}$$

or, since $\sin (180° - \theta) = \sin \theta$, this yields

$$\sin \phi = \frac{b}{R} \sin \theta \qquad 4.4$$

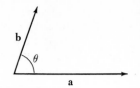

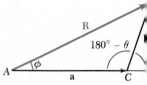

FIGURE 4.6 Computing the resultant trigonometrically.

Both R and ϕ are thus determined.

In a problem where there are more than two given vectors, the calculations can be carried out repeatedly by computing the resultant of two of the vectors, then combining this with a third, etc. This process can be cumbersome, and a better scheme will be worked out in Sec. 4.8.

In many situations arising in practice, the problem is to find the resultant of two vectors that meet at *right angles.* Then the magnitude R is given simply by the right-triangle rule, and the angle ϕ is expressed as a trigonometric function of the sides of the triangle:

$$R = \sqrt{a^2 + b^2}$$
$$\tan \phi = \frac{b}{a} \qquad 4.5$$

It often happens that a mechanics situation involves two vectors that lie in the same line. If they also have the *same direction,* the magnitude of their resultant is merely the *sum* of the two separate values, and its direction is that of the given vectors. If the two original vectors are in *opposite* directions, the resultant has a magnitude equal to the *difference* of the two, and its direction is that of the larger of the given vectors.

4.6 FORCES IN EQUILIBRIUM

A situation frequently encountered in mechanics is that of a body in equilibrium. **Equilibrium** is the condition in which the resultant of all the forces applied to the body is zero. If we can trust the polygon construction, the existence of equilibrium means that, when all the given vectors have been laid down end to end, the last vector itself will just succeed in closing the polygon; no resultant can be, or need be, drawn. This expresses the idea that the acting forces are such that they add up vectorially to no force at all.

Simple laboratory experiments show that forces can be arranged to hold each other in check in exactly the way described by the closed-polygon condition. One of the main concerns of a structural engineer is to design a bridge, crane, or framework so that every particle can be in equilibrium under the action of the applied forces at all times.

The term *equilibrium* does not imply that a body is *at rest*. When we say that a body is at rest, we assume that the ground is the particular reference system being used. It will be shown (page 92) that an object at rest on the surface of the earth can be said actually to have a number of different kinds of motion if other reference frames are used. Experience shows that mechanics experiments performed in a uniformly moving vehicle have the same outcome as when conducted in a "stationary" laboratory. Objects stay in place on a ship sailing a straight, steady course on a calm sea, and games can be played on deck with customary ease. Questions of this nature are extremely fundamental and will be discussed further in Chap. 6. For the present, we must be careful not to confuse equilibrium with rest or to assume that one of these conditions implies the other.

The following are examples of typical equilibrium situations that show how to apply the general methods developed in the preceding section. It is important to work through these examples in detail, step by step, in order to get an adequate appreciation of the broad scope of the equilibrium principle.

Worked example 4.1

A 50-lb weight hangs from a cord that is firmly tied to a given point on a rope suspended between two hooks (Fig. 4.7a). Both the cord and the rope are ideally assumed to be negligible in weight and perfectly flexible. Find the tension force in each of the portions AB and AC of the rope.

Solution   In order to make clear what is meant by the tension force in a structural member, consider the simplest case—that of a cord or rope, ideally so flexible that no appreciable bending forces exist in it. The **tension** is defined as the magnitude of either of the pair of equal but oppositely directed forces tending to pull the cord apart. If the cord is considered to be a string of particles, the tension is the force that any one particle exerts on its neighbor on either side. The magnitude of the tension is the same at all points of any segment of a weightless cord acted upon only by forces at each end. This is not as obvious as it seems, and the principle will be enlarged upon later.

The situation described in this example is a system in equilibrium under the action of several forces. Since the weight of the cord itself is neglected, the hanging mass alone is responsible for the existence of tension forces in the cord. These forces all pass through the point A, the knot in the rope, and so this knot may be considered a particle whose equilibrium will be examined.

First, we insert in the sketch of Fig. 4.7a vectors representing all the forces exerted on A by other parts of the system. The first of these is a force vector drawn vertically downward from A to represent the weight of the suspended body. This may be labeled "50 lb." Then tension vectors $\mathbf{T}_1$ and $\mathbf{T}_2$ are drawn; each one runs outward from A along its segment of rope. The length of each of these vectors is not yet known, but they are sketched provisionally nevertheless.

The magnitudes of $\mathbf{T}_1$ and $\mathbf{T}_2$ can be found graphically by the polygon method, since these two forces together with $\mathbf{W}$ constitute a set of vectors

FIGURE 4.7 A weight supported by two cords. The tension in each segment of the cord can be found by the polygon method.

(a)

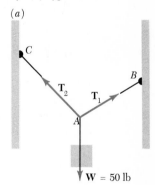

(b)

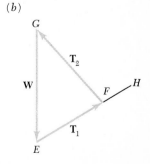

in equilibrium and should form a closed polygon, in this case a triangle. This is drawn in Fig. 4.7b, starting with **W,** whose magnitude and direction are both known. Some convenient scale, say 1 in. equivalent to 20 lb, is used. Next, a line is drawn from the end point of **W** in the direction of AB in the first sketch. We do not yet know where to terminate this vector **T₁**; if this were known, we would then draw the remaining force **T₂** from the end of **T₂**, making it parallel to AC, and should then have to end at G, the initial point of **W,** in order to close the polygon.

The way to proceed is now evident: Simply construct the force triangle by working from both ends toward the middle. Draw a line from G parallel to CA until it crosses EH. This determines the point F, which is marked with an arrow as the end point of **T₁**. Put an arrow on **T₂** as shown, and the force diagram is complete. Finally, by measurement in terms of the scale used in drawing **W,** the magnitudes of **T₁** and **T₂** can be found.

Make your own sketch of such an arrangement and determine the magnitudes of the two tensions by the above procedure.

Worked example 4.2   Forces of 10 lb in a direction 30° west of north and 15 lb toward the northeast act on a particle. Find by calculation a third force that must be added to produce equilibrium.

Solution   The pair of given forces is shown in Fig. 4.8a. Since the problem is to be solved by computation rather than by measurement on a scale drawing, only a freehand sketch of the force polygon is required (Fig. 4.8b). Such a sketch should not be done too roughly. Try to make the directions and relative lengths of the vectors fairly true to the given description; this often helps to show up a wrong result arising from an error in calculation. Label the proper magnitudes and angles on this sketch, taking care to identify angles and sides with the standard arrangement of Fig. 4.6.

In Fig. 4.8b, the resultant of the two given forces has been labeled R. Its magnitude can be computed by using Eq. 4.3:

$$R^2 = (10)^2 + (15)^2 + 2(10)(15) \cos 75°$$

and with the value of the cosine taken from the trigonometric tables on page 633, this becomes

$$R^2 = 403 \qquad R = 20.1 \text{ lb}$$

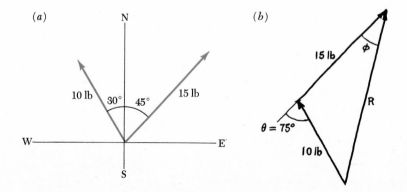

FIGURE 4.8 In order to hold the given forces in equilibrium by means of a single force, the latter must be equal and *opposite* to the resultant of the given forces.

The direction of R can now be found by using Eq. 4.4:

$$\sin \phi = \frac{10}{20.1} \sin 75° = 0.483 \qquad \phi = 29°$$

The vector that is wanted is one that is equal in magnitude to R but *opposite* in direction (see Prob. 4.7).

4.7 COMPONENTS OF A VECTOR

It often happens that mechanical circumstances do not allow a vector to act in its most effective direction. Suppose a tugboat is towing a freighter by means of a line, as shown in Fig. 4.9. If the tension in the line amounts to 800 lb, the force applied to the freighter can be represented by the vector AB, drawn to some convenient scale. Since the towline slopes downward, it is not pulling on the freighter in the most effective direction to produce forward motion. We are not interested so much in the tension in the line as in the force tending to move the freighter forward. This force can be found by means of a geometric construction whose validity will be justified below:

First draw a line AM in the forward direction, through the point of attachment of the towline. Then from B, the end of the tension vector, *drop a perpendicular* onto AM. The segment AC thus cut off will then represent the effective magnitude of **T** in the forward direction. With an arrowhead inserted as shown, it becomes what is called the **component** of the vector **T** in the direction AM. In this instance, AC may be referred to as the horizontal component of the tension. Its magnitude is given by its length, measured on the same scale as the one used in drawing **T.**

Meaning of component of a vector.

This component is obviously smaller in magnitude than **T** itself. What has become of "the rest" of **T?** To explore this question, consider the *vertical* component of **T.** This vector is determined in the same way as the horizontal component, by dropping a perpendicular from B onto the direction in question. This gives AD as the vertical component of **T.** Thus, because of its inclination, the towline does two things: It pulls the freighter through the water with an effective force AC and, at the same time, pulls down on the bow of the ship with a force AD. Only the former force is of practical interest in this problem.

The original force is said to be *resolved into its horizontal and vertical*

FIGURE 4.9 The tension force in the towline has been resolved into horizontal and vertical components. Only the horizontal component is effective in producing a forward pull.

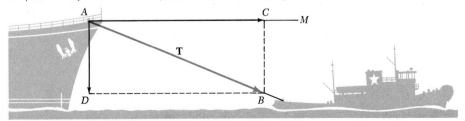

components. If the angle of inclination is known, the magnitudes of these components are (Fig. 4.10*a*)

$$F_x = T \cos \theta$$
$$F_y = T \sin \theta$$

4.6

The process of resolving a vector into two components amounts to finding a pair of vectors whose resultant is the given vector. Thus the resolution into components is the direct opposite of the process of adding vectors to obtain their resultant. There is one point of difference: Addition gives a unique resultant; resolution into components can be done in many ways (see Fig. 4.10*b*), depending on the choice of the two directions. It is possible to carry out a similar resolution process even when the two directions are not perpendicular to each other; however, in all the examples with which we shall deal, only components at right angles will be used. This simplifies matters since, from the way the construction of components was defined, one such force has zero component in the direction of the other.

It should be emphasized that, although only force vectors appeared in the foregoing discussion, the conclusions apply to vectors of any kind.

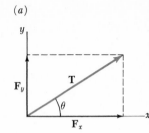

(*a*)

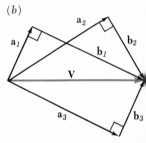

(*b*)

FIGURE 4.10   Resolving a vector into components at right angles.

4.8   SOLVING EQUILIBRIUM PROBLEMS BY USING COMPONENTS

In Sec. 4.6, the polygon method was used to solve problems concerning the equilibrium of forces. In many instances, especially those involving a considerable number of vectors, resolution into components proves to be a more straightforward procedure. This method is based on the idea that, *when a set of vectors is in equilibrium, the net component* (algebraic sum) *of all the vectors along any direction must be zero.*

To apply this scheme, any convenient pair of rectangular coordinate axes is chosen, and each of the given vectors is resolved into components along these axes. In Fig. 4.11, **a, b, c,** and **d** are the given vectors and a_x, a_y, b_x, b_y, etc., are their components along the two axes. Each component may be computed by using Eqs. 4.6. Then, for equilibrium, it is necessary that the algebraic sum of all the x components be equal to zero and, independently, the algebraic sum of the y components be zero. Written in symbols, these statements are

$$a \cos \theta_a + b \cos \theta_b + \cdots = 0$$
$$a \sin \theta_a + b \sin \theta_b + \cdots = 0$$

4.7

where the dots indicate that as many terms are to be summed as there are original vectors.

It is important to remember that, in taking these sums, proper attention must be given to the algebraic sign of each term. This can be handled simply by placing an arrow on each component in the directional sense indicated by the vector that is being resolved. Then all the x components pointing toward the right are counted positive and all those to the left, negative. Similarly, count the upward y components positive and the downward ones negative.

What has been gained by resolution into components is that vectors

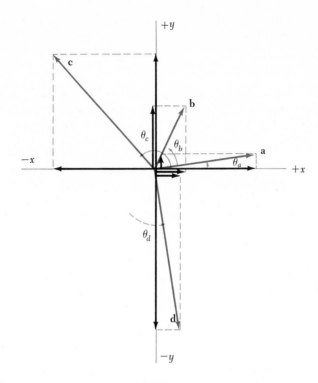

FIGURE 4.11 Equilibrium described in terms of vector components. The algebraic sum of the components along each of the two axes must be zero.

having various directions have been replaced by vectors lying along only one line, and these can be combined by simple algebraic addition.

For Stevin's highly original solution of the problem of an inclined plane, see Ref. 4.2.

Resolution of the acting forces into components is useful in dealing with the most diverse mechanical situations. A case in point is that of an object on a ramp, or inclined plane. Although it was undoubtedly used even in prehistoric times for the raising of heavy loads (the wedge and the screw are merely other forms of this device), its fundamental principle was first worked out by the talented scientist Simon Stevin of Bruges (1548–1620).

Worked example 4.3

A ramp makes an angle θ with the horizontal direction. If the frictional forces are assumed to be negligibly small, how large a force in the uphill direction is needed to hold a car of weight W stationary on this slope? How large a force does the roadway exert on the wheels of the car?

For this problem, the car may be considered a particle.

Solution   As in the previous examples, the first step is to identify all the forces exerted *on* the car by its environment. It must be emphasized again that this procedure of isolating the body from its surroundings is an essential part of the analysis.

One of the most obvious forces acting on the car is its weight, which is represented by a downward vector **W** in a rough sketch of the system (Fig. 4.12). The required uphill force **F** is another force applied to the car. Finally, the roadway acts on the car with a force **P,** as shown. Since the contact between wheels and road is assumed frictionless, **P** is drawn

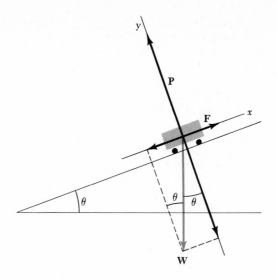

FIGURE 4.12 The problem of a car on a ramp is solved by resolving the weight into two components.

perpendicular to the slope. Thus **P** is a force that *supports* the car but exerts no drag on it (see page 77).

The problem is to find the magnitudes of **F** and **P.** This can be done by considering the equilibrium of the car under the combined action of **W, F,** and **P.** If we choose to use the component method, we must set up a pair of axes. The fact that two of the vectors are already perpendicular to each other suggests the convenience of selecting the axes along and perpendicular to the direction of the slope, rather than horizontally and vertically. After the x and y axes are inserted as indicated in Fig. 4.12, it remains to resolve **W** into components along these directions. The acute angle in each of the triangles is equal to the slope angle θ of the ramp. According to Eqs. 4.7, the condition for equilibrium of the x components is then $F - W \sin \theta = 0$, and for the y components $P - W \cos \theta = 0$, so that the complete solution becomes

$$F = W \sin \theta$$
$$P = W \cos \theta$$

4.8

Practice set*

1. In Worked example 4.3, what is the ratio of F to W when the slope angle is 30°?
2. What is the ratio of F to P when the slope is 45°?
3. The example can also be solved by using the polygon construction. Sketch the appropriate self-closing polygon, starting with the side **W.** How does this polygon compare with the triangles in Fig. 4.12?

4.9 FRICTION

It is impossible to eliminate frictional forces completely in constructing any kind of mechanism involving moving parts. Fortunately, in many arrangements, these forces can be made so small compared with the other

* *Ans.:* $\frac{1}{2}$; 1; identical.

acting forces that they can be neglected. This kind of idealization was used in the situation of Worked example 4.3. Where it is not possible to do this, the frictional forces must be determined, usually by experiment, and included in the problem.

Solid friction   This is the kind of frictional force that exists at the dry surface of contact between solid objects. It resists the sliding or rolling of one surface on the other.

Sliding friction cannot be thought of merely as the interlocking of roughnesses on the two surfaces. In most cases, the drag is due directly to forces of attraction between the molecules of the two bodies and to the breaking of minute welds that occur at high spots as a result of pressure between the surfaces.

Experience shows that the magnitude of the frictional drag between two solid surfaces depends mainly on the force pressing the two objects together, as well as on the nature of the materials and the smoothness of the surfaces. However, it also depends slightly on the area of the common surface of contact and on the relative speed of sliding.

In rolling friction, the resistance to motion is caused by the deformation (usually only temporary) of the two objects at their point of contact. A wheel is slightly flattened and the rail on which it rolls is minutely deformed where they press together. The effect of rolling friction is to oppose the motion, just as in sliding friction, except that rolling friction is usually much smaller in magnitude. The use of ball or roller bearings is highly effective in reducing solid friction in machines.

Figure 4.13 shows two solid bodies in contact. Let **R** be the total contact force exerted by one surface on the other at any instant, regardless of whether the objects are relatively at rest or sliding or rolling on each other. This force may be resolved into two components, one along the common plane of contact and the other perpendicular to it. The first is the frictional (drag) force; the other is the normal (supporting) force. The smaller the friction between the surfaces, the smaller will be the component **F** and

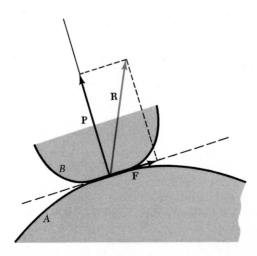

FIGURE 4.13 Contact forces between two solid objects.

the nearer the total contact force will be to the perpendicular. The limit is the ideal case represented by Worked example 4.3.

Fluid friction   This kind of friction shows itself typically as the resistance experienced by an object as it moves through a liquid or a gas. It is associated with the viscosity (page 47) of the fluid. The magnitude of the resistance force depends on the nature of the fluid and on the size, shape, attitude, and speed of motion of the body. Because of the complexity of the phenomena, most of the practical questions concerned with fluid friction have to be solved experimentally by tests made in wind tunnels or towing channels or by means of ballistic trials or rocket flights.

The efficiency of a mechanism is often increased considerably by replacing solid friction by fluid friction, as in the lubrication of machinery. A recent development is the use of air-supported surface vehicles in place of those with wheels (see page 122).

The most general statement that can be made about friction of any kind is that it gives rise to a *resistive* force, one that always acts in a direction *opposite* to the motion (or incipient motion) of the body concerned.

4.10 GENERAL PROCEDURE FOR SOLVING EQUILIBRIUM PROBLEMS

A general scheme will now be summarized for solving problems on the equilibrium of a particle acted upon by forces. This procedure will always lead to a straightforward solution. Its use is recommended in solving problems given at the end of the chapter.

1. *Draw a simple sketch* representing the situation described in the statement of the problem. The drawing may be freehand, but the lines should be reasonably straight and the angles approximately the right size.
2. *Mark the symbols and values* of the given quantities on the sketch alongside the items to which they apply. Put in symbols for the unknown quantities.
3. *Sketch vectors* on the diagram to represent all the forces that act on the particle whose equilibrium is to be investigated. Be sure these forces are given their proper directions. They must be the forces exerted *on* the particle *by* its surroundings. Always draw force vectors extending *outward* from the particle, rather than pushing *in* on it. Make sure that all vectors are labeled.
4. *Apply the principle of equilibrium* of forces. This may be done in either of two ways:
   a. *The polygon construction.* Draw a force polygon to scale and measure from it the parts to be determined. If the geometry of the polygon happens to be simple, the results can be found by calculation, using only a rough sketch of the polygon instead of a scale drawing.
   b. *The component method.* This is especially convenient when there are many forces. Resolve all forces into their components along two mutually perpendicular axes, making a separate sketch for this purpose. Label the components. It is advisable to cross off each of the given vectors as soon as it has been replaced by its two components. Finally, compute the required quantities by the method given in the preceding text.

4.11 TORQUE; ROTATIONAL EQUILIBRIUM

The above discussion of equilibrium has been confined to cases where all the acting forces are, in effect, applied to a single particle. However, there are many instances in practice where forces are applied to various points of an extended object. This introduces something new: the possibility of rotation. In order to ensure the *complete* mechanical equilibrium of an extended body, another condition must be imposed on the acting forces in addition to the one already described by $\mathbf{R} = 0$. This new principle concerns not only the force vectors themselves but the places at which they are applied.

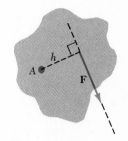

FIGURE 4.14 The magnitude of the torque exerted by **F** around an axis drawn through A and perpendicular to the page is given by $L = Fh$. This torque is in the clockwise direction.

First, it is assumed that the body is *perfectly rigid,* the distances between its particles remaining constant at all times. If we consider only cases where the forces all lie in one plane (Fig. 4.14), any line perpendicular to this plane can serve as a prospective axis of rotation of the body. The magnitude L of the **torque** (turning effect) exerted by any force **F** around any axis A is defined as

$$L = Fh$$

where h is the length of the perpendicular drawn between the line of action of **F** and the axis A. This is inherently a reasonable definition since, for instance, experience shows that the most effective way to push open a door is to exert our force in a direction perpendicular to the door and at a point as far from the line of hinges as possible.

If there are several applied forces, the quantity L belonging to each force must be given an algebraic sign: positive if it tends to turn the body in one rotational sense; negative, if in the other. The condition for *rotational equilibrium* (equilibrium of torques) is then found to be

$$F_1 h_1 + F_2 h_2 + F_3 h_3 + \cdots = 0$$

where the sum is understood to be an *algebraic* one in the sense explained above. This equation merely states that the net tendency of the acting forces to produce rotation about any axis is zero. With Eqs. 4.7 (page 74), it furnishes information for determining some of the forces required for complete equilibrium of the body.

No further details need be given here except to remark that the principle of the lever, discovered empirically by Archimedes in the second century B.C., is a special case of the equilibrium of torques.

4.12 WHY STUDY EQUILIBRIUM?

This chapter has devoted considerable space to a discussion of the equilibrium of vectors, particularly of forces. Since very many mechanical situations do *not* involve a condition of equilibrium, it may be questioned whether such attention to this case is deserved and whether the topic had better be left entirely to structural engineers.

Although equilibrium will be seen later to be a special case of the more general considerations of Chap. 6 (Force and Motion), there are good reasons for discussing this topic at this stage. The equilibrium principle is an excellent illustration of the wide applicability of a scientific generalization. The principle as developed above is a simple and plausible rule that governs an enormous variety of special situations, from seesaws to suspension bridges. Also, consideration of the polygon rule and the component method provided an opportunity to see the advantage of making operational definitions of physical quantities—in this case, of a vector. Another advantage is an increased appreciation of the usefulness of idealization. A large number of mechanical situations become manageable through the use of simplified conceptual models. This involves replacing the actual situation, with all its complexities, by an idealized one using such concepts as particles, perfectly flexible and massless strings, frictionless surfaces, etc.

Programmed review

Instructions: See page 18.

1. Give a provisional description of what is meant by a force.

    Any agency capable of deforming bodies or changing their motion. An example is the push or pull exerted by means of muscles.   [4.2]

2. Distinguish between scalar and vector quantities.

    A scalar is completely described by stating a number, expressed in appropriate units. A vector requires, in addition, the specification of its direction in space.   [4.3]

3. What is meant by a particle?

    A material body whose extension in space can be ignored.   [4.4]

4. Define the resultant of a set of vectors.

    The single vector that produces the same effect as the combined action of the given vectors.   [4.4]

5. State the polygon rule for finding the resultant of a set of vectors.

    Lay off the vectors, tail to head, in any order. Then the vector drawn from the initial point of the first vector to the end of the last will be the required resultant.   [4.5]

6. What is meant by saying that a given set of forces, acting on a particle, is in equilibrium?

    The resultant of the set is zero.   [4.6]

7. Explain the meaning of the phrase "component of a vector in a given direction."

    The effective value of the vector in that direction. It is found by projecting the original vector onto a line running in the given direction.   [4.7]

8. In terms of components along two mutually perpendicular axes, what must be true if a set of vectors is in equilibrium?

    The algebraic sum of all the component vectors along each of these axes must be zero.   [4.8]

9. Define torque.

> The turning effect of a force applied to a body. The magnitude of the torque L around a given axis is Fh, where F is the magnitude of the force and h is the perpendicular distance of F from the axis.   [4.11]

10. What is the condition for rotational equilibrium of a rigid body?

> The algebraic sum of the torques due to all the forces, taken about any selected axis, must be zero.   [4.11]

For further reading

4.1. *Crew, H.* ''The Rise of Modern Physics.'' Read the account of the early work on the combination of forces.

4.2. *Taylor, L. W.* ''Physics, The Pioneer Science.'' Stevin's treatment of the inclined plane is found in chap. 6; torque and the lever principle are described in chap. 7.

4.3. *Chalmers, T. W.* ''Historic Researches.'' Chapter 1 traces the development of our understanding of friction.

4.4. *Rabinowicz, E.* Stick and Slip, *Sci. Am.,* May, 1956. Describes an interesting aspect of solid-friction phenomena.

Questions and problems

In any problem requiring the determination of a vector, find the result by computation rather than by measuring a scale drawing, except where otherwise directed.

4.1. A bug placed at the center of a turntable 8 ft in diameter crawls outward along a radius and arrives at the rim just as the table completes a full turn. If both turntable and bug move at constant rates, describe the total displacement of the bug, referred to his initial position and direction of motion, after the table has made (*a*) a quarter turn; (*b*) five-eighths of a turn.

4.2. Verify by means of a careful drawing the fact that the order in which vectors are added is immaterial. Draw a set of four arbitrary vectors applied at a point. Combine them by means of the polygon construction and draw their resultant. Combine them a second time, taking the vectors in some other order. Satisfy yourself that the resultant so obtained is approximately parallel to and of the same length as the first one.

4.3. Can two forces of magnitude 7 and 10 kg, respectively, have a resultant of 12 kg? What are the greatest and least values that the magnitude of their resultant can have? Explain.

4.4. Determine the resultant of three vectors of magnitudes 1, 2, and 3 whose directions are those of the sides of an equilateral triangle, taken in order.

4.5. A reconnaissance plane leaves its base and flies 80 mi due east, 150 mi northwest, 100 mi northeast, and finally 150 mi due south. Find by means of a scale drawing how far the plane is then from its base and in what direction (specify angle) it must fly to return.

4.6.   *Lami's theorem.* If three forces, all in the same plane, hold a particle in equilibrium, the magnitude of each is proportional to the sine of the angle between the other two. Prove this by sketching a force triangle and applying the trigonometric law of sines to it.

4.7.   In Worked example 4.2 (page 72), determine the geographical direction of the vector required to produce equilibrium. Also, draw it, together with the two given vectors, to scale and check to see if the polygon formed by them is self-closing.

4.8.   Find by computation the resultant in the Practice Set on page 69 and compare it with the graphical solution.

4.9.   The polygon rule is not limited to vectors in the same plane but can easily be extended to three dimensions. Compute the size of the resultant of the following three vectors: 3 units toward the east, 4 units to the north, and 12 units vertically upward (make a sketch). What angle does the resultant make with the vertical direction?

4.10. Equation 4.4 for the direction of the resultant of two vectors makes use of the value of R, which had to be computed first by means of Eq. 4.3. Show that the direction can also be found from the following formula, which employs only the given quantities a, b, and θ:

$$\tan \phi = \frac{b \sin \theta}{a + b \cos \theta}$$

4.11. A baseball thrown straight upward is momentarily at rest when at the top of its path. Is it in equilibrium at that instant?

4.12. Discuss the statement: If a set of forces is known to be in equilibrium, any one of them must be equal and opposite to the resultant of all the remaining forces.

4.13. Prove that, if three vectors applied to a point are in equilibrium, they must all lie in the same plane. HINT: Consider the statement in Prob. 4.12.

4.14. A helicopter flies a straight course of 80 km in a direction 30° north of west. How far is it then (a) to the north of its takeoff point; (b) to the west?

4.15. The inclined portion of an escalator that runs from the basement to the ground floor of a department store is 16 ft long. If the distance between floors is 10 ft, describe the displacement vector of a man who has ridden from the basement to the ground floor. How far was he transported horizontally?

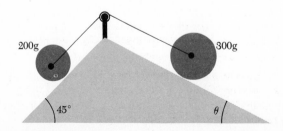

FIGURE 4.15 Stevin's roller paradox.

4.16. *Stevin's roller paradox.* Two rollers of different mass, when connected by a cord passing over a pulley as in Fig. 4.15, are able to hold each other in balance on the double slope. Find the angle θ for which this is possible. Friction is negligible.

4.17. A 250-lb cake of ice is observed to slide down a sloping board at a constant speed. It will be shown in Chap. 5 that this means the resultant force acting on the block is zero. If the board makes an angle of 2° with the ground and there are no other restraints, how large is the force of sliding friction between the ice and the board?

4.18. A spring-driven toy car is able to exert a tractive force equal in magnitude to half its weight, when placed on a certain sloping surface. Will the car be able to climb the slope if it makes an angle of 35° with the horizontal? Explain your result.

4.19. A 200-lb box placed on a rough slope that makes an angle of 20° with the horizontal experiences a frictional drag of 23-lb force, assumed constant. How hard must a man pull in the uphill direction on a rope attached to the box in order to have it just on the point of slipping upward?

4.20. The tension in each band of a slingshot is 12 lb when the rubber is stretched to the extent shown in Fig. 4.16. Neglecting the weight of the stone, compute what net force (give magnitude and direction) acts on it the instant after it is released. Note that, although the tensions in the two bands are applied to the stone at the points A and B, they effectively act at the common point C.

4.21. An iron ball of mass 10 kg hangs from a rope. It is pulled aside by a *horizontal* force of 4 kg. What angle does the rope make with the vertical when equilibrium is attained, and how large is the tension force in the rope at that time?

4.22. How large a horizontal force would be needed to pull the ball of Prob. 4.21 aside by an angle of 75°? What is then the tension in the rope?

4.23. Keeping in mind the results of Probs. 4.21 and 4.22, determine how large a horizontal force would be needed to pull the ball aside until the rope assumed an exactly horizontal position. Confirm your conclusion by sketching the force vectors. In view of the result, make a conjecture as to the possibility of stretching a clothesline perfectly straight.

4.24. Instead of the cord being tied firmly to a given point on the rope in Example 4.1, suppose the weight had been hung on by means of a frictionless pulley, as shown in Fig. 4.17. How would this change the equilibrium configuration of the setup? What statement can now be made about the tensions in the two segments of the rope? Support your answer by sketching the force triangle.

4.25. When a slack-wire performer at the circus balances at the center of a wire 20 ft long, the midpoint of the wire is depressed 3 ft below the level of the ends and the tension in the wire amounts to 275 lb. How much does the man weigh?

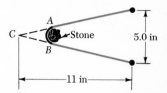

FIGURE 4.16 Force exerted by a slingshot.

FIGURE 4.17 The suspended weight rides on the cord by means of a roller.

4.26. A length of rope that can sustain a maximum tension of 400-lb force is stretched taut between two posts 32 ft apart, and a force of 50 lb is applied to the midpoint of the rope in a direction perpendicular to the line joining the posts. How far to one side can this point be moved before the rope breaks? Assume that the rope does not stretch appreciably.

4.27. There is a tension of 30 lb in the supporting chain of the street lamp shown in Fig. 4.18. Find the weight of the lamp and the thrust force in the horizontal strut, which is of negligible weight.

4.28. Two perfectly smooth glass balls, each 4 cm in diameter and of mass 0.5 kg, are hung from a hook by cords 5 cm long (Fig. 4.19). How hard, in kilograms force, does one sphere press against the other?

4.29. Two boys of weight 100 lb and 80 lb, respectively, improvise a seesaw by pivoting a board of negligible weight across a fence. If the boys are effectively 9 ft apart, how far from the heavier boy should the pivot be located for balance?

4.30. In Prob. 4.27, find the magnitude of the torque exerted by the tension force in the chain, computed around the left-hand end of the strut. Equate this torque to that of the weight of the lamp around this same point and thus find the value of the weight. Compare with the result obtained by using equilibrium of forces. Why is it unnecessary to consider the torque due to the thrust force in the strut?

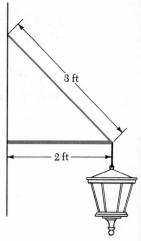

FIGURE 4.18 The weight of the lamp causes a tension in the slanting chain and a compressive force in the strut.

FIGURE 4.19 Glass balls suspended from a fixed point. How great a force presses them together?

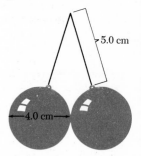

Chapter 5
THE MOTION
OF A PARTICLE

In the preceding chapter the vector concept was introduced by considering displacement because this is a readily visualized kind of vector. A displacement vector serves to describe the transfer of a particle from one location to another, and for certain purposes this is all that is of interest about the process. However, in many practical problems concerning motion, the *time* element is of great importance. A person traveling by train or plane expects to be transported a given distance in a given direction. He also expects it to be done in a stated interval of time, within reasonable limits.

This chapter will explore the ways of describing the motion of a particle, and concepts such as speed and velocity will be developed. Quantitative relations will be derived for certain important types of motion, including the case of a particle moving freely under the influence of gravity near the surface of the earth.

5.1 SPEED AND VELOCITY

A vehicle or any other moving body travels along some kind of path in space. If we disregard the spin of a baseball, the twirling motion of a baton, or the gyrations of a skidding car, each of these bodies may be replaced by a particle moving along some sort of curved path in space. Thus, for some purposes, a particle can substitute for an extended, rotating body. The rate, in time, at which a particle traverses a path in space is called the **speed** of the motion.

Motion along a path, in contrast with spinning motion (rotation), is called **translational** motion.

Speed is a derived quantity, representable as a combination of two of the fundamental quantities distance and time. The basic method of measuring speed is to use a measuring tape and a clock. Thus if a car moves at a uniform rate over a path 80 mi long and takes 2 hr to do so, its speed is 80 mi divided by 2 hr, or 40 mi/hr (miles per hour). This means that, if the car continued to move at that rate, it would cover 20 mi in the next half hour, 120 mi in 3 hr, etc.

This intuitive representation of speed can be expressed as a simple algebraic relation. Let x stand for the distance covered, t for the elapsed time, and v for the speed, assumed constant here. Then, by using a particular value of v, the above can be stated in general form as

Relations for motion at constant speed.

$$v = \frac{x}{t} \quad \text{or} \quad x = vt \quad \text{or} \quad t = \frac{x}{v} \qquad\qquad 5.1$$

Speed is recognized as a scalar quantity having the dimensions L/T (Sec. 3.8). For its measurement, any convenient pair of distance and time units may be used. Some common units of speed are miles per hour, kilometers per hour, feet per second, and knots.

A knot is one International nautical mile (6,076 ft) per hour.

Even if the car mentioned above had not moved at a steady rate during its 80-mi journey, 40 mi/hr would still represent a useful fact—the **average speed.** Moreover, the total distance covered in the given time would be the same as that of a car moving at a *constant* speed of this magnitude. Equation 5.1 in any of the three forms could still be used; we need only replace the constant speed v by the average $\bar{v}$, where the bar over the letter indicates an average value.

Worked example 5.1

A motorist sets out on a trip at 8:00 a.m., travels 200 mi at a constant speed of 50 mi/hr, makes a 45-min stop for lunch, and then does an afternoon run of 5 hr at a constant speed of 40 mi/hr. What time does he get to his destination, and what was his average speed for the trip?

Solution   Using the third form of Eq. 5.1, we find the time for the morning run was

$$\frac{200 \text{ mi}}{50 \text{ mi/hr}} = 4 \text{ hr}$$

To find the total time for the trip, the 45-min lunch period must be expressed in the same units as the other time intervals and so must be written 0.75 hr. Then the total time is 4 + 0.75 + 5, or 7.75 hr, and the time of arrival is 3:45 p.m.

The average speed for the trip is the total distance, 400 mi, divided by the total time, 7.75 hr, or about 52 mi/hr.

If the driver of the car in Worked example 5.1 had looked at his speedometer at any time during the morning run, he would have seen the needle standing steady at the 50-mi/hr mark. Throughout the afternoon run it would have indicated 40 mi/hr and during the lunch interval, zero. The average speed computed above, 52 mi/hr, differs from all these figures. What, then, does a speedometer show?

To answer this question, consider a vehicle (or particle) moving over a path PQ (Fig. 5.1). Along a part of this path, markers A, B, C, . . . and a, b, c, . . . have been set down. The markers were not spaced in any special way; but once they were put in place, their distances from, say, A were carefully measured and recorded. Riding in the vehicle are a number of observers, one for each pair of markers. Each observer has a stopwatch.

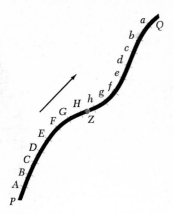

Interval	$\triangle x$, m	$\triangle t$, sec	$\bar{v}=\triangle x/\triangle t,$ m/sec
Aa	16.5	1.51	10.9
Bb	13.0	1.32	9.88
Cc	10.5	1.04	10.1
Dd	8.60	0.915	9.40
Ee	6.78	0.720	9.41
Ff	4.40	0.470	9.37
Gg	2.85	0.304	9.37
Hh	1.04	0.111	9.37

FIGURE 5.1 The history of a trip. Each measured distance interval divided by the corresponding time gives the average speed over that interval. The numbers approach a steady value as the intervals are taken smaller and smaller.

The vehicle now moves over the path PQ, and as it does so, each observer sets himself the task of measuring the time it takes to travel the distance between his assigned pair of markers: One observer is given A and a, the next has B and b, etc. After the run, the observed times and distances are recorded, as in the table in Fig. 5.1, and the values of distance divided by time are computed and entered as well. When these values are examined, it is seen that the numbers vary considerably for the first few entries, where the intervals are large, but soon approach a constant value as the intervals become smaller.

The delta notation.

The delta is a mathematical operator; it prescribes the operation of taking a piece of the quantity to which it is applied.

Some symbols will now be introduced that will prove useful through the whole study of physics. In mathematical notation, a change, or interval, in the value of any quantity is represented by writing the Greek letter Δ (delta) directly before the symbol for the quantity. Thus Δx means "a change, or difference, in the value of x." It is a *single* quantity and must *not* be taken to mean "Δ times x." It should be read: "delta eks." Moreover, an interval designated with a Δ is usually considered to be a *small* interval.

If the delta notation is applied to the experiment described above, each of the time intervals measured by the observers would be the difference between two times, t_2 at the end of the measured interval and t_1 at the beginning. Each such measured interval could be called a Δt, and

$$\Delta t = t_2 - t_1$$

If the stopwatch has been reset to zero before making a reading, t_1 will be zero.

Similarly, each of the corresponding distance intervals is the difference between two distances measured from P. If we call the distance to the end of the interval x_2 and that to the beginning x_1, we can write

$$\Delta x = x_2 - x_1$$

Then the quantity listed in the last column of the table in Fig. 5.1 is $\Delta x/\Delta t$; as we read down the column we come to values of $\Delta x/\Delta t$ corresponding to smaller and smaller values of Δt. But we see that, as Δt decreases, the quotient $\Delta x/\Delta t$ appears to come closer and closer to a steady value. We are justified in expecting no surprises as the time and distance intervals are made even smaller, and it is reasonable to assume that, as both intervals approach zero, $\Delta x/\Delta t$ would take on *a definite value* which, however,

Read Eq. 5.2
this way:
"The limit,
as Δt
approaches
zero, of Δx
divided by
Δt is equal
to v."

generally differs from zero. In symbols, this statement can be written

$$\lim_{\Delta t \to 0} \frac{\Delta x}{\Delta t} = v \qquad\qquad 5.2$$

where v is called the **instantaneous speed.** In words: To find the instantaneous speed of a moving particle at any moment or as it passes a given
point in space, proceed operationally as follows: Find an average speed
by measuring the length of any segment of the path that includes the point
in question and divide it by the corresponding time interval. Repeat the
process, taking smaller and smaller intervals. Then the limit of these values
of the average speed, as the interval is allowed to approach zero, is the
instantaneous speed of motion.

The instantaneous speed is, in fact, the value that would be indicated
by an ideal speedometer. For a body moving in a completely general way,
this reading might be expected to change from one moment to another,
but at each instant it would have a perfectly definite value.

It is instructive to represent the history of a trip, such as the one in the
above example, by means of a graph of distance against time. This is done
in Fig. 5.2a. The plotted times and distances are cumulative values, measured onward from some chosen starting point.

*Notice that
the curve in
the graph of
Fig. 5.2 is
not the
same as the
actual path
of which a
part is
shown in
Fig. 5.1.*

Figure 5.2b shows an enlarged view of the part of 5.2a in the dashed
enclosure. D,d represents the ends of one of the intervals assigned to
an observer in the experiment. Its Δx is the segment dS on the graph,
and its Δt is DS. We draw the chord Dd and note that $\Delta x/\Delta t$ represents
what mathematicians call the **slope** of the chord line. But we have previously identified $\Delta x/\Delta t$ with the *average speed* over the interval, and so
it appears that *the average speed over a segment of the path is represented
on the distance-time graph by the slope of the chord* drawn through the
end points of the segment.

It must be remembered that, in measuring a slope, each Δ interval must
be read off along the scale of the corresponding axis of the graph. For
example, suppose that, for the larger of the two intervals shown in Fig.
5.2b, Δx is 8.60 m and Δt is 0.92 sec. Then the slope of the chord is
$\bar{v}$ = 8.60 m/0.92 sec = 9.35 m/sec, and this is the average speed over
this interval. The instantaneous speed at Z is somewhat smaller in this
example, a fact that is evident from the smaller inclination of the tangent
drawn to the curve at that point.

Now we take one of the smaller intervals, Ee, included within the previous
one. The slope of the chord is again given by the appropriate Δx divided
by the corresponding Δt, and it is evident that this slope differs slightly
from the previous one: The average speed of the moving body was slightly
different over this interval than over the larger one. Continuing in this way,
using smaller and smaller intervals, we close down on the point Z by letting
both Δt and Δx approach zero. In the limit, the chord has become the
tangent line to the curve. Then, since this limit has been identified with
the instantaneous speed, we can say that *the slope of the tangent to the
distance-time curve at any point represents the instantaneous speed of the
body at that point.*

In seeking a complete description of the motion of a particle, it is not
enough to know its speed at every instant. We must also know something
of its direction of travel in space, and a vector is needed for this purpose.

FIGURE 5.2 Graphical representation of average speed as the
slope of the chord on a distance-
time diagram. In the limit, the chord
approaches the tangent line and
its slope is the instantaneous
speed at the point in question.

(*a*)

x

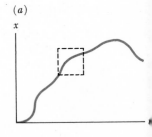

(*b*)

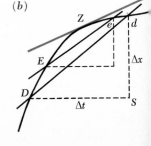

A vector drawn tangent to the path of a moving particle at any point and having the same direction as the motion of the particle and a magnitude equal to the instantaneous speed is called the **instantaneous velocity** of the particle at the point in question. Its symbol is **v**. Thus the *magnitude* of the velocity vector **v** is the speed v, which is a scalar.

The Latin velocitas, from which velocity derives, means "swiftness"; it does not imply direction.

Although the terms "speed" and "velocity" are often used interchangeably in ordinary affairs, the distinction between them must be carefully maintained. For example, a point on the rim of a wheel that turns at a uniform rate has constant speed, but its instantaneous velocity is *not* constant because the direction of this vector is continually changing. Only in the case of motion with constant speed in a straight line do both the speed and velocity remain constant.

5.2 RELATIVE VELOCITY

The discussion of the preceding chapter showed that it is possible to apply several displacements or several forces to a particle at the same time and that in each case there is a perfectly definite resultant. Since velocity, like force or displacement, is a vector, it might be asked: Is it possible for a particle to have two or more velocities at the same time? What would this mean?

Consider a familiar example: A car overtakes and passes another car going in the same direction on a straight road (Fig. 5.3). The vectors represent the velocities of the two cars. These velocities are referred to the ground; but if each driver concentrates on how the other car moves *with respect to him,* quite different velocities are involved. The slower driver would say that the other one is passing him at the rate of $65 - 50 = 15$ mi/hr: *Relative to the slower car,* the faster has a velocity of 15 mi/hr toward the *east.* At the same time, the faster car sees the other one falling behind at this rate: *Relative to the faster car,* the slower one has a velocity of 15 mi/hr toward the *west.* Similarly, if the two cars were traveling in opposite directions at the given speeds, their relative speed would be $65 + 50$, or 115 mi/hr. Even these simple examples show how strongly the velocity of a body depends on the reference frame relative to which it is observed.

Practice set*

The guns of a bomber have a muzzle speed of 2,800 ft/sec. If the airplane is flying through still air at a speed of 400 ft/sec, find the initial *velocity relative to the ground* of the bullets from (*a*) the nose gun, (*b*) the tail gun,

*See footnote on page 90.

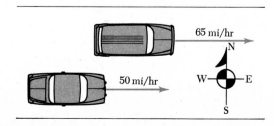

FIGURE 5.3 An example of relative velocity: one car passing another.

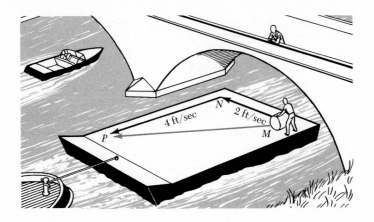

FIGURE 5.4 Resultant of two velocities. The resultant velocity of the sailor, relative to the earth, is given by the vector *MP*.

(*c*) the waist gun, assuming that these fire directly forward, backward, and sideward, respectively.

A vector can be multiplied or divided by a scalar. The result is a new vector having a different magnitude and dimensions but the same direction as the original vector. The magnitude has merely been scaled up (or down) by a certain factor.

We take next a more general case of relative motion where the velocities are not in the same line. A sailor crosses the deck of a barge that is moving along a river (Fig. 5.4). He walks in the direction shown with a speed of 2 ft/sec, and the vector *MN* represents his velocity relative to the deck. But the barge itself—and everything on it, including the sailor—has a forward speed of 4 ft/sec, as represented by the vector *NP*. This is the velocity of the barge and its contents relative to the earth. Any *displacement* of the sailor over the deck will add to the displacement of the barge to give the displacement of the sailor relative to the earth. If each of these displacement vectors is divided by the time interval in which the displacement takes place (the same for all three), the quotient is a velocity. Thus the velocity of the sailor with respect to the earth is the resultant *MP* of his partial velocities *MN* and *NP*. The sailor moves in the direction *MP* with a speed of about $4\frac{1}{2}$ ft/sec in this example. This would be his motion as seen by an observer on the bridge overhead (which is stationary with respect to the earth), provided that the observer can refrain from thinking of the two partial motions of the sailor and concentrate on their resultant.

To avoid confusion when combining velocities, displacements, or other vectors that are given with respect to different reference frames, it is advisable to label their extremities, as in the example of the barge. The vector that is to represent the velocity of *A* relative to the frame *B* should be labeled with an "*A*" at its tail and a "*B*" at its head (Fig. 5.5), and similarly for the velocity of frame *B* with respect to another frame *C*. Then, to get

Ans.: 3,200; 2,400; 2,830.

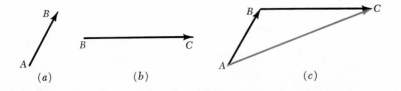

(*a*)   (*b*)   (*c*)

FIGURE 5.5 Method of combining vectors that are referred to different frames.

the resultant, the two vectors are joined so that the ends having the label B in common coincide, and the resultant is drawn. In accordance with the scheme used for naming the given vectors, this resultant will be called "the velocity of A relative to C," which is the vector required:

$$\mathbf{v}_{AC} = \mathbf{v}_{AB} + \mathbf{v}_{BC} \qquad\qquad 5.3$$

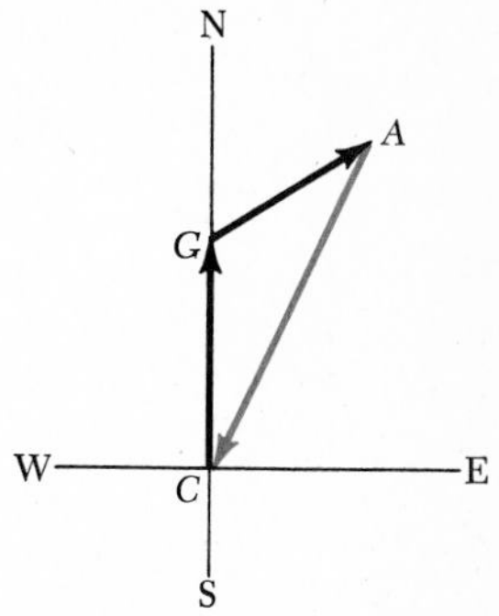

Worked example 5.2 A motorcyclist rides due north at 50 mi/hr while a wind blows from the northeast at 30 mi/hr. Find the direction and speed of the wind as they appear to the cyclist.

Solution Call the velocity of the cyclist relative to the ground $\mathbf{v}_{CG}$, the velocity of the air relative to the ground $\mathbf{v}_{AG}$, and the velocity of the air relative to the cyclist $\mathbf{v}_{AC}$. Then, in accordance with the above scheme, $\mathbf{v}_{CG}$ and $\mathbf{v}_{AG}$ are drawn in succession as shown in Fig. 5.6, with the two ends G coinciding.

Note that in order to do this the vector GA in Fig. 5.6 had to be drawn *opposite* to the direction of the wind. According to the above procedure for naming the vectors, it represents the velocity of the *ground* relative to the *air*. Each of these vectors is the *negative* of the other; in vector notation,

$$\mathbf{v}_{GA} = -\mathbf{v}_{AG}.$$

For this problem, the resultant is the vector $\mathbf{v}_{AC}$ shown in Fig. 5.6. From either a scale drawing or by computation, this vector is found to have a magnitude of about 74 mi/hr, and a direction of about 17° west of south.

FIGURE 5.6 Finding the velocity of the wind relative to a moving vehicle.

Worked example 5.3 The engine of a motorboat can give it a forward speed of 10 ft/sec in still water. The boat operates on a stream that is 500 ft wide and has a uniform current of 4 ft/sec. A number of interesting questions can be asked about this situation:

a. If the boat is headed straight across the river, how long will the crossing take, and at what point on the opposite bank will the boat land?
b. How must the boat be headed in order to land exactly opposite its starting point, and how long does this trip take?
c. If, instead of crossing the stream, the boat makes a round trip by going downstream for 500 ft and back again, how long will this take?

Solution Here again the problem is that of a particle having two component velocities: The boat has a stated velocity relative to the water, and the entire body of water has a certain velocity relative to the earth. We wish to know the resultant velocity of the boat relative to the earth.

a. Figure 5.7a shows the motorboat headed directly across the stream. Its keel is kept perpendicular to the banks, but because of the current there is a sidewise drift (leeway) during the crossing. As long as the current and the propeller rate remain constant, the boat moves on the straight course AC at a speed of $\sqrt{(4)^2 + (10)^2} = 10.8$ ft/sec. The cross-stream component of the boat's velocity has the magnitude 10 ft/sec, and so the time for the trip is

$$\frac{500 \text{ ft}}{10 \text{ ft/sec}} = 50 \text{ sec}$$

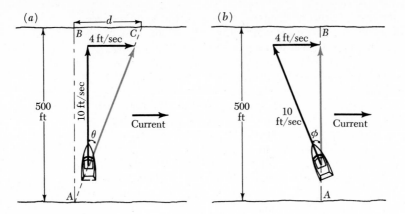

FIGURE 5.7 The velocity, relative to the earth, of a motorboat crossing a stream is the resultant of the boat's velocity in still water and the velocity of the current.

The downstream angle of the course is such that

$$\tan \theta = \frac{4 \text{ ft/sec}}{10 \text{ ft/sec}}$$

so that $\theta = 21.8°$. The distance d is found from $d = 500 \tan \theta = 100$ ft.

b. An intuitive answer to the first part of this question might be: Head the boat upstream at 21.8°. This is not correct. The heading must now be such that the resultant of a 10-ft/sec vector in some direction and the 4-ft/sec downstream vector should lie exactly in the cross-stream direction (Fig. 5.7b). If ϕ is the upstream angle, $\sin \phi$ must equal 4/10, or $\phi = 23.6°$, which differs somewhat from the angle computed in part a. The keel of the boat keeps pointed in the direction shown, but the boat follows the course AB directly across the river. The crossing speed is 10 cos 23.6°, or about 9.17 ft/sec. At this rate, the crossing takes

$$\frac{500 \text{ ft}}{9.17 \text{ ft/sec}} = 54.5 \text{ sec}$$

c. Going directly downstream, the motorboat has a resultant speed of 14 ft/sec, so that this half of the trip takes 500/14 = 35.7 sec. The resultant speed on the return trip upstream is 6 ft/sec, and so this part takes 500/6 = 83.3 sec.

The times for each round trip are (a) 100 sec; (b) 109 sec; (c) 119 sec.

5.3 VELOCITY ADDITION IN THE THEORY OF RELATIVITY

Most examples of motion are ordinarily referred to the surface of the earth, so that it is scarcely necessary even to mention the frame of reference. However, this is not the end of the matter if other frames are taken into account. The ground itself is in motion because of the earth's rotation. Relative to a reference frame fixed at the center of the earth, the eastward speed amounts to some 700 mi/hr at middle latitudes. The laboratory and everything in it possess this motion, of which we are not even aware. Further, the earth itself moves in its orbit around the sun at an average speed of about 67,000 mi/hr; and the entire solar system, including the earth, is observed to be moving in the direction of the constellation Hercules at about 43,000 mi/hr. In addition, we move around the center of the galaxy at about 600,000 mi/hr and share in the motion of our galaxy among the others.

At any instant, the combination of such motions, each in its own direction, gives our resultant velocity. Velocity referred to what? Newton believed in the possibility of determining the "absolute" motion of bodies with respect to space itself. He and his contemporaries even went so far as to assume the existence of an all-pervading medium, the "ether," as the carrier of gravitational force and light. Nineteenth-century physicists assumed that the hypothetical ether was responsible for transmitting electric forces and radiation. Some even considered atoms to be mere whirlpools in the ether.

If such a medium existed, it could presumably serve as an absolute frame of reference. Since the earth is continually changing the direction of its motion as it travels along its orbit, it would have to show motion with respect to the ether. This should have an effect on the propagation of light and could conceivably be tested experimentally.

Albert A. Michelson (1852–1931), the first American Nobel Prize winner in science, was noted for his ingenious and extremely accurate experiments in optics and especially for his unsurpassed measurements of the speed of light.

An investigation to detect such effects was begun by the American physicist A. A. Michelson in 1880 and continued by him and E. W. Morley in the following years. The experimenters used an optical device called the Michelson interferometer, to be described in detail in Chap. 15. In this instrument, a beam of light is split into two beams that travel out equal distances along perpendicular paths and are then reflected back to their starting point (Fig. 5.8). One of the paths is placed in the direction of the orbital motion of the earth; the other, crosswise to it. This arrangement is exactly like the motorboat situation of Worked example 5.3: The two boat trips are replaced by the passage of two beams of light, and the current in the river is analogous to the presumed ether drift. Such a drift would show up as a difference in the transit times for light going along the stream and crosswise to it. Instead of returning to the starting point in the same stage of vibration, the two beams would be slightly out of step. This would result in a shift in the pattern of bright and dark bands seen in the interferometer. The purpose of the experiment was to see whether such a shift existed.

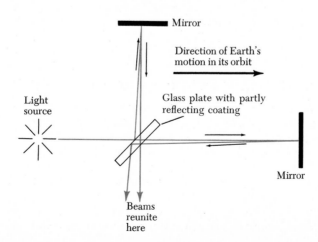

FIGURE 5.8 The Michelson-Morley experiment (schematic). The situation is analogous to the example of the motorboat crossing a river.

In the boat example, the time for the round trip parallel to the bank was nearly 10 percent greater than for the trip across the stream and back. In the corresponding ether-drift experiment the effect is very small because the earth's orbital speed (18.5 mi/sec) is very small compared with the speed of light (186,000 mi/sec). The result is that the two times of passage could be expected to differ by only half a millionth of a percent. However, the interferometer was capable of detecting a difference 100 times smaller than the computed one.

A repetition of the experiment, using masers, proved (1959) that if an ether drift exists, it must be less than 10 m/sec.

The astonishing result of the Michelson-Morley experiment was that *no time difference was found.* The test has since been repeated, at various times of the year and at different locations, always with a *null* outcome. No motion through the ether could be detected. After examining all reasonable alternatives, many physicists were compelled to conclude that *an observable universal frame of reference does not exist.*

Einstein had already come to the conclusion that, for various reasons, the classical ideas of space and time were in need of reformulation. The null result of the Michelson-Morley experiment lent further support to this conviction.

Others, among them the French mathematician Henri Poincaré, arrived independently at similar conclusions from another point of view.

In 1905 Einstein drafted his **special theory of relativity,** which he was able to base on two fundamental assumptions:

1. *Principle of relativity.* All the laws of physics are the same in all inertial systems (see below). No one system of this kind is more fundamental than any other. This assumption is also basic to pre-relativity physics.
2. *Principle of constancy of the speed of light.* The speed of light in a vacuum is a constant, independent of the motion of the source relative to the observer. This principle is something new; it is not in harmony with pre-relativity physics.

It is necessary at this point to define an inertial system. An **inertial system** may be described as any reference frame in which a body subject to no force will be observed to move with constant vector velocity. Einstein's first formulation of relativity concepts, designated the *special* theory, applies only to inertial systems. All other cases come under the heading of what is called the *general theory of relativity.*

Starting with the two fundamental postulates, Einstein derived conclusions regarding the effects of relative motion on the measurement of space, time, mass, energy, etc. Some of the consequences will be discussed later. Where high relative speeds are involved, the relativity formulas give results that differ markedly from those of classical physics. However, for low speeds the relativity formulas reduce to the familiar classical ones. Relativity has not *overthrown* classical physics; it has merely generalized it, extending it to include a wider range of experience.

Velocity addition according to the theory of relativity.

For the present, the discussion will be confined to what special relativity has to say about the measurement of relative velocity. Consider two inertial systems, designated as reference frames B and C (Fig. 5.9), moving with relative speed v_{BC}. Arbitrarily, we have shown system B moving to the right, relative to C, with this speed. A particle A is seen by an observer in B to be moving with a speed v_{AB}. For simplicity, both of these motions have been taken to be in the same line. How does A move as seen by an observer who is stationary in the system C?

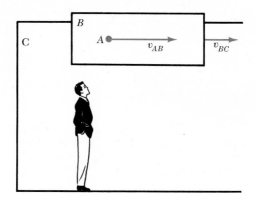

FIGURE 5.9 Relative velocity in the theory of relativity.

In classical theory, the answer would be given at once by Eq. 5.3:

$$v_{AC} = v_{AB} + v_{BC}$$

but special relativity replaces this by the quite different relation

$$v_{AC} = \frac{v_{AB} + v_{BC}}{1 + v_{AB}v_{BC}/c^2} \qquad \textbf{5.4}$$

where c represents the *speed of light in vacuum,* numerically about 3×10^8 m/sec.

It may seem surprising to find the speed of light appearing in a formula

Michelson (left) with Albert Einstein and Robert A. Millikan (right). (*Courtesy of Niels Bohr Library, American Institute of Physics.*)

that describes a purely mechanical situation apparently having nothing to do with the phenomenon of light. However, some of the arguments of relativity make use of the concept of transmitting signals from one place to another, and c is the greatest speed with which this can be done. This constant appears in many other relativity formulas as well, including the now-familiar $E = mc^2$.

When either v_{AB} or v_{BC} approaches zero, Eq. 5.4 reduces to Eq. 5.3, but when either of these speeds is *not* small compared with c, the numerical results given by the two formulas are widely different. The following example illustrates this point.

Worked example 5.4   A rocket ship approaching an unfriendly space station at a relative speed of $\frac{3}{4}c$ fires a projectile in the forward direction at a speed of $\frac{1}{2}c$ relative to the ship.

a. What is the speed of the projectile as seen by an observer on the space station?

b. This observer measures the speed of light coming from a searchlight on the rocket ship. What value does he get?

Solution a. Here $v_{AB} = \frac{1}{2}c$ and $v_{BC} = \frac{3}{4}c$. If the classical formula 5.3 is used, the result is $v_{AC} = \frac{1}{2}c + \frac{3}{4}c = \frac{5}{4}c$, which is greater than the speed of light. On the other hand, substitution into the relativity equation 5.4 yields

$$v_{AC} = \frac{\frac{1}{2}c + \frac{3}{4}c}{1 + 3c^2/8c^2} = \frac{10}{11}c$$

which is *less* than the speed of light. It is an interesting consequence of the relativity formula that *no material object can have a speed, relative to any observer, that is as great as c.* Note, however, that it is proper to consider the projectile and the space station to be closing in on each other at the rate $\frac{5}{4}c$. This relative speed of two objects, although greater than c, could be measured by an observer in a third system, with no violation of relativity principles.

b. In this case $v_{AB} = c$ and $v_{BC} = \frac{3}{4}c$. Substituting into Eq. 5.4,

$$v_{AC} = \frac{c + \frac{3}{4}c}{1 + 3c^2/4c^2} = c$$

The resulting value is c, *independent of the motion of the source*, as required by the principle of the constancy of the speed of light.

Examination of such ordinary concepts as speed and velocity has led to some surprising results, suggesting again that intuitive ideas about familiar aspects of nature cannot be trusted blindly but must always be open to investigation from new points of view. For the present, we return to a discussion of simpler ideas connected with the study of the motion of a particle under more familiar conditions where it is not necessary to use relativity formulations. The question of reference frames will be taken up again in Sec. 6.3, page 122.

ACCELERATED MOTION

5.4 ACCELERATION

Most of the types of motion usually observed are of a complicated nature and so are difficult to describe quantitatively. As examples, think of the motion of a car in city traffic, the flight of a butterfly, or the progress of a football player running for a touchdown. Even some apparently more regular motions such as the swinging of a pendulum or the motion of a planet in its orbit are not easy to analyze. The reason is that in all such cases the instantaneous velocity continually changes, either in direction, magnitude (speed), or both. To describe the way in which the change takes place in a particular kind of motion, Galileo introduced a special physical quantity, **acceleration.** The term derives from a Latin word *accelerare:* to cause to move faster.

Difference of two vectors.   In Fig. 5.10*a*, A and B are two successive positions of a particle moving along a path, and $\mathbf{v}_1$ and $\mathbf{v}_2$ are their instantaneous velocities at these two places. It is seen that the two vectors may differ in both direction and magnitude. Draw both velocity vectors from a common point (Fig. 5.10*b*) and construct the vector $\Delta\mathbf{v}$, as shown. What does this vector represent? From the diagram, $\Delta\mathbf{v}$ is obviously a vector which, when added to $\mathbf{v}_1$, yields $\mathbf{v}_2$. Expressed as a vector equation,

$$\mathbf{v}_1 + \Delta\mathbf{v} = \mathbf{v}_2$$

or, transposing,

$$\Delta\mathbf{v} = \mathbf{v}_2 - \mathbf{v}_1 \qquad\qquad 5.5$$

$\Delta\mathbf{v}$ may be called the **vector difference** between $\mathbf{v}_1$ and $\mathbf{v}_2$.

Instantaneous acceleration.   The change in velocity of the particle in going from A to B took place in some interval of time Δt. Divide $\Delta\mathbf{v}$ by Δt and pass to the limit in which Δt approaches zero; this is the same process as used on page 88. Here the quantity obtained as a limit is a *vector* called the **instantaneous acceleration** of the particle at the point in question:

$$\mathbf{a} = \lim_{\Delta t \to 0} \frac{\Delta\mathbf{v}}{\Delta t} \qquad\qquad 5.6$$

The vector $\mathbf{a}$ is shown in Fig. 5.10*c*. It has the same direction as $\Delta\mathbf{v}$, but its scale has been changed by dividing it by a time interval (see marginal note, page 90).

In Fig. 5.10*c* the vector $\mathbf{a}$ is shown resolved into two components, one

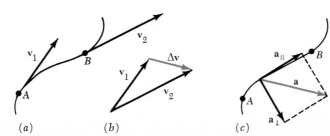

(a)   (b)   (c)

FIGURE 5.10 The instantaneous acceleration may be resolved into the two components $\mathbf{a}_\parallel$ and $\mathbf{a}_\perp$. The first changes the magnitude of the velocity vector; the second changes its direction.

along the path and the other perpendicular to it. In driving a car along a winding road, the motorist is free to change $\mathbf{a}_\parallel$ by manipulating the accelerator and the brake, but $\mathbf{a}_\perp$ is governed by the twisting and turning of the road. Thus $\mathbf{a}_\parallel$ changes the *magnitude* of the vector $\mathbf{v}$ and $\mathbf{a}_\perp$ changes its *direction*. In some situations soon to be discussed, the path is straight and the total acceleration lies along it; the perpendicular component is then zero. In these special cases, the analysis is simplified.

There is no specific word for the scalar part of the acceleration vector, corresponding to speed *for the scalar part of velocity.*

The motorist is concerned mainly with the magnitude of the instantaneous acceleration *along* the road, which is the instantaneous rate of change in his speed. From the way in which acceleration was defined we could say, for instance, that a certain vehicle is changing its speed at the rate of 3 ft/sec each second, or at the rate

$$\frac{3 \text{ ft/sec}}{\text{sec}}$$

This is usually written even more compactly as 3 ft/sec² and is to be read: "Three feet per second squared." The "sec²" has no meaning by itself but indicates merely that acceleration has the dimensions of length divided twice by time, or L/T^2. One of the T factors is part of the quantity of *speed.* Permissible units for the magnitude of acceleration are ft/sec², cm/sec², m/sec², etc. A mixed unit such as mi/hr/sec may be useful in automotive engineering but should be avoided in physics; it is not good policy to have more than one unit of a given kind in any problem.

Even in instances where the vector nature of acceleration is not important and only numerical values are wanted, it must still be carefully specified whether one is dealing with increase or decrease of speed. A rate of increase is called a *positive* acceleration; a rate of decrease is a *negative* acceleration, or, sometimes, a *deceleration.* However, no separate name is needed, and a minus sign written before the number is sufficient. For example, if a car has a speed of 40 ft/sec at the instant when the brakes are applied, and a speed of 10 ft/sec after the brakes have been in operation for 5 sec, its average acceleration over this interval will be

$$\bar{a} = \frac{v}{t} = \frac{10 - 40}{5} = -6 \text{ ft/sec}^2$$

The minus sign shows that the acceleration vector would be *opposite* in direction to the velocity vector.

Practice set*

The speed of a car is 10, 14, and 12 ft/sec at times $t = 0$, 1, and 2 sec, respectively. What is the magnitude, in ft/sec², of its average acceleration during (*a*) the first second; (*b*) the second second; (*c*) the first two seconds?

5.5 MOTION WITH CONSTANT ACCELERATION

The problem of describing the motion of a body when accelerations are present is sometimes complicated. However, the problem of motion with *constant acceleration,* which is of great practical interest, is comparatively easy. This is the type of motion possessed by a block sliding down an

*Ans.: 4; −2; 1.

inclined board, by a stone falling freely near the surface of the earth or, strictly, by any particle acted upon by a constant force.

The detailed quantitative description of motion with constant acceleration was worked out by Galileo in connection with the example of free fall, using an ingenious blend of intuition and geometrical reasoning. We can pursue his arguments for the general case of motion with constant acceleration, returning later to the particular application to falling bodies.

Consider, for simplicity, a particle moving along a straight path so that the displacement, velocity, and acceleration vectors are all in the same line. Then we need deal only with the magnitudes of these vectors. However, we must still allow for the directional *sense* of each quantity: whether it is toward one end of the line or the other. Suppose the particle is moving with an initial speed v_0 when we begin to apply a constant acceleration of magnitude a. The problem is to describe the subsequent motion quantitatively and to be able to predict where the particle will be at any later instant, how fast it will then be going, etc. This means that we must agree on where to begin measuring the time, and the simplest procedure is to start at the instant the acceleration begins.

When the time $t = 0$, the instantaneous speed v of the particle is v_0, and since the acceleration is assumed to be constant, v will increase in proportion to the time. This is represented graphically in Fig. 5.11a, where v is plotted against t. The curve turns out to be a straight line for the following reason: The average acceleration of the particle over a stated interval is given by $\Delta v/\Delta t$, and since the average acceleration is constant, this ratio is constant. On the graph, this ratio represents the slope of the tangent to the curve, and a curve of constant slope can only be a straight line. Notice that this line does not pass through the origin but through the point $t = 0$, $v = v_0$. The physical interpretation of the graph is that, starting with an initial speed v_0, the body picked up speed at a constant rate a as time passed. Figure 5.11b shows the speed-time graphs for two particles having different initial speeds and different values of a. The line corresponding to the larger a has the steeper slope.

Deducing general formulas. The regularity expressed by the speed-time graph can be put in the form of a useful algebraic equation. From the definition of acceleration, an object that changes its speed by an amount Δv in a time interval Δt can be said to have an average acceleration during that interval of $\bar{a} = \Delta v/\Delta t$. Apply this to the entire time interval from $t = 0$ to $t = t$, corresponding to a change in speed from $v = v_0$ to $v = v$. Since we are here interested in motion with constant acceleration, the average value $\bar{a}$ can be replaced by the constant value a. Then

$$a = \frac{v - v_0}{t} \quad \text{or} \quad v - v_0 = at$$

Transposing,

$$v = v_0 + at \qquad \qquad 5.7$$

This is an algebraic representation of the straight-line speed-time graph; it gives values of v, the speed attained, as a function of the elapsed time t.

The question of the *distance* covered by the particle in a given time is slightly more complex. In the discussion of motion with *constant speed*

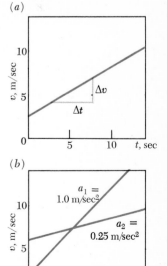

FIGURE 5.11 The magnitude of the acceleration is given by the slope of the vt curve; this curve is a straight line for motion with constant acceleration.

(a)

(b)

(page 86), it was found that equal distances were covered in equal intervals of time, and the whole story was told by Eq. 5.1. Here, however, the moving object is constantly changing its speed, and this affects the subsequent distances covered. Galileo was able to resolve the difficulty by considering the *average* speed of the accelerated body over a given interval. His method is applied in the following.

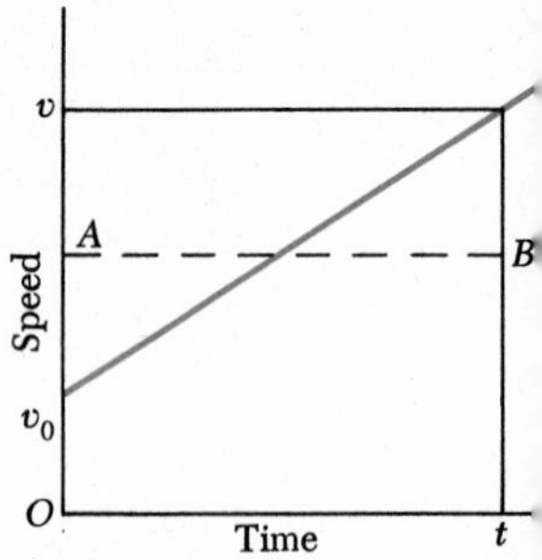

In the speed-time graph of Fig. 5.12, examine any interval of the motion, starting with time $t = 0$. The speed at the beginning of this interval is v_0 and at the end it is v. Because the rate of increase of speed is uniform in this case of constant acceleration, the average speed over the interval is merely $(v_0 + v)/2$. This value is shown on the graph by the dashed line *AB*. At any instant during the first half of the motion the instantaneous speed is below this line; at a corresponding point during the second half, it is an equal amount above. Consequently, the average represents the motion exactly, as far as distances are concerned. Stated in another way, the accelerated particle is replaced by one moving at a constant speed equal to the average speed of the former. The distance covered is given by this average speed multiplied by the elapsed time, or

The average value of any pair of numbers is half their sum.

$$x = \bar{v}t = \frac{v_0 + v}{2} t \qquad \text{5.8}$$

Substitute the value of v from Eq. 5.7:

$$x = \frac{v_0 + (v_0 + at)}{2} t$$

or

$$x = v_0 t + \frac{at^2}{2} \qquad \text{5.9}$$

Figure 5.13 shows the distance-time graphs for the same two values of a that appeared in Fig. 5.11b. Notice that, whereas the v,t graphs were straight lines, the x,t graphs are curved upward. The term in t^2 in Eq. 5.9 is the main cause of the rapid increase in x as time passes.

[Figure 5.12 graph: Speed vs Time]

FIGURE 5.12 For motion with constant acceleration, the average speed over an interval is halfway between the initial and final values.

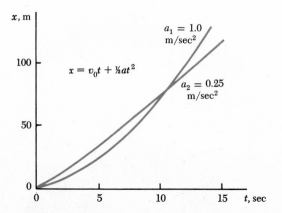

FIGURE 5.13 Distance-time curves for two examples of motion with constant acceleration. The graph corresponding to the greater value of a curves upward more sharply.

Equations 5.7 and 5.8 contain the complete description of motion with constant acceleration. However, Eq. 5.9, a combination of the other two, is frequently useful. Another equation of some importance can be obtained by eliminating t between Eqs. 5.7 and 5.9: From Eq. 5.7, we get $t = (v - v_0)/a$. When this is substituted into Eq. 5.9,

$$x = \frac{v_0(v - v_0)}{a} + \frac{a(v - v_0)^2}{2a^2}$$

After multiplying out and collecting terms, the result is

$$v^2 = v_0^2 + 2ax \qquad \textbf{5.10}$$

It is now advisable to bring together Eqs. 5.7 to 5.10 and see what has been accomplished:

MOTION WITH CONSTANT ACCELERATION

$v = v_0 + at$	5.11a
$x = v_0 t + \frac{1}{2}at^2$	5.11b
$v^2 = v_0^2 + 2ax$	5.11c
$x = \frac{1}{2}(v_0 + v)t$	5.11d

Notice that there are five quantities involved in the discussion of this type of motion: a, v_0, x, v, and t. The first two are constants; the other three are variables. Recall the exact meaning of the variables: t is the time measured from the instant the acceleration begins, v is the speed attained at the end of this time, and x is the distance covered in this same time. Since only two of the above equations are independent, only two quantities can be calculated; the other three must be known or given. However, the above relations are sufficient for answering any legitimate question concerning motion with constant acceleration.

Such questions can take various forms. Suppose that the values of a and v_0 are given, as is often the case. It may then be asked: How fast will the body be going after a specified time? The answer can be obtained at once by substituting into Eq. 5.11a, which gives v as a function of t. Solved for t, this equation also answers the reciprocal question: How long does it take the object to attain a given speed? In the same way, Eq. 5.11b tells how far the object moves in a given time, or how long it takes to go a given distance. Similarly, Eq. 5.11c relates speed and distance, and Eq. 5.11d connects the three variables: distance, speed, and time. In seeking to answer a specific question, we can select an equation from the above set that involves the given items and the unknown quantity, and solve for the latter.

There is, however, always a way of solving the problem without the use of these equations and that is to *reason through the situation by using only the definition of acceleration*, which is Eq. 5.11a, *and the expression for the average speed*, which is 5.11d. This procedure was used in deriving the general equations. Both methods will be illustrated by the numerical examples that follow. Work through them carefully.

Worked example 5.5   An elevator starts upward from rest with a constant acceleration of 1.5 ft/sec². How far does it rise in 4 sec?

Solution   It is always a good idea to list the quantities involved, both known and unknown:

$$v_0 = 0 \qquad\qquad x = ?$$
$$a = 1.5 \text{ ft/sec}^2$$
$$t = 4 \text{ sec}$$

The equation connecting all four of these quantities is Eq. 5.11*b*. Substituting the numbers, $x = 0 + \frac{1}{2}(1.5)(4^2)$, or $x = 12$ ft. Without recourse to a formula, the problem can be reasoned through in this way: If the speed of the car is initially zero and increases at the rate of 1.5 ft/sec² for 4 sec, the final speed will be $(1.5)(4) = 6.0$ ft/sec. This means that the average speed over the stated interval was $(0 + 6)/2 = 3$ ft/sec. If an object were to move at a constant speed of 3 ft/sec for 4 sec, it would cover a distance of $3 \times 4 = 12$ ft, as found above.

Worked example 5.6   A car increases its speed uniformly from 30 to 70 ft/sec in 8 sec. Find the acceleration and the distance traveled in that time.

Solution   List the known and unknown quantities:

$$v_0 = 30 \text{ ft/sec} \qquad a = ?$$
$$v = 70 \text{ ft/sec} \qquad x = ?$$
$$t = 8 \text{ sec}$$

From Eq. 5.11*a*, $a = (v - v_0)/t = (70 - 30)/8 = 5$ ft/sec². Then, to compute x, we have a wide choice of formulas; we can use Eq. 5.11*b*, 5.11*c*, or 5.11*d*.

By Eq. 5.11*b*: $x = (30)(8) + \frac{1}{2}(5)(64) = 400$ ft
By Eq. 5.11*c*: $x = (v^2 - v_0^2)/2a = (4{,}900 - 900)/10 = 400$ ft, as before
By Eq. 5.11*d*: $x = \frac{1}{2}(30 + 70)(8) = 400$ ft again

Notice that the questions can be answered almost at once without formulas by using the definition of acceleration and the concept of average speed over an interval. Equation 5.11*a*, solved for a, is nothing more than the definition of acceleration, and Eq. 5.11*d* states only that the distance covered is the average speed $\frac{1}{2}(v_0 + v)$ multiplied by the time.

Worked example 5.7   A train is coasting along at 60 ft/sec when the brakes are applied, bringing it uniformly to rest at a station 1,200 ft ahead. Calculate the acceleration (assumed constant) and the total time in coming to a stop. Find also the speed of the train 12 sec after the brakes were applied and how far it went in this time interval.

Solution Stating the given and the unknown quantities,

$$v_0 = 60 \text{ ft/sec} \qquad a = ? \qquad v_{12} = ? \atop v = 0 \qquad\qquad t = ? \qquad x_{12} = ? \}\qquad \text{for first 12 sec of braking}$$
$$x = 1{,}200 \text{ ft}$$

To find a, substitute in Eq. 5.11c: $0 = (60)^2 + 2a(1{,}200)$, or $a = -1.5$ ft/sec^2. The minus sign shows that the object is slowing down. Next, use Eq. 5.11a to compute the time: $0 = 60 - 1.5t$, $t = 40$ sec. To find the speed after 12 sec of braking, make use of Eq. 5.11a again: $v_{12} = 60 - (1.5)(12) = 42$ ft/sec. The corresponding distance of travel can be found from Eq. 5.11d: $x_{12} = \frac{1}{2}(60 + 42)(12) = 612$ ft.

As in the preceding examples, the same results can be reached by a less formal argument. To find the total time, notice that the average speed of the train while coming to rest is 30 ft/sec, so that the time will be $1200/30 = 40$ sec. The acceleration will be the entire change (decrease) in speed divided by this time, or $-60/40 = -1.5$ ft/sec^2. After 12 sec, the speed of the train will have decreased by $(1.5)(12) = 18$ ft/sec, making the speed at that instant $60 - 18 = 42$ ft/sec. The distance traveled in this 12-sec interval will be equal to the time multiplied by the average speed over this interval, or $12(60 + 42)/2 = 612$ ft.

Units of a result. In using the set of equations 5.11, the question of the units to be attached to the result of a computation may be puzzling. As long as a consistent set of units is used throughout, this presents no problem. For example, if distances are measured in feet and times are measured in seconds, than all speeds must be expressed in ft/sec, all accelerations in ft/sec^2, etc. However, the principle of dimensional consistency explained on page 53 always provides a further check on the units.

As an example, take Eq. 5.11b. Since the quantity on the left is a length, each of the terms on the right side of the equation must be a length also. Specifically, in the English system, the first term on the right will use feet per second as the appropriate units for v_0 and seconds for t, giving this term the units

$$\left(\frac{\text{ft}}{\text{sec}}\right)(\text{sec}) = \text{ft}$$

where "sec" was canceled in numerator and denominator just as for any common factor. This term of the equation is, then, a length expressed in feet, as it should be.

In the second term on the right side of the equation, the factor $\frac{1}{2}$ is a pure number (no units), a is measured in ft/sec^2, and t is in sec. Thus the combination will have the units

$$\left(\frac{\text{ft}}{\text{sec}^2}\right)(\text{sec})^2 = \text{ft}$$

which is again the correct unit.

Practice set Check Eq. 5.11c for dimensional consistency, using units based on meters and seconds.

MOTION UNDER GRAVITY

5.6 FREE FALL UNDER GRAVITY

The motion of an object that is allowed to fall to the ground was described by Aristotle as being "quicker in proportion to its size," implying that doubling the weight of a falling body will make it cover a given distance in half the time. It is likely that he reached this conclusion by his customary method of seeking knowledge by pure reason, but he may have had in mind the observed behavior of such things as falling leaves and dropped stones. In any case, systematic quantitative experiments would have shown the above statement to be false. Aristotle's influence and authority were so great that these notions went unchallenged for a long time.

Although the story is often quoted, there is no definite evidence that Galileo actually experimented with bodies dropped from the Leaning Tower of Pisa.

In the sixteenth century, Galileo questioned the ideas of Aristotle on this matter, having undoubtedly observed that the times of fall of two objects of different weight did not differ noticeably as long as both were quite heavy. With remarkable insight, Galileo was able to see that the difference in rate of fall of a leaf and a stone is merely an effect of the disturbing influence of the surrounding air and that if two compact, fairly heavy bodies are compared the difference will be quite small.

Galileo had no way of eliminating the effects of air resistance, for the air pump was not invented until some years later. His conclusions were probably the outcome of one of the thought experiments (page 17) for which he was noted. He was able to idealize the phenomenon of the motion of a falling body, replacing the actual situation by a model—a particle falling in a vacuum, near the earth's surface. Then he identified the ideal motion with the theory of motion with constant acceleration which he next developed. The new point of view that Galileo introduced was the recognition of the necessity of making simple models and of appealing to experiment to justify them.

Since the direct observation and measurement of free fall require timing devices which were not available to him, Galileo experimented instead with a ball rolling down an incline. He reasoned that the movement of the ball down the slope would be of the *same type* as that of free fall, except that it would be in "slow motion" and so could be measured with the means at his disposal. The results of such experiments showed that a ball descending a given incline picked up speed at a constant rate, and Galileo defined the word acceleration especially for this purpose, to be used exactly as we do today. He could then project his experience to conclude that the motion of a freely falling object is motion with constant acceleration, since free fall is simply the limiting case of motion down an incline whose slope angle is 90°.

Having identified free fall with motion with constant acceleration, we have only to determine the magnitude of the acceleration by experiment and then put this value into the equations we have been using for this type of motion. The magnitude in question is called the **acceleration due to gravity** and its symbol is g. Although the numerical value varies slightly from place to place on the earth's surface, the greatest spread is only about 0.4 percent of the average. Part of this variation is caused by differences in altitude, so that if we use a constant numerical value of g in our computa-

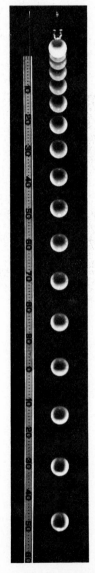

A falling ball photographed at $\frac{1}{30}$-sec intervals. The scale at the left is in centimeters. (*From PSSC Physics, D. C. Heath & Co., 1965.*)

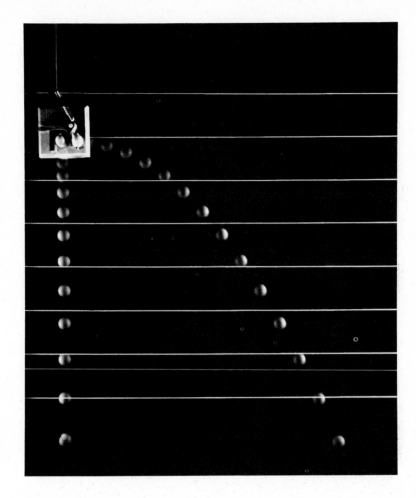

One ball was projected horizontally to the right at the same time that the other was dropped from rest. Their vertical motions are identical. (*From PSSC Physics, D. C. Heath & Co., 1965.*)

tions, we are assuming that the distance of fall is small compared with the radius of the earth. For most of the calculations in this book, the following two-digit values can be used:

$$g = 9.8 \text{ m/sec}^2 \quad \text{or} \quad 32 \text{ ft/sec}^2$$

The official International standard values are 9.80665 and 32.1740, respectively, at 45° latitude and sea level. A value correct to 1 part in a billion can now be obtained.

In applying Eqs. 5.11 to the case of free fall, not only must the special value g be used in place of a, but the directions of the various quantities must be carefully noted. It is convenient to take the upward direction as positive. Then all displacements and velocities in the vertical upward direction will be positive, and those in the downward direction will be negative. Also, a negative sign must be attached to g, so that the acceleration due to gravity becomes -32 ft/sec^2. The algebra will then indicate the correct direction. Any quantity that turns out to have a plus sign will be directed upward, and one with a minus sign will be directed downward. The following examples will illustrate this.

Worked example 5.8   A ball is thrown upward from the top of a 144-ft tower with a speed of 64 ft/sec. On the way down, the ball misses the tower and falls to the ground below.

a. How long did the ball rise?
b. How high above the top of the tower did it go?
c. How long was it in flight?
d. How fast was it moving when it hit the ground?

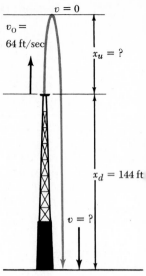

FIGURE 5.14   Ball thrown upward from the top of a tower.

Solution   Figure 5.14 is a sketch of the circumstances. With the upward direction chosen as positive, the given and unknown quantities are

$$v_0 = 64 \text{ ft/sec} \qquad\qquad t_u = ?$$
$$a = -g = -32 \text{ ft/sec}^2 \qquad x_u = ?$$
$$x_d = 144 \text{ ft} \qquad\qquad t = ?$$
$$v = ?$$

Notice, first, that although the ball is rising during part of its flight its motion is properly called free fall. This term means only that gravity acts to *decrease the upward speed* of the object while on the way to the top and to *increase its downward speed* on the descent, both at the rate *g*.

a. The ball rises until its upward speed becomes zero, so that Eq. 5.11a applied to the upward part of the path becomes

$$0 = 64 - 32t_u \qquad \text{or} \qquad t_u = 64/32 = 2.0 \text{ sec}$$

b. Equation 5.11d applied to the upward part is

$$x_u = \tfrac{1}{2}(64 + 0)(2.0) \qquad \text{or} \qquad x_u = 64 \text{ ft}$$

c. Equation 5.11b applied to the entire path yields $-144 = 64t - \tfrac{1}{2}(32)t^2$, which reduces to $t^2 - 4t - 9 = 0$, a quadratic equation in *t*. The two values of *t* are found to be 5.6 sec and −1.6 sec. The positive value, 5.6 sec, is the total time of flight of the ball, from the instant it was thrown upward from the top of the tower until it struck the ground. The other solution, −1.6 sec, also has a meaning. It is the time at which the ball would have had to be thrown upward from the ground in order to pass the top of the tower with an upward speed of $v_0 = 64$ ft/sec. That is to say, the formal algebraic equation has no way of knowing whether the ball actually started from the ground or from the top of the tower, and so it gives both answers.

d. Use Eq. 5.11a again, this time for the entire flight starting upward from the tower top and finally arriving at the ground. The appropriate time to use is 5.6 sec:

$$v = 64 - (32)(5.6) \qquad \text{or} \qquad v = -115 \text{ ft/sec}$$

The minus sign shows that this is a *downward* speed.

Worked example 5.9   A stone is thrown straight upward with a speed of 14.7 m/sec from a stationary balloon. Neglecting air resistance, find the position, velocity, and acceleration of the stone (*a*) after 1.5 sec; (*b*) after 4.0 sec.

Solution   Here the quantities are

$$v_0 = 14.7 \text{ m/sec} \qquad x = ?$$
$$g = -9.8 \text{ m/sec}^2 \qquad v = ?$$
$$t = 1.5 \text{ sec; } 4.0 \text{ sec} \qquad a = ?$$

a. By Eq. 5.11a, $v = 14.7 - (9.8)(1.5) = 0$. Then, from Eq. 5.11d, $x = \frac{1}{2}(0 + 14.7)(1.5) = 11.0$ m.

The object is at the top of its path, 11.0 m above the starting point. The instantaneous *velocity* of the stone is zero there, yet its *acceleration* is -9.8 m/sec², the same as for any unsupported object near the earth. At certain times the instantaneous velocity and acceleration may even be in opposite directions, as on the upward part of the path. It must be remembered that in general *velocity and acceleration are independent quantities*.

b. Repeating the calculations for $t = 4.0$ sec, we substitute into Eq. 5.11a again: $v = 14.7 - (9.8)(4)$, or $v = -24.5$ m/sec. The minus sign indicates that the body is moving in the *downward* direction at that time. From Eq. 5.11d, $x = \frac{1}{2}(14.7 - 24.5)(4)$, or $x = -19.6$ m, where the minus sign shows that the stone is 19.6 m *below* its starting point. Notice that x is the momentary *displacement* of the particle, not necessarily the total distance it has traveled.

5.7   THE FLIGHT OF A PROJECTILE

A simple projectile is any inert particle that is thrown or shot from some point near the earth's surface and is subsequently acted upon only by gravity. It is assumed, also, that (1) the projectile travels only a limited distance over the earth's surface, this part being considered a horizontal plane, and (2) effects of the rotation of the earth can be disregarded.

In general, a projectile may be sent out at any angle, including directions below the horizontal if it is projected from an elevated position. A pitched baseball or a stone thrown into the Grand Canyon are good general examples of a simple projectile, whereas Worked examples 5.8 and 5.9 presented the special case of a projectile launched vertically upward.

If it were not for gravity, a projectile would continue to move with its initial speed in the direction in which it was launched. Actually, the earth's attraction makes the object drop below the original line of projection because it is a free, unsupported body. Its position after a stated time will be vertically below this line by an amount equal to the distance a particle released from rest would fall in this interval. The result is that the projectile follows a curved path, known geometrically as a **parabola** (Fig. 5.15).

For a projectile shot vertically upward or downward, the parabolic path degenerates into a straight line, as in Worked examples 5.8 and 5.9.

A convenient way of analyzing the motion of a projectile is to resolve the velocity vector at each point into its horizontal and vertical components, as shown in Fig. 5.16. Since by definition gravity acts vertically, it can change only the vertical-velocity component, decreasing its magnitude at the rate g when the particle is on the way up and increasing its magnitude at this same rate on the way down. At the top of the path the vertical component is zero, and the total velocity of the particle is $v_{0,x}$ at that instant. The horizontal component is unaffected because it is perpendicular to the total acting force (the weight of the projectile). This component remains constant throughout the flight.

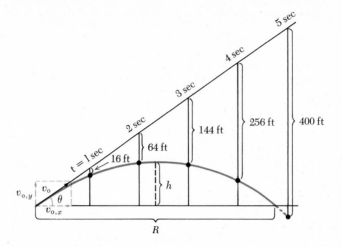

FIGURE 5.15 The projectile drops below its original line of projection in the same manner as any freely falling object.

The general motion of a projectile may be looked upon as the resultant of the motion of a particle shot straight upward with an initial speed $v_{0,y}$ and at the same time moved horizontally with a constant speed $v_{0,x}$. An example will show how to compute the details of the motion of a projectile from this point of view.

Worked example 5.10 A projectile leaves a gun which is elevated at an angle of 30° above the horizontal with an initial speed (muzzle speed) of 196 m/sec. Find the maximum height reached by the projectile, the time of flight, and the distance from the gun at which it hits the ground. Assume the muzzle of the gun to be essentially at ground level.

Solution Resolve v_0 into its horizontal and vertical components (see Fig. 5.16):

$$v_{0,x} = v_0 \cos \theta = (196)(0.866) = 170 \text{ m/sec}$$
$$v_{0,y} = v_0 \sin \theta = (196)(0.500) = 98 \text{ m/sec}$$

A particle projected upward at a speed of 98 m/sec will ascend a distance h given by Eq. 5.11c as $0 = (98)^2 - 2(9.8)h$, yielding

$$h = 490 \text{ m}$$

as the maximum height attained.

The time of rise can be calculated most simply by using Eq. 5.11a:

$$0 = 98 - 9.8t \qquad \text{or} \qquad t = 10.0 \text{ sec}$$

The total time of flight, T, is just twice this amount, or

$$T = 20.0 \text{ sec}$$

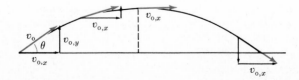

FIGURE 5.16 Velocity components of a projectile at various points in its flight.

BALLISTICS

The parabolic path of an ideal projectile can be shown to be symmetric around a vertical line through the highest point, which also marks the midpoint of the range on a horizontal plane. Further, the particle comes into the ground at the end of its flight at the same angle as it was initially projected and strikes with a speed equal to its launching speed. These facts were not known before Galileo pointed them out. In the sixteenth century, a common view was that the path of a missile consists of three parts: a "violent mode" during which the line of projection is followed exactly, a "mixed mode" when gravity begins to make the projectile deviate from this line, and a final "natural mode" in which gravity takes complete control and the object plunges straight to earth. This idea seems to have enough plausibility to make people sometimes offer it, even today, as an intuitive description of what happens.

The results calculated in this chapter for falling bodies and projectiles are strictly valid only in the ideal circumstances that were assumed, and it may be questioned why we do not compute what happens in a practical case. The reason is that the ever-changing effects of the surrounding air make the mathematical treatment of an actual case very complex. In addition, for long-range missiles, the change in the value of g with height, the change in direction of g from one place to another on a curved earth, and the effects of the earth's rotation must be taken into account as well. As a consequence, the science of ballistics must rely on theory plus results derived from test firings in order to describe satisfactorily the performance of an actual missile. However, in physics we are more concerned with the idealized model, which in many cases gives a good approximation to actual ballistics.

FIGURE 5.17 Path of a projectile according to Galileo's predecessors.

FIGURE 5.18 The effect of air resistance is to make the actual path of the projectile fall short of the ideal parabola (broken line). The difference is especially prominent near the end of the flight, producing a "snub-nosed" path. Also, the total time of flight is increased.

All during the time the projectile is in flight it is moving horizontally at a speed $v_{0,x}$. Then the distance covered along the ground, the range of the projectile, will be

$$R = v_{0,x}t = (170)(20) = 3{,}400 \text{ m}$$

Programmed review

Instructions: See page 18.

1. What is meant by the average speed of motion of a particle? By the instantaneous speed?

   Average speed over an interval is the distance covered divided by the elapsed time. The instantaneous speed is the limit of the average speed as the size of the interval approaches zero.   [5.1]

2. Define instantaneous *velocity*.

A vector whose direction is tangent to the path of a particle at any point and whose magnitude is equal to the instantaneous speed at that place.   [5.1]

3. If the velocity of a particle in one reference frame is known, how can its velocity in another frame, moving relative to the first, be found?

By combining the vectors representing the velocity of the particle in the first frame and that representing the velocity of the first frame with respect to the second:

$$\mathbf{v}_{AC} = \mathbf{v}_{AB} + \mathbf{v}_{BC}$$ [5.2]

4. For what purpose was the Michelson-Morley experiment conducted, and what was its result?

For determining the motion of the earth in space. The null result showed that such a quantity cannot be determined because it has no operational meaning.   [5.3]

5. Describe how the speed of light in empty space depends on the motion of the source.

The speed is completely independent of the motion of the source.   [5.3]

6. Give an expression for the instantaneous acceleration of a particle.

The instantaneous acceleration is given by

$$\mathbf{a} = \lim_{\Delta t \to 0} \frac{\Delta \mathbf{v}}{\Delta t}$$ [5.4]

7. What is meant by the acceleration due to gravity? What is its approximate value?

The acceleration experienced by a freely falling object in the neighborhood of the surface of the earth. Approximately, $g = 9.8$ m/sec^2 or 32 ft/sec^2.   [5.6]

8. How does the distance fallen by a particle released from rest vary with the time of fall?

It is proportional to the square of the time: $g = \frac{1}{2}gt^2$.   [5.5]

9. For a particle projected horizontally from an elevated position, what is true of (a) the vertical component of its motion; (b) the horizontal component?

The vertical component is identical with that of a freely falling body; it has constant acceleration. The horizontal component has constant speed.   [5.7]

10. Describe the effect of air resistance on the range and on the time of flight of a projectile.

Range is decreased; time of flight is increased.   [Box, p. 109]

For further reading

5.1.   *Holton, G.,* and *D. H. D. Roller*   "Foundations of Modern Physical Science." Read chaps. 2 and 3 on Galileo's work on falling bodies and projectiles.

5.2.   *Galilei, G.*   "Two New Sciences." (Transl. by Crew and de Salvio.) Read excerpts in J. R. Newman (ed.), "The World of Mathematics," vol. 2.

5.3.   *Shapiro, A. H.*   "Shape and Flow." A highly interesting and readable account of the effect of the surrounding fluid on the motion of bodies.

5.4.   *Jaffe, B.*   "Michelson and the Speed of Light." Read chap. V on the Michelson-Morley experiment.

5.5.   *Shankland, R. S.*   The Michelson-Morley Experiment, *Sci. Am.,* November, 1964, p. 107.

Questions and problems

In the examples on falling bodies and projectiles, use $g = 9.8 \text{ m/sec}^2 = 32 \text{ ft/sec}^2$, and disregard the effects of air resistance.

5.1.   A car runs at constant speed on a circular track of radius 1,000 ft, taking 61 sec on each lap. What is (*a*) its average speed; (*b*) its average *velocity* on each complete lap?

5.2.   A train goes 100 mi in 2.5 hr, makes a 30-min station stop, and then goes an additional 90 mi in 2.0 hr. Find the average speed for (*a*) the first 2.5 hr; (*b*) the last 2.0 hr; (*c*) the entire trip.

5.3.   A car maintains a speed of 20 mi/hr for 3 hr and then suddenly changes to a speed of 40 mi/hr, which it maintains for the next 3 hr. How far does it go altogether? What is its average speed over the 6-hr interval?

5.4.   A motorist is compelled to restrict his speed to 20 mi/hr while passing through a populated zone 10 mi long. How far must he then go at 50 mi/hr in order to make his average speed for the whole trip 30 mi/hr?

5.5.   The graph in Fig. 5.19 shows the variation of displacement with time for a particle moving along a straight path. (*a*) Did the particle start from rest? (*b*) Was it at rest at any place or over any part of the path other than the beginning? If so, where? (*c*) On which part of the path was the speed constant but different from zero? (*d*) At what place was the speed greatest? (*e*) Did the particle reverse its direction of motion at any time? If so, at what place or places?

5.6.   If, in the Practice set on page 89, the tail gun had a muzzle speed numerically less than the forward speed of the airplane, would the projectile emerge from the gun? What would be seen by a stationary observer located at one side?

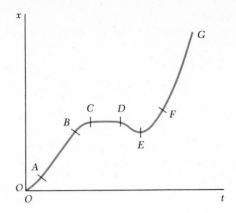

FIGURE 5.19 Displacement-time graph of the motion of a certain particle. Various points on the curve are labeled for reference.

5.7. A hurricane is drifting at a speed of 15 mi/hr in a direction 30° north of east. Compute how fast it is moving (a) eastward; (b) northward.

5.8. Raindrops striking the side windows of a car moving along a straight, level track at constant speed make streaks whose angle with the vertical is 60°. If the rain is falling vertically with a constant speed of 20 ft/sec, how fast is the car moving?

5.9. Can a homing pigeon whose maximum flying speed is 17 km/hr fly directly back to a city lying 24 km due south of its point of release if there is a wind blowing from the west at (a) 15 km/hr; (b) 18 km/hr? Make a sketch of the velocity vectors, and decide how long each trip will take, explaining the answers fully.

5.10. A freighter steams due north at a speed of 15 km/hr and a steady wind blows from the southeast at 30 km/hr. Find by means of a scale diagram the direction taken by smoke from the funnel, as seen by a man on the ship. HINT: The smoke particles immediately take up the motion of the surrounding air.

5.11. An atom moving through a vacuum tube with a speed of 2.0×10^8 m/sec ejects an electron in the backward direction with a speed, relative to the atom, of 2.5×10^8 m/sec. How fast and in which direction is the electron moving, referred to the laboratory? Note that the relativistic formula (page 95) must be used at these high speeds.

5.12. An electron is moving through a vacuum tube at a speed of 2.5×10^8 m/sec when it hits a plastic disk which stops it in a distance of 0.05 mm. Calculate the magnitude of its acceleration, assuming it to be constant, while coming to rest. How many times g is this?

5.13. At a certain instant, a vehicle is heading due north at 100 km/hr, and 20 sec later it is heading 30° north of east with a speed of 50 km/hr. Describe its average acceleration vector over this interval (direction and magnitude).

5.14.   A vehicle starts from rest and accelerates uniformly at the rate $a = 1.5\,\text{m/sec}^2$. Find (a) its speed after 4.0 sec and (b) the distance moved in the third second.

5.15.   A train coasting at 56 ft/sec was brought uniformly to rest by its brakes in 8.0 sec. Compute (a) its acceleration, (b) how far it traveled while coming to a stop, and (c) how far it went during the first 4.0 sec after the brakes were applied.

5.16.   A freight train traveling at a constant speed of 30 ft/sec passes a station at the instant when a passenger train starts out in the same direction on an adjoining track. If the passenger train moves with a constant acceleration of 2 ft/sec^2, (a) how far beyond the station will it overtake the freight train? (b) How do the speeds of the two trains compare at this time? Explain.

5.17.   One method being considered for driving an interplanetary space vehicle is ion propulsion, in which electrified atomic particles are ejected from the rear of the craft, driving it forward. Computations show that such a vehicle could be given a steady acceleration of about 1 mm/sec^2 by this means. (a) Starting from rest, how far would it have to travel in order to attain a speed of 2 km/sec? (b) How long would it take to reach this speed?

5.18.   The motors of a subway train can give it an acceleration of 1.0 ft/sec^2, and the brakes can give it an acceleration of -3.0 ft/sec^2. What is the shortest time in which the train can make the run between two stations 1,215 ft apart? HINT: Make a sketch of speed against time for the entire trip.

5.19.   The tallest man-made structure is a 2,063-ft television mast at Blanchard, North Dakota. How long would it take an object to fall to the ground from the top of the tower? How fast would it strike the earth in miles per hour?

5.20.   Work out a method of finding the numerical value of g from the strobe photo on page 104. For your data, read the positions on the scale, to the nearest 0.5 cm, of alternate images of the ball.

5.21.   A stone is thrown vertically upward, reaching the top of its path 1.5 sec later. (a) How high above the ground, in meters, will it be 2.5 sec after it is thrown? (b) How fast will it then be moving?

5.22.   Exactly 1 sec after dropping a stone over a cliff, a boy throws another stone downward with a speed of 40 ft/sec. Does the second stone overtake the first one? If so, where and when?

5.23.   Show that, for a particle projected straight upward, the speed on passing any higher level on the way up is the same as the speed on passing the same level on the way down. Use general algebraic symbols rather than specific numerical values.

5.24.   Describe the motion of a shell fired from a mortar (a) as seen by an observer in a helicopter hovering a considerable distance directly above the line of fire and (b) as seen by an observer standing a considerable distance directly behind the mortar.

5.25.   A boy riding in a wagon that moves with a constant speed of 10 ft/sec along a straight, level road throws a stone straight upward with a speed of 25 ft/sec. Is he in danger of being hit by the stone as it falls back? HINT: Does the stone have any forward speed with respect to the ground as it leaves the thrower's hand? If so, how does this compare with the speed of the wagon?

5.26.   Construct a diagram similar to Fig. 5.15 for a ball thrown at an angle of elevation of 30° with an initial speed of 120 ft/sec. Find an approximate value for the range on level ground by scaling off the diagram.

5.27.   A hunter points his gun directly at a monkey hanging from the limb of a distant tree and fires a shot. At the instant the monkey sees the flash, he lets go of the limb, falling freely. Will he be hit by the bullet? Does the answer depend on the angle of elevation or the speed of the bullet? Justify your answers.

5.28.   A ball rolls straight down a sloping roof that makes an angle of 30° with the horizontal. The ball leaves the edge of the roof, which is 16 ft above the ground, with a speed of 24 ft/sec. How far beyond the edge of the roof does it land?

5.29.   At a certain point in the flight of a thrown ball, its total velocity vector has a magnitude of 19.6 m/sec and is directed at an angle of 30° above the horizontal. (a) What vector would represent the *change* in velocity of the ball during the next second? (b) Combine this with the previous vector to find the total velocity vector at the end of that time (give direction and magnitude). (c) If the ball was thrown with an angle of elevation of 45°, find the maximum height to which it rose.

5.30.   Following the procedure of Worked example 5.10, page 108, show that the maximum height h, the time of flight T, and the range R of a projectile having initial speed v_0 and angle of elevation θ are given, respectively, by

$$h = \frac{(v_0 \sin \theta)^2}{2g}$$

$$T = \frac{2v_0 \sin \theta}{g}$$

$$R = \frac{2v_0{}^2 \sin \theta \cos \theta}{g}$$

Part Two
DYNAMICS AND ENERGY

Chapter 6
FORCE AND MOTION

6.1 FROM ARISTOTLE TO GALILEO AND NEWTON

The preceding chapter was concerned with the quantitative description of motion, including motion with constant speed or with constant acceleration, as well as special instances such as free fall or the flight of a projectile. The study of the description of motion in the abstract is a branch of mechanics called **kinematics.**

The term comes from the Greek word kinema, meaning motion.

Aristotle and the scholars of the Middle Ages speculated about the causes of motion, basing their conclusions on pure reasoning or on qualitative observations that were often wrong. The results did little to further an understanding of the phenomena. Galileo, by contrast, saw the importance of trying to discover an accurate description of motion before attempting to find the causes, realizing that progress could be made only by idealizing the situations presented by nature. He succeeded in finding quantitative relationships for motion with constant acceleration and thus established the foundations of kinematics.

Galileo was fully aware of the fact that there are slight differences in the observed rates of fall of light and heavy objects in air. Nevertheless, he was able to surmise what the motion of all objects would be in the absence of disturbing influences and to identify this as motion with constant acceleration, without troubling about what produces it.

Galileo went further than this, preparing the way for Newton's attainment of a rational explanation of the causes of motion. The branch of mechanics dealing with motion and the agencies that produce or change it is known as **dynamics.** It is the subject of this and several of the following chapters.

6.2 INERTIA

"Dynamics" is derived from the Greek word dynamikos: powerful.

Most of the inanimate objects in our surroundings are at rest relative to the earth, and it is a matter of experience that they remain so, even if they are not constrained or fastened down. A stone will not of itself suddenly begin to move about or take off into space. Whenever something of this sort is actually observed to happen we immediately suspect—unless we believe in magic—that some undisclosed physical influence is at work. Similarly, bodies initially set into motion are observed eventually to come to rest.

A ball rolling along the floor, a car coasting in neutral along a level road, or a ship drifting through the water with engines shut off will finally come to rest. So far, everything appears to be in harmony with the idea that the natural state of all objects is that of rest. Aristotle did, however, exempt the planets from this rule. For some reason he believed they traveled in "perfect orbits" (circles), and that their natural condition was motion rather than rest. He could not know that the presence of another body, the sun, and the absence of resistance were responsible for the observed motion.

Galileo was not satisfied with the aristotelian view, which still persisted almost unchallenged. He felt there must be a simpler way to describe this phase of nature, if only certain outside influences were absent. Since the complete elimination of such forces was as impossible as in his work on falling bodies, Galileo resorted to a thought experiment. From observation, he knew that a ball allowed to run down an inclined plane of any chosen slope has a constant positive acceleration. It will be remembered (page 104) that by extrapolating to the case of a vertical board he concluded that a freely falling object would also have constant acceleration. Further, he knew that a ball thrown onto an incline in the upward direction moves with a constant negative acceleration and that the magnitude of this acceleration decreases as the slope angle is made smaller.

Galileo was now ready to extend these experiences, and he devised a thought experiment of the following nature: Imagine a ball released from rest at some point P on the incline at the left in Fig. 6.1. It gathers speed at a constant rate until it reaches the bottom, after which it will move up another incline A which is joined smoothly to the first one. Ascending A, the ball loses speed at a constant rate and comes to rest momentarily at some point Q.

Now let A be lowered to a new position A', and start the ball again from rest at P. This time the ball will be observed to have a numerically smaller deceleration on the second slope, and it will roll farther along this slope before stopping. The more the second slope is lowered, the farther the ball will roll before coming to rest. What would be expected to happen if, in the extreme case, the second slope were made exactly *horizontal?* Galileo concluded that the negative acceleration would become zero altogether, and the motion would go on indefinitely at a constant rate.

This view was directly opposed to Aristotle's pronouncement that motion persists only as long as some propulsive force acts. Galileo pointed out that motion is as natural a state of affairs as rest and that the observed

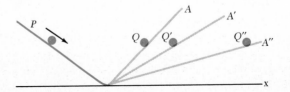

FIGURE 6.1 Galileo's double-incline experiment. The smaller the slope, the farther the ball rolls before coming to rest.

slowing down and eventual stopping of actual moving bodies are merely evidence for the existence of outside influences such as friction. Thus he recognized that, if a material body is at rest it tends to remain at rest, but if it is in motion it tends to continue moving.

Newton gave this idea a more specific form and introduced the term **inertia.** Inertia is the attribute of all objects that apparently makes them persist in their state of rest or motion. He conceived of inertia as a universal property of all material bodies, regardless of their size, composition, physical state, etc. This strongly suggests a connection with another general characteristic of matter already discussed: mass. The exact nature of this connection will be described below.

Building on Galileo's ideas, Newton was able to formulate his **law of inertia,** the first of three general laws of motion which he published in 1687: *Every body continues in its state of rest or of uniform motion in a straight line, unless it is compelled to change that state by forces impressed upon it.*

The concept of a body that is moving and yet is free of all forces is almost as difficult for most people to accept today as it was for the aristotelians. Our experience and intuitions are influenced by the inevitable presence of frictional forces in all familiar mechanical situations. Nevertheless, we recognize that with successive refinements a mechanism can be brought closer and closer to the condition of zero external force, bringing the observed behavior nearer to the description given by the law of inertia. As for direct observation, perhaps the best approximation to the ideal force-free case is the motion of a star, which is so far removed from other bodies that their gravitational influence is virtually absent. Also, the negligible amount of matter in interstellar space should not be able to retard a star's motion appreciably. Observations extending over many years show that such motion is truly uniform.

In earthbound experiments frictional forces can sometimes be reduced

Statements almost exactly equivalent to Newton's law of inertia had been made some time earlier by both Descartes and Huygens. The idea was evidently "in the air" around the middle of the seventeenth century.

This air-cushion vehicle can attain a speed of 80 mi/hr. (*Textron's Bell Aerosystems Co.*)

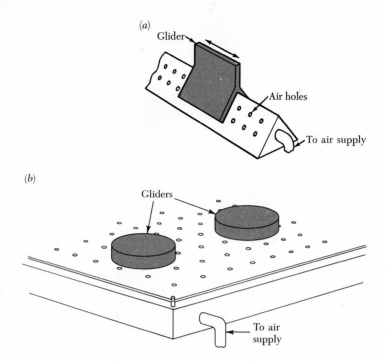

FIGURE 6.2 Portions of (a) a linear air track and (b) an air table. The gliders in (b) can slide over the table in any direction, like a hockey puck on ice.

to a surprising extent. One of the best means of reducing such forces is to support a moving object on a cushion of air or some other gas. This principle is used in "ground effect" vehicles such as the hovercraft, capable of skimming over land or water. A laboratory version is the **air supported glider** (Fig. 6.2) which shows quite convincingly the tendency of motion to persist when friction is greatly reduced. A glider of this kind, given an initial speed of 1 m/sec, would go a distance of the order of 100 m before coming to rest.

6.3 INERTIAL SYSTEMS

In Newton's statement of the law of inertia, the phrase "in a straight line" requires further discussion. The kinematic description of any kind of motion, including the shape of the path, depends on the frame of reference that is used. Imagine a small ball rolled radially outward from the center of a smooth, rotating turntable (Fig. 6.3). To an observer located on the turning surface, the ball appears to move along a spiral curve, but to an observer on the ground, the ball moves uniformly along a straight line. The observer riding on the turning platform would conclude that the law of inertia does *Mechanical* not hold for the rolled ball, although he would find nothing to which he *effects in* could ascribe the existence of any resultant force acting on it. Of all con- *rotating sys-* ceivable reference frames, which one is assumed in the statement of the *tems will be* law of inertia?

discussed in   The type of reference system appropriate to this statement has already *more detail* been mentioned on page 94 as an **inertial frame,** hence the name. The *in Chap. 8.*

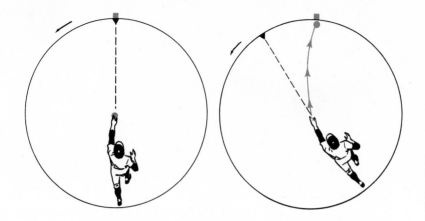

FIGURE 6.3 With respect to the earth, the ball moves in a straight line at constant speed, but to the observer on the rotating platform it has a more complicated motion.

test of whether a given frame is inertial is to find out by experiment if the first law correctly describes the motion in that system. This can be done by launching a particle, arranging that the net force acting on it is zero, and observing whether or not it pursues a straight path at constant speed. This kind of test would have disclosed that the rotating platform is definitely *not* an inertial system. The same test with a ball thrown out along a smooth, level floor would show that the earth is very nearly an inertial system, but for an object allowed to travel far enough, a slight but unmistakable deviation would be observed (Fig. 6.4). This sidewise drift, called the **Coriolis effect,** has many interesting consequences. See Ref. 6.7.

A simple laboratory experiment showing that the earth is not an inertial frame was devised by the French physicist J. L. Foucault (1819–1868). He hung a massive metal ball from a long wire and set it swinging, as a pendulum, in a vertical plane. After a few minutes he noticed that the plane of swing was apparently turning slowly in a clockwise direction. Foucault correctly interpreted this to mean that, because of its inertia, the pendulum bob maintained its direction of motion while the earth literally rotated under it. For a Foucault pendulum at either pole, the rate of turning amounts to one complete rotation in a day, or 15° per hour, with smaller

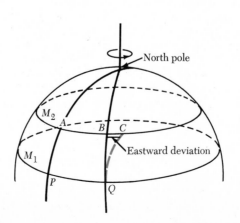

FIGURE 6.4 A body projected northward from P also has the eastward speed of the ground corresponding to the latitude circle M_1. A little later, when the body arrives at latitude M_2, it will be east of A by a distance $AC = PQ$, which is greater than the distance AB through which the meridian has turned meanwhile. For any direction of motion, the deviation is always toward the *right* in the Northern Hemisphere and toward the *left* in the Southern Hemisphere, as seen when facing in the direction of motion.

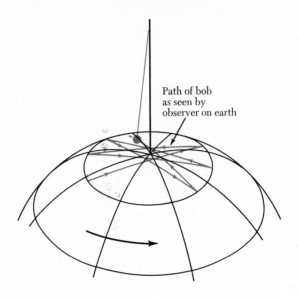

Path of bob
as seen by
observer on earth

FIGURE 6.5 A Foucault pendulum swinging at the North Pole.
The pendulum continues to swing
in a plane that is stationary in the
primary inertial frame, while the
earth turns beneath it.

values at lower latitudes (Fig. 6.5). The turning of the plane of swing may
be considered direct evidence for the rotation of the earth, independent
of any outside observation.

Newton's belief that motion could be determined with respect to space
itself was attacked by the Irish theologian and philosopher George Berkeley
(1685–1753). He suggested that the ultimate reference frame is determined
by the amount and distribution of all the matter in the universe, and this
idea was taken up nearly two centuries later by Mach (page 12). According
to this way of thinking, the distribution of all the matter in the universe
determines a definite system of reference called the **primary inertial frame.**
It is presumably the primary inertial frame in which the plane of a Foucault
pendulum retains a constant direction or in which an isolated star moves
with constant vector velocity.

The principle of relativity (page 94) asserts that any system that moves
with constant vector velocity relative to an inertial frame is itself an inertial
frame. This means that, in addition to the primary inertial system, there
are infinitely many other such systems. The magnitude and direction of
the velocity of an object depend on which of these frames is used. In
particular, a body having a constant vector velocity $\mathbf{v}$ in a given inertial
system S_1 will be *at rest* in another inertial system S_2 that has a constant
velocity $\mathbf{v}$ with respect to S_1. This is still another way of saying that rest
and motion are indistinguishable as long as one is dealing with inertial
systems.

6.4 EFFECT OF A FORCE; NEWTON'S SECOND LAW

The law of inertia is a statement of what is to be expected in the special
dynamic case where the resultant force applied to a particle is zero. As
already pointed out, this condition is not often encountered in practice and

special arrangements must be made to ensure even a good approximation to the force-free condition. In most cases, the forces acting on a body do not cancel out. The central problem of dynamics then becomes: What happens when a body *is* acted upon by a resultant force that differs from zero?

The answer to this question was supplied by Newton in his **second law of motion,** which he stated in this way: *Change of motion is proportional to the impressed force and takes place in the line in which the force acts.*

In order to use the law in this form, it must be made clear exactly what Newton meant by the word *motion*. Is it to be interpreted as velocity, acceleration, or something still different? This question will be answered in the next chapter. Further, Newton's laws are of the nature of hypotheses, and historians of science have no definite evidence of how he arrived at his conclusions. It is likely that they came both from reflection (thought experiments) and from direct experimentation. In any case, the laws conform with the results observed in a multitude of mechanical situations. In addition, they satisfy the criterion for a good theory because they can predict the outcome of particular experiments. In what follows, it will be shown how Newton's second law can be deduced from systematic experiments in which forces are applied to bodies.

The ideal procedure for such experiments would be to conduct them on a body far out in space, at a great distance from any other bodies. This would make certain that the selected object is free of any forces except the ones that the experimenter chooses to apply. A constant force could be exerted by means of a stretched spring, and the effect on the motion of the body could be determined with the help of a clock and a linear measuring stick (Fig. 6.6). However, since a space laboratory of this nature is not yet available, such experiments must be carried out in an ordinary earthbound laboratory. The object whose motion is to be examined can then be an air-supported glider (page 122). The supporting effect of the film of air balances the weight of the object so that no net force acts on it except any that the experimenter wishes to apply. A permanent record of the ensuing motion of the object can be obtained either (1) by lighting the apparatus with a strobe lamp and recording a series of images in a photograph or (2) by arranging for a spark to pass from the object to a sensitized strip of paper at regular intervals during the motion.

Figure 6.7 shows such a record schematically. The glider is seen to move in the direction in which the constant force was applied and, with the body

FIGURE 6.6 A mechanics experiment in space.

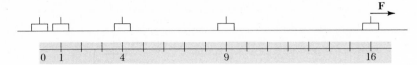

FIGURE 6.7 Successive positions of a glider, at equal intervals of time, when it is pulled to the right by a constant force.

The spacing of the images is of the same kind as in the multiflash photograph of a freely falling body (page 104).

starting from rest, the total distances it moves in 1, 2, 3, 4, . . . equal intervals of time are found to be proportional to 1^2, 2^2, 3^2, etc. This proportionality between distance and the square of the time is characteristic of motion with constant acceleration, starting from rest. If the experiment is repeated with a different body and a new value of the applied force, the motion is again found to be of this kind. The numerical value of the acceleration may be different from the previous one, but it is constant throughout the motion. Thus it may be concluded from such experiments that *a body acted upon by a constant force will move with constant acceleration in the direction of the applied force.*

6.5 EXPERIMENTAL BASIS FOR THE SECOND LAW

Having shown that a constant force applied to an object gives it a constant acceleration, we can now consider a systematic series of experiments leading to a quantitative general relation between the acceleration and the force producing it. For this, two sets of experiments are needed.

In the first set, forces of various magnitudes are applied to a glider in successive runs. Each time, the positions of the glider at given times are recorded. For simplicity, let the magnitudes of the applied forces be F, $2F$, $3F$, etc. These forces may be exerted by means of a spring balance that has been calibrated by hanging a succession of equal weights from it. According to Hooke's law (page 44), the divisions on such a balance will be uniformly spaced. Let the positions of the glider be marked at the end of 1, 2, 3, etc., units of time, starting from rest at $t = 0$.

The results of such a series of experiments are shown schematically in Fig. 6.8a. For each run, the total distances are found to be in the ratio 1^2, 2^2, 3^2, etc., demonstrating that the motion takes place with constant acceleration, as already noted. But when the records for the various runs are compared with each other, something additional is observed: The spacing of the images is discovered to be proportional to the magnitude of the force, being twice as great for $2F$ as for F, three times as great for $3F$, and so on. This regular increase is shown by the dashed lines in Fig. 6.8a. The conclusion is that, *for a given object, the acceleration is directly proportional to the applied force,* as shown graphically at the right of the figure.

In a second series of experiments the applied force is held *constant* throughout, but in successive runs the single metal disk that forms the heavy base of the glider is replaced by stacks of 2, 3, 4, etc., identical disks. The results of this series of experiments are shown in Fig. 6.8b. As additional disks are loaded onto the glider, the distances traveled in a given

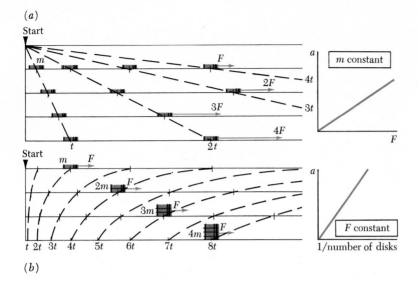

FIGURE 6.8 Positions of the gliders at equal intervals of time, starting from rest.

time are seen to decrease in a regular manner with increased loading. However, the dashed lines connecting the corresponding positions in the various records are no longer straight. It is found by trial that a straight line is obtained when the acceleration is plotted against the *reciprocal* of the number of disks, as indicated by the graph at the right in Fig. 6.8*b*. The conclusion is that, *with a given applied force, the acceleration is inversely proportional to the number of disks* used.

What is the mechanical significance of the number of disks making up the glider in each run? Suppose the disks are made of a pure, elementary material such as refined copper. Then it is a truism that a pair of such identical disks will contain just twice as many copper atoms as a single disk, that three disks will contain three times as many such atoms, and so on. Moreover, experiment would show that if an added disk is melted and cast into another shape or broken into small fragments or hammered out into a thin sheet before being loaded onto the glider, the observed motion would be exactly the same as if the disk had been left intact. Such attributes are precisely the ones we associated earlier (page 49) with the *mass* of a body, and we conclude that adding disks is merely a convenient way of adding measured amounts of mass. The results of this set of experiments may be summarized by saying that, *with a given applied force, the acceleration is inversely proportional to the mass of the body*. Thus, from the role it plays here, the mass becomes the definite measure of what we have called the inertia of a body: *The greater the mass of an object, the more difficult it is to change its existing state of motion or rest.*

Experience shows that a change in other properties of a body, such as its color, state of division, phase, etc., will not affect its motion, but a change in mass will do so. The fundamental question of how a body moves under the influence of forces is therefore one that involves the three variables force, mass, and acceleration. Of these, mass is the only one that is a fundamental property of the object.

The two series of experiments described above are enough to reveal the exact connection between these quantities; however, for completeness, the outcome of the remaining set of possible thought experiments is shown in Fig. 6.9.

The results of the first two series, as summarized in the graphs of Fig. 6.8a and b, were as follows:

1. With m constant, $a \propto F$
2. With F constant, $a \propto 1/m$

where the standard symbol $\propto$ means "is proportional to." Combining these two statements, we can say

$$a \propto \frac{F}{m}$$

which may also be written

$$F = kma$$

where k is a constant of proportionality, a pure number whose value depends on the units used to measure F, m, and a.

It is convenient to choose the units in this relation so that k becomes numerically equal to 1. This would allow us to write the above equality simply as

$$F = ma \qquad\qquad 6.1$$

How can this be done? If, as is usual, we retain L, M, and T as primary dimensions (page 52), then m is already one of the primary quantities, and the dimensions of a are expressible as the simple combination L/T^2. This compels us to choose, for use in Eq. 6.1, a new unit of force which satisfies the equation dimensionally. Since the right side of this equation has the dimensions ML/T^2, the quantity F on the left side must also have these dimensions. The appropriate unit of force is one that will impart one unit of acceleration to an object of unit mass. Specifically, in the mks system, we define a new force unit as follows: A force of such size that it can give an acceleration of 1 m/sec² to a body of 1-kg mass is called one **newton.** Thus the term *newton* is a substitute for *kilogram meters per second per second.* The newton is approximately equivalent in magnitude to the weight of a large-sized egg. A 157-lb man weighs 700 newtons.

So far we have been using a makeshift force unit that we called "one kilogram of force," equal to the earth's gravitational pull on the standard kilogram when determined at a standard location, and this was acceptable for problems of equilibrium. But whenever accelerations are involved the

A new kind of force unit.

The abbreviation for the newton is N. In the past nt was sometimes used.

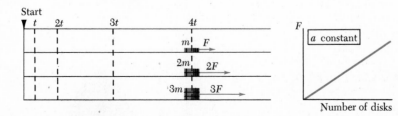

FIGURE 6.9 The acceleration remains constant if F increases in proportion to the number of disks.

use of Eq. 6.1 is called for, and all forces must be measured in newtons when the mks system is used. In order to show that such a unit is more fundamental than one described by referring to the pull of the earth on a standard mass, a force unit defined by means of Eq. 6.1 is sometimes referred to as an *absolute* unit. Absolute units of force have also been defined in other systems of measure. In the metric cgs system, the absolute unit of force is the **dyne,** defined as the force needed to impart an acceleration of 1 cm/sec^2 to a body of 1-g mass. Since 1 kg = 1,000 g and 1 m = 100 cm, Eq. 6.1 shows that 1 newton = 10^5 dynes. Throughout the remainder of this book we shall use only the newton for the measurement of force.

Practice set* A force F, acting on a particle of mass m, gives it an acceleration a. What will be the magnitude of the (a) acceleration if the force acting on m is tripled; (b) force needed to give a mass $\frac{2}{3}m$ an acceleration $\frac{1}{2}a$; (c) mass that could be given an acceleration $4a$ by a force $2F$?

If Eq. 6.1 is written in vector form, it reads

$$\mathbf{F} = m\mathbf{a} \qquad\qquad 6.2$$

and, in this form, includes the statement that the acceleration takes place in the direction of the resultant acting force. A simple example will show how this relation can be used in a particular case.

Worked example 6.1 An object of mass 100 kg, initially at rest and free of all other forces, is acted upon by a steady pull of 500 newtons directed toward the north. Describe the resulting motion. Where will the body be and what will be its instantaneous velocity after 20.0 sec?

Solution Since the mass and the force are constant, Eq. 6.1 shows that the motion will have constant acceleration—constant in both amount and direction. Substituting the numbers,

$$500 \text{ newtons} \left(\text{or } \frac{\text{kg-m}}{\text{sec}^2} \right) = (100 \text{ kg}) \,(a)$$

or

$$a = 5.00 \text{ m/sec}^2$$

This acceleration will be in the same direction as the force, toward the north.

Having determined, through the application of the second law, that the motion has constant acceleration, and having computed its magnitude, we can now turn to the relations developed in Chap. 5 for describing this type of motion in order to compute the details. Since the object starts from rest, its final speed will be given (page 99) by

$$v = at \qquad \text{or} \qquad v = (5)(20) = 100 \text{ m/sec}$$

The body will be moving northward at this speed.

*Ans.: $3a$; $\frac{1}{3}F$; $\frac{1}{2}m$.

The distance covered is

$$x = \tfrac{1}{2}at^2 \qquad \text{or} \qquad x = \tfrac{1}{2}(5)(20)^2 = 1,000 \text{ m}$$

The object will be this distance directly north of its starting point.

Equation 6.2 has just been applied to a problem where the force was constant in magnitude, but in many mechanical situations the acting force is variable. If this is the case, then Eq. 6.2 must be looked upon as a relation between the *instantaneous* values of F and a. The physical nature of the system may show that F depends in a certain way on other variables. For example, the oscillations of an object hanging from a spring are governed by an elastic force that depends on the momentary displacement of the body from its rest position. The motion of a stone sinking in a liquid is affected by its weight, which is constant, and by the force of fluid resistance whose magnitude depends on the instantaneous speed (see Box below).

Problems dealing with variable forces can become quite complicated. Their solution requires the use of what mathematicians call **differential equations,** which must be handled by the methods of the calculus. Such problems will not be treated in this book.

The force represented by F in the second law is understood to be the resultant of all the individual forces that may be acting on the object in question. It has already been mentioned (page 69n) that, according to experience, each partial force produces its own effect, so that it would be correct

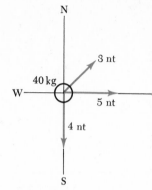

FIGURE 6.10 A body acted upon by forces in various directions.

MOTION OF A BODY FALLING IN A FLUID

The great difference between the motion of a body falling in air and falling in a vacuum has already been noted in Chap. 5. This difference is caused by the resistance of the surrounding fluid, in this case, air. The resistance offered to the motion is a function of the speed as well as of the size, shape, and attitude of the body and the properties of the fluid. Reasonable assumptions can sometimes be made about the dependence on speed over limited ranges. Then the details of the motion can be worked out by calculus methods, leading to formulas for speeds, distances, times, etc.

A body falling in air has an acceleration of g at the start, where its speed is low. However, the resisting force is found to increase with speed, and this makes the acceleration decrease as the motion continues. When, ultimately, the resisting force becomes essentially equal to the weight of the body, the acceleration becomes zero because the net force acting on the body is then zero. From that time on, the body continues to fall with its constant **terminal speed** (see Fig. 6.11).

Objects with a large ratio of cross-section area to weight usually have low terminal speeds. For example, a parachutist in a delayed-opening jump may attain a speed of more than 150 mi/hr, but his landing speed with the parachute open is less than 25 mi/hr. The reduction is due largely to the increase in cross-section area of the falling object. Raindrops of moderate size have a terminal speed of about 25 ft/sec, but a fog droplet may settle with a speed of only a small fraction of an inch per second.

Other circumstances affect the motion of any object moving through the air when its speed exceeds that of sound waves ("Mach 1"). For an interesting account of many aspects of fluid dynamics, consult Ref. 6.8.

FIGURE 6.11 The speed-time curves for an object falling in air and in a vacuum diverge markedly as time passes.

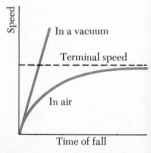

to use Eq. 6.2 to compute an acceleration due to each such force. Vector addition of these partial accelerations would then give a resultant acceleration. Instead of doing this, however, it is more natural first to combine the acting forces vectorially and then to use Eq. 6.2 to calculate at once the resultant acceleration. A numerical illustration follows.

Worked example 6.2   A body of mass 40 kg is acted upon by three forces: 5 newtons toward the east, 3 newtons northeast, and 4 newtons southward (Fig. 6.10). How does it move?

Solution   The 3-newton force can be resolved into two components, 2.12 newtons toward the north and 2.12 newtons toward the east. Combining all east-west forces yields a net eastward component of 7.12 newtons. Similarly, there is a net southward component of 1.88 newtons. The vector sum of these two forces has a magnitude $F = \sqrt{(7.12)^2 + (1.88)^2} = 7.37$ newtons. Its direction is determined by $\tan \phi = 1.88/7.12 = 0.268$, or $\phi = 15°$.

The value of this resultant force can now be substituted into Eq. 6.2, yielding $7.37 = 40a$, or $a = 0.184$ m/sec^2. The body will move with this acceleration in a straight line directed 15° south of east. Once the acceleration has been found, further details of the ensuing motion can be computed from kinematics, as in Worked example 6.1.

After jumping from 10,000 ft the parachutists managed to join hands in a circle. Their terminal speed was about 82 mi/hr. (*London Daily Express—Pictorial Parade.*)

6.6 THE CONCEPT OF WEIGHT

Weight is one of the most universal forces with which we have experience. Weights are often used in practical devices through the employment of pulleys, levers, or other mechanical elements, to exert required forces. The fact that we "heft" an object in order to get some idea of its weight is evidence of the intuitive connection between muscular effort and force in general.

A relation between the mass of an object and its weight is embodied in Newton's second law. The weight of a body is simply a particular force that can act on it—the force of the earth's gravity—and this is the only force acting on the body when it is in free fall. In addition, we know by observation that any freely falling body has an acceleration $\mathbf{g}$. Substituting into $\mathbf{F} = m\mathbf{a}$ the particular value $\mathbf{w}$, the weight of the body, in place of $\mathbf{F}$, and the corresponding value $\mathbf{g}$ for $\mathbf{a}$, the relation becomes

$$\mathbf{w} = m\mathbf{g} \qquad\qquad 6.3$$

This is a formal definition of what is meant by the weight of an object. At the same time it points up the distinction between mass and weight: The mass of a body is a *scalar* constant that gives a quantitative measure of its inertia and is linked to the amount of matter making up the body (number of atoms of each kind in it). The weight of the body, on the other hand, is a *vector*. It is the force with which the earth attracts the object gravitationally and is sensitive to changes in location with respect to the earth. Equation 6.3 asserts that, *at any given place* (constant value of $\mathbf{g}$), *the magnitude of the weight of an object is proportional to its mass.* Thus we can compare the masses of two bodies by weighing them at the same location. If one of the bodies is selected as the standard (kilogram), this procedure determines the mass of the other. The relationship of mass and weight will be discussed further in Chap. 7.

Practice set*

1. What is the mass, in kilograms, of a body whose weight is 49 newtons?
2. What acceleration, in m/sec^2, would an applied force of 20 newtons give to this object?
3. Answer question 2 for the same object when located at a place where $g = 0.9$ of the standard value.

6.7 SOME FURTHER APPLICATIONS OF THE SECOND LAW

Worked examples 6.1 and 6.2 illustrated the application of Newton's second law to simple dynamic cases. Before showing how this law can be used to investigate a variety of other situations, we examine a general method of attacking such problems.

1. The first step is to select one object (particle) whose behavior is of interest and to concentrate on it to the exclusion of all others.
2. The next step is to determine all the forces acting *on* the chosen body, carefully excluding all other forces such as any that the body in question exerts on other things. In doing this, we mentally enclose the selected

*Ans.: 5; 4; 4.

body in a shell and ask: What forces are exerted on this body by its environment? At this point it is useful to make a separate sketch showing the selected object and all the forces applied to it from its surroundings.

3. The resultant of all the acting forces should then be determined and the second law, **F** = m**a**, applied.
4. The procedure may be repeated for any other particle of the system.

We shall now show how this method is used in specific instances.

Worked example 6.3 A particle of mass m hangs from an unstretchable cord of negligible weight whose upper end is held in the hand. Find the tension in the cord when the mass is (a) held at rest; (b) rising with constant speed v; (c) descending with constant speed v; (d) rising with constant acceleration a; (e) descending with constant acceleration a.

Solution There are two forces acting on the particle from its environment: the weight **w** (gravitational pull exerted on it by the earth) and the tension **T** in the cord. These are shown in Fig. 6.12.

First, we note that in the three cases (a), (b), and (c) there is no acceleration; the body is in equilibrium. This means that the resultant force is zero, and so **T** and **w** are equal and opposite. This makes $T = w = mg$; the tension in the cord is just equal to the weight of the suspended body.

The remaining cases, however, are different. In case (d), the fact that the body has an upward acceleration shows, according to the second law, that there must be a net upward force acting on it. This means that the tension (T_u) now exceeds w in magnitude, the resultant being $T_u - w$. Equating this force to the product of mass and acceleration yields $T_u - w = ma$, and putting mg in place of w makes this

FIGURE 6.12 Tension in a cord from which a massive body hangs.

$$T_u - mg = ma \quad \text{or} \quad T_u = m(g + a) \qquad 6.4$$

Similarly, in case (e), the existence of a downward acceleration implies that there is a resultant force on the particle in the downward direction. Now w is greater in magnitude than the tension (T_d), and the second law states that

$$mg - T_d = ma \quad \text{or} \quad T_d = m(g - a) \qquad 6.5$$

Thus, in the last two instances the tension differs from the weight of the suspended body, mg. The reason is that accelerations are present. It is only when $a = 0$ that both Eqs. 6.4 and 6.5 reduce to $T = mg$.

As a numerical illustration of the above ideas, consider the firing of a Saturn 5 rocket whose total mass at takeoff is 2.82×10^6 kg and whose jets exert a vertical thrust of 3.33×10^7 newtons. This arrangement is mechanically equivalent to the particle accelerated upward by the pull in a cord, with the thrust of the jets taking the place of the tension in the string. Equation 6.4 can be applied to find the initial upward acceleration. Substituting,

$$3.33 \times 10^7 = 2.82 \times 10^6(9.8 + a) \quad \text{from which} \quad a = 2.0 \text{ m/sec}^2$$

This acceleration is about one-fifth as great as g.

The importance of isolating one body at a time when applying the second law is shown by the following example.

Worked example 6.4 A block of mass M resting on a smooth, horizontal table (Fig. 6.13) is pulled by a cord of negligible mass that passes over a frictionless pulley and has a second block, of mass m, tied to it. Find (a) how the system moves and (b) the magnitude of the tension in the cord during the motion.

Solution First examine all the forces acting from the outside on the block that rests on the table. This object may be considered a particle to which the following forces are applied:

1. The weight, Mg, acting downward.
2. The force exerted on the block by the table. Since the table is perfectly smooth, this force will have no component *along* the table top and so must be directed straight up. We shall call this force N.
3. The tension in the string, T.

Since the block does not accelerate in the vertical direction, the net vertical force $N - Mg$ must be zero, or $N = Mg$. This leaves T as the resultant force acting on the block. Observation shows that T remains constant in magnitude during the experiment, and so the block will have a constant acceleration to the right. If we call its magnitude a, the second law states that

$$T = Ma \qquad\qquad 6.6$$

This single equation is not sufficient to determine both a and T. In order to find the values of these two quantities, another relation is needed. This is obtained by shifting attention to the other body, the suspended block of mass m.

The forces acting on this block from its surroundings are shown in Fig. 6.13. One of these forces is the weight of the block, mg; the other is the upward-directed tension in the cord. Since the magnitude of the tension is the same at all points on this cord (page 71), the value of this force is again T. Also, the cord has a constant length so that, if the first block moves with a constant acceleration a, the other will also have an acceleration of this amount. The only effect of the pulley, assumed to be light in weight, is to make the directions of motion of the two objects different.

The resultant force acting on the suspended mass is the difference between its weight and the force applied to it by the cord. We write this difference as $mg - T$, rather than $T - mg$, for the following reason: Since the block is observed to be accelerating *downward*, the resultant force on it must be in this direction, which means that mg must be greater than T.

Now, applying the second law to the hanging block,

$$mg - T = ma \qquad\qquad 6.7$$

Equation 6.7 is the second independent relation between a and T that is needed in order to solve for both quantities. Substituting the value of T given by Eq. 6.6 into Eq. 6.7 and solving for a, the result is

$$a = \frac{m}{M + m}\, g \qquad\qquad 6.8$$

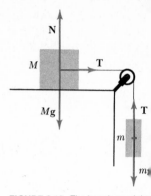

FIGURE 6.13 The hanging weight provides the tension in the string, and this is the force tending to move the block.

and putting this value of a back into Eq. 6.7 yields

$$T = \frac{Mm}{M + m} g \qquad\qquad\qquad \textbf{6.9}$$

Thus, by applying the second law independently to each block in turn, the values of both the acceleration and the tension have been found.

A closer look at the results brings out some interesting points:

1. The tension in the cord is *not* equal to the weight of the suspended block but is smaller than this amount. This can be seen immediately by rewriting Eq. 6.9 in the form

$$T = mg \frac{M}{M + m}$$

The fraction $M/(M + m)$ is always less than 1, since the denominator is greater than the numerator. The equation then says that T is less than the weight mg, as stated above.
2. For the same reason, the value of a as given by Eq. 6.8 is seen to be less than g, the full acceleration of gravity.
3. If the mass of the block on the table were zero, Eqs. 6.8 and 6.9 would reduce to $a = g$ and $T = 0$, respectively. This result is reasonable, since the suspended block, relieved of any responsibility for accelerating M, would become a freely falling object. At the same time, the cord would no longer be under any tension.

In order to see how Eqs. 6.8 and 6.9 apply in a specific case, let $M = 5.0$ kg and $m = 2.0$ kg. Then, substituting in the two equations,

$$a = \tfrac{2}{7}g = 2.8 \text{ m/sec}^2$$
$$T = \tfrac{10}{7}g = 14 \text{ newtons}$$

Worked example 6.5 A block of wood sliding along the floor at a speed of 4.0 m/sec encounters a straight slope that rises 2.0 m for each 10 m measured along the incline (Fig. 6.14). How far up the slope does the block go before coming to rest? Frictional forces are assumed to be negligible.

Solution   The block, considered a particle, is the object of interest here. The forces applied to the block are its weight w and the supporting force N exerted by the plane. In the absence of friction, N is perpendicular to the plane. It is convenient to choose a pair of axes as shown, resolving w into components along these axes. As in Worked example 6.4, there is no acceleration in the y direction, and so N is equal and opposite to P, the component of w that is perpendicular to the surface. This leaves F, the x component of w, as the resultant force on the block. It is directed downhill (in the negative x direction), and so the associated acceleration will be in that direction.

From the force triangle, $F = -w \sin \phi$, so that the statement of the second law becomes

$$-mg \sin \phi = ma \qquad\qquad\qquad \textbf{6.10}$$

The factor m cancels out, showing that the resulting acceleration is independent of the mass of the block; the motion will be the same for a heavy

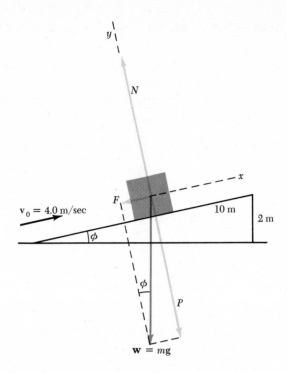

FIGURE 6.14 Block projected upward along a slope.

block as for a light one. Also, the angle ϕ in the force triangle is the same as the slope angle of the inclined plane, from which we see that $\sin \phi = \frac{2}{10}$. Then Eq. 6.10 becomes

$$a = -0.2g = -1.96 \text{ m/sec}^2$$

The remainder of the calculation is a question of kinematics: Starting onto the slope with an initial speed of 4.0 m/sec, how far along the incline will the block travel before coming to rest if it has an acceleration of -1.96 m/sec^2 all the while? We can use the formal relation 5.11c, putting the final speed $v = 0$ and using the above values of v_0 and a:

$$0 = (4.0)^2 + 2(-1.96)x \qquad \text{from which} \qquad x = 4.1 \text{ m}$$

What happens to the block after that?

This is the distance traveled by the block along the slope before reaching its highest point.

The preceding example showed the importance of getting the correct directions of forces, velocities, and accelerations in dynamics problems. Since the positive x axis was chosen in the uphill direction, the force and the acceleration of the block were in the negative direction, whereas the initial velocity was in the positive direction. This point is emphasized again in the example that follows.

Worked example 6.6

A 50-kg boy stands on a platform spring scale in an elevator that is going down with a constant speed of 3.0 m/sec. (a) What is the reading, in newtons, of the scale? (b) If the elevator is now brought to rest with constant acceleration in a distance of 9.0 m, what does the scale read during this period?

Solution The forces acting on the boy are his weight and the supporting force of the scale platform, as indicated in Fig. 6.15. In case (*a*), there is no acceleration, so that the scale reading is the same as the static weight:

$$S = w = mg = (50)(9.8) = 490 \text{ newtons}$$

In case (*b*), we take the upward direction to be positive. Then the initial velocity, being downward, is written $v_0 = -3.0$ m/sec, and the distance traveled to rest is $x = -9.0$ m. Using the kinematic formula $v^2 = v_0^2 + 2ax$,

$$0 = (-3.0)^2 + 2a(-9.0) \quad \text{or} \quad a = 0.50 \text{ m/sec}^2$$

This quantity has a positive sign, which means that the acceleration of the elevator is directed *upward*. Using $F = ma$,

$$S - w = ma \qquad S - 490 = (50)(0.50)$$

and finally

$$S = 515 \text{ newtons}$$

This exceeds the static weight, 490 newtons, by 25 newtons. The scale platform may be looked upon as exerting a 490-newton force to support the weight of the boy plus a 25-newton force to slow down his motion as described.

Programmed review

Instructions: See page 18.

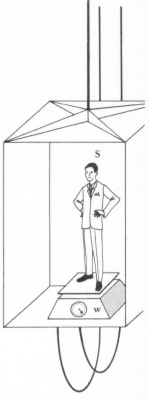

FIGURE 6.15 Forces on a body in an elevator.

1. Distinguish between *kinematics* and *dynamics*.

    Kinematics is a description of motion, whereas dynamics deals with force as producing or changing the motion of a body.   [6.1]

2. Contrast the views of Aristotle with those of Galileo on motion.

    Aristotle believed a force was necessary to keep a body moving; Galileo held that motion is just as natural as rest.   [6.2]

3. What is meant by the term *inertia*?

    The property possessed by all bodies that causes them to persist in their state of rest or motion.   [6.2]

4. State the law of inertia (Newton's first law).

    Every body continues in its state of rest or of uniform motion in a straight line, unless acted upon by a force applied from the outside.   [6.2]

5. What is an inertial frame of reference?

    A reference frame in which Newton's laws of motion are observed to hold.   [6.3]

6. State Newton's second law in words.

Change of motion is proportional to the applied force and takes place in the direction in which the force acts. An algebraic statement is preferable (see following question). [6.4]

7. State the second law in algebraic form.

$F = ma$ [6.5]

8. Name and define the mks unit of force.

One newton is the force that can give an acceleration of 1 m/sec^2 to a body of 1-kg mass. [6.5]

9. How can the acceleration of a particle be found when more than one force is applied to it?

Find the resultant of all the acting forces and use this value in $F = ma$. The acceleration will have the same direction as this resultant. [6.5]

10. Distinguish between *mass* and *weight*.

The weight of a body is the force of the earth's attraction for it. It is related to the mass by $\mathbf{w} = m\mathbf{g}$, where m is the mass. Mass is scalar, whereas weight is a vector. [6.6]

For further reading

6.1. *Galilei, G.* "Two New Sciences." Read about Galileo's work on motion with constant acceleration in the section called "Third Day."
6.2. *Taylor, L. W.* "Physics, the Pioneer Science." Chapters 9 and 10 of this work should be read at this point.
6.3. *Andrade, E. N.* "Sir Isaac Newton." A short biography depicting Newton's personality as well as his scientific work.
6.4. *Cohen, I. B.* Newton, *Sci. Am.,* December, 1955, p. 73.
6.5. *Sciama, D.* Inertia, *Sci. Am.,* February, 1957, p. 99. Presents Mach's ideas on the nature of inertia.
6.6. *Armitage, A.* The Foucault Pendulum, *Discovery,* vol. 12, February, 1951, p. 50.
6.7. *McDonald, J. E.* The Coriolis Effect, *Sci. Am.,* May, 1952, p. 72.
6.8. *Shapiro, A. H.* "Shape and Flow." Good descriptions and explanations of fluid dynamics phenomena.

Questions and problems

Use the value $g = 9.8$ m/sec^2 in all calculations where needed.

6.1. On the basis of Newton's first law, explain why passengers in a car lurch forward when the brakes are suddenly applied. In what reference frame may they be said to do this? Is this an inertial frame?

6.2. A 1-kg body is at rest, in a given inertial frame, at a point out in space, far from all other objects. How big a force is needed to set the body in motion?

6.3. Which has the greater inertia, a 1-cm cube of aluminum or 2,200 cm³ of air at standard temperature and pressure? For relevant numercial data, see Table 3.4 on page 55.

6.4. A ball hangs from a string inside a railroad car moving along a straight, level track. The string is observed to be inclined toward the rear of the car, making a constant small angle with the vertical. Describe how the car is moving.

6.5. Because of the rotation of the earth, an object dropped from a tower will not land exactly beneath the point of release. In what geographical direction with respect to the vertical will the path be deflected? Explain.

6.6. A fly crawls at constant speed straight outward along a radius of a uniformly rotating phonograph turntable. Does Newton's first law apply to the fly? Give a reason for your answer.

6.7. A parcel falls off a shelf in a train moving at constant speed along a straight track. Describe the path of the parcel as seen (a) by a passenger on the train and (b) by an observer on the ground.

6.8. An object of mass 10.0 kg is acted upon by two equal forces that make an angle of 120° with each other. Find the magnitude and direction of the resultant acceleration of the body if each force has a magnitude of 5.00 newtons.

6.9. Three forces, each of magnitude f, act simultaneously on a particle of mass m. One force is directed 60° north of east, another due east, and the third 60° south of east. Find the direction and magnitude of the acceleration of the particle.

6.10. When a golf ball of mass 0.062 kg is hit by a club, it acquires a speed of 84 m/sec. The impact lasts 3.0×10^{-3} sec. Assuming the force that the club exerts on the ball is constant, what is its magnitude?

6.11. A proton, whose mass is 1.67×10^{-27} kg, is acted upon by a force of 2.40×10^{-14} newton while traversing one section of a particle accelerator. What is the magnitude of its acceleration, and by what factor does this exceed g?

6.12. A steady force changes the speed of body A by 2.0 m/sec in a time interval of 0.5 sec. The same force applied to a second object, body B, changes its speed by 12 m/sec in 2.0 sec. Find the ratio of the masses of A and B.

6.13. Three trailers, each of mass 100 kg, form a train that is pulled along the floor of a factory by a tractor exerting a forward force of 750 newtons. Find (a) the acceleration of the train; (b) the tension in the coupling between the first and second cars; (c) the tension in the coupling between the second and third cars. Frictional forces may be neglected.

6.14. The takeoff speed of a 60,000-kg airplane is 200 km/hr, and it acquires this speed after a run lasting 50 sec. (a) What is the average thrust force exerted by the jets, and (b) how long a runway is needed?

6.15. An electron leaves the cathode of a vacuum tube with negligible speed and reaches the anode, 0.050 cm away, with a speed of 5.0×10^6 m/sec. Find the magnitude of the electric force that acted on this particle during its flight. (Take the mass to be 9.1×10^{-31} kg.)

6.16. A 4-kg projectile going 800 m/sec hits a cliff and penetrates to a depth of 1.5 m before coming to rest. Compute (a) the time of stopping and (b) the average force that opposed the motion.

6.17. When a hammer of mass 0.90 kg moving with a speed of 6.0 m/sec strikes a small nail, it drives it 0.8 cm into a heavy plank. Find (a) the average acceleration of the hammer head and (b) the average force it exerted on the nail.

6.18. Find the effective braking force needed to bring a car of mass 1,200 kg, moving initially at a speed of 70 km/hr, to rest in a distance of 70 m. Find also the braking force that would be needed to bring the car to rest in half this distance.

6.19. A 60-g bullet going 400 m/sec encounters a plank 4.0 cm thick. The bullet emerges from the other side with a speed of 100 m/sec. Assuming that the plank offered a constant resistance force to the bullet during penetration, compute the magnitude of this force.

6.20. A heavy block is dragged along a rough table by means of a spring scale attached to it (Fig. 6.16). When the block is moving with constant velocity the scale reads 0.50 newton, and when it is dragged with a constant acceleration of 0.10 m/sec^2 the scale reads 1.70 newtons. Find (a) the retarding force of friction and (b) the mass of the block.

6.21. A rocket has a total mass of 20,000 kg, and its jets exert a thrust of 250,000 newtons. (a) With what acceleration does the rocket start when fired vertically upward? (b) A 20-kg instrument is mounted on the inside wall of the rocket. What upward force is exerted on this instrument at that time?

6.22. An object of mass 2.0 kg hangs from a cord whose upper end is held in the experimenter's hand. When it is just on the point of

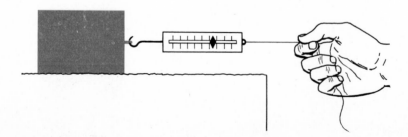

FIGURE 6.16 Block dragged along a rough surface.

breaking, a piece of this cord can support a 5.0-kg mass at rest. What is the maximum acceleration with which the 2-kg mass can be raised without breaking the cord, expressed as a multiple of g?

6.23. A 75-kg man carrying a 12-kg suitcase rides in an elevator that has an upward acceleration of 0.50 m/sec². Find (a) the pull on his arm and (b) the force with which his feet press on the floor of the car, each expressed as a multiple of the weight of the supported object or system.

6.24. With what minimum acceleration can a fireman slide down a rope whose breaking strength is three-fourths of his weight?

6.25. To what do Eqs. 6.8 and 6.9 reduce (a) when M becomes infinite; (b) when $m = 0$? Justify these results on physical grounds.

6.26. *Atwood's machine.* A system consisting of two unequal masses attached to a cord and hung over a pulley (Fig. 6.17) is sometimes used in laboratory experiments on dynamics. If the two masses are 2.2 and 2.7 kg, respectively, find (a) the acceleration of either object and (b) the tension in the cord when the system is allowed to run. HINT: Use the method of Worked example 6.4, page 134.

FIGURE 6.17 Atwood's machine.

6.27. A 5-kg block of wood sliding on a long, rough board is subject to a constant frictional drag of 20 newtons. The board is inclined at an angle of 30° to the horizontal. If the block is set into motion on the slope in a downhill direction, will it come to rest eventually? If not, describe its motion.

6.28. Calculate the total pull in the drawbar of the end coach of a train (mass of coach is 28,000 kg) if the train has an acceleration of 0.30 m/sec² as it goes up a grade rising 1 m for each 10 m along the track. Friction is to be neglected. HINT: There are two contributions to the total force.

6.29. A stone is thrown vertically upward with an initial speed of 25 m/sec, reaching a maximum height of 30 m. Find the average force of air resistance during the upward motion, expressed as a fraction of the weight of the stone.

Chapter 7
MOMENTUM

The first two laws of motion characterize force as an agency capable of changing the motion of material bodies. In Chap. 6, we found it convenient to consider one object at a time and to investigate the effect of the forces impressed on this object from the outside. There was no need to ask how the forces were brought to bear by the environment nor to study what effect these forces may have had on any other object. The applied forces were assumed to originate in very massive objects such as the whole earth or something fixed on earth—objects having so much inertia that any slight changes of motion they experience can be ignored. Is it nevertheless true in principle that even the earth itself is affected when it exerts gravitational attraction on a falling stone?

Newton addressed himself to such questions. In this chapter we follow his lead in considering *interactions* between objects and describe what happens to a collection of material bodies as a result of these mutual influences. This line of investigation has interesting and far-reaching results.

7.1 NEWTON'S THIRD LAW OF MOTION

A useful physical law summarizes a great variety of individual observations and experiences and predicts what will happen in particular situations not yet directly observed. Newton's third law is an excellent example of such a generalization. In his own words, it reads: ''To every action there is always opposed an equal reaction; or the mutual actions of two bodies upon each other are always equal and directed to contrary parts.''

In many situations it is convenient to interpret Newton's use of the terms *action* and *reaction* as referring to *forces*. With this in mind, his original statement may be recast in this form: *Whenever one object exerts a force on a second object, the second exerts an opposite force of equal magnitude on the first.*

Algebraically, the assertion is simply

$$\mathbf{F}_{AB} = -\mathbf{F}_{BA} \qquad\qquad 7.1$$

where $\mathbf{F}_{AB}$ is the force exerted by body A on body B, and $\mathbf{F}_{BA}$ is the force exerted by body B on body A. The oppositeness in direction is shown by the minus sign. The statement made by Eq. 7.1 is true at any instant, even if the forces vary in time, and it does not depend in any way on the state of motion of either body.

The third law has implications that are provocative and sophisticated. It says that forces arise only as the result of *interactions* between pairs of objects, so that there must be more than one object in order for any force to be exerted. The second member of such a pair may be any other object, even another part of the same body. Further, during such inter-action, one of the objects pushes or pulls on the other just as hard as it, in turn, is pushed or pulled by its partner. The relationship is perfectly reciprocal, and either force may be termed the action force. One of these forces does not come into existence before the other, in the nature of cause and effect; rather, the two forces arise at the same time. The tail *does* wag the dog and, as will be seen later, eventually stirs the distant stars.

One of the most important points to remember in this connection is that *the two forces always act on different bodies.* This distinguishes the third law from a statement about equilibrium. Moreover, it will be shown below that the third law applies universally, even to nonequilibrium conditions. In any example of the equilibrium of a pair of forces, the two equal and opposite forces act on the *same* object and their resultant is zero. It is meaningless to compute the resultant of a pair of action-reaction forces, because they act on *different* bodies.

In some of the mechanical situations discussed in earlier chapters we tacitly recognized the applicability of the third law, as in Worked example 6.4. However, it will be useful to analyze a simple case in detail at this point. Figure 7.1 represents an apple hanging from the limb of a tree. Altogether, there are four forces acting:

1. $\mathbf{F}_{EA}$, exerted by the earth on the apple. This is the weight of the apple, which appropriately may be taken to be 1 newton.
2. $\mathbf{F}_{TA}$, exerted by the tree on the apple.
3. $\mathbf{F}_{AE}$, exerted by the apple on the earth through gravitational attraction.
4. $\mathbf{F}_{AT}$, exerted by the apple on the tree.

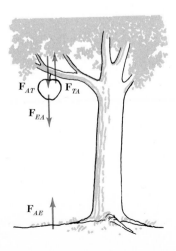

FIGURE 7.1 All four forces have the same magnitude, but for different reasons.

From their names, we recognize that forces 1 and 3 form an action-reaction pair, and so do forces 2 and 4. Thus

$$\mathbf{F}_{EA} = -\mathbf{F}_{AE} \qquad\qquad 7.2$$

and

$$\mathbf{F}_{AT} = -\mathbf{F}_{TA} \qquad\qquad 7.3$$

Each of these forces has a magnitude of 1 newton.

Notice, however, that because they act on the *same* body, $\mathbf{F}_{EA}$ and $\mathbf{F}_{TA}$ are *not* an action-reaction pair. They are equal and opposite because the apple is in equilibrium, and so the *second law* yields

$$\mathbf{F}_{EA} = -\mathbf{F}_{TA} \qquad\qquad 7.4$$

Each of these forces has a magnitude of 1 newton in this example.

Suppose, now, that the stem breaks away from the branch to which it is attached. The apple, although falling, is still attracted by the earth with a force equal to its weight; that is, $\mathbf{F}_{EA} = 1$ newton. Also, by Eq. 7.2, $\mathbf{F}_{AE}$ continues to exist, with a magnitude of 1 newton.

Both $\mathbf{F}_{AT}$ and $\mathbf{F}_{TA}$ became zero at the moment the stem came loose, and now $\mathbf{F}_{EA}$ and $\mathbf{F}_{TA}$ are no longer equal. One has vanished while the other is still a downward force of a magnitude of 1 newton. This means that equilibrium no longer exists.

The force $\mathbf{F}_{EA}$ acting on the apple accelerates it toward the earth, but it is equally true that the force $\mathbf{F}_{AE}$ acting on the whole earth accelerates it toward the apple. The earth's mass is so great that we are completely unaware of the infinitesimal motion it receives and are conscious only of the falling motion of the apple.

It is interesting to note that the third law requires the earth to react continually to countless effects of natural and man-made actions. No wind can blow, no airplane take off, no volcano erupt without calling up a reaction on the part of the earth. Even the act of walking must cause the earth to "roll back" in response to the reaction force.

As a further example, consider the forces acting on an airplane in flight. These are shown in Fig. 7.2 and identified along with their reaction forces in tabular form below:

FORCES ACTING ON AIRPLANE	REACTIONS OF AIRPLANE ON ITS ENVIRONMENT
L, the *lift* force	Downward push on surrounding air
W, the *weight*	Upward gravitational pull on earth
T, the *thrust* of propellers or jets	Backward push against the air or against ejected gases
D, the *drag* force	Forward force pulling along some of surrounding air

Newton's second law gives relations between these forces:

1. If the airplane is in steady, level flight, it is in equilibrium both vertically and horizontally, and

$$\mathbf{L} = -\mathbf{W} \qquad \text{and} \qquad \mathbf{D} = -\mathbf{T}$$

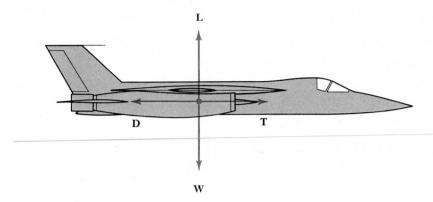

FIGURE 7.2 The forces exerted on an airplane. In steady, level flight, **L** and **W** are equal and opposite and so are **T** and **D**.

2. If the airplane is climbing at *constant speed* it is still in equilibrium, but an additional drag force **R** will come into play as a result of the upward motion. **R** will be in the downward direction, and then

$$\mathbf{L} = -(\mathbf{W} + \mathbf{R})$$

3. If the plane is climbing at increasing speed, L will be numerically greater than $W + R$. Similarly, if the plane is accelerating *along* its flight path, T must be numerically greater than D.

Practice set

Name the reaction force corresponding to each of the following. This includes naming the object to which the reaction force is applied.

a. Force with which a book presses on the table on which it rests
b. Horizontal force exerted by the feet of a sprinter at the start of a race
c. Force on a baseball while in contact with the bat
d. Force acting on a bullet while it is gathering speed inside a gun barrel

7.2 NEWTONIAN FORM OF THE SECOND LAW; MOMENTUM

Newton's third law may be combined with his second law to yield one of the most useful and general laws of mechanics. First, we review Newton's original statement of the second law, as given on page 125: "Change of motion is proportional to the impressed force and takes place in the line in which the force acts." With this as a starting point and with the help of the results of experiments described in the preceding chapter, we were able to reduce this statement to the quantitative relation $\mathbf{F} = m\mathbf{a}$.

But there is another way of expressing the second law. Newton interpreted the word "motion" in the above statement to mean a quantity that measures jointly the velocity of a body and the amount of matter that is moving. This quantity is known as the **momentum** of a body. The momentum of a particle is the product of its mass and its velocity. This definition is arbitrary, but it proves exceedingly useful. In symbols,

$$\mathbf{p} = m\mathbf{v} \tag{7.5}$$

The product of the mass and velocity of a particle is specifically referred to as the linear momentum *in order to distinguish it from a quantity called angular momentum, to be defined later.*

where **p** represents the momentum. This is a vector having the same direction as the velocity vector **v,** since multiplying **v** by the scalar quantity

m simply increases its magnitude by this factor. The dimensions of p are seen to be ML/T, and the appropriate units in the mks system are kg-m/sec.

The quantity **p** has important properties. Newton himself must have recognized this fact by defining momentum essentially as above on the very first page of the *"Principia."* If we interpret the phrase "change of motion" occurring in his statement of the second law to mean *rate of change of momentum,* then this statement can be shown to be equivalent to **F** = m**a,** a form that we have already found useful in a variety of problems in dynamics. The reasoning is as follows:

For another interpretation, see the discussion on page 163.

Introduce the quantity **p,** as defined by Eq. 7.5, into the above formulation of the second law. If the momentum of a particle changes by an amount Δp in a time interval Δt, the average rate of change of momentum will be $\Delta p/\Delta t$. According to the interpretation of the second law given above, this is proportional to the magnitude of the applied force. With the simplification introduced by using absolute force units, this can be set numerically *equal* to the average force:

$$\bar{F} = \frac{\Delta p}{\Delta t} = \frac{\Delta(mv)}{\Delta t} \qquad 7.6$$

Now we make a significant assumption. Suppose the mass of the particle remains constant. Then the factor m can be placed in front of the Δ symbol because it is constant and contributes nothing to any change in mv, any such change being due entirely to a change in v. Equation 7.6 may now be written

$$\bar{F} = \frac{m\,\Delta v}{\Delta t} = m\frac{\Delta v}{\Delta t}$$

But the fraction in this expression is recognized to be the average *acceleration* over the interval in question. On passing to the limit, $\bar{F}$ becomes the instantaneous force F, and $\Delta v/\Delta t$ becomes the instantaneous acceleration a, so that the relation takes the familiar form $F = ma$. Thus (using the vector forms of the two expressions) **F** = m**a** is merely a way of writing

$$\mathbf{F} = \lim_{\Delta t \to 0} \frac{\Delta \mathbf{p}}{\Delta t} \qquad 7.7$$

for the special case where the mass remains constant.

As mentioned above, the momentum-change formulation of the second law expressed in Eq. 7.6 or 7.7 is more general than **F** = m**a** because it takes into account the possibility that the mass m may not stay constant during the motion. For example, a rocket propels itself by the reaction of gases ejected in the backward direction (see Box on page 150). Since the rocket vehicle is continually losing material in this way, its mass is changing all the time. In the same way, a snowball rolling down a hill constantly adds to its total mass. Problems connected with such mass changes are best handled by applying conservation of momentum to the complete system consisting of the main body together with the mass that is ejected or taken on in a given time.

The case of variable mass.

A more fundamental fact is that, according to the theory of relativity, the mass of an object changes with its speed (page 49), and this effect becomes more and more pronounced as the speed approaches that of light. In any case, Eq. 7.7 remains correct as it stands and has not required modification in the light of any subsequent development up to the present time. It is indeed remarkable that, although Newton could not have antici- pated relativity, he was led to express the second law in the more general form that could be taken over unchanged into this theory.

A steady rain starts to fall on the deck of a ship that moves at a speed of 7.0 m/sec. The rain is falling straight down at the rate of 5×10^{-2} kg of water per second on each square meter, and the total deck area of the ship is 6,000 m². What additional propulsive force is needed to keep the ship moving at its former speed?

Solution   The water striking the deck must be set into forward motion with a speed of 7.0 m/sec, and a force is needed to give this mass of water the required forward momentum. In 1 sec, the mass set into motion is $(5 \times 10^{-2})(6,000) = 300$ kg. This is given a forward speed of 7.0 m/sec, so that the resulting rate of momentum change of the water that is swept up is $(300)(7.0) = 2,100$ kg-m/sec. According to Eq. 7.7, this requires a force of magnitude $F = 2,100$ newtons, or about 470-lb force.

7.3 INTERACTIONS AND MOMENTUM CHANGES

When two bodies interact, the force between them may vary in a compli- cated way, as it does, for example, during a collision. However, the second law, particularly when stated in the momentum form, furnishes a method of measuring the net effect of an interaction without troubling about the way the force may vary in the process. Newton himself experimented with colliding pendulums in order to observe the changes in motion produced by the impact. In another experiment, he placed a magnet and a piece of iron on separate floats in a vessel of water and noted that the pair of objects was at rest after drawing together and colliding.

An even better ar- rangement is to replace the cars with air- supported gliders.   A simplified experiment on the interaction of two bodies is shown in Fig. 7.3. Two small cars, A and B, of equal mass are placed on a level, virtually frictionless track. A small spring of negligible mass is placed between the cars, and a loop of thread is tied between them to hold the spring in a compressed position.

With the cars at rest, the thread is snipped with a scissors. The com- pressed spring relaxes, thrusting the cars apart. A multiple-flash photograph reveals equally spaced images of both cars, showing that, from the moment

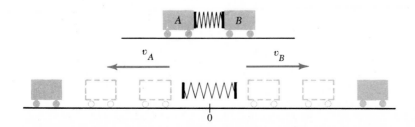

FIGURE 7.3 Cars thrust apart by releasing a compressed spring. A multiflash photograph shows equally spaced images when the cars have the same mass.

the spring is completely relaxed, the cars move with constant speed in opposite directions. An equivalent way of saying this is that the cars acquire equal and opposite momenta, or

$$mv_A = -mv_B$$

where v_A and v_B are the velocities of the respective cars and m is the mass, which is the same for both. The vector nature of momentum is taken into account in this simple case of motion along a single line by designating velocity toward the right as positive and toward the left as negative.

The above result comes as no surprise in view of the perfect right-left symmetry of the arrangement. The same result would be expected if each half of the system had been replaced by a single car that was made to back away from a rigid wall by releasing a spring made of half the original spring and compressed to the same extent as before. This is a useful idea to which we shall return later.

We now make the experiment more general by placing *unequal* loads on the two cars so that their total masses have different values, m_A and m_B. This time, the flash pictures (Fig. 7.4) show that the cars recoil with *different* speeds, the more massive car having the lower speed; in fact, a series of trials will show that the two *speeds are inversely proportional to the masses:*

$$-\frac{v_A}{v_B} = \frac{m_B}{m_A}$$

or

$$m_A v_A = -m_B v_B \qquad\qquad 7.8$$

This observed result may be considered a verification of Newton's third law. The reasoning is as follows:

Write Eq. 7.8 for a small change taking place in a time Δt:

$$m_A \Delta v_A = -m_B \Delta v_B \qquad\qquad 7.8a$$

Divide both sides by Δt:

$$m_A \frac{\Delta v_A}{\Delta t} = -m_B \frac{\Delta v_B}{\Delta t} \qquad\qquad 7.9$$

But each fraction in Eq. 7.9 represents the average acceleration over the interval. In the limit, when Δt approaches zero, each of these in turn becomes the instantaneous acceleration, as defined on page 97, so that 7.9 is now

$$m_A a_A = -m_B a_B$$

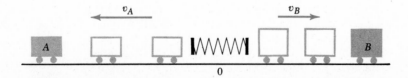

FIGURE 7.4   Cars of unequa mass. The heavier car acquires th smaller speed.

According to the second law, this can be written

$$F_{BA} = -F_{AB}$$

which is the same as Eq. 7.1, the analytical statement of Newton's third law. Thus the third law is established by experiments done with the reaction cars.

Worked example 7.2   A shell of mass 2 kg is fired horizontally forward at a speed of 200 m/sec from a gun mounted on a car that is free to roll on a smooth level track. The car and gun together have a mass of 500 kg. With what speed does the car recoil?

Solution   The two interacting bodies are the projectile and the gun mounted on the car. Substituting into Eq. 7.8,

$$500v_A = -2(200) \qquad \text{from which} \qquad v_A = -0.80 \text{ m/sec}$$

This is the recoil speed of the car and gun. The negative sign shows that the motion is to the left. In the absence of any other forces, the bullet would continue to move with a speed of 200 m/sec to the right and the car would continue to move with a speed of 0.80 m/sec to the left. The effect of the exploding charge has been not only to propel the bullet forward but, unavoidably, to give the gun a backward kick. In principle, the explosion has succeeded in blasting the gun and bullet apart. The gun and carriage are much more massive than the projectile, and so moved more slowly, but the two objects came away with *equal* amounts of momentum. If, instead of rolling freely, the gun were firmly fixed in the ground, it would have to share its momentum with the whole earth and only the motion of the bullet would be perceptible.

Worked example 7.2 dealt with two objects that were originally joined together, effectively forming a single particle, and then thrust apart by an internal action. This interaction resulted in equal and opposite momentum changes in the two objects. This is just as true for the reverse process in which two moving objects collide and stick together. A simple case of a completely **inelastic** (no rebound) collision will be illustrated next:

Worked example 7.3   A lump of soft clay of mass 200 g moving to the right with a speed of 4.5 m/sec collides with a block of wood of mass 500 g moving to the left at 2.5 m/sec. The collision is assumed to be *central* (head-on), so that no rotational effects are produced by the impact. If the clay sticks to the block, how does the combined body move after the collision?

Solution   Call the final speed of the combined object V. This speed will be in the original line of motion of the two objects, since their interaction forces are entirely in this line, but the direction of V (right or left) is not known in advance. Both the direction and magnitude of V can be found by using the relation 7.8a, which states that the momentum change of the lump of clay must be equal and opposite to the momentum change of the wood block. Using the above data and taking the positive direction toward the right, this becomes

$$200(V - 4.5) = -500[V - (-2.5)] \quad \text{or} \quad V = -0.50 \text{ m/sec}$$

Thus the combined body moves to the *left* at 0.50 m/sec just after the collision. Notice that the masses were inserted in grams (as originally given) and not in kilograms. It is not necessary to convert as long as the same mass unit is used throughout, since the conversion factor would be present in every term of the equation and so could be canceled out.

This example differs from Worked example 7.2 about the cannon. The cannon-bullet system was at rest to begin with, whereas the clay-block system of the present example was in motion at all times. This difference

JET PROPULSION AND ROCKET ENGINES

The forerunner of all jet propulsion devices was probably the reaction engine constructed by Hero of Alexandria in the second century B.C. Steam was produced by boiling water in a hollow sphere equipped with two nozzles bent in opposite directions. The jets of escaping steam caused the sphere to rotate. A rotary lawn sprinkler operates in much the same way.

In recent years several kinds of jet engines have been developed for propelling aircraft, missiles, and space vehicles. Most of them use chemical fuels whose combustion produces a large volume of gas. The rearward ejection of these gases at high speed produces a reaction against the nozzles that drives the vehicle forward.

The ram jet   This is a simple form of reaction engine that has no moving parts. It is used for the propulsion of guided missiles at altitudes up to about 15 mi. When an engine of this kind moves through the air at high speed, air is "rammed" into

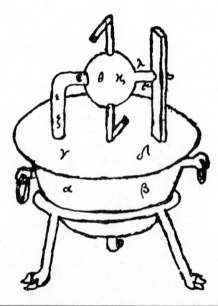

Reaction steam engine devised by Hero of Alexandria. (*Courtesy of the Granger Collection.*)

is not an essential one, since Newton's laws are valid in any inertial system. For instance, in Worked example 7.3 we might have chosen to refer all velocities to another inertial frame moving to the left at 0.50 m/sec with respect to the laboratory, and in this frame the combined body would be at rest. This will be illustrated in Worked example 7.6.

The above calculations show the usefulness of the momentum concept in determining the subsequent motions even when no details of the interaction forces are known. In order to make use of the constancy of linear momentum (see Sec. 7.5), we must first make sure that we are working with an isolated system and that all velocities are measured in the same inertial frame of reference.

a duct where it mixes with the ignited fuel continuously. The hot combustion gases are ejected at the rear at high speed. A ram jet works best at speeds above 500 mi/hr.

The turbojet engine   In this engine, air is drawn mechanically into the duct by means of a compressor (Fig. 7.5a). Besides furnishing thrust, the escaping gases drive a turbine which operates the compressor. Turbojets are especially effective for driving high-speed high-altitude aircraft.

The rocket engine   Ram-jet and turbojet engines use the oxygen of the air for burning their fuel. The rocket engine carries its own oxygen supply and so can operate even beyond the atmosphere (Fig. 7.5b). Calculations show that the final speed of a rocket can be made very great by having the ratio of mass of fuel to mass of vehicle as large as possible. The effectiveness can be increased by using multistage rockets, the method used in launching artificial satellites and space vehicles.

Other reaction engines   Various forms of rocket propulsion that do not use chemical processes are under development or have been proposed. In *plasma (ion) rockets,* atoms or molecules are broken up into electromagnetic particles which are then expelled at high speed by means of electric forces. A *photon rocket* would operate by the reaction of a strong beam of radiation. All these engines have low thrust, and regular rocket propulsion would be needed to move the vehicle to a great distance from the earth before the other forms could take over.

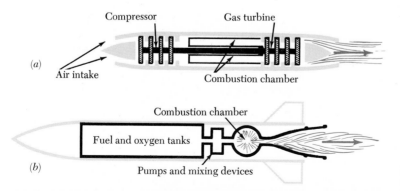

FIGURE 7.5 (a) Turbojet engine. (b) Rocket engine.

7.4 IMPULSE FORCES

When two fairly hard objects collide, each exerts a large force on the other, but the interaction usually lasts a very short time. For instance, when a nail is struck by a hammer, a very great force is exerted for a small fraction of a second. The force exerted by the hammer head on the nail drives the nail into the wood. The equal and opposite reaction force exerted by the nail on the hammer head reduces the motion of the latter to zero. Violent interactions of this kind deform both bodies during the impact, and permanent deformations sometimes result. Any situation in which a very large force is exerted for a very short time is called an **impulse** interaction.

Think of a golf ball that is suddenly set into motion by a blow from a club. It is possible to find out something about the effect of this force by relating it to the observed net change in motion of the ball. To do this we make use of the momentum-change form of Newton's second law,

$$\mathbf{F} = \frac{\Delta \mathbf{p}}{\Delta t}$$

where $\mathbf{F}$ is assumed to remain essentially constant during the short interval Δt. Cleared of fractions, this becomes

$$\mathbf{F}\,\Delta t = \Delta \mathbf{p} \qquad\qquad 7.10$$

The quantity $\mathbf{F}\,\Delta t$ on the left is called the **impulse** of the force $\mathbf{F,}$ and the equation says that *the change in momentum of the body is equal to the impulse of the applied force.*

Impulse and momentum are measured in the same units. In the mks system, the left side of Eq. 7.10 has the units N-sec; the right side has units corresponding to mass multiplied by speed, or kg-m/sec. These two combinations are, however, identical, since newtons can be expressed as kg-m/sec² through the use of $F = ma$.

Equation 7.10 shows that a given change in the momentum of a body can be brought about by a large force acting for a short time or by a small force acting for a correspondingly longer time. In an actual case such as the impact of a hammer on a nail, the instantaneous value of the force may vary in some continuous but complex way whose details are not easy to determine. Let such a force be represented by the curve in Fig. 7.6. The variable force may be replaced by its *time average,* which is of such magnitude as to make the area of the dashed rectangle equal to the total area under the actual force curve. This can be seen in the following way:

Area under a curve.

Imagine the entire area under the curve to be divided into a large number n of narrow, vertical strips, all of the same width (Fig. 7.6). The height of each strip represents a value of F that may be considered constant over the very short time interval corresponding to the width $\Delta t/n$ of each strip. If the areas of all these strips are added, we have

$$\Delta A = (F_1 + F_2 + F_3 + \cdots + F_n)\,\frac{\Delta t}{n}$$

But $(F_1 + F_2 + \cdots)/n$ represents the arithmetical average of all the F's, denoted by $\overline{F}$. Then the above equation may be written

$$\Delta A = \overline{F}\,\Delta t$$

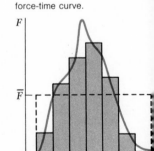

FIGURE 7.6 The impulse of a force is measured by the area under the force-time curve.

If, now, the number of strips is made indefinitely large, their individual widths become very small and their total area approaches the area included between the curve and the t axis. Then we can say that the value of $\bar{F}$ is such that, when multiplied by Δt, the result is exactly the area under the force curve. Thus, if we draw the dashed rectangle so that its area is the same as that under the curve, the height of this rectangle will represent the magnitude of the average force $\bar{F}$.

Notice that the areas under discussion are not ordinary surface areas of dimension L^2. Rather, the dimension is given by the product of the two plotted quantities, in this case force multiplied by time, which is momentum. Also, it is not essential that the curve cut the axis at each extremity of the interval. The average value over a given interval, multiplied by the width of the interval, will still equal the area between the curve, the horizontal axis, and the two ordinates drawn at the beginning and end of the interval.

A special case of this averaging process was used in Chap. 5, page 100, where the area under a *straight* sloping segment was replaced by a rectangle whose height in that particular instance was just halfway between the heights of the two end points. The general method of averaging values of a variable represented by a curve will be found useful on later occasions in this book.

Worked example 7.4

A baseball of mass 0.15 kg, pitched with a speed of 50 m/sec, is struck by a bat and sent back roughly in the opposite direction with a speed of 70 m/sec.

a. What is the total change in momentum of the ball, and in what direction does this change take place?
b. Find the average force exerted on the ball if it was in contact with the bat for 0.08 sec, and state the direction of this force.
c. What were the magnitude and direction of the average force exerted on the bat by the ball during the impact?

Solution a. The change in momentum of the ball is given by $\mathbf{p} = m\mathbf{v'} - m\mathbf{v}$, where $\mathbf{v}$ is the velocity before impact and $\mathbf{v'}$ is the velocity after impact. We have seen that in computing such changes the directions of the velocities are important. To avoid confusion, assume that the ball is moving initially to the right, and let this be the positive direction. Then $\mathbf{v}$ will be positive and $\mathbf{v'}$ will be negative. The above equation becomes

$$\Delta p = 0.15 \, (-70 - 50) = -18 \text{ kg-m/sec}$$

and since this number is negative, the momentum change is directed toward the left, the direction in which the bat pushed on the ball.

b. According to Eq. 7.10 the impulse is equal to the change in momentum. If the magnitude of the average force is $\bar{F}$, then $\bar{F} \, \Delta t = \Delta p$, or $\bar{F}(0.08) = -18$ and so $\bar{F} = -225$ newtons. This amounts to about 50-lb force. The negative sign shows that this force, like the momentum change, is directed toward the left.

c. According to the third law, the force exerted by the ball on the bat is equal and opposite to the force exerted by the bat on the ball. Thus the bat experiences a force of 225 newtons to the right.

In the preceding example, the bat was not a free object but was held in the hands of the batter. This being so, are we justified in saying that the force exerted on the bat amounted to 225 newtons as computed above? The answer is that in an impact the impulse forces are usually very large compared with the other forces acting, so that these other forces do not produce appreciable effects during the time the impact lasts. The force that the batter applied to the bat is of this nature.

In the same way, the retarding force of friction acting on a hockey puck *during impact* is insignificant compared with the forces produced by the blow of the stick. After the impact is over, the frictional force can influence the motion of the puck, but during the time of impact this force can be neglected completely.

7.5 CONSERVATION OF LINEAR MOMENTUM

Equation 7.8a (page 148), which reads

$$m_A \, \Delta v_A = -m_B \, \Delta v_B$$

was deduced from observations of two interacting objects initially at rest. However, the statement is also valid if the combined body is in motion to begin with, for this involves merely shifting to another inertial frame of reference.

Equation 7.8a says that the momentum change experienced by one part of the system is equal and opposite to that of the other. This is equivalent to saying that the total linear momentum of the pair of particles remains constant. In the example of the pair of cars initially at rest, the total momentum was zero before the spring was released and it was (algebraically) zero afterward as well.

Experience shows that the above statement can be generalized to include isolated systems of more than two particles, moving and interacting in any way. Thus there is a general **principle of the conservation of linear momentum:** *In any isolated mechanical system, the total vector linear momentum remains constant.*

This is a physical law of extreme generality. It is valid for an isolated system of any nature, whether it is made up of individual particles, extended objects in any state of aggregation, or any combination of these. The components of the system may undergo the most diverse and violent inter-actions—collisions, explosions, change of phase, and so on. There is no need to follow the detailed motion of each particle; the conservation of momentum makes a simple statement about the behavior of the whole collection. It is necessary only that the system be uninfluenced from the outside and that all momenta be measured in the same inertial frame of reference. It will be shown in Chap. 18 that the law is valid even for subatomic particles.

Perhaps the most important point to remember in using the momentum principle is that, since momentum is a vector, the direction in space of the momentum of each particle must be taken into account.

Our previous examples of momentum conservation involved bodies mov-ing in a single line, where the vector nature of momentum required merely choosing plus and minus directions along this line. In a more general case the parts of the system may be considered to move in two dimensions, like balls on a billiard table, or even in three dimensions like the molecules of a gas confined in a box (Chap. 10). For the present, we illustrate the vector nature of the conservation of momentum by the following two-dimensional example.

Momentum conservation in two dimensions.

Two balls A and B of unequal mass are rolling on a smooth, level plane (Fig. 7.7), with velocity vectors as shown in Fig. 7.7a. Also shown, alongside each ball, is its momentum vector. This vector has a magnitude equal to the product of the mass and the speed of the ball, and the direction is the same as that of the velocity vector. The diagram on the right shows the polygon construction for finding the resultant (total) momentum vector P of the system, which is assumed to be isolated.

The symbols for velocities after impact are to be read "V A, prime" and V B, prime."

The two balls collide and bound off obliquely in different directions with velocities $\mathbf{v}'_A$ and $\mathbf{v}'_B$, respectively, as shown in Fig. 7.7b. Once more the momentum vectors are constructed. When they are combined by means of the polygon scheme it is seen that their resultant is identical with the vector **P** found in Fig. 7.7a: *The total momentum vector remains unchanged by the collision.*

Now suppose that some time after the collision ball B breaks into two pieces, C and D, as shown in Fig. 7.7c. The momentum vectors of these two fragments just after the break are $\mathbf{p}_C$ and $\mathbf{p}_D$, respectively. If all the momentum vectors are combined we again get a resultant identical with **P,** as shown at the right in Fig. 7.7c.

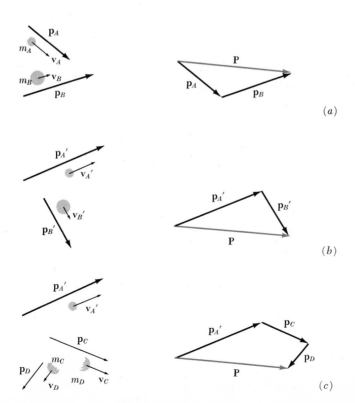

FIGURE 7.7 Regardless of what happens to the system internally, the total momentum vector remains constant.

Instead of using the polygon construction it is sometimes advantageous to resolve all the individual momentum vectors into x and y components, as was done earlier for forces and velocities. Then a statement of the conservation principle can be made for each set of momentum components independently:

$$p_{Ax} + p_{Bx} + p_{Cx} + \cdots = P_x = \text{constant}$$

and

$$p_{Ay} + p_{By} + p_{Cy} + \cdots = P_y = \text{constant}$$

where p_{Ax} stands for the x component of the linear momentum vector of body A, p_{Bx} for that of B, etc. All the particles in the system must be included in taking these sums.

Worked example 7.3 (page 149) showed what happens in a two-body collision in which the objects stick together after impact. A collision in which the two objects rebound after impact will now be analyzed. Here again, in order to keep the computations simple, all motions will be restricted to a single line.

The situation before impact is shown in Fig. 7.8. According to the principle of momentum conservation,

Total vector momentum before impact = total vector momentum after impact

If the speeds of the balls after collision are designated v'_A and v'_B, respectively, this becomes

$$(50)(90) - (80)(70) = 50\ v'_A + 80\ v'_B \qquad\qquad 7.11$$

The minus sign of the second term reflects the fact that v_B is toward the left, or in the negative x direction.

The equation simplifies to

$$5\ v'_A + 8\ v'_B + 110 = 0 \qquad\qquad 7.12$$

Since there are two unknowns in this equation, it cannot be solved explicitly for their numerical values unless further information is available. To see that there is no unique solution, notice that the total momentum of the system is given by the left side of Eq. 7.11 as $(50)(90) - (80)(70) = -1{,}100$ g-cm/sec and is represented by a vector of magnitude 1,100 directed toward the left. Any two momentum vectors that combine to produce this resultant give a possible solution of the problem. But there is an infinite number of such vector pairs. How can the correct one be found?

Colliding bodies are real objects, and the outcome of an impact experiment depends on how much of the original motion is dissipated by being converted to heat. The extent to which this happens can often be approxi-

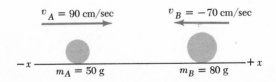

$v_A = 90$ cm/sec $v_B = -70$ cm/sec

$-x$ ——————————————————————— $+x$

$m_A = 50$ g $m_B = 80$ g

FIGURE 7.8 Central collision of two balls. The velocities before collision are shown.

The momentum principle sometimes is not sufficient for a complete solution.

mated by measuring a characteristic of the colliding bodies called the coefficient of restitution. In the special case of the completely **inelastic** (no rebound) collision of two particles, we saw (Worked example 7.3) that because both objects move as a single body after impact the conservation of momentum provides enough information to determine the motion of this body. At the other extreme—perfectly elastic impact—virtually none of the original motion is dispersed. In this case, momentum conservation can be supplemented by a principle that deals with energy, leading to a definite solution for the subsequent motion of both bodies. This will be discussed in Chap. 9.

If the motion of either body is *observed* after impact, the momentum equation is sufficient to determine the motion of the other. In the example, suppose that B is observed to be moving to the left at 5.0 cm/sec after impact. Then v'_B can be set equal to -5.0 cm/sec in Eq. 7.11, yielding

$$5\, v'_A + (8)(-5) + 110 = 0 \qquad \text{from which} \qquad v'_A = -14 \text{ cm/sec}$$

Thus both balls will move to the left after colliding, A with a speed of 14 cm/sec and B with a speed of 5 cm/sec.

The principle of the conservation of momentum is of great importance in connection with collisions and disintegrations ("explosions") that involve atomic and subatomic particles. Some of these particles record their paths by leaving visible trails in a cloud chamber or bubble chamber or in photographic emulsions. Measurement of the lengths and directions of such tracks, together with the law of momentum conservation, permits the calculation of an unknown mass or speed. The momentum principle seems to hold universally for such events. In one case, where this appeared *not* to be true, physicists provisionally assumed the existence of an unseen particle, the **neutrino** (page 583), in order to maintain the validity of con-

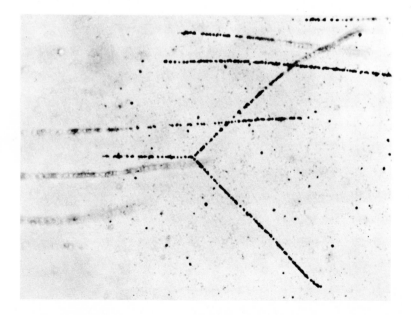

Photographic traces of a subatomic collision. A proton coming from the left hits a proton originally at rest. For all such encounters except head-on impact, the two particles move off at right angles, in conformity with conservation of momentum and of mechanical energy. (*Courtesy of Powell and Occhialini, "Nuclear Physics in Photographs," Clarendon Press.*)

servation of momentum. In due course, direct observational evidence for the neutrino was found, justifying their confidence in the conservation of momentum.

7.6   CENTER OF MASS

When an explosion occurs we expect to see fragments shooting out to all sides. It would be surprising if all the pieces flew off in much the same direction. When the compressed spring between the two cars described on page 147 was released, it was natural to expect the cars to move off to *opposite* sides. Observations of this kind lead to the following conclusion. There is a special point that represents the motion of a mechanical system as a whole, regardless of any internal motions that may be present. This point is called the **center of mass** of the system. It can be shown that the system moves as if (1) its whole mass were concentrated at the center of mass and (2) the resultant of all the *external* forces were applied at that point.

According to Newton's third law, internal forces cancel each other in pairs.

In several previous examples, we replaced an extended object such as a ball or a vehicle by a single particle. This particle was assumed to be located at the center of mass of the actual object and to have a mass equal to the total mass of the object. Thus a way is provided for using Newton's second law to describe the motion of a complicated system as a whole: If the resultant of all the external forces is F and the total mass of the system is m, then the center of mass of the system will move with an acceleration given by $F = m\mathbf{a}.$

In an isolated system, the net external force is zero and so in this instance the center of mass continues to move with constant velocity (zero acceleration). This is illustrated in the multiflash photograph of a wrench that was flung out along a very smooth floor. The wrench rotated as it moved along, but its center of mass (indicated by a cross in the picture) moved with constant speed in a straight line. This can be readily checked by measuring the photograph.

The center of mass of the wrench was located by trial. For a body of regular geometric shape, made of homogeneous material, the location of this point can be computed. Figure 7.9 shows the results for a few solids.

Motion of a wrench spinning as it slides along a smooth horizontal surface.

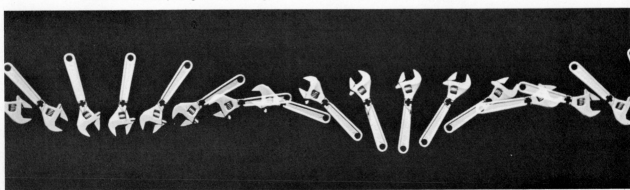

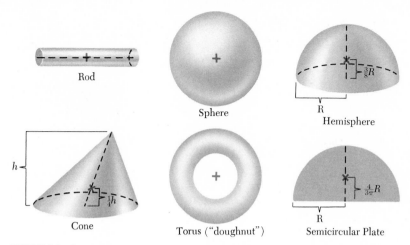

FIGURE 7.9 Center of mass of some common shapes. The solids are assumed to be homogeneous. In addition, the semicircular plate has constant thickness. The center of mass of a body is not within the material itself in every case, but its position is fixed with respect to the body.

The center of mass of a system consisting of just two particles lies on the line joining the particles and divides this line in the inverse ratio of the two masses. For instance, if one of the particles has three times the mass of the other, the center of mass will divide the distance between them into two segments, one of which is three times as long as the other. The shorter segment will be nearer the larger mass (Fig. 7.10).

The center of mass has another property that can be illustrated as follows: Think of a pair of particles joined by means of a rigid but massless rod (Fig. 7.11). Apply a force to the midpoint of the rod in any direction except along the rod. Then the system will be observed to rotate, as shown in the upper part of the figure. The more massive particle will lag behind because of its greater inertia. However, if the force is applied quite close to the larger mass, the rod will rotate in the opposite direction. By successive trials, a point can be found where the applied force will not produce any rotation of the rod but merely a forward motion of the system in which the rod continues to move parallel to itself. This point is the center of mass.

The center of mass is useful in describing the interaction of a pair of particles, as in the following situation:

Worked example 7.5 Two skaters of mass 50 and 70 kg, respectively, stand facing each other 6 m apart (Fig. 7.12). They then pull, hand over hand, on a light rope stretched between them. How far has each moved when they meet? Friction is negligible.

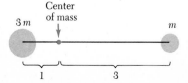

FIGURE 7.10 The center of mass of two particles divides the line joining them in the inverse ratio of their masses.

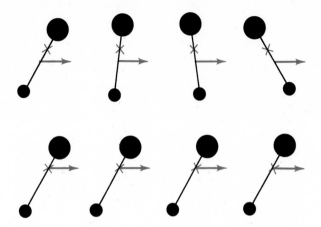

FIGURE 7.11 If the object is free, it will move without rotating when pulled at its center of mass.

Solution The center of mass of the pair is at rest to begin with; since there is no net external force, this point remains at rest, and the skaters eventually meet there. The center of mass divides the distance between them in the ratio 50:70, and so it is located a distance of

$$\frac{50}{50 + 70}\, 6 = 2.5 \text{ m}$$

from the heavier man. Thus he travels 2.5 m, while his lighter companion travels 3.5 m. The result is in no way dependent on how each man takes in rope. It would be the same if either man did all the pulling and the other merely held fast to his end of the rope.

The inelastic collision of a lump of clay and a block of wood described in Worked example 7.3 (page 149) can also be treated by using the center-of-mass concept:

Worked example 7.6   Assume the same data as before. Figure 7.13 shows the pair of bodies at two specific times: (*a*) 2 sec before collision and (*b*) 1 sec before collision. The initial positions of the two objects are chosen to make the impact occur at the origin (see Fig. 7.13).

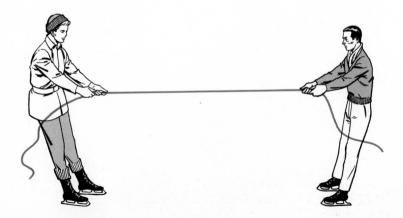

FIGURE 7.12 The lighter man moves farther before they meet.

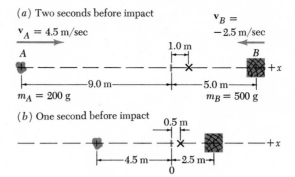

FIGURE 7.13 Head-on collision of a lump of clay and a wood block.

Solution   In case *a*, the distance of the center of mass of the pair of bodies from A is

$$\frac{500}{200 + 500} (9.0 + 5.0) = 10 \text{ m}$$

that is, it is located at

$$x = +1.0 \text{ m}$$

In case *b*, this distance is

$$\frac{5}{7} (4.5 + 2.5) = 5.0 \text{ m}$$

or the center of mass is at

$$x = +0.5 \text{ m}$$

This change in position takes place in 1 sec so that, before collision, the center of mass is moving at a speed of -0.5 m/sec. Since there is no external net force, this speed must be the same after the impact as before, so that the speed of the combined objects is also -0.5 m/sec, as found on page 150.

Figure 7.14 shows what happens in a head-on elastic collision between two objects by what amounts to a series of snapshots taken at intervals of 0.2 sec. Notice that the dashed lines are *not* the actual paths of the bodies. The two objects move along a fixed line. At any instant the segment of this line between the bodies is divided in inverse ratio to their masses by the position of the center of mass, which continues in uniform motion before, during, and after the impact.

Figure 7.14*a* shows how the motions appear when referred to the fixed frame of reference of the laboratory, and Fig. 7.14*b* shows them as viewed from the center of mass. The fact that the center of mass moves according to Newton's first law enables us to couple an inertial frame to it, thereby referring the motions of the parts of the system to this point. *The total momentum of the system, referred to the center of mass, is zero at all times.*

Consider a shell which explodes into two parts at some point of its trajectory (Fig. 7.15), each of the fragments going off on its own parabolic

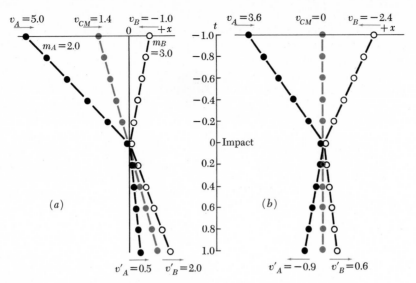

FIGURE 7.14 The history of a central collision of two objects, from 1 sec before to 1 sec after impact (see vertical time scale). In (a) the positions are shown relative to the laboratory system, in (b) relative to the center of mass of the pair. In (b) the total vector momentum is zero at every instant. All distances are given in meters, times in seconds, and speeds in meters per second.

trajectory. What can be said about the subsequent motion of the pieces? The answer is that the center of mass of the pair continues along the original parabola just as the shell would have done had it not exploded.

7.7   NEWTON'S LAWS IN RETROSPECT

Apart from certain modifications introduced by the theory of relativity for the extreme case of speeds approaching that of light, Newton's laws constitute a complete basis for mechanics, including the practical situations faced by engineers. Nevertheless, the exact meaning of and logical basis for Newton's laws are still debated. It is not feasible to go into such matters extensively here, but it is profitable to raise certain questions of interpretation at this time.

Newton offered his laws as axiomatic statements that were in harmony with observations made on mechanical systems. With regard to the second law, some authorities, following the example of Ernst Mach, believe that $F = ma$ does not represent a law describing the behavior of nature but

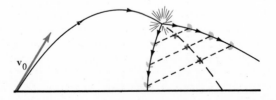

FIGURE 7.15 Shell exploding in flight. The center of mass of the fragments continues on the original trajectory as if nothing had happened.

is only a quantitative statement of what is meant by force. On this basis, the operational way of measuring a force is to cause it to accelerate a material body, and force becomes merely a name for the product of mass and acceleration.

By introducing additional mechanical concepts such as momentum and energy, it is possible to discuss mechanical systems without explicitly using the idea of force. Some historians of science maintain that what Newton called "motive force" in his statement of the second law is, in modern terminology, *impulse,* and "change in motion" is *change of momentum.*

Nevertheless, there are good reasons for retaining force as a physical concept of great intuitive value. For example, a wealth of experience shows that forces, as we think of them, can do many other things besides altering motion. Specifically, the effects of forces in deforming bodies were described in Chap. 3. L. W. Taylor (Ref. 7.3) remarks: "Whatever may be the degree of ultimate reality possessed by forces, they are useful as a means of unifying the explanation of the breaking of hawsers, the crushing of stones, and the pushing of the pens of those who write about their non-existence."

Considering the other two laws of motion, we find at the outset that the first law, the law of inertia, is included in the second law as a special case. To see this, set **F** equal to zero in $\mathbf{F} = m\mathbf{a}$. Then **a** must be zero, since the factor m always has a finite value. Thus, in the absence of a net applied force, a body will be unaccelerated and will have a constant vector velocity, as the first law states. It may happen that the velocity measured in a given inertial system is zero. Then the body is at rest in this reference frame and remains so. Although the first law follows from the second in this way, it is by no means trivial. It serves to define in a direct way the kind of reference system in which all newtonian mechanics is valid—the inertial frame.

Despite the disarming simplicity of its statement, the third law is probably the most subtle and meaningful of all three. With one swift, sure stroke it eliminates the necessity of having to consider the myriad internal forces present in a mechanical system. The third law leads directly to the conservation of momentum for an isolated system, one of the most general and powerful of all physical principles.

Conservation of momentum provides a logical and consistent procedure for measuring mass. Mach pointed out that in a system made up of two interacting particles the relative masses can be measured by observing the velocity changes. If one of the particles happens to be the standard mass, this amounts to *measuring* the mass of the other particle by what may be termed a purely inertial method, that is, one that does not involve the weight. In this connection, it is highly probable that Newton's experimental and theoretical work on impact did much toward clarifying his ideas of mass and momentum.

Apparent violations of Newton's third law.

Does the third law hold for nonmechanical forces such as those that arise when electrically charged particles interact? Later it will be shown that, when electrified particles are in motion, additional forces called magnetic forces arise. The result is that the total force on each of the electrified particles is no longer in the line joining them. Also, since electromagnetic effects move with the finite speed of light, they take a finite time to travel across the space separating the objects, and each body feels the effect of the other only after some delay. The total momentum of the system is not constant. It is as if two players were throwing an invisible ball back and forth between them. During the time the ball is in the air, the total momentum of the two players is not conserved, because some of it is concealed in the motion of the unseen ball. In electromagnetism simplicity is preserved in this way by recognizing that some of the momentum of the system is stored in space in what is called the electromagnetic field of the bodies. In this instance the sum of the mechanical and electromagnetic momentum is constant.

Newton's laws of motion represent one of the greatest syntheses in the history of human thought. Their validity is unchallenged except at extreme conditions in which bodies move at speeds close to that of light, in which case the theory of relativity introduces a necessary modification. The classical relation $F = ma$ implies that the acceleration of a particle depends only on two factors—the applied force and the mass. There is no indication, for example, that the instantaneous speed of the particle is involved. However, relativity points out that there is an effect connected with the speed of the body relative to the frame of reference of the observer. It was mentioned earlier (page 49) that the mass of a body is not strictly constant but is a function of its speed. As a result, the factor m in the second law increases with the speed of motion, particularly at speeds approaching that of light, and a constant force can no longer be said to produce a constant acceleration. For such cases—for instance, the motion of high-speed protons in a cyclotron—the equation $F = ma$ must be modified. This will be discussed more fully later.

Programmed review

Instructions: See page 18.

1.   Make a statement of Newton's third law.

Whenever one object exerts a force on a second object, the second exerts an equal and opposite force on the first.   [7.1]

2.   Define the linear momentum of a particle.

A vector given by the product of the mass and velocity: $\mathbf{p} = m\mathbf{v}$.   [7.2]

3.   State Newton's second law in momentum form.

The instantaneous force is equal to the instantaneous rate of change of momentum.   [7.2]

4.   Define the impulse of a force.

A vector given by the product of the force and the interval of time during which it acts.   [7.4]

5.   What is the relation of impulse to momentum?

The change in momentum of a body is equal to the impulse of the applied force: $\mathbf{F}\,\Delta t = \Delta\mathbf{p}$.   [7.4]

6. How can the impulse of a variable force be obtained from the force-time curve?

The impulse is equal to the area under the curve. [7.4]

7. What is meant by the principle of the conservation of linear momentum?

In any isolated mechanical system, the total vector linear momentum remains constant. [7.5]

8. What determines the location of the center of mass of a system of particles?

It is the point where the entire mass of the system can be considered concentrated and where all the external forces can be considered to act. [7.6]

9. What is true of the total linear momentum of all the particles of an isolated system, figured with reference to the center of mass?

It is zero. [7.6]

10. Why must $F = ma$ be modified for bodies moving at high speed?

Because the theory of relativity shows that the mass of a body changes with its speed. [7.7]

For further reading

7.1. *Newton, I.* *"Principia Mathematica"* (Motte's translation). Read the first 18 pages.
7.2. *Magie, F. W.* "A Source Book in Physics." Excerpts from the *"Principia"* on pp. 30–46.
7.3. *Taylor, L. W.* "Physics, the Pioneer Science." Read chap. 11 on action and reaction.
7.4. *Holton, G.,* and *D. H. D. Roller* "Foundations of Modern Physical Science." Read pp. 81–87 on the third law and chap. 17 on conservation of momentum.
7.5. *Millikan, R. A., D. Roller,* and *E. C. Watson* "Mechanics, Molecular Physics, Heat and Sound." Chapter 5 deals with impact phenomena.

Questions and problems

7.1. Show that Newton's first law may be rephrased in terms of momentum in this way: If a body of constant mass is subject to no external force, its linear vector momentum remains constant.

7.2. What is the rate of change of momentum of a ball of mass m at the instant it reaches the top of its path? Does the presence of air resistance affect the answer?

7.3. A cowboy and a steer pull on a rope in opposite directions, neither one moving. The tension in the rope is 200 newtons. What is the magnitude of the total force acting on (*a*) the cowboy; (*b*) the steer?

(c) What force acts on the rope at each end? (d) Which of the above forces form action-reaction pairs?

7.4.   What interactions occur when a cannon fixed in the earth fires a shell into the side of a cliff some distance away? Describe in detail.

7.5.   A 70-kg man is pushing a 300-kg hand truck along the floor of a warehouse with a constant acceleration of 0.50 m/sec². There is a frictional drag of 400 newtons acting on the truck. (a) What forward force is exerted on the man by the floor, through friction on his shoes? (b) How hard does the man push on the truck? (c) How hard does the truck push on the man? (d) Two of the above forces are equal in magnitude. Is this because they are the only two forces acting on a body in equilibrium, or for some other reason? Explain.

7.6.   A horse hitched to a wagon starts it into motion. Is it true that during this process the rearward force exerted by the wagon on the horse is exactly equal to the forward pull of the horse on the wagon? If so, how does the horse manage to set the wagon into motion? It is assumed that the horse has adequate traction with the ground. Compare your statement with the one on p. 142 of Ref. 7.3.

7.7.   An untethered astronaut finds that he has drifted several meters away from his earth-orbiting space station. Attempting to use his rocket pistol to propel him back, he finds that it has run out of fuel. If the pistol is expendable, how can he use it to get him back to the space station?

7.8.   A pendulum mounted on a glider (Fig. 7.16) is pulled to one side while the glider is held stationary and then both are released simultaneously. Describe the resultant motion. HINT: The momentum of the pendulum changes as it swings from one side to the other. As a result, what other momentum changes occur?

7.9.   On a calm day, is it possible to propel a sailboat by means of a large motor-driven fan mounted on the boat? Give reasons for your

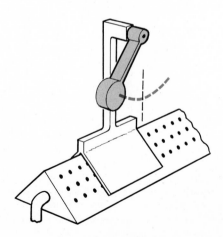

FIGURE 7.16 Pendulum mounted on a glider.

answer. What happens if the fan is pointed toward the stern of the boat instead of toward the sail?

7.10. A 200-g steel ball moving northward with a speed of 40 cm/sec collides (not necessarily head on) with an identical ball that is at rest. What is the amount of the total momentum of the pair after the collision? What is the direction of the total momentum vector?

7.11. Water streams horizontally with a speed of 10 m/sec from a fire hose at the rate of 220 kg/sec. The water strikes the side of a house and flows down to the ground without rebounding. What force is exerted on the wall? HINT: The wall effectively changes the forward momentum of the water. The rate of this change can be computed from the given information.

7.12. A shell is acted upon by an average force of 2.1×10^5 newtons while being fired from a cannon. If the shell spends 7.0×10^{-3} sec inside the gun barrel, how fast is it going when it emerges? The mass of the shell is 1.8 kg.

7.13. When a tennis ball, mass 60 g, moving at a speed of 8 m/sec is struck by a racket, it leaves in the opposite direction with a speed of 12 m/sec. Compute the impulse applied to the ball by the racket.

7.14. A golfer makes a chip shot, the ball taking off at an angle of elevation of $45°$ and landing 10 m away on level ground. If the mass of the ball is 36 g, calculate the impulse given the ball by the club. HINT: Use the range formula given in Prob. 5.30 (page 114) to compute the initial speed of the ball.

7.15. The center of mass of a uniform rectangular strip of matter of constant thickness is at its geometric center. Show how to find the center of mass of a triangular sheet of metal of constant thickness by dividing it, in imagination, into a large number of narrow strips parallel to one side. How must the centers of mass of all these strips lie? If the same thing is done by taking strips parallel to one of the other sides of the triangle, where must the center of mass of the triangle be located?

7.16. Two particles of mass 1 and 3 kg, respectively, are connected by a spring of negligible mass. The spring is stretched until the particles are 80 cm apart. (a) Locate the center of mass of the system, referred to the heavier object. (b) The particles are released. Find the separation of the pair after the heavier object has moved a distance of 10 cm from its rest position.

7.17. A lumberjack of mass 75 kg stands at one end of a uniform log of mass 600 kg that floats at rest in a lake. The man then walks along the log to the other end, a distance of 4.0 m away, and stops. If fluid resistance is assumed to be negligible, (a) how fast is the log moving after the man stops? (b) How great is the total displacement of the log in the process? HINT: What happens to the center of mass of the system consisting of man and log?

7.18.   Which has greater momentum, a 100-g bullet moving with a speed of 800 m/sec or a 500-kg catboat drifting with a speed of 12 cm/sec? What would happen if the bullet embedded itself in the boat's mast when the two were moving directly toward each other at the above speeds?

7.19.   A hovercraft of total mass 480 kg is drifting over level ground toward the east at 0.50 m/sec when a stone of mass 2.0 kg is thrown out sideways from it. A man on the ground sees the stone moving due north over the ground with a speed of 50 m/sec just after it is projected. Find the speed and direction of motion of the hovercraft at that time, assuming that all external horizontal forces that act on it are negligible.

7.20.   When a steel ball collides centrally with an identical ball originally at rest, the first one is observed to stop dead. Prove that the struck ball goes forward with the same speed as the first ball had initially; that is, one ball appears to have passed through the other.

7.21.   The nucleus of a radioactive atom, initially at rest, simultaneously emits an electron with momentum 1.0×10^{-22} kg-m/sec and a neutrino whose momentum is 6.0×10^{-23} kg-m/sec, the two particles moving in roughly opposite directions. Compute the speed of the residual nucleus if its mass is 3.7×10^{-25} kg.

7.22.   A 20-g ball moving to the right with a speed of 60 cm/sec collides centrally with a 50-g ball at rest. After the impact the two balls rebound, and the heavier one is observed to be moving to the right with a speed of 20 cm/sec. Find the velocity (direction and speed) of the lighter ball after the collision.

7.23.   A proton, mass 1.67×10^{-27} kg, having a speed of 1.20×10^8 m/sec collides centrally with a deuteron that is originally stationary. After the collision the proton moves in the reverse direction at 8.00×10^7 m/sec and the deuteron moves forward with a speed of 1.00×10^8 m/sec. Compute the mass of a deuteron.

7.24.   A 30-g bullet moving horizontally at 400 m/sec strikes and embeds itself in a 2,970-g wood block that is free to slide on a rough floor. The frictional force is constant and amounts to 6 newtons. How far does the block slide before coming to rest?

Chapter 8

ORBITAL MOTION AND GRAVITATION

The work begun by Galileo and brought to a successful conclusion by Newton in the form of his three laws of motion furnished a basis for dealing with mechanical problems of the most varied and general kinds. Newton was able to take a tremendous step further by showing that the range of his laws could be extended from everyday mechanical systems of ordinary size to the mechanics of the solar system itself—to **astromechanics.** He was able to do this through a law *supplementary* to the laws of motion, one of unlimited applicability in the realm to which it refers.

This chapter outlines the circumstances of Newton's discovery of his law of universal gravitation and reveals some of its many consequences. Finally, an even more accurate formulation given by the theory of relativity will be described. As a necessary preliminary to an understanding of these developments, we next study the dynamics of a particle moving in a curved path.

8.1 MOTION WITH CONSTANT SPEED IN A CIRCLE

Let a particle be moving at constant speed v around a circle of radius R (Fig. 8.1). When the particle is at P, its velocity vector is **v**, which is drawn tangent to the circle at that point. At a slightly later instant the particle has moved to P' and its velocity vector is **v'**. The two vectors have the same *length* (magnitude), for the particle is assumed to move along the circle at constant speed; however, these two vectors have slightly different *directions,* and this means that in going from P to P' the *velocity* of the particle has changed.

To compute this change, **v** and **v'** have been drawn from a common point in Fig. 8.1*b*. We want to find the *difference* of these two vectors, which is represented by the vector **Δv** drawn as shown. It becomes apparent that **Δv** is the correct difference if the diagram in Fig. 8.1*b* is viewed as a vector polygon for which $\mathbf{v} + \Delta\mathbf{v} = \mathbf{v'}$. Transposing yields $\mathbf{v'} - \mathbf{v} = \Delta\mathbf{v}$, so that **Δv** represents the change in the velocity vector, as stated above.

Computing the change in velocity.

The next step is to compute the magnitude of **Δv**. By geometry, the angle

FIGURE 8.1 Particle moving with constant speed in a circle. The velocity vector is continuously changing its direction, although its magnitude (the speed) stays constant.

(*a*)

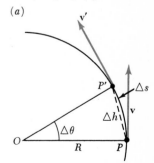

(*b*)

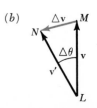

between the two radii in Fig. 8.1*a*, designated by $\Delta\theta$, is the same as the angle between **v** and **v'** in Fig. 8.1*b*, since **v** is perpendicular to one of these radii and **v'** is perpendicular to the other. As a result, the two equilateral triangles OPP' and LMN are similar, and we can write a proportion between the lengths of corresponding sides:

$$\frac{\Delta v}{v} = \frac{\Delta h}{R} \qquad \text{from which} \qquad \Delta v = \frac{v\,\Delta h}{R} \qquad\qquad 8.1$$

Suppose that the particle moves along the circle from P to P' in a time Δt. Dividing both sides of Eq. 8.1 by this time interval,

$$\frac{\Delta v}{\Delta t} = \frac{v}{R}\frac{\Delta h}{\Delta t} \qquad\qquad 8.2$$

Now let this time interval approach zero. Then P and P' move closer and closer together, θ becomes very small, and in the limit the chord Δh joining P and P' becomes indistinguishable from Δs, the arc of the circle between these two points. Thus we can replace Δh in Eq. 8.2 by Δs:

$$\lim_{\Delta t \to 0}\frac{\Delta v}{\Delta t} = \frac{v}{R}\lim_{\Delta t \to 0}\frac{\Delta s}{\Delta t}$$

Passing to the limit, the left side becomes the instantaneous acceleration a_c of the moving particle (page 97), and the second factor on the right becomes the instantaneous speed v (page 88). The equation then reads

$$a_c = \frac{v^2}{R} \qquad\qquad 8.3$$

The word centripetal, meaning "toward the center," was introduced by Newton.

Equation 8.3 asserts that, while traveling around the circle at constant speed, the particle is subject to an *acceleration* of *constant* magnitude a_c. What is the direction of this acceleration? Since acceleration is the change in vector velocity divided by the time interval, which is a scalar, the acceleration **a**$_c$ must have the same direction as Δ**v**. But in approaching the limit, as Δt becomes very small, Δ**v** becomes more and more nearly perpendicular to **v**, as Fig. 8.1*b* shows. Since **v** is tangent to the circular path, **a** must be directed *toward the center of the circle* (Fig. 8.2). It is called the **centripetal acceleration** of the particle. The formula was deduced by Huygens in 1673, using much the same reasoning as that given here.

It may seem paradoxical that a particle moving at *constant speed* in a circle nevertheless has an acceleration. This is another example of the fact that velocity and acceleration are different and independent physical quantities. The velocity vector of the particle moving uniformly around the circle does not change its magnitude, but it does change its direction at a constant rate, and this means that the particle experiences an acceleration of constant magnitude. The direction of the acceleration vector is, at any instant, perpendicular to the path and so has no component along the path (Fig. 8.2). Therefore the *speed* of the particle is unchanged by the centripetal acceleration. Although the particle may, through other causes, have a variable speed along its path, the centripetal acceleration at any instant is always calculated by using the instantaneous speed.

It may be said that the function of the centripetal acceleration is only

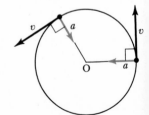

FIGURE 8.2   The acceleration vector is always directed toward the center and so has no component in the direction of motion.

to change the *direction* of motion of the particle, so that it rounds the circle instead of following a straight path (along the tangent) at constant speed, as it would if it were subject to no acceleration. Although the particle remains at a constant distance from the center of rotation, it is always "falling" toward the center in the sense of veering inward from the momentary tangent direction.

A body moving in a curved path other than a circle is nevertheless subject to a centripetal acceleration. In such case, however, the magnitude as well as the direction of this acceleration may vary from point to point on the path. At any place, a small segment of the curve approximates a circle of a certain size. This is called the *circle of curvature* at that point, and its radius is the radius of curvature (Fig. 8.3). Thus Eq. 8.3 is valid for the instantaneous magnitude of the centripetal acceleration, with R representing the radius of curvature at the point in question. Note that the center of curvature always lies on the perpendicular to the tangent and is on the concave side of the curve.

The word *trajectory* was used in Chap. 7 in connection with the path of a projectile. A trajectory is any path described by a particle under the action of given forces. Later in this chapter, the more specific word *orbit* will be used. An orbit is a path described by a particle under a *gravitational* force.

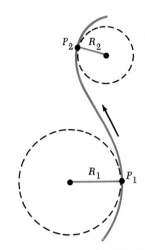

FIGURE 8.3 There is a definite radius of curvature at each point of the path.

Practice set*

1. A particle is moving uniformly around a circle. If both the speed and the radius are doubled, how does the centripetal acceleration compare with its former value?
2. A particle at rest on a circular track begins to move with uniformly increasing speed. After 2 sec, what will be the ratio of its centripetal acceleration to the value after 1 sec?
3. Verify the dimensional consistency of Eq. 8.3.

Worked example 8.1

Calculate the magnitude of the centripetal acceleration of a body at the equator resulting from the earth's rotation.

Solution A point on the equator describes a circle of circumference $2\pi R$ in a time T, where R is the radius of the earth and T is its period of rotation relative to the stars. Then the linear speed will be given by $2\pi R/T$ and, by Eq. 8.3, the centripetal acceleration will be

$$a_c = \frac{(2\pi R/T)^2}{R} = \frac{4\pi^2 R}{T^2} \qquad 8.4$$

Substituting the numerical values $R = 6.4 \times 10^6$ m and $T = 8.6 \times 10^4$ sec, we get $a_c = 0.034$ m/sec², or about $1/300$ of g. The values for locations at higher latitudes are even smaller, and the effect of the earth's motion in its orbit is only about one ten-millionth of g. These results show why, for most purposes, the surface of the earth can be considered an inertial frame.

* *Ans.:* Doubled; 4.

8.2   CENTRIPETAL FORCE

Since a particle moving in a circle is subject to a centripetal acceleration, there must also be a corresponding **centripetal force,** given by $\mathbf{F} = m\mathbf{a}$. If the mass of the particle is m, the magnitude of the centripetal force is

$$F_c = \frac{mv^2}{R}$$

8.5

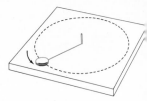

According to the second law, this force vector must have the same direction as the associated acceleration; the force is directed inward along the radius of curvature. In the mks system, F_c is expressed in newtons, like any other force. It should be noted that centripetal force is not a new *kind* of force but merely one that is used for the purpose of constraining a particle to move in a curved path. Centripetal force may be exerted by means of springs, cords, gravitational attraction, or other means.

Figure 8.4 shows a glider moving uniformly in a circle or an air table. The glider is tied to a fixed peg by a string, and the tension in the string provides the centripetal force necessary to keep the object in its circular path. If the fixed point of the string is held in the hand instead of tied to the peg, a steady, outward-pulling force is felt. This is sometimes referred to as a **centrifugal force.** The use of this term implies that there is a force that tends to pull the glider away from the center of the circle. Strictly speaking, there is no such applied force. The object, because of its inertia, is merely exhibiting its natural tendency to move in a straight line, as prescribed by Newton's first law. The centrifugal force is the reaction to the inward-directed centripetal force, which in this example is applied to the glider by the string. Thus the centripetal force is exerted by the string on the glider, whereas the centrifugal reaction is the force exerted by the glider on the string. Like any pair of action-reaction forces, they are not applied to the same object. If the string suddenly breaks, both forces immediately become zero, and the body reverts to straight-line uniform motion along the direction of the tangent to the circle at that instant.

The whole question of centripetal and centrifugal force often causes serious misunderstandings, and it is worthwhile to discuss this matter in more detail.

The concept of inertia (Chap. 6) describes what is observed to happen when, for example, a car makes a sudden start. A passenger in the car finds himself pressed back in his seat, although nothing appears to exert a force in the rearward direction. We readily recognize, however, that what the passenger experiences is the momentary absence of a force that would give him the forward acceleration of the car. In just the same way, centrifugal force is merely the reaction to whirling an object in a curved path—that is, giving it a centrally directed acceleration.

Such situations are clarified by recognizing that neither the accelerating car nor the reference frame turning with the revolving body is an inertial system, and that the use of noninertial frames is responsible for bringing in these fictitious inertial or centrifugal forces. Only one type of reference frame should be used at a time, and usually the simplest is an inertial frame. In making calculations about motion in a curved path, it is best to avoid the term centrifugal force altogether.

The word centrifugal means "fleeing from the center."

Centrifugal reaction.

When an automobile rounds a curve on a level road, the required centripetal force must be furnished by friction between the tires and the pavement. The forces acting on the car are shown in Fig. 8.5a. The car is *not* in equilibrium. As in the case of the glider described above, the weight is counterbalanced by the supporting force exerted by the underlying surface. This leaves F_c as the net acting force. If friction is unable to supply a force as large as mv^2/R, the car skids off the road.

Banking of road curves.

In order to avoid dependence on friction for the needed centripetal force, modern high-speed roads are designed with *banked* curves, as shown in Fig. 8.5b. The external forces applied to the car are its weight **W** and the reaction of the road, **N.** The latter is drawn perpendicular to the road surface, since we are assuming complete absence of friction. Because the pavement is banked at an angle θ with the horizontal, **N** is inclined to the vertical by the same angle. The force **N** may be resolved into a horizontal and a vertical component, as indicated. The vertical component must be equal and opposite to the weight **W** since there is no acceleration of the car in the vertical direction. There remains the horizontal component of **N,** and this furnishes the centripetal force needed to keep the car moving in the curved path.

It is a straightforward matter to compute how much the road must be banked. From the equilibrium of *vertical* forces,

$$N \cos \theta = mg \quad \text{or} \quad N = \frac{mg}{\cos \theta} \qquad 8.6$$

Further, the horizontal component of **N** must be the centripetal force, or

$$N \sin \theta = \frac{mv^2}{R} \qquad 8.7$$

Substituting the value of N from Eq. 8.6 into 8.7, the result is

$$\frac{mg}{\cos \theta} \sin \theta = \frac{mv^2}{R} \quad \text{or} \quad \tan \theta = \frac{v^2}{Rg} \qquad 8.8$$

Note that the mass of the car has canceled out. The required angle of banking does not depend on the mass of the vehicle but it does depend on the speed, and so the road can be banked correctly only for cars

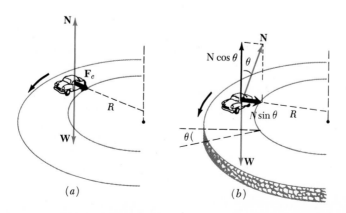

(a)

(b)

FIGURE 8.5 Car rounding a curve (a) on a level road, with friction furnishing the needed centripetal force, and (b) on a suitably banked track, with the inward component of the supporting force of the roadway supplying the centripetal force.

traveling at a particular speed. In practice, a reasonable range of speeds can be safely accommodated by making some use of road friction or restraining rails.

The problem of an airplane making a turn in a horizontal circle is exactly the same as the above, with the aerodynamic lift force on the wing replacing the supporting force of the roadway. Another equivalent example is that of the *conical pendulum* (Fig. 8.6), in which a particle hanging from a fixed point by a string is set into uniform motion in a horizontal circle.

Interesting mechanical effects arise when a body describes a curved path in a vertical plane. A high-speed aircraft can fly an "outside loop" in such a way that objects inside the airplane lose contact with the floor and hover in the air, a condition popularly described as "weightlessness." This term is not an accurate one. The *weight* of a body is the force mg of the earth's gravitational attraction for it, and despite what happens to the body this force continues to act on it as long as it remains in the general neighborhood of the earth. Thus "weightless" is intended to imply *completely unsupported.* A high diver is very nearly "weightless" while in flight, but a parachute jumper is not. "Weightless" is a word whose use is best avoided.

In the case of the conical pendulum, what supplies the centripetal force? See Prob. 8.9.

"Weightlessness."

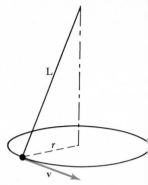

FIGURE 8.6 The conical pendulum corresponds mechanically to a car or an airplane making a banked turn.

Worked example 8.2

To what extent is a passenger "weightless" (unsupported) if he is in an elevator descending with an acceleration of 4.2 m/sec²?

FIGURE 8.7 "Superweightlessness": The elevator has a downward acceleration greater than g

Solution If the mass of the passenger is m, the two forces acting on him are his weight $m\mathbf{g}$ downward and the supporting force of the floor $\mathbf{S}$ upward. Equating the net force acting on the passenger to the product of his mass and acceleration,

$$mg - S = ma \quad \text{or} \quad S = mg - ma = mg\left(1 - \frac{a}{g}\right)$$

Substituting the numbers, this becomes

$$S = mg\left(1 - \frac{4.2}{9.8}\right) = \frac{4}{7}\,mg$$

which indicates that the supporting force of the elevator floor amounts to four-sevenths of the passenger's normal weight.

Worked example 8.3 When an airplane executes an outside loop, what must the minimum flying speed be in order to leave interior objects unsupported when the airplane rounds the top of a loop whose radius of curvature at that point is 1,500 m?

Solution An object resting on the floor of the airplane has two forces applied to it, $\mathbf{N}$ and $m\mathbf{g}$ (Fig. 8.8). The resultant of these forces must be the centripetal force needed to make the object describe its curved path. Then $F_c = mg - N$, and if N is to become zero this reduces to $F_c = mg$.

$$F_c = mg \quad \text{or} \quad \frac{mv^2}{R} = mg \quad \text{and so} \quad v = \sqrt{gR}$$

Substituting numerical values, $v = \sqrt{(9.8)(1,500)}$, or $v = 120$ m/sec. If the flying speed is less than this, objects will remain in contact with the supporting surface. If it is greater, they will drift upward until stopped by the ceiling. The critical condition has been maintained for as much as a minute in actual U.S. Air Force test flights (see photograph).

A device called a **centrifuge** is an important application of the ideas discussed in this section. Chemists and biologists make use of this instru-

FIGURE 8.8 Flying a "zero-gravity" path.

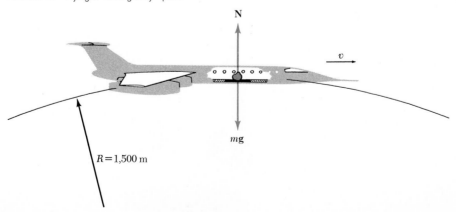

"Weightlessness" in an airplane during "zero-gravity" flight. (*United States Air Force.*)

Cream separators and centrifugal laundry dryers are examples of a centrifuge.

ment to separate two liquids of different densities or to remove suspended solid matter from a liquid. When such mixtures are spun rapidly, the difference between the centripetal force on equal volumes of the two materials causes the denser matter to collect toward the outer edge of the vessel. The separation takes place in much the same way as the settling out of a sediment under gravity but much more rapidly because enormously greater accelerations can be obtained by rotation. By using special experimental arrangements, centripetal accelerations amounting to more than a billion times g have been attained. See Ref. 8.1.

8.3 KEPLER'S LAWS

Section 1.4 (page 7) should be reread at this point.

The development of ideas about the motions of the planets and other objects in the sky was traced very briefly in Chap. 1. Copernicus, by assuming the central body of the planetary system to be the sun rather than the earth, was able to bring about a great simplification of the picture of astronomical motions. Galileo's construction of a telescope enabled him to discover direct visual evidence supporting Copernicus' theory. However, the most convincing confirmation of this idea did not come from observation but from calculations made by Kepler.

Kepler did not arrive at his solution to the problem of the structure of the solar system by mathematical analysis, in which one step follows logically from the previous one by standard methods. Instead, he attempted to find order among the many thousands of observations left by Brahe; and although his procedures were tinged with mysticism, he represented a modern point of view in his conviction that a relatively simple analytical description of planetary motion must exist. In 1609, after years of work, he was able to announce two generalizations about the motion of the planets, and a third one followed nearly 10 years later. Kepler's three laws of planetary motion are:

1. The orbit of each planet is an ellipse, with the sun occupying one focus.
2. A planet moves in its orbit in such a way that the line joining the planet and the sun sweeps out equal areas in equal times.
3. The squares of the periods of orbital revolution of the planets are proportional to the cubes of their average distances from the sun.

Observation shows that, except for some minor deviations whose causes are well known, Kepler's laws give an accurate description of the motions of the planets around the sun. They hold as well for the motions of satellites about the planets and for the mutual revolution of double stars. We shall now examine briefly the content of each of these laws.

Elliptical orbits. The earlier astronomy was characterized by a complicated system of circles turning upon circles. Kepler's first law replaces it with a well-known geometric curve, the ellipse. This type of curve has the shape shown by a circle when it is inclined to the line of sight. The greater the inclination, the greater is the *eccentricity* of the ellipse. An ellipse has two special points called the *foci,* located in symmetric positions on the long axis (Fig. 8.9). According to Kepler's first law, the sun occupies one of these locations in a planetary orbit, the other being vacant.

For the planets of our solar system, the orbits are very nearly circular; the difference between the extreme distances from the sun amounts at most to just under 3 percent.

The law of areas. The second of Kepler's laws is an ingenious way of specifying how the instantaneous speed of a planet varies with its position on the orbit. In Fig. 8.10, the shaded regions represent areas swept over in equal intervals of time by the line connecting the planet and the sun. The second law says that *all these areas are equal.* As a result, the orbital speed of a planet is greatest when it is nearest the sun, and least when farthest away. For the earth's orbit, which is very nearly circular, these two speeds have almost the same value—about 18.8 mi/sec and 18.2 mi/sec.

Kepler's third law. The first law specifies the shape of a planetary orbit and its position with respect to the sun, and the second law describes how each planet moves in its orbit. The function of the third law is to relate the various orbits to each other. An understanding of this law requires that the meaning of two terms appearing in its statement (see above) be made clear. These terms are "period" and "average distance."

The motion of a planet around its orbit is *periodic;* that is, it is repeated

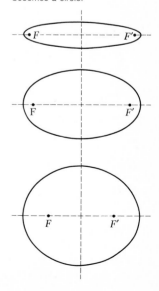

FIGURE 8.9 As the foci of an ellipse are brought closer together, the eccentricity decreases. In the limit, when they coincide, the ellipse becomes a circle.

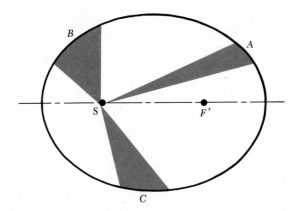

FIGURE 8.10 As a planet moves along its orbit, the line connecting it to the sun sweeps over equal areas in equal times.

at equal intervals of time. The time for one complete trip is called the **period.** For the earth, this time has been named one *year,* equivalent to about 3.16×10^7 sec.

Kepler used the phrase "average distance" in a special way to mean the arithmetical average of the largest and the smallest distances of the planet from the sun. Interpreted in this way, the average distance is equivalent to one-half the long axis of the ellipse.

The third law then states that, for any two planets,

$$\frac{T_1^2}{T_2^2} = \frac{R_1^3}{R_2^3}$$ 8.9

where T represents the period and R the average distance. Thus, if the period and size of orbit are known for one planet, the size of the orbit of another planet can be calculated from its observed period.

Worked example 8.4   The period of Jupiter in its orbit is 11.9 yr. What is its average distance from the sun, expressed in terms of the earth's average distance?

Solution   Transposing Eq. 8.9,

$$\frac{R_J}{R_E} = \left(\frac{T_J}{T_E}\right)^{2/3}$$

where the subscript J refers to Jupiter and E to the earth. Substituting, $R_J/R_E = (11.9/1)^{2/3}$, or $R_J/R_E = 5.20$. Jupiter is more than five times as far from the sun as the earth is.

8.4   THE LAW OF GRAVITATION

Kepler's accomplishment was the discovery of the *kinematics* of the solar system, but it remained for Newton to supply a *dynamical* explanation of the observed motions. Through his great insight he was able to synthesize the content of Kepler's laws and to perceive that these laws could be accounted for by making the assumption that there is a force of attraction between each planet and the sun. Newton was not the first scientist to

William Gilbert (1540–1603) was court physician to Queen Elizabeth I, a philosopher, a strong advocate of the experimental method, and one of the founders of modern physical science.

suggest the presence of such a force. Others had suspected its existence, stimulated perhaps by the observations of Gilbert on electric and magnetic forces—other instances of "action at a distance." However, Newton was able to go much farther by deducing how the magnitude of this gravitational force depends on the circumstances of the situation.

With the help of Kepler's second law (page 177), Newton was able to prove that the force acting on a planet must always be directed toward a *fixed point*. His reasoning follows:

Suppose that a particle subject to no net force is moving at speed v along a line MN (Fig. 8.11). Let A, B, C, etc., be the positions of the particle at equal intervals of time, Δt. According to Newton's first law, these positions will be equally spaced.

Now assume that the particle is not allowed to continue along MN but, as it passes through A, is struck a sharp blow directed toward a fixed point O. The impulse of this blow changes the vector velocity of the particle by a definite amount that is independent of whatever motion the particle had at the time it was struck. Instead of going on to B as it would have if undisturbed, the particle experiences an additional displacement Δr_1 in the time interval Δt and arrives at a new point, B_1.

Left to itself, the particle would now continue along $B_1 N_1$ at a new but constant speed v_1. However, as the particle passes through B_1 it is given another blow, directed again toward O. This sends it abruptly off in a new direction $B_1 N_2$, and it arrives at the point C_2 after a time Δt.

The triangles AOB_1 and B_1OC_1 have equal areas, since the base of each is $v_1 \Delta t$ and the altitude is h. Also, B_1OC_1 and B_1OC_2 have the same area because they have the common base OB_1 and the same altitude k. As a result, the areas of triangles AOB_1 and B_1OC_2 are equal.

The same process can be continued: If, instead of moving along B_1D_2, the particle is given another blow directed toward O when it reaches C_2, then the line joining the particle to O will again sweep over an area of the same magnitude in the next time interval Δt.

Newton next concluded that, if the time interval between blows applied to the particle were made smaller and smaller, it would amount to applying

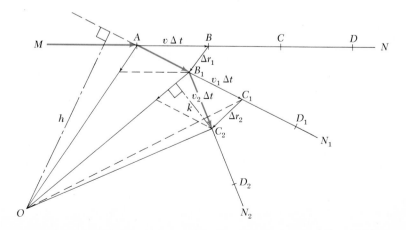

FIGURE 8.11 A particle subjected to a series of centrally directed blows.

a continuous force always directed toward the center O, and the path of the particle would become a continuous curve instead of a succession of straight segments. The line from O to the particle would still sweep over equal areas in equal times in conformity with Kepler's second law. Newton then reversed the chain of reasoning and drew the conclusion that *a particle moving in the way described by Kepler's second law must be acted upon by a force which is always directed toward a fixed center.*

The orbit lies in a plane. The identification of the force governing the motion of a planet as a *central* force leads to an important conclusion about the orbit. In Fig. 8.12, a particle P is moving in a closed orbit under the influence of a central force. Since the force is along PO, the acceleration **a** is also along this line, and so is the change in velocity of the particle. Thus, after any short interval of time the velocity vector will still be in the plane determined by PO and **v.** The location, velocity, and acceleration of the particle remain in the same plane during the motion.

In the solar system, each planetary orbit is observed to be in a plane of its own. The various planes happen to have very nearly the same position in space, with the exception of that of Pluto, which is inclined at nearly 20° to the others. Facts such as these must be taken into account in attempting to develop a theory of the origin of the solar system.

Form of the force law. In order to get more specific information about the nature of the orbit, such as its shape, one must know how the magnitude of the central force depends on the circumstances of the problem. Newton next asked himself the question: Given the fact that the orbit is an ellipse (Kepler's first law), what form must the force law take in order to yield an orbit of this kind? His calculations finally led to the answer that *the force of attraction of the sun for a planet must be inversely proportional to the square of their distance apart.*

To give a general proof of this statement requires advanced methods, but the result is readily derived for the special case of circular orbits. The force acting on a planet in a circular orbit is the centripetal force, given by

$$F = \frac{mv^2}{R} \qquad\qquad 8.10$$

where m is the mass of the planet, v its speed, and R the radius of its orbit. For v we may substitute $2\pi R/T$, where T is the period, i.e., the time for one trip around the orbit. Inserting this value, Eq. 8.10 becomes

$$F = \frac{4\pi^2 mR}{T^2} \qquad\qquad 8.11$$

Now, according to Kepler's third law,

$$T^2 = KR^3 \qquad\qquad 8.12$$

where K is a constant. If this value of T^2 is substituted into Eq. 8.11, the result is

$$F = \frac{4\pi^2 m}{KR^2} \qquad\qquad 8.13$$

which shows that F is inversely proportional to R^2.

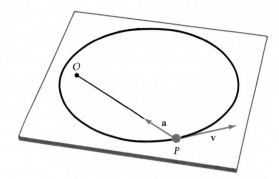

FIGURE 8.12 Under a central force, the particle describes an orbit that lies in a plane.

The value of the constant of proportionality K remains to be determined. The force of mutual attraction of the sun and a planet can be considered to be either one of two equal and opposite (action-reaction) forces: that of the sun on the planet, or of the planet on the sun. Newton concluded that the force law must therefore contain not only the mass m of the planet but also the mass M of the sun, and that these two masses must enter into the force law symmetrically. Accordingly, he proposed that the law of force must be of the form

$$F = \frac{GMm}{R^2} \qquad\qquad 8.14$$

where G, a new constant, always has the same value regardless of which two attracting bodies are involved. This factor is called the **constant of gravitation,** not to be confused with **g,** the acceleration due to gravity at the earth's surface. The general phenomenon of the attraction of one body for another is called *gravitation;* if one of the objects is a planet or other astronomical body and the other an object at or near its surface, the term *gravity* is used.

Equation 8.14 is a statement of Newton's **law of universal gravitation.** In words, it states: Every particle in the universe attracts every other particle with a force that is directly proportional to the product of the masses of the particles and inversely proportional to the square of their distance apart. This statement, by extending the law of gravitation to *any* two particles, represents a bold generalization on Newton's part, and one with which he felt somewhat uneasy pending more direct confirmation. He therefore de-cided to test the law on the problem of the motion of the moon. Knowing the radius of the moon's orbit, r_M, and the period of its motion around the orbit, he computed the centripetal acceleration of the moon $a_M = v^2/r_M$ and compared the result with g, the acceleration of an object at the surface of the earth. The resulting value of a_M was about $1/3600$ of g. It was known that the ratio of the moon's distance to the radius of the earth is about 60, which is $\sqrt{3600}$, so that the acceleration actually does fall off with the square of the distance, as required by the law of gravitation. Thus it appeared justified to conclude that the same force that makes a stone fall to the ground holds the moon in its orbit.

Even with this confirmation at hand, Newton retained some doubts and cautiously withheld publication of his results for more than 20 years. In

Hundreds of thousands of stars, held together by gravitation, make up this globular cluster in the constellation Hercules. (*Photograph from the Mount Wilson and Palomar Observatories.*)

The calculus was conceived independently at about the same time by the German philosopher and mathematician G. W. von Leibnitz (1646–1716).

his calculations he had assumed that the earth and moon attract each other as if the entire mass of each were located at the center. This is by no means evident. The gravitational law refers to the attraction between particles, not extended objects, and in principle it should be necessary to compute the force of attraction between every particle of the earth and every particle of the moon and add all these forces vectorially to get the total force. Newton discovered how to effect this summing up by inventing an entirely new branch of mathematics, the integral calculus. He was finally able to prove that any spherical object whose density variation is the same in all directions from the center will attract an outside body as if the entire mass of the sphere were concentrated at its center.

The moon does not describe its orbit around the center of the earth. The fixed point in the earth-moon system is the *center of mass* (page 158) of the two bodies, which is located almost 2,900 mi from the earth's center (Fig. 8.13). Both the earth and the moon revolve in nearly circular orbits

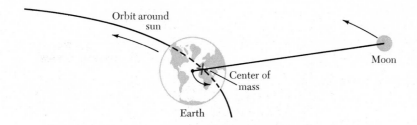

FIGURE 8.13 The point that moves along the earth's orbit in conformity with Kepler's second law is the center of mass of the earth-moon system.

around this point, the earth in a circle of the above radius and the moon in a circle about 81.6 times as big. It is the center of mass of earth and moon that follows the "earth's" orbit around the sun in conformity with Kepler's laws. As a result, the earth itself has a slight wobble as it moves about the sun. The irregularity can be measured, and from it the ratio of the mass of the moon to the mass of the earth can be computed.

Practice set   From the information given in the above paragraph, calculate the distance between the center of the earth and the center of the moon.

8.5 GRAVITATION AND GRAVITY

See Worked example 8.5.   The magnitude of the attractive forces between the sun and the planets is enormous, whereas that between objects of ordinary size is so small that it escapes notice. Newton established the whole concept of gravitation without being able to detect or measure the attraction between bodies in a laboratory.

Cavendish measured G.   It was not until over a century after Newton's announcement of the law that the English scientist Henry Cavendish (1731–1810) succeeded in measuring G by a laboratory experiment. Figure 8.14 is a simplified diagram of his apparatus. A light rod with a small metal ball at each end is hung from a fixed point by means of a thin wire. When two massive lead spheres are brought close to the small balls, the gravitational forces of attraction make the suspended system turn slightly to a new position of equilibrium. The torque with which the suspending wire opposes twisting can be measured in a separate experiment. Then, if the masses of the objects and the

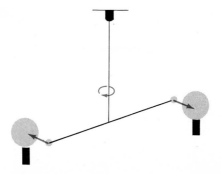

FIGURE 8.14 Schematic diagram of Cavendish's apparatus for measuring G.

distance between their centers are known, the value of G can be computed from Eq. 8.14.

The best present value of the gravitational constant is

$$G = 6.673 \times 10^{-11} \text{ N-m}^2/\text{kg}^2$$

Notice that the units, as stated here, come directly from the form of Eq. 8.14.

With the numerical value of G available, it becomes possible to compute the actual force of attraction in a specific situation.

Worked example 8.5   Calculate the mutual force of gravitational attraction of (a) two bowling balls, each of mass 7.2 kg, when placed with their centers 25 cm apart; (b) the earth and the moon: $m_E = 6.0 \times 10^{24}$ kg, $m_M = 7.3 \times 10^{22}$ kg, and $R_{EM} = 3.8 \times 10^7$ m.

Solution   a.   Substitution into Eq. 8-14, with the above value of G, yields

$$F = \frac{(6.7 \times 10^{-11})(7.2)(7.2)}{(0.25)^2} = 5.5 \times 10^{-8} \text{ newton}$$

or about the weight of a gnat's wing.

b.   Here the substitution yields

$$F = \frac{(6.7 \times 10^{-11})(6.0 \times 10^{24})(7.3 \times 10^{22})}{(3.8 \times 10^7)^2} = 2 \times 10^{22} \text{ newtons}$$

or a force of more than 2 billion billion tons.

Newton applied his law of gravitation to a number of interesting problems. He explained ocean tides as an effect of the gravitational attraction of the moon and the sun, and he showed how small irregularities observed in the motions of the planets could be ascribed to the attractions exerted on them by the other planets. The two outermost planets of the solar system, Neptune and Pluto, were later discovered by tracking down possible sources of the disturbances ("perturbations") in the observed motion of the other planets.

Relation between g and G.   The force of attraction of the earth for a particle of mass m at its surface is given by $F = GM_E m R_E^2$. This force may also be expressed by $F = mg$, the weight of the particle. When these two values are equated, m drops out and the result is

$$g = \frac{GM_E}{R_E^2} \qquad \qquad 8.15$$

a relation between the gravitational constant and the acceleration due to gravity at the earth's surface. For positions above the surface, this relation should be written $g_R = GM_E/R^2$, where R is the distance from the center of the earth. The equation shows that the value of g falls off rapidly with increased altitude. For places at middle latitudes and within a mile or so of the surface, the numerical value of g decreases only about 0.03 percent per kilometer increase in altitude. This is equivalent to saying that a kilogram mass will weigh less by about one milligram of force (10^{-5} newton) when raised from the floor to the ceiling of a room.

TABLE 8.1 VALUE OF g AT VARIOUS LOCALITIES

PLACE	LATITUDE, deg	ELEVATION, m	g, m/sec^2
Equator	0	0	9.78039
Panama Canal	9	5	9.78243
Key West, Fla.	25	1	9.78970
San Francisco, Calif.	38	114	9.79965
Washington, D.C.	39	14	9.80112
Denver, Col.	40	1,638	9.79609
Standard station (mid-latitude, sea level)	45	0	9.80665
S.E. Alaska	55	4	9.81524
Greenland	70	0	9.82534
North Pole	90	0	9.83223

The value of g varies from one point to another on the earth's surface because of a combination of circumstances: (1) Different distances from the axis of rotation lead to different values of the centripetal acceleration; (2) the earth is not perfectly spherical; (3) there are local inhomogeneities in the crust. Measurements made at several locations are given in Table 8.1.

Worked example 8.6 The earth is 81.6 times as massive as the moon, and its radius is 3.67 times as big. Calculate the acceleration, due to its own gravity, at the moon's surface.

Solution   Equation 8.15 gives the value of g at the surface of the earth. Writing a similar equation for the moon and dividing one by the other, we get

$$g_M = g_E \frac{M_M}{M_E} \left(\frac{R_E}{R_M}\right)^2$$

Substituting,

$$g_M = \frac{9.80}{81.6} (3.67)^2 = 1.62 \text{ m/sec}^2$$

or about one-sixth of g_E.

8.6 EARTH SATELLITES

The moon, because of its transverse motion, maintains its distance from the earth. Can an object be given a suitable transverse speed that will enable it, too, to circle the earth? A projectile fired horizontally from a gun located on a mountain top (Fig. 8.15) would follow a curved path and hit the ground some distance away. Increasing the initial speed of the object would move the point of impact farther away. At some critical speed, the object would not hit the ground at all but would circle the earth as a satellite. Still higher initial speeds would result in elliptical orbits and, if the object were launched from the surface with a speed of about 7 mi/sec or more, the orbit would open up at the far end and the body would escape permanently from the earth (page 207).

In 1957 the first artificial earth satellite, *Sputnik 1,* was successfully launched. Astromechanics became an experimental science, splendidly confirming the laws of mechanics on a cosmic scale. A considerable number of satellites are now in orbit, serving as communications relays and as observation stations for collecting information about various aspects of the earth's environment in space.

To place a satellite in orbit, it is carried to a height of several hundred miles by one or more stages of rocket propulsion and then is given transverse motion by the final rocket stage. The velocity thus imparted determines the size and shape of the orbit taken up by the satellite.

A satellite in a circular orbit just above the surface of the earth must have a speed given by equating its centripetal acceleration to the acceleration due to gravity. Thus

$$\frac{v^2}{R_E} = g \qquad \text{or} \qquad v = \sqrt{gR_E} \qquad\qquad 8.16$$

When numerical values are inserted, we get $v = 7,850$ m/sec, or about 17,500 mi/hr. The period of revolution about the earth is given by $T = 2\pi R_E/v$ and is about 5×10^3 sec = 1 hr 23 min. Satellites traveling in larger orbits will have longer periods, as given by Kepler's third law (see Prob. 8.17, page 193).

FIGURE 8.15 Newton's anticipation of artificial earth satellites. When the launching speed is great enough, the projectile does not hit the ground at all but continues around the earth in a complete orbit.

8.7   RELATIVITY AND GRAVITATION

Gravitational and inertial mass.

It has been pointed out that the mass of a body can be determined in two distinct ways. One method consists in weighing it at a standard location, the other in observing its motion when a known force is applied or when the body interacts with a standard mass. It may well be asked whether these two diverse kinds of experiments really measure the same thing. There seems to be no reason to assume that a single quantity could serve to represent both the property of gravitational attraction and that of inertia. Conceivably, an object may have a *gravitational mass* m_G and a different *inertial mass* m_I.

Some familiar observations may throw light on this question. Galileo discovered that all bodies fall at the same rate, regardless of difference in mass; that is, at a given place the acceleration of gravity g is the same for all freely falling bodies. Writing Newton's second law for the case of free fall, we have

$$F = w = m_I g$$

This indicates at least that the *weight of a body is proportional to its inertial mass.*

Whatever meaning may be assigned to the term "gravitational mass," it is taken to be proportional to weight. Hence if both m_I and m_G are proportional to the weight of the body, they must be proportional to each other. If we take the constant of proportionality to be g in both cases, as we have been doing, then m_G and m_I may be considered *numerically* equal.

We conclude that a single quantity m is sufficient for specifying both the inertial and gravitational properties of a material body.

There is ample support for this idea. First, the center of mass (page 158) of an object, determined dynamically, is found to coincide with its **center of gravity,** the point from which the body must be suspended in order to remain in static equilibrium in any position. Further, the theory of the motion of a simple pendulum shows that, at a given locality, the period of swing of a pendulum of fixed length should be directly proportional to the inertial

A GRAVITY-ASSISTED "GRAND TOUR" OF THE OUTER PLANETS

Although unmanned spacecraft have already sent back information on neighboring planets Mars and Venus, the power requirements for direct flights to the outermost planets are not available.

At this time, only a fly-by of Jupiter, using combinations of existing launch vehicles, is feasible. A spacecraft passing near Jupiter would have its speed increased by the gravitational pull of this planet and at the same time would be deflected into a new orbit. If the more distant planets happen to be in proper positions, the spacecraft could be similarly accelerated on passing them, making what is called the "Grand Tour" of the solar system (Fig. 8.16). It could do this with much less expenditure of energy than direct flights would require and in much shorter time. By this "crack-the-whip" method the travel time would be reduced from 6 years to less than 2 years for Saturn, from 16 to 5.7 years for Uranus, and from 31 to 9.1 years for Neptune.

There are plans for a spacecraft to be launched by a moderately powerful chemical rocket vehicle, with additional propulsion provided by a solar-powered ion engine (see Box, page 151). The spacecraft is designed to weigh about $\frac{3}{4}$ ton, with an instrument load weighing 100 lb. Measurements of various kinds of radiation in space, the solar wind, magnetic fields, etc. are to be made, and the results will be transmitted to earth.

The most favorable set of planetary positions for a "Grand Tour" occurs about every 175 years, the nearest one to the present time being November 1977.

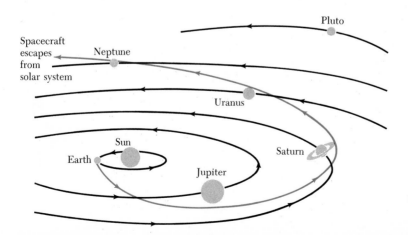

FIGURE 8.16 The "Grand Tour" of the outer planets. Distances are not shown to scale.

mass of the bob and inversely proportional to its gravitational mass. On the assumption that $m_I = m_G$, these quantities cancel out, leaving an expression that does not involve the mass. Very careful pendulum experiments confirm that the period is independent of the mass of the pendulum bob. Such tests were made by Newton and later by others. Recent highly refined experiments have verified the proportionality between m_I and m_G to an accuracy of a few parts in a hundred billion.

The connection between these two characteristics of mass was accepted merely as an experimental fact for two centuries. In 1914, Einstein reasoned that there must be a fundamental basis for this correspondence. He became convinced that *the effects produced by accelerated motion and by gravitational action are not distinguishable from each other.* This **principle of equivalence** became the starting point for the general theory of relativity.

Principle of equivalence.

The principle of equivalence can be demonstrated in the example of an astronaut inside a space vehicle that moves with constant acceleration relative to an inertial frame (Fig. 8.17). The vehicle is far from any massive body such as the earth, yet the astronaut finds that his feet press on the floor of the cabin in a way that seems normal to him. He performs mechanics experiments such as dropping simultaneously two objects of different mass and observing that they hit the floor together. He throws a ball in a direction parallel to the floor and sees it follow a parabolic path until it strikes the floor across the cabin. Accustomed to life on earth, he concludes that the cause of all such observed phenomena is the existence of a massive gravitating body located somewhere beneath his feet. An outside observer, however, sees the two dropped objects simply continuing to move along uniformly in an inertial frame while the accelerated floor of the cabin comes up to meet them. Both descriptions are correct, as the principle of equivalence asserts.

Bending of light near a gravitating body.

In developing the special theory of relativity, Einstein saw that its applicability went beyond mechanics to include optical and electrical phenomena as well. He realized that the principle of equivalence might be valid over a wider range that included the behavior of light. In the example of the space ship, a beam of light sent across the cabin in a direction parallel to the floor might be expected to follow a curved path like that of the thrown ball, and if the astronaut thinks that acceleration phenomena are due to gravity he will say that gravitational force is the cause of the bending.

Turning to a particular case, Einstein reasoned that light rays from distant stars, in passing close to the sun on their way to an observer on earth, should bend slightly because of the sun's gravitation (Fig. 8.18). This effect was soon verified by direct observation: Stars photographed near the edge of the sun's disk at the time of total eclipse were found to have about the computed outward displacement. In 1970, observation of radio signals sent back from space probes traveling beyond the sun confirmed the deflection predicted by relativity to within 2 percent.

Another experimental test of the general theory of relativity.

After several years of work, Einstein was able to formulate a relativity theory of gravitation along the above lines. This theory gives virtually the same results as the classical theory, so that the calculations of astromechanics remain very nearly unchanged. There is another important case where relativity and classical mechanics predict somewhat different results. According to classical theory the orbit of a planet, because of small per-

FIGURE 8.17 The principle of equivalence.

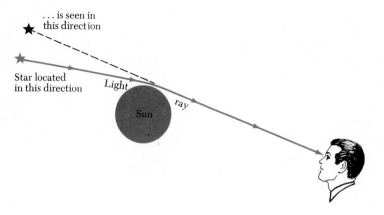

FIGURE 8.18 Light rays are bent as they pass close to a massive body such as the sun. In this schematic diagram the observer does not see the star in its true position but in the direction in which the ray enters his eye. This direction is shown by the dashed line.

turbations due to other planets, turns in its own plane at the extremely small rate of one rotation in about 3 million years. Observation of the planet Mercury shows that the rate is not in agreement with newtonian mechanics but is almost exactly what relativity predicts, so that this observation is considered one of the best experimental verifications of the general theory of relativity at present.

More detailed descriptive discussions will be found in Refs. 8.3 and 8.5.

The classical and relativistic theories of gravitation may be compared in this way: The classical theory invokes the action of a force exerted between bodies, whereas the relativity formulation ascribes such effects to a *change in the properties of space* in the neighborhood of material bodies. The latter assumes that the presence of a massive object produces a curvature or warping of space itself, so that the natural path of a moving body or a light beam is no longer a straight line but is curved (Fig. 8.19).

Theories of gravitation.

It may appear that neither the newtonian nor the relativistic theory offers a satisfactory explanation of the nature of gravitation. The former assumes the existence of a **field** of force in the neighborhood of gravitating bodies whereas the latter relies on the idea of an equally mysterious curvature of space. Can both these descriptions be valid?

In Newton's time, the concept that bodies completely isolated from each other should be able to interact was puzzling and distasteful. However, this same idea of "action at a distance" arose later in the description of electric and magnetic effects. In this way it gained acceptance through increasing

FIGURE 8.19 Visualizing the warping of space produced by the presence of a massive body. A marble rolled across a tightly stretched rubber membrane follows a straight path, but if a heavy object distorts the rubber sheet the marble will be deflected on passing by.

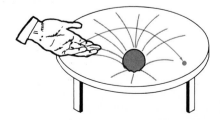

familiarity. It must be remembered that the relativity theory of gravitation is more accurate and complete than the newtonian, as in its correct prediction of the motion of the orbit of Mercury and the gravitational bending of light rays.

Nevertheless, neither theory can offer any *inherent* explanation of gravitation, for the question of the ultimate nature of any phenomenon is completely outside the scope of science. Newton himself expressed this view in a famous passage:

But hitherto I have not been able to discover the cause of those properties of gravity from phenomena, and I feign no hypotheses. For whatever is not deduced from the phenomena is to be called an hypothesis; and hypotheses, whether metaphysical or physical, whether of occult qualities or mechanical, have no place in experimental philosophy.

And again:

To us it is enough that gravity does really exist and act according to the laws which we have explained, and abundantly serves to account for all the motions of the celestial bodies, and of our sea.

Newton must have realized that such pronouncements sounded too pessimistic and that even though scientists must accept the unattainability of absolute truth, they must continue to press their efforts at explanation as far as seems justified. From this point of view, Einstein's theory of gravitation must be regarded as an improvement over Newton's.

Programmed review

Instructions: See page 18.

1. Explain why an object moving uniformly along a curved path has an acceleration.

    There is an acceleration because the *direction* of the velocity vector is changing.   [8.1]

2. What is meant by centripetal force, and what is the formula for computing its magnitude?

    The inward-directed force needed to make an object follow a curved path. $F_c = mv^2/R$.   [8.2]

3. What is the relation of *centrifugal* to *centripetal* force?

    One may be considered the reaction to the other, but the primary force applied to a moving body is the centripetal one.   [8.2]

4. Why is "weightlessness" an objectionable term?

    Because it falsely implies that the body in question has no weight—that is, attraction for the earth. A "weightless" object is one that is unsupported, or in free fall.   [8.2]

5. Which of Kepler's laws gives a kinematic description of how a particular planet describes its orbit?

The second law (law of areas).   [8.3]

6. Which of Kepler's laws expresses a relation between the orbits of the various planets?

The third law relates the periods to the average distances.   [8.3]

7. State Newton's law of universal gravitation in algebraic form. Identify each symbol.

$F = GMm/R^2$, where F is the attractive force, G the constant of gravitation, M, m the masses of the two bodies, and R the distance between them.
                                                                  [8.4]

8. Distinguish between gravitation and gravity.

Gravitation is the name for the general phenomenon of attraction between bodies; *gravity* is the manifestation of this force on an object near a planet, star, etc.   [8.4]

9. What is the relation between the gravitational and inertial mass of a given object?

They are proportional to each other and, when measured in suitable units, identical in value.   [8.7]

10. How is the answer to Question 9 connected with the general theory of relativity?

This principle of equivalence is the basis for the Einstein theory of gravitation (general relativity).   [8.7]

For further reading

8.1. *Beams, J. W.* Ultrahigh-speed Rotation, *Sci. Am.,* April, 1961, p. 134.
8.2. *Taylor, L. W.* "Physics, the Pioneer Science." Read chaps. 12 and 13 on the development of Kepler's laws and gravitation.
8.3. *Gamow, G.* "Gravity." An interesting paperback book treating the classical and modern ideas of the nature of gravitation.
8.4. *Whipple, F. L.,* and *J. A. Hynek* Observations of Satellite I, *Sci. Am.,* December, 1957, p. 37. How the orbit of Sputnik was determined and what information this gave astronomers.
8.5. *March, A.,* and *Ira M. Freeman* "The New World of Physics." Read pp. 94–110 on the relativity theory of gravitation.

Questions and problems

8.1. In rounding a sharp curve, a passenger in a car traveling on a level road finds that he is being pressed against the side of his seat. Is the force exerted on his body by the armrest a centripetal or a centrifugal force?

8.2. By what factor would the earth's rotational speed have to increase in order to make objects at the equator require no support from the ground?

8.3. At what angle with the horizontal must a turn on a bobsled run be banked if a sled moving at 30 m/sec is to round a turn of radius 11 m without depending on friction?

8.4. A freshly opened bottle of soda water is grasped by the neck and swung briskly in a vertical circle. Do the gas bubbles collect near the neck or near the bottom of the bottle? Explain.

8.5. An instrument package is released from an artificial earth satellite by detaching it from the outer wall of the satellite. Will the object plunge to earth? Assume that air resistance is negligible.

8.6. A boy rides a roller coaster. With what force does he press against the seat when the car goes over a crest whose radius of curvature is 8.0 m at a speed of 4.0 m/sec? Express the result as a fraction of his weight.

8.7. A plane is flying along a parabolic path at the speed with which a projectile would follow that path, maintaining a "zero gravity" condition for some time so that objects on board are "weightless." If, in a certain flight of this kind, the radius of the path at its highest point is 1,000 m, what is the plane's speed at this point?

8.8. A stone tied to a string is whirled in a vertical circle of radius 0.75 m at a speed of 3.0 m/sec. (a) Where is the tension in the string greatest; least? (b) If the difference between these two extreme values amounts to 98 newtons, what is the mass of the stone?

8.9. For the conical pendulum (Fig. 8.6), find an expression for the linear speed with which the ball must travel in order to follow a circle of radius r, the pendulum length being L. Hint: Sketch a diagram showing the forces acting on the ball, and relate the situation to that of a car traveling on a banked turn (page 173).

8.10. Two ice skaters of mass 50 and 70 kg, respectively, hold hands and whirl about in circles at the rate of one turn per second. If their mass centers are effectively 1.2 m apart, how great is the tension force in the arm of either skater? Hint: First decide where the axis of rotation is located with respect to the center of mass of the pair; then determine the value of the centripetal force on either skater.

8.11. It was stated in a newspaper account that a certain artificial earth satellite was "free from the pull of the earth's gravity" when in orbit. Comment on the accuracy of this statement.

8.12. A storekeeper in a mountain village uses a spring scale that was calibrated at sea level with standard weights. To whose advantage is this—his or the customers'?

8.13. Calculate the average rate at which the line joining the sun to the earth is sweeping out area, in mi^2/sec. The average linear speed in the orbit is 18.5 mi/sec, and the average radius of the orbit is 9.3×10^7 mi.

8.14.   The planet Pluto is, on an average, 39.5 times as far from the sun as the earth is. What is Pluto's orbital period, in years?

8.15.   The orbits of comets that are permanent members of the solar system are extremely elongated ellipses so that a comet near the distant end of its orbit is too remote to be visible. Halley's comet made its most recent close passage to the sun in 1910 and is expected to return in 1986. It passed 8.9×10^7 km from the sun. Using Kepler's third law and the definition of "average distance" given on page 178, compute the greatest distance from the sun reached by this comet. Compare this with the radius of the orbit of Neptune, 4.5×10^9 km.

8.16.   By equating the centripetal force to the gravitational attraction, show that the speed of a satellite in a circular orbit about a planet is inversely proportional to the square root of the radius of the orbit.

8.17.   What must be the radius of the circular orbit of an earth satellite that will hover over a given point on the equator, that is, will have the same rate of turning as the earth? Also, express the height of the satellite above the ground, in miles. HINT: Use Kepler's third law in the form given in Prob. 8.22, replacing GM by $gR_E{}^2$ as in Eq. 8.15. For numerical data use $R_E = 6.4 \times 10^6$ m and take the time of rotation of the earth with respect to the stars to be 8.62×10^4 sec.

8.18.   Show that G has the dimensions L^3/T^2M.

8.19.   How would the greatest distance you could throw a ball on the moon compare with the greatest distance you can throw it on earth?

8.20.   At what altitude above the earth's surface would the numerical value of g be half that at the surface?

8.21.   Using Eq. 8.15, compute the mass of the earth. Take $g = 9.8$ m/sec^2 and $R_E = 6.4 \times 10^6$ m.

8.22.   By equating the forces given by Eqs. 8.13 and 8.14, show that Kepler's third law (Eq. 8.12) can be written

$$T^2 = \frac{4\pi^2}{GM} R^3$$

This provides a means of finding the mass M of a central body if it has a satellite or planet whose period and distance can be observed.

8.23.   Use the relation given in Prob. 8.22 to compute the mass of the sun. Use the following data: Average distance of the earth from the sun is 1.5×10^8 km, number of seconds in a year $= 3.2 \times 10^7$, and $G = 6.7 \times 10^{-11}$ mks units.

8.24.   Compute the average density of the earth, assuming it to be a sphere of radius 6.4×10^6 m and mass 6.0×10^{24} kg. The result is roughly twice the average density of the materials in the outer crust. What does this indicate about the presence of denser materials in the interior? Newton obtained a first approximation to a value for G by assuming the average density of the whole earth to be about five times that of water.

8.25. From the observed angular size and distance of the sun, its radius is computed to be 7.0×10^5 km. Using this figure and the value of the sun's mass found in Prob. 8.23, calculate the value of the acceleration due to gravity at the surface of the sun and compare it with the value of g on earth.

8.26. Two stones of equal mass rest on the pans of an equal-arm balance in an elevator. Does the scale remain balanced when the elevator accelerates upward? Explain.

Chapter 9
MECHANICAL ENERGY

9.1 THE NATURE OF ENERGY

The energy concept was introduced in a preliminary way (page 50) as an agent capable of causing changes in matter. Energy occurs in a variety of forms: mechanical, thermal, electromagnetic, radiative, chemical, and others.

Experience shows that a hot iron, although it looks the same as before it was heated, has properties it did not have when cold. A bullet in flight differs in some way from the same bullet at rest, and an electric cable takes on important new attributes after the switch is closed. All such changes involve some variety of energy.

Since energy cannot be handled and does not take up space, it is somehow different in character from matter itself. However, there are processes in which energy and matter are converted, one into the other. For the present, at least, energy may be considered a condition rather than a substance.

Although it is the central unifying idea in all physical science and perhaps in biology as well, the energy concept is of relatively recent origin. Newton's work on the foundations of mechanics had been completed for more than a century before the energy concept was given a name, but some of its aspects were recognized much earlier. In this chapter the nature of the energy connected with purely mechanical processes will be examined; other forms of energy and their interrelations will be taken up later on.

9.2 ENERGY OF A MOVING PARTICLE: KINETIC ENERGY

Think of a particle at rest in an inertial frame located far out in space so that it is subject to no appreciable forces of any kind. If a force is intentionally applied to the particle, motion results, and because of this acquired motion the particle is able to produce certain effects that it could not produce when at rest. For instance, it can deform another body by impact. It may shatter the struck object and set the fragments into motion. It may encounter a spring and compress it, or it may plunge into a fluid and churn it up; there are many ways in which a moving body can show that it possesses energy. Energy connected with the motion of a body is called **kinetic energy.**

In order to obtain a definite measure for kinetic energy, consider a particle of mass m subject to a resultant force **F.** If the particle is already in motion relative to the observer's reference frame with a velocity **v,** then consider the case where **F** is *in the direction of motion.* This force produces an acceleration **a** given by the second law:

$$\mathbf{F} = m\mathbf{a} \qquad\qquad 9.1$$

Suppose the force acts for a short time Δt, during which **F** may be considered constant. Write the above equation in scalar form and multiply both sides by Δs, the distance covered in this time:

$$F\,\Delta s = ma\,\Delta s \qquad\qquad 9.2$$

Since the force may be thought of as constant during the short interval in which it acts, the acceleration will be constant, and the kinematic formulas of Chap. 5 apply. In particular, Eq. 5.10 (page 101) may be written as

$$v^2 = v_0{}^2 + 2a\,\Delta s$$

and can be rearranged to read

$$a\,\Delta s = \frac{v^2 - v_0{}^2}{2} \qquad\qquad 9.3$$

Substituting on the right side of Eq. 9.2,

$$F\,\Delta s = m\,\frac{v^2 - v_0{}^2}{2}$$

or

$$F\,\Delta s = \tfrac{1}{2}mv^2 - \tfrac{1}{2}mv_0{}^2 \qquad\qquad 9.4$$

The quantity $F\,\Delta s$, the magnitude of the force multiplied by the distance moved in the direction of the force, is defined as the quantity of mechanical **work** ΔW done on the particle. The right side of the equation represents the consequent *change* in the kinetic energy of the particle.

Equation 9.4 has the form

$$\Delta W = K_{\text{final}} - K_{\text{initial}} \qquad\qquad 9.5$$

The contrac-tion KE may be used for convenience to designate kinetic energy; in equations, the symbol K is sufficient.

This expression will now be shown to be valid even if the resultant applied force is variable and if the particle is confined to a prescribed path of any form whose direction at any point may not necessarily be the same as that of the force.

There are many instances where a body is not moving in the direction of the applied force, so that the force and the displacement are not in the same direction. For example, a sled is pulled along over a frozen pond by means of a rope. The sled moves in a horizontal direction, while the force applied by the rope slopes upward at some angle θ (Fig. 9.1). To find the work done, the force **F** may be resolved into two components,

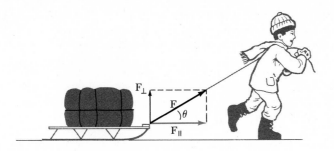

FIGURE 9.1 Work is done only by the component $\mathbf{F}_\parallel$ of the force in the rope because the motion takes place entirely in the direction of $\mathbf{F}_\parallel$.

$\mathbf{F}_\parallel$ along the direction of motion and $\mathbf{F}_\perp$ perpendicular to this direction (Fig. 9.2). The latter component does no work, does not alter the speed of motion, and does not alter the KE of the sled. *Work is done only by* $\mathbf{F}_\parallel$, *the component in the direction of motion.* Since the magnitude of this component is $F \cos \theta$, the work amounts to

$$W = Fs \cos \theta \qquad\qquad\qquad \textbf{9.6}$$

In order to discuss the motion of a particle moving along a prescribed path, think of a bead made to slide along a smooth, curved wire under the action of a variable force (Fig. 9.3). For each short segment of the path the work done is given by an expression of the form of Eq. 9.6, and so Eq. 9.4, written for the successive segments, yields

$$F_1 \cos \theta_1 \, \Delta s_1 = \tfrac{1}{2}mv_1^2 - \tfrac{1}{2}mv_0^2$$
$$F_2 \cos \theta_2 \, \Delta s_2 = \tfrac{1}{2}mv_2^2 - \tfrac{1}{2}mv_1^2$$
$$F_3 \cos \theta_3 \, \Delta s_3 = \tfrac{1}{2}mv_3^2 - \tfrac{1}{2}mv_2^2$$
$$\cdots \qquad \cdots \qquad \cdots \qquad\qquad \textbf{9.7}$$
$$\cdots \qquad \cdots \qquad \cdots$$
$$F_f \cos \theta_f \, \Delta s_f = \tfrac{1}{2}mv_f^2 - \tfrac{1}{2}mv_{f-1}^2 \qquad \text{(for the final segment)}$$

When these equations are added, all but two of the terms on the right cancel in pairs, as indicated by the colored strokes, and the result can be written

$$\Delta W = F_1 \cos \theta_1 \, \Delta s_1 + F_2 \cos \theta_2 \, \Delta s_2 + \cdots = \tfrac{1}{2}mv^2 - \tfrac{1}{2}mv_0^2 \qquad \textbf{9.8}$$

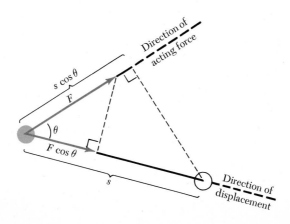

FIGURE 9.2 The work done can be written as $(F \cos \theta)\ s$ or as $F\ (s \cos \theta)$, that is, the product of either F or s and the projection of the other upon its direction line.

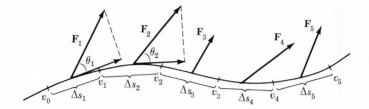

FIGURE 9.3 Work done along a curved path. The force may be considered constant in direction and magnitude over each short interval, and so the work done on such an interval is $\Delta W = F \Delta s \cos \theta$.

where the subscript f is now omitted from the final speed v. Equation 9.8 is the general form of Eq. 9.4 and shows that a relation of this kind is valid even when the force varies in amount and direction.

The theory of relativity shows that it is necessary to modify the form of the right side of Eq. 9.8. The difference is important only at very high speeds.

9.3 MECHANICAL WORK AND ENERGY TRANSFER

It might seem that in attempting to find an expression for the newly defined quantity KE we have succeeded only in bringing in the equally unfamiliar concept of mechanical work. However, the idea of work is simple, natural, and widely useful. The following facts about this quantity should be noted:

1. Work, as defined in physics, is done only when the acting force succeeds in *moving* the point to which it is applied. Here again, a common word is used for scientific purposes in a limited, special sense. In this meaning of *work,* no force, however large, does any work unless it produces a displacement. A workman holding up one end of a heavy plank while it is being fitted into place would consider himself to be working, but in the mechanical sense no work is accomplished. It is true, however, that he is working (expending energy) merely to keep his muscles tensed.

2. Work may be thought of as mechanical energy *exchanged* between bodies. Work is always done *by* something *on* some other thing. For example, in setting an object into motion, the acting force may be said to do work *on* the object or impart (kinetic) energy to the object. This work is done *by* the agency that exerts the force.

3. Notice that work and energy are *scalar* quantities; no element of direction is involved in their specification.

Practice set*

1. How much work is done by a force of magnitude F applied in a direction perpendicular to the line of motion of a particle of mass m when it moves a distance x?

2. What change, if any, takes place in the linear momentum of a particle of mass m moving with a speed v when its direction of motion is reversed?

3. What change, if any, takes place in the KE of the particle? HINT: Replace v by $-v$.

*Ans.: none; $2mv$; none.

9.4   UNITS; NUMERICAL EXAMPLES

The unit is named for the British physicist J. P. Joule (see marginal note on page 229).

According to its definition, work has the dimensions of force times distance. By an equation such as 9.5, KE must have the same dimensions. The mks unit of work (or energy) is the *newton-meter,* which is called the **joule** (pronounced "jool"), whose symbol is J. In the cgs system, the basic energy unit is the *erg,* which is 10 million times smaller than a joule:

$$10^7 \text{ ergs} = 1 \text{ joule} \qquad\qquad 9.9$$

In the English system, the unit commonly used is the foot-pound, of the same order of size as the joule:

$$1 \text{ ft-lb} = 1.356 \text{ joules} \qquad\qquad 9.10$$

Worked example 9.1

A block of mass 30 kg is dragged along a rough floor at constant speed by means of a rope that slopes upward at an angle of 30°. If the tension in the rope is 50 newtons, how much work is done in moving the block a distance of 20 m along the floor?

Solution $W = Fs \cos \theta = (50)(20)(0.87) = 870$ joules. The mass of the block does not enter into the problem. Note that work is done only by the horizontal component of the force—the component in the line of motion—as long as the front end of the block does not lift up under the action of the force.

Worked example 9.2

It has been proposed to use the radiation pressure of sunlight to propel a space ship equipped with a large "sail." If solar radiation pressure at the earth's distance amounts to about 4×10^{-6} newton on each square meter, what speed could be acquired by a ship of mass 2,000 kg and sail area 400 m² in moving from rest at the earth's orbit directly away from the sun a distance of 4×10^5 km (about a quarter of a million miles)? Disregard gravitational effects on the ship caused by other bodies.

Solution   The total propulsive force will be $F = (4 \times 10^{-6})(400)$ newton. Substituting into Eq. 1.4,

$$(4 \times 10^{-6})(400)(4 \times 10^8) = \tfrac{1}{2}(2,000)v^2$$

leading to $v = 25$ m/sec, just under 60 mi/hr.

9.5   POTENTIAL ENERGY

Work done on a body does not always appear in the form of increased kinetic energy. Consider a weight hanging freely from a rope. If the weight is hauled upward at constant speed, the tension in the rope is equal to the value of the weight. When the object is pulled up through a finite distance, work is done on it. This work has given the body a kind of latent energy, for it now in turn has an ability to do work: In descending again to its former level, it can be made to raise another weight, or it can stretch a spring in the process or, if allowed to fall freely, it can demolish something that it hits.

In the process of raising the hanging weight, KE need not be considered. The weight may have been resting on the floor, and the hauling operation could have been carried out very slowly, leaving the object at rest again at the higher level. The entire energy imparted to the body is energy of position. Energy of position is called **potential energy.**

"Potential energy" in general may be referred to by the contraction PE; in equations, the symbol P is sufficient.

There are other forms of energy of position besides the PE of a raised weight. When a watch is wound, work is done in moving the particles of the spring slightly farther apart against the elastic forces of attraction acting between them. The wound-up spring thus has a type of energy of (relative) position: **elastic PE.** A stick of dynamite has **chemical PE,** and so has a lump of coal.

Other forms of PE, such as electric, magnetic, atomic, etc., will be taken up later in this book.

9.6 GRAVITATIONAL POTENTIAL ENERGY

The particular kind of PE associated with gravitational forces is called **gravitational potential energy.** It may be referred to conveniently as GPE. In equations, the symbol P_g is used.

In many cases of practical interest, one of the gravitating bodies is the earth. The GPE of a mass in the neighborhood of the earth must always be specified with respect to some reference level. The location of this level is entirely a matter of choice, since only *differences* of GPE are of interest in practical situations. A suitable reference level may be the floor, sea level, or even the center of the earth, which is obviously the lowest point from which a weight can be raised or to which it can descend. Once a zero level of GPE is selected, it must be retained through the entire problem.

GPE of a raised weight.

The increase in GPE of a weight that is raised a moderate distance is readily computed. The force needed to support the weight is mg, where m is the mass of the body. To pull it up through a vertical distance Δh requires an amount of work given by Eq. 9.6 as

$$P_g = mg\,\Delta h \qquad\qquad 9.11$$

since $\cos\theta = 1$ in this case. In the mks system, P_g is expressed in joules, m in kilograms, g in meters per second per second, and h in meters. Notice that the equation in this form assumes that h is small compared with the radius of the earth, so that g may be considered constant over the distance h.

Work done is independent of the path.

Equation 9.11 is valid even if the mass is not raised straight upward but is carried to the higher level by any other path, as long as the beginning and the end of the path are a *vertical distance h* apart. This can be shown as follows: In Fig. 9.4, the curved line represents a smooth slope along which the particle of mass m can be moved from A to a higher point B. The force needed to slide the body in the uphill direction over any short segment Δs of the slope, without acceleration, is equal to the component along the path of the weight of the object. Its magnitude is $F_\parallel = mg\sin\theta$. The work done along this segment Δs is then given by $\Delta W = (mg\sin\theta)\,\Delta s$. But, from Fig. 9.4, $\Delta s\sin\theta = \Delta h$, and so $\Delta W = mg\,\Delta h$, which is the same

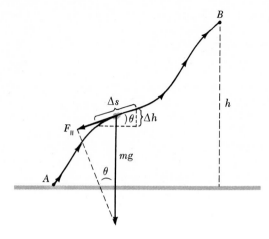

FIGURE 9.4 The work done is the same as if the weight had been raised vertically between the levels of the end points and is independent of the path.

as the work done (GPE gained) in pulling the object straight up a distance Δh. This is true for any short segment of the path so that, when the effects are added up, it is valid for the entire path AB.

The work done against gravity on an extended object may be calculated from Eq. 9.11 by taking Δh to be the vertical distance through which the center of gravity of the body (page 187) is raised.

Worked example 9.3   A uniform brick of mass 2.25 kg and dimensions 5.70 by 9.54 by 20.4 cm rests with its largest face on a level table. Find how much work must be done to stand it on end.

Solution   The center of gravity is located at the geometric center of the block. Originally it is $\frac{1}{2}(5.70)$ cm above the table, and in the final position $\frac{1}{2}(20.4)$ cm, making $\Delta h = 7.35$ cm, or 0.0735 m. Then $\Delta P_g = mg\,\Delta h = (2.25)(9.80)(0.0735) = 1.62$ joules. The result is independent of the particular way the brick was upended.

The energy resides in the field of force of the system.   The discussion above has referred to the work done *on the raised object,* or to the increase in GPE *of this object.* More precisely, the reference should be to the work done in *separating* the earth and the object, or to the GPE stored in the *system* composed of these two bodies. The work is done against the attractive force between the two, and the resulting PE does not reside exclusively in one or the other but is shared by both. In the case of the raised weight, only a matter of circumstance makes it possible to ascribe the GPE to the weight alone because the earth is so massive compared with the raised object that the amount of the earth's response is entirely negligible (Fig. 9.5).

9.7 TRANSFORMATIONS OF MECHANICAL ENERGY

In the course of most mechanical processes, neither the PE's nor the KE's remain constant, and even a rough check on the values of these quantities will suggest that the changes are related in some way.

Consider a stone of mass m that has been raised slowly to a height h above its original position on the ground. In the process, an amount of work mgh was done. For simplicity, take the ground as a reference level. Then mgh represents the GPE of the object in its elevated position.

Now let the stone fall freely. On the assumption of no resisting forces, the entire force acting on the stone is its weight mg. Under the action of this force, the stone acquires KE, just as in the case discussed in Sec. 9.2. Meanwhile, its height above the ground—and consequently its GPE—is continuously getting smaller while its speed—and consequently its KE—is getting larger. It may be said that as the stone falls, its initial GPE is being transformed into KE.

By the time the stone is about to hit the ground it will have lost all its initial GPE, but it will then have a certain amount of KE. How great will this KE be? From kinematics, the speed just before striking the ground is given by $v^2 = 2gh$. Multiplying both sides of this relation by $\frac{1}{2}m$ gives

$$\tfrac{1}{2}mv^2 = mgh \qquad\qquad 9.12$$

The left side of this equation is the KE acquired in the fall; the right side is the GPE lost during the process. It then seems reasonable to say that the initial GPE of the stone has been transformed into an equal amount of KE in the course of falling to the lower level.

Follow the process a step further. Assume that the stone strikes a spring (Fig. 9.6) on its way down. It will do work in squeezing the coils of the spring together. When the object reaches its lowest position it will momentarily be at rest and its KE will be zero. At the same instant, the spring will have its maximum compression and will have stored within it a certain amount of elastic PE.

The work done in compressing the spring by this amount can be measured in a separate experiment. It is found that, no matter what the mass and speed of the striking body or whether it compresses a stiff spring a short distance or a weaker spring a greater distance, the work done in the compression is equal to the maximum KE acquired by the striking body

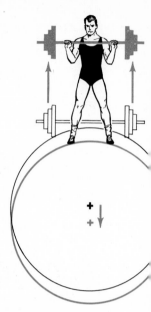

FIGURE 9.5 A weight lifter on a small planet. The work he does is stored up in the form of mutual GPE of the system.

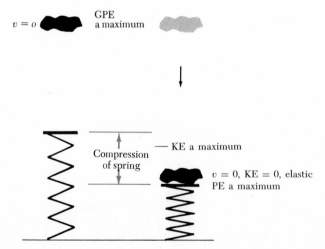

$v = 0$ GPE a maximum

Compression of spring

— KE a maximum

$v = 0$, KE $= 0$, elastic PE a maximum

FIGURE 9.6 The initial GPE of the stone is converted into KE during the fall and finally into elastic PE of the compressed spring. A detailed analysis shows that the KE reaches a maximum value some distance below the point where the stone first makes contact with the spring.

A transformation of mechanical energy: GPE into KE. (*State of New Hampshire; photo by Dick Smith.*)

plus the additional decrease in GPE as the spring compresses. This KE, in turn, is equal to the initial GPE of the stone in its highest position. Thus there has been a **transformation** of the GPE of the raised weight into KE of the moving object and then into elastic PE of the spring.

This is not the end of the process. After it brings the stone to rest, the compressed spring will again flex itself, once more imparting KE to the stone by projecting it upward. In an ideal case, the stone would return to the level from which it originally dropped, and the sequence of events would repeat itself indefinitely. The reason this does not happen in an actual situation will be brought out below.

A **simple pendulum** is a particle hanging from a fixed point by a perfectly flexible thread of negligible mass. It furnishes a good example of mechanical energy conversion. If the particle is pulled aside, keeping the thread taut, work must be done. This work is stored in the form of GPE, since the displacement has raised the particle to a higher level. If the pendulum is now released from rest, it swings back toward its initial position. As it does so, the particle descends, acquiring KE at the expense of its initial GPE until, on passing through the lowest point of its swing, its mechanical energy is all in kinetic form.

The pendulum bob does not stop dead at the bottom of its arc. Its inertia carries it through this point and it now rises, losing KE and gaining GPE until, in the ideal case, it reaches a height on the far side just equal to its height before release. The process then repeats, time after time. Figure

The lowest level attained by the particle is taken as the reference (zero) level for GPE.

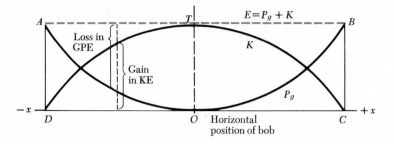

FIGURE 9.7 As the pendulum bob swings down its arc, the loss in GPE equals the gain in KE, so that the total mechanical energy remains constant.

9.7 shows graphs of the energy changes as the pendulum swings from one extreme end of its arc to the other. For any displacement of the pendulum, the decrease in GPE is just equal to the gain in KE, or the other way around. The dashed line represents the total mechanical energy of the pendulum bob, which is constant.

9.8 DISSIPATION OF MECHANICAL ENERGY; FRICTION

In practice, a weight falling onto a spring as discussed in Sec. 9.7 actually behaves somewhat differently from the way described above. Instead of regaining its original height exactly, the weight reaches a slightly lower position after each rebound until it finally comes to rest. The same is true of a real pendulum as contrasted with the ideal model discussed in the preceding section. The swings become shorter and shorter and eventually stop altogether. Similarly, the height of rebound of a dropped ball is a little short of the previous rebound, until the ball is finally at rest on the floor. What becomes of the energy initially given to such systems?

It is impossible to build a working mechanism that does not involve frictional or other resistive forces of some kind. Such forces are said to **dissipate** the mechanical energy of the system. This means that the system performs work against the resistive forces. Whenever material bodies collide, mechanical energy is dissipated by work done against *internal* friction, the friction between parts of the bodies.

In any case where work is done against dissipative forces, the work is not converted into PE or KE but into another form called heat energy, to be described later. When a block of wood is thrown out along the floor and finally comes to rest, it is observed to be slightly warmer than before. Similarly, although harder to detect, heat develops in the spring hit by the falling weight and in the surrounding air that was churned up by the moving object. In Chap. 11 it will be shown that a general statement of great importance and simplicity can be made about the *total* energy of any isolated system: It remains constant.

Friction, as well as any other dissipative forces that affect the motion of a body as a whole, always acts in a direction *opposite* to that of the motion (Fig. 9.8). A ball thrown straight upward in the air is acted upon by a *downward* resistive force on the way up and by an *upward* resistive force while falling back. Work is done against dissipative forces on all parts of the round trip, and the ball returns to the hand of the thrower with less energy (smaller speed) than it had when thrown. There is a net loss of

FIGURE 9.8 Frictional forces arising from the motion of a body are always in the direction opposite to the motion.

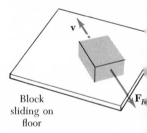

Block sliding on floor

Drop falling through air

mechanical energy on the round trip. Unlike work done against gravity, the work done against friction on the way up is not recoverable on the way down.

9.9 CONSERVATION OF MECHANICAL ENERGY

An isolated system in which no appreciable dissipative forces exist is called a **conservative system.** When such a system goes through any sequence of changes that brings it back to its original state, no net amount of work is done. This may be expressed as follows: *In any conservative system, the sum of the PE and KE remains constant.* This statement is known as the principle of the **conservation of mechanical energy.** It follows from the laws of mechanics and the definitions of PE and KE. A similar statement can be made about the conservation of linear momentum (page 154) and for several other conservation principles to be discussed later.

Conservation of mechanical energy implies that, in a conservative system, any change in one form of mechanical energy is accompanied by an equal and opposite change in another form. Thus, if P_1, K_1 and P_2, K_2 are the PE and KE of such a system at two different instants, the assertion is that

$$P_2 - P_1 = -(K_2 - K_1)$$

where the minus sign before the parentheses indicates the oppositeness of the two changes. Transposing, this equation becomes

$$P_1 + K_1 = P_2 + K_2$$

which expresses the constancy (conservation) of mechanical energy as stated above.

The principle of the conservation of mechanical energy is very useful because it provides the solution of a great variety of mechanics problems without dealing with accelerations, intermediate forces, speeds, etc. It relates initial and final states without requiring any knowledge of intermediate states.

In a sense the scientist himself is a black box to most people. See Ref. 9.5.

A system whose detailed construction is not revealed is called a "black box" in the vernacular of technology. A black box usually has *input* and *output* connections linking it to its surroundings. Adding machines, television sets, and diesel engines are black boxes to all but experts. The casual user knows what these devices do, but he may not understand their internal construction or operation. This does not prevent people from learning by experience how to use them.

Figure 9.9 represents a black box containing a concealed mechanism. Two cords come from holes in the bottom. Inside the box there is assumed to be some arrangement of direct-acting mechanical elements involving levers, pulleys, etc., but no such things as springs, compressible gases, or flywheels, where energy could be stored. Friction is assumed to be negligible.

It is not necessary to convert the distances to meters since a conversion factor would cancel throughout.

Suppose that when cord A is pulled down 10 cm, a 25-cm length of cord B is pulled up into the box. The question is: What pull must be applied to cord A to lift slowly a 2.0-kg mass attached to B? According to the energy principle, the work done by the pull on A must equal the work done

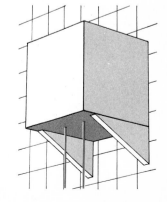

FIGURE 9.9   A "black box," or "input-output" mechanism.

by B on the suspended mass. Since all motions are to be performed slowly, no appreciable amount of KE is involved, and the amount of work done by B will be simply the gain in GPE of the hanging mass. If F is the applied force in newtons, $F \times 10 = 2.0 \times 9.8 \times 25$, or $F = 49$ newtons. The required force has been found, in spite of the fact that no details are known concerning the mechanism inside the box.

In many instances the use of the energy concept makes it possible to bypass Newton's second law, because the energy conservation principle really represents an overall application of this law. The energy method is especially direct in situations involving the *distance* moved.

Worked example 9.4   A 10,000-kg rocket takes off vertically, propelled by a constant thrust of 118,000 newtons. What will be the speed of the rocket when it reaches an altitude of 900 m? Neglect air resistance and any decrease of mass caused by consumption of fuel.

1. Solution by Newton's second law   The resultant upward force is the thrust of the motors minus the weight of the rocket, and this must be equated to the mass of the rocket times its upward acceleration: $118{,}000 - (10{,}000)(9.8) = 10{,}000a$, from which $a = 2.0$ m/sec². Then, from kinematics, the speed attained may be computed by using $v^2 = 2ax$:

$$v^2 = (2)(2.0)(900) \qquad \text{or} \qquad v = 60 \text{ m/sec}$$

2. Solution by the energy principle   The work done by the resultant upward force must equal the sum of the GPE and KE given to the rocket, or

$$Fh = mgh + \tfrac{1}{2}mv^2$$

Substituting numbers, $(118{,}000)(900) = (10{,}000)(9.8)(900) + \tfrac{1}{2}(10{,}000)v^2$. After canceling common factors and reducing, this again yields $v = 60$ m/sec.

9.10 THE WORK-ENERGY PRINCIPLE

The principle of the conservation of mechanical energy applies rigorously only to a conservative system. This is an ideal that can only be approximated, but it is possible to generalize the principle so that it is valid even when dissipative forces are present, provided only that these forces themselves or the total work done against them can be measured. The general statement then reads

> (*Total mechanical energy of a system at any time*)
> $=$ (*total mechanical energy at a later time*)
> $+$ (*work done against dissipative forces in the interval*)

This may be called the **work-energy principle.** A numerical example will show how to apply it.

Worked example 9.5   A boy coasting on a sled comes to a hill that rises 1 m for each 5 m measured along the slope (see Fig. 9.10). The combined mass of boy and sled is 30 kg, and the constant frictional drag of the runners amounts to

FIGURE 9.10   Coasting uphill.

8.0 newtons. If the sled has a speed of 10 m/sec just before encountering the hill, how far up the slope will it go before coming to rest?

Solution   Consider the sled (*a*) when it is about to start up the slope and (*b*) when it reaches its highest point. If, for convenience, the ground is chosen as the reference level for GPE, the sled will have only KE at the bottom. At the top point, when it is momentarily at rest, it will have only GPE. Then the statement of the work-energy principle for this case becomes

Initial KE = (final GPE) + (work done against friction along the way)

If the distance of ascent along the slope is designated s, then the vertical height of rise of the sled is $h = s/5$. Substituting numbers,

$$\tfrac{1}{2}(30)(10)^2 = (30)(9.8)\,\frac{s}{5} + (8.0)s \qquad \text{from which} \qquad s = 22.5 \text{ m}$$

To solve the problem without the use of the work-energy principle, it would first be necessary to compute the downhill component of the weight of the sled and add the force of friction to get the total force opposing the motion of the sled. Then Newton's second law would have to be used to calculate the (negative) acceleration that the total force would impart to the given mass. Finally, the kinematic relation $v^2 = 2as$ would be needed to find the distance traveled along the slope. The relative simplicity and directness of the work-energy idea are apparent.

9.11   ESCAPE FROM THE EARTH

In Chap. 8 it was shown that a particle, given the proper initial velocity, can be launched into a closed orbit around the earth or some other gravitating body. The work-energy principle shows something further: By giving the particle sufficient velocity it can be made to escape from the central body altogether. The critical speed, called the **speed of escape,** would carry the particle to an infinite distance and leave it there with no appreciable remaining motion.

The increase in GPE of a particle when raised to a higher level is given by Eq. 9.11 (page 200): $\Delta P_g = mg\,\Delta h$. As already pointed out, this expression is strictly valid only for relatively small increases in height, where g may be considered constant over the whole distance. If h is not small, allowance must be made for the decrease in the gravitational attraction of the earth with increased altitude. By calculus methods, it is found that the GPE of a particle at a distance r from the center of the earth is given by

$$P_g = -\frac{GmM}{r} \qquad\qquad 9.13$$

where m is the mass of the particle and M the mass of the earth. The minus sign appears because, for convenience, P_g is taken to be *zero* at an infinite distance. At a finite distance, it is negative since GPE decreases when the height is reduced.

According to the energy principle, a particle shot away from the earth will experience a decrease in KE just equal to its increase in GPE, air resistance being neglected. This means that, if it starts out from the surface of the earth ($r = R$) with a speed v_0 and reaches an infinite distance with zero speed,

$$\tfrac{1}{2}mv_0{}^2 - 0 = -\frac{GmM}{\infty} - \left(-\frac{GmM}{R}\right) = 0 + \frac{GmM}{R}$$

Cancel m and solve for v_0:

$$v_0 = \sqrt{\frac{2GM}{R}}$$

According to Eq. 8.15 (page 184), the last expression can be written

$$v_0 = \sqrt{2gR} \qquad\qquad 9.14$$

This is just $\sqrt{2}$ times the speed in a sea-level circular orbit (page 186), and so it amounts to 1.41 × 17,500, or about 24,600 mi/hr. The value of v_0 depends only on G, M, and R, and so this is the escape speed for *any* direction of launching from the earth's surface. A *rocket,* as opposed to a projectile, can start with a smaller speed, even zero, because its propulsion continues after the start, whereas a true projectile must get all its KE at the point of launching.

Some of the lighter molecules in the upper atmosphere of a planet attain escape speed because of favorable collisions with other molecules. This accounts in part for the absence of an appreciable atmosphere on certain planets and on the moon.

9.12   ROLES OF KE AND MOMENTUM COMPARED

Is the proper measure of the quantity of motion of a body its KE or its momentum? A brisk scientific controversy arose over this question among prominent scientific men during the latter part of the seventeenth century. It was finally recognized that the question is not a valid one. Any scientific quantity is "proper" if it provides a useful description of phenomena and if simple and comprehensive statements can be made about it. In this respect both KE and momentum are useful, each having its special function in mechanics.

According to Sec. 9.2, the KE stored in a free mass m by the action of a force F that moves it from rest a distance s in the direction of F is measured by the product Fs. The acquired KE is a measure of the effect of the force F expressed directly in terms of the *distance* moved. On the other hand, the momentum p imparted to a body of mass m by a force F acting for a time t is given by $p = Ft$ (page 152). Thus the momentum is a measure of the effect of the force expressed directly as a function of the *time* it acts.

An important difference between mechanical energy and momentum is shown by the conservation principles. Conservation of linear momentum holds for *any* isolated system, regardless of the nature or violence of the interactions of its parts. But conservation of mechanical energy is maintained only for special systems where no dissipative forces are present (conservative systems). A fundamental difference in the nature of the two quantities is that linear momentum is a vector, whereas mechanical energy in any form is a scalar.

Worked example 9.6   A 20-g ball moving to the right with a speed of 60 cm/sec collides centrally with a 50-g ball which is at rest. After impact, the heavier ball is observed to be moving to the right with a speed of 25 cm/sec.

a.   Find the velocity (direction and speed) of the lighter ball after impact.
b.   Is mechanical energy conserved in the process?

Solution   *a.*   The entire process takes place in a single line. If the line is assumed to be horizontal, with the positive direction toward the right, conservation of linear momentum yields

$$(20)(60) = (50)(25) + 20v$$

where v is the velocity of the lighter ball after impact. Solving, $v = -2.5$ cm/sec, the minus sign indicating that this ball is moving to the *left* after impact.

b.   The total KE before impact is $K_1 = \frac{1}{2}(0.020)(0.60)^2 = 3.6 \times 10^{-3}$ joule; the total KE afterward is $K_2 = \frac{1}{2}(0.050)(0.25)^2 + \frac{1}{2}(0.020)(0.025)^2 = 1.57 \times 10^{-3}$ joule, which is considerably less. There has been a fractional loss of $(3.6 - 1.57)/3.6$, or about 56 percent of the initial KE. The "lost" KE has been converted largely into heat, with perhaps a small amount changed to energy of sound waves produced in the impact.

9.13 POWER

The work-energy principle deals only with the *total amount* of work done on a system or the *total energy* stored in it. In many actual situations another factor—the time element—is of interest, and the *rate* of doing work is sometimes of more concern than the total work done.

In mechanics, the term **power** is defined as the time rate of doing work. If a quantity of work ΔW is done in a time Δt, the instantaneous power is given by

$$L = \lim_{\Delta t \to 0} \frac{\Delta W}{\Delta t} \qquad \qquad \text{9.15}$$

Power may be measured in any units that express work divided by time. The mks unit is the **watt,** equal to a rate of working of one joule per second. A larger unit, useful in rating practical machines (engines, motors, or other power sources), is the **kilowatt** (kW), equal to 1,000 watts. Engineers in English-speaking countries use the **horsepower** (hp) as a practical unit. According to an estimate originally made by James Watt himself, 1 hp represents a rate of working equal to 550 ft-lb/sec. One horsepower is equivalent to 0.746 kW.

Worked example 9.7 Each second, 3.0 m³ of water passes over the brink of a waterfall 8.0 m high, starting essentially from rest. If the water is sent through a turbine, how much useful mechanical power is obtainable, assuming that 20 percent is dissipated in the machine?

Solution   Referred to the bottom of the fall, the GPE of 3 m³ of water is (3,000)(9.8)(8) joules. This GPE of the water at the top of the fall is converted into KE as it descends, and 80 percent of this KE is delivered as useful work. Since the above amount of work is done each second, the available power is given by

$$L = (0.8)(3,000)(9.8)(8) = 1.9 \times 10^5 \text{ watts or 190 kW}$$

The power relation 9.15 may be written in another form, useful in many situations. Suppose a force F is exerted on a body while the body is displaced a distance Δs in the direction of F. Then we may substitute $F \, \Delta s$ for ΔW in Eq. 9.15, which yields

$$L = F\left(\lim_{\Delta t \to 0} \frac{\Delta s}{\Delta t} \right)$$

But, in the limit, the second factor is simply v, the instantaneous speed with which the force F is moving the body acted upon, so that the above expression becomes

$$L = Fv \qquad\qquad\qquad 9.16$$

The instantaneous power is equal to the acting force times the speed with which the body is moving.

Practice set* At what rate is the earth's gravitational attraction doing work (*a*) on an object of mass m, falling freely near the earth's surface, at the instant its speed is v? (*b*) On a satellite moving around the earth in a circular orbit? (*c*) What is the rate, averaged over one revolution, for a satellite moving in an elliptical orbit?

Worked example 9.8 A spider of mass 1×10^{-5} kg is climbing straight up his web at a speed of 6.0 cm/sec.
  a.   How much KE does he have?
  b.   What power is he expending while climbing at this rate?

Solution   *a.* $KE = \frac{1}{2}mv^2 = \frac{1}{2}(1 \times 10^{-5})(0.06)^2 = 1.8 \times 10^{-8}$ joule
      b. $L = Fv = (1 \times 10^{-5})(9.8)(0.06) = 6 \times 10^{-6}$ watt

Notice that the two questions are completely independent of each other. The first deals with the *work* done initially by the spider in setting himself into motion; the second refers to his steady *rate of working* after that. If a complete graph were made of the power expended, right from the start, it might resemble Fig. 9.11. After the initial stage, the value of L remains constant at 6×10^{-6} watt (6 µwatts).

Ans.: mgv; 0; 0.

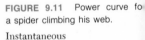
FIGURE 9.11   Power curve fo a spider climbing his web.

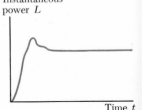
Instantaneous power L

Time t

Programmed review

Instructions: See page 18.

1. Define mechanical work, and give an expression for its amount.

    Work is the force component in the direction of motion multiplied by the distance moved. $W = Fs \cos \theta$.   [9.2]

2. Define kinetic energy, and give the expression for the KE of a moving particle.

    KE is the energy of motion of a body. $K = \frac{1}{2}mv^2$.   [9.2]

3. What are the mks and English units of work or energy? How is the mks unit defined?

    The joule and the foot-pound. One joule is the work done by a force of 1 newton when it moves through a distance of 1 m.   [9.4]

4. Describe, in words, the meaning of potential energy.

    PE is energy possessed by a system because of the position or configuration of its parts.   [9.5]

5. What is meant by the gravitational potential energy of a stone located near the surface of the earth?

    The mutual PE of the stone and the earth, resulting from the gravitational attraction between them.   [9.6]

6. What is meant by dissipation of mechanical energy?

    Conversion into irrecoverable forms such as heat through friction or other resistive forces.   [9.8]

7. What is conservation of mechanical energy?

    In a system where dissipative forces can be neglected, the sum of the PE and KE remains constant.   [9.9]

8. State the work-energy principle.

    The total mechanical energy of a system at any time equals the total mechanical energy at a later time plus any work done against dissipative forces in the interval.   [9.10]

9. Define mechanical power.

    Power is the rate of doing work. $L = \Delta W / \Delta t$.   [9.13]

10. How is the mks unit of power specified?

    A rate of working of 1 joule/sec is called 1 watt. A larger unit, 1 kilowatt, is 1,000 watts.   [9.13]

For further reading

9.1.   *Taylor, L. W.* ''Physics, the Pioneer Science.'' Read chap. 17.
9.2.   *Holton, G., and D. H. D. Roller* ''Foundations of Modern Physical Science.'' Read chap. 18.
9.3.   *Ruhemann, M.* ''Power.'' Chaps. 1–4.
9.4.   *Swann, W. F. G., and I. M. Freeman* ''Physics.'' Chap. V.
9.5.   *Snow, C. P.* ''The Two Cultures; and a Second Look.'' The scientist as a ''black box.''

Questions and problems

9.1.   State how much work is done by the weight of a block that rests on a level table when the block is (*a*) moved from one place to another; (*b*) pulled along the table with constant acceleration.

9.2.   A boy pulls a sled across a frozen pond at constant speed, using a force of 20 lb. The rope makes an angle of 30° with the ground. How much work is done in pulling the sled 600 ft?

9.3.   A slender, uniform pole 12 ft long and weighing 10 lb lies on the ground. A man picks up the pole and stands it in a vertical position with its lower end resting on a fence 3.0 ft high. How much work, in foot-pounds, was done in the process?

9.4.   Show that a 600-kg sports car moving 3.0 m/sec has more momentum but less KE than a 0.50-kg projectile shot with a speed of 300 m/sec.

9.5.   A steel ball is dropped onto a steel plate and rebounds with essentially the same speed it had before impact. Has there been any change in (*a*) its momentum and (*b*) its KE? Explain.

9.6.   Using the expression $K = \frac{1}{2}mv^2$, show that, in the mks system, the units of KE are newton-meters (joules).

9.7.   Design a suitable mechanism for the inner parts of the black box shown in Fig. 9.9.

9.8.   A sled starts from rest and coasts down a frictionless hill. Using the energy principle, find how fast it is going as it passes a point 16 m lower than the starting point, measured vertically. Does the answer depend on the mass of the sled or on the shape of the hill?

9.9.   A small ball is allowed to roll from rest down each of the two slopes *AB* and *AC* in turn (Fig. 9.12). Compare the speeds on passing the level *DE*. Neglect friction.

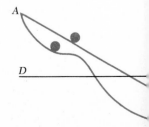

FIGURE 9.12 Ball rolling down different slopes.

9.10.   From energy considerations, show that an ideal projectile has the same speed at any level on the way up as on the way down.

9.11.   Compute the energy loss in the collision shown in Fig. 7.14 (page 162). Will the answer depend on whether the data are taken from Fig. 7.14*a* or from Fig. 7.14*b*? Explain.

9.12.   In the energy graph of a pendulum (Fig. 9.7), show that *AOB* is a circular arc. What is the form of *DTC*?

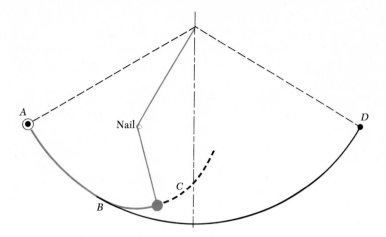

FIGURE 9.13 The swing of the pendulum is interrupted when the cord strikes a nail.

9.13. *Galileo's interrupted pendulum.* The bob of a simple pendulum is released from rest at A (Fig. 9.13). The cord strikes a nail and the bob continues its swing along the arc BC. To what height will it rise before coming to rest? Explain. Set up such a pendulum in front of a wall, mark a horizontal line AD, and try the experiment.

9.14. As the earth goes around the sun on its elliptical orbit, its distance from the sun varies (Fig. 8.10, page 178). The ratio of the greatest to the least distance is 1.035. Using Kepler's second law, show that the earth's KE at A is about 7 percent greater than at B. What is the source of the difference in KE?

9.15. A helper tosses bricks up to a mason a vertical distance of 3.0 m so that they reach him with a speed of 2.0 m/sec. What fraction of the helper's work is wasted?

9.16. A simple pendulum 3.43 m long is pulled back through a 60° angle and released from rest. (*a*) What fraction of its initial GPE, measured from the bottom, remains after it has descended to 30°? (*b*) How fast is the bob then moving?

9.17. An alpha particle is ejected from the nucleus of an atom of radon whose *initial* mass is 55.5 times that of the alpha particle. The nucleus is free to recoil. Comparing the alpha with the remaining nucleus, find the ratio of their (*a*) recoil speeds; (*b*) linear momenta; (*c*) KE's.

9.18. A coal car of total mass 2.0×10^4 kg, coasting at a speed of 2.5 m/sec, collides with a similar car of mass 3.0×10^4 kg moving in the opposite direction at 2.0 m/sec. The cars couple on contact. How much energy is lost in the collision?

9.19. A 20-g ball moving to the right at 40 cm/sec collides centrally with a ball of equal mass that is at rest. Show that, if the collision is perfectly elastic (no dissipation of mechanical energy), the two balls merely interchange velocities: The incoming ball stops dead and the second ball goes to the right with a speed of 40 cm/sec. HINT: Call the speeds after collision v_1 and v_2, respectively, and use conservation of momentum and conservation of mechanical energy.

9.20. A fielder catches a fly ball, mass 150 g, moving horizontally at a speed of 20 m/sec. If his glove moves back a distance of 30 cm while bringing the ball to rest, what average force was exerted on his hand? Solve by the work-energy method.

9.21. A warehouseman gets a 200-kg cart moving at a speed of 0.25 m/sec along a level floor. How hard must he pull back on the cart to stop it in a distance of 4.0 m? Frictional forces may be neglected.

9.22. The barrel of a field gun has a mass of 400 kg. When a 5.0-kg projectile is fired horizontally with an initial speed of 430 m/sec, the barrel recoils a distance of 45 cm. Find the average force developed by the antirecoil device in bringing the gun barrel to rest.

9.23. Solve Prob. 6.19 (page 140), using the work-energy principle.

9.24. With each blow of a hammer a carpenter drives a nail 1.0 cm into a board. If the hammer head has a mass of 1.0 kg and a speed of 8.0 m/sec just before striking, what is the average force acting on the nail during each blow?

9.25. A car of mass 1,500 kg has brakes capable of exerting a retarding force of 20,000 newtons. The driver has a reaction time of 0.40 sec. If, when traveling 72 km/hr, he suddenly becomes aware of an obstacle in the road 29 m ahead, can he stop in time to avoid a collision? Work out the details.

9.26. In Worked example 9.5 (page 206), what will be the speed of the sled on returning to ground level if the frictional drag is assumed to be of the same magnitude as on the ascent?

9.27. A projectile is fired with a speed of 70 m/sec from the top of a tower 77 m high. Use the energy principle to find the speed of the projectile on hitting the ground. Does the result depend on the mass of the projectile or the angle at which it is fired? Air resistance is negligible.

9.28. An acrobat standing on a high platform swings forward and downward on a trapeze. At the lowest point of the swing, his center of gravity is 2.5 m below its starting position. At this instant the acrobat lets go and lands in a net below, at which time his center of gravity is 6.0 m lower than at the start. If there is no change in the attitude of his body during the stunt, how far horizontally beyond the point of suspension of the trapeze does he land?

9.29. A bullet of mass 10 g leaves the gun at a speed of 900 m/sec and returns to level ground with a speed of 400 m/sec. How much energy was dissipated against air resistance?

9.30. A test rocket of mass 13,500 kg rose vertically and attained a height of 170 km above the ground. (a) Compute its GPE at the top point, relative to the starting level. (b) If the average thrust of the jets amounted to 3.15×10^5 newtons and the charge burned until the rocket reached an altitude of 80.0 km, find the work done against air resistance on the upward flight. Assume g to have its standard value over the entire path.

9.31. A projectile having an initial speed v_0 is launched at an angle of elevation θ. Use the energy principle to find an expression for the maximum height reached. HINT: Make use of the fact that the horizontal component of the velocity remains constant (page 107).

9.32. A certain watch, when fully wound, has 1.0 joule of elastic PE stored in the spring. This energy is released at the rate of 1×10^{-5} watt as the watch runs down. How many hours can it run on one winding?

9.33. A tractor exerts a steady forward force of 2,200 lb when pulling a trailer at a constant speed of 15 ft/sec on a level road. What horsepower is being expended by the motor? What happens to the work done by the motor?

9.34. A 10-ton plane is going 200 mi/hr while climbing steadily at an angle of 20°. If the motors are exerting a total of 4,800 hp, how great is the force of air drag at this speed?

9.35. A pilot bails out of a disabled airplane. After his parachute opens, he finds himself descending at a steady speed of 27 ft/sec. If the total weight of pilot and parachute is 220 lb, at what rate is energy being imparted to the surrounding air?

9.36. A crewman in a racing shell exerts an average force of 300 lb on his oar, working at the rate of 33 strokes per minute. If his hands move back a distance of 28 in. with each stroke, at what rate, in horsepower, is he working? If the shell moves at an average speed of 12 mi/hr, how much work, in ft-lb, is done by the eight-man crew in a race lasting 6 min?

9.37. What average power (kW) is expended by a locomotive whose constant pull imparts a speed of 20 m/sec to a train of mass 4.0×10^5 kg (including that of the locomotive) after moving it on a level track a distance of 500 m from rest? Assume a constant resistive force of 4×10^4 newtons during the run.

Chapter 10
THE KINETIC MODEL OF MATTER

The molecular structure of matter was described in Chap. 3, and an intuitive idea of energy was briefly introduced. Subsequent chapters traced the development of the laws of mechanics, culminating in the concept of mechanical energy. In this chapter it will be shown how mechanical principles, especially that of energy, can be related to the microscopic structure of matter.

In the previous discussion of certain mechanical situations it was stated that mechanical energy is converted into heat when dissipative forces are present. The implication was that heat itself is a form of energy. This can now be confirmed, beginning with an examination of the temperature concept and going on to an interpretation based on a model of the structure of matter.

10.1 TEMPERATURE AND THERMOMETERS

Each of us has an intuitive qualitative appreciation of the meaning of temperature as a measure of the apparent "coldness" or "hotness" of objects that we touch. We also know by experience that when a hot body is placed in good contact with a cold one, the two eventually attain what is called **temperature equilibrium,** where no further changes associated with hotness or coldness occur. This is what would be expected if, as it was once believed, heat were a *fluid* that could flow from a hotter to a colder body until there was no longer any difference of "level" (temperature). Even today, long after the fluid theory had to be abandoned, we continue to speak of the flow of heat from one place to another.

Crude judgment of temperature based on the response of nerve endings in the skin must be replaced by a more objective method for measuring this quantity. From experience, we know that certain other changes in the condition of matter can be correlated with changes in temperature. Most obvious is the fact that almost all bodies, when free to do so, expand when heated and contract when cooled. This expansion and contraction is the basis for the construction of the commonest and most convenient tempera-

All gases behave in this way, but not all solids and liquids do.

ture-measuring devices **(thermometers),** notably the mercury-in-glass and the bimetallic forms (Fig. 10.1). Other properties that change with temperature and on which the construction of thermometers can be based are phase (physical state), pressure, electric resistance, character of radiated energy, etc. (See Ref. 10.1, pp. 5–9.)

Thermometers based on the expansion of a liquid were constructed in the seventeenth century, but it was some time before a standard calibration (affixing of a scale) was developed. Of the many temperature scales that were devised, two remain in general use today. The one employed in all scientific work is called the **Celsius scale,** after the Swedish astronomer Anders Celsius who suggested this scheme in 1742. Until recently, it was called the *centigrade* scale. The other is called the Fahrenheit scale (Fig. 10.2). Since it is not of scientific interest, it will not be used here.

Two conveniently reproducible temperatures, called **fixed points,** can be used in calibrating a Celsius thermometer. They are (1) the **ice point:** the equilibrium temperature of a mixture of pure ice and water in contact with the air at standard atmospheric pressure; and (2) the **steam point:** the equilibrium temperature of steam from pure water boiling at standard atmospheric pressure.

Note that the expression "°C" refers to a particular point on the temperature scale; "C°" refers to a temperature interval.

In calibrating a liquid-in-glass thermometer, the bulb is immersed in water at the ice point. The place along the stem where the top of the thread of liquid comes to rest is marked 0°C (degrees, Celsius). Then the bulb is placed in steam and the position of the liquid thread is marked 100°C. The intervening space along the stem is divided into 100 equal parts, each being 1 C° (Celsius degree). The scale may be extended in both directions beyond the fixed points.

In 1954, by international agreement, the use of two fixed points for determining the temperature scale was replaced by a scheme based on a single fixed point, the **triple point** of water (page 254) and the absolute zero. This change has no effect on practical thermometry.

(*a*)

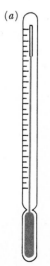

(*b*)

FIGURE 10.1 (*a*) *Mercury-in-glass thermometer.* The slight changes in the volume of the mercury are magnified by letting the liquid expand into the fine-bore tube. (*b*) *Bimetallic thermometer.* The two metals making up the spiral have different rates of expansion. As a result, the spiral twists when the temperature changes, and this moves the pointer over the scale.

In spite of its convenience, the liquid-in-glass thermometer is not the best instrument to use from the point of view of accuracy or fundamental significance. One difficulty is that, even if a number of such thermometers using different liquids are calibrated to agree at the two fixed points, there is no assurance that they will agree at other places on the scale. For example, Fig. 10.3 shows the expansion curves of a number of liquids, starting with unit volume at the ice point. Different liquids expand to different extents as the temperature changes, and this is to be expected. However, one of the substances, water, actually contracts with elevation of temperature over a short range just above the ice point before beginning a regular expansion above that range. In order to compare the expansion of the various substances, the ordinates of the curves should be replaced by a factor that will make them agree at the steam point. When this is done there still are discrepancies between the curves. Although these departures are small, it is desirable to find a way of reducing them or completely eliminating them.

10.2 THE GAS LAWS; ABSOLUTE TEMPERATURE

A thermometer whose behavior is independent of any particular substance can be made by using a *gas* rather than a liquid as a thermometric material. In particular, when gases confined at low pressure are used, they are found to agree very well among themselves and to have a constant rate of expansion. Another advantage of using a gas is that its rate of expansion in

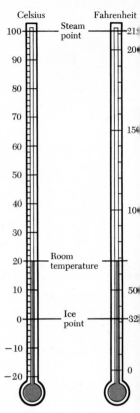

FIGURE 10.2 Comparison of Celsius and Fahrenheit thermometers. Temperatures on these scales are related by $t_F = 1.8\ t_C + 32$ where t_F and t_C are corresponding readings on the two thermometers.

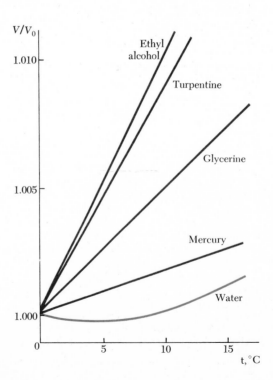

FIGURE 10.3 Volume expansion of some liquids that are used as thermometric substances. V/V_0 the ratio of the volume at any temperature t to the volume at $0°C$. Beginning at its melting point, water contracts slightly with rise of temperature and begins to expand beyond about $4°C$. In constructing a liquid thermometer, the expansion of the glass must be taken into account. See Prob. 10.1, page 24.

volume is 30 to 40 times as much as that of typical liquids and over 700 times as much as that of most metals.

The constant-volume gas thermometer.

The standard instrument is the **constant-volume gas thermometer.** In principle it is a rigid enclosure containing a gas at low pressure. The apparatus is equipped with a pressure gauge and with some means of heating and cooling the vessel. This arrangement is represented schematically in Fig. 10.4, although the actual construction of the instrument is very different from that shown. In use, the vessel is brought to a particular temperature, which may be one of the fixed points mentioned above, and the pressure gauge is read. The temperature is then changed and the new pressure is recorded. In this way, a series of corresponding values of pressure and temperature is obtained.

When these data are plotted on a graph of pressure versus temperature, the points are found to fall quite accurately along a straight line, as in Fig. 10.5, showing that the pressure changes are proportional to the temperature changes. For practical reasons, data can be obtained from the

TEMPERATURE AND THE NATURAL ENVIRONMENT

Ecologists consider temperature the main factor controlling life on earth. In recent years, the increasing discharge of waste heat into natural bodies of water threatens to disturb the balance of nature. The environmental effect is referred to as thermal pollution.

The main sources of this waste heat are electric power stations, especially those using nuclear energy. In the United States, the rate of use of electric power is doubling every 8 to 10 years. It is estimated that by the year 2000, heat energy equivalent to more than 2×10^{19} joules per day will have to be disposed of.

Cold-blooded animals such as fish cannot adjust to abrupt temperature changes. According to the Federal Water Pollution Control Administration, temperatures above 93°F (34°C) would make a body of water uninhabitable for nearly all fishes. This temperature can easily be reached if several power plants are located on the same river, since the heat discharge from each can raise the temperature 10 to 30 Fahrenheit degrees.

At elevated temperatures the amount of oxygen dissolved in water is reduced. This cuts down the amount of organic waste (sewage) that a stream can assimilate and also reduces the oxygen available to fish and shellfish. At the same time, the higher temperature raises the metabolic rate of these animals; further increasing their need for oxygen. The preferred temperature ranges are about 5 Fahrenheit degrees for most commercially important fish, and the creatures on which they feed are even more sensitive to temperature changes.

There are feasible alternatives to the present practice of discharging waste heat directly into streams and lakes. Cooling water can be stored temporarily in artificial ponds, or the heat can be dissipated into the air by means of cooling towers. The latter method, however, requires an expensive installation.

Increased public awareness of the heat pollution problem is necessary in order to influence effective legislation. Some progress is already evident. In 1970, government agencies devised a "heat quota" system under which a legal limit will be set on the amount of heat that can be discharged into a given body of water. The location of new power plants will have to be guided by these regulations.

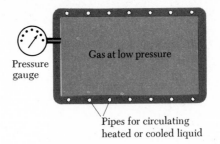

Pressure gauge

Gas at low pressure

Pipes for circulating heated or cooled liquid

FIGURE 10.4 Schematic representation of a constant-volume gas thermometer. The container has a fixed volume and is equipped with a pressure gauge and a means of changing the temperature.

constant-volume gas thermometer only over a restricted range of temperature. All gases eventually liquefy if cooled sufficiently, but even before this happens, slight departures from the straight-line relationship are observed (see Fig. 10.6). A gas that obeys the linear relationship between temperature and pressure exactly is called an **ideal gas.** All gases at sufficiently low pressure, particularly hydrogen and helium, are excellent approximations to this ideal model.

Now we make the assumption that an ideal gas continues to follow the linear relationship, no matter how far the temperature is lowered. If the straight line that represents the behavior of the gas at moderate temperatures is extended to the left, it will eventually cut the temperature axis at some point marked A in Fig. 10.6. Moreover, if the whole series of measurements is carried out again with a different sample of gas, the extended line will again cut the temperature axis at the *same* point (Fig. 10.6). This is found to be true regardless of the kind, amount, and initial pressure of the sample, as long as the gas is confined at low pressures.

There appears to be some fundamental significance to the common point of intersection of all the temperature-pressure lines. On extending the Celsius scale markings down this far, the point of intersection is found to lie at $-273.15°C$. It is called the **absolute zero** of temperature. Unlike the Celsius zero, which depends on the behavior of a particular material (water), the absolute zero represents something more general and more fundamental. In addition, the entirely different line of reasoning of the science of thermodynamics (Chap. 11) leads to precisely the same point. Experimentally, temperatures down to within a millionth of a degree of absolute zero have been reached.

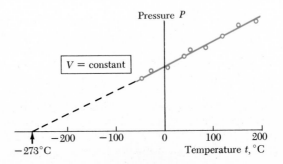

Pressure P

$V = $ constant

$-273°C$

-200 -100 0 100 200

Temperature t, $°C$

FIGURE 10.5 The pressure-temperature curve of a rarefied gas at constant volume is a straight line. Projected backward, this line cuts the temperature axis at $-273°C$.

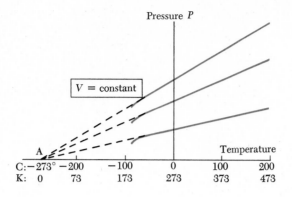

Pressure P

V = constant

Temperature

C: $-273°$ -200 -100 0 100 200
K: 0 73 173 273 373 473

A

FIGURE 10.6 Extensions of the lines for various samples of gas all intersect the temperature axis at the same point, the absolute zero. In practice, departures begin to occur, as shown, before this point is reached.

William Thomson (1824–1907), Lord Kelvin, was one of the greatest British physicists of his time. Besides his fundamental researches in heat, electricity, and radiation, his work made possible the first undersea cable.

Each of the lines in Fig. 10.6 has an intercept on the P axis. It was suggested by Kelvin that the algebraic equation of such a line could be written more simply if the P axis were moved to the absolute zero point. Then the line would pass through the origin, so that P would be directly proportional to the temperature as measured to the right from that place. If this shift is made, the result is called the **absolute** or **Kelvin scale** of temperature. Its starting point is the absolute zero, and it uses the same-sized degree as the Celsius scale. Temperatures on the absolute scale are designated algebraically by the symbol T and are measured in **kelvins** (K). For example, the ice point is at $T_i = 273.15$ K and the steam point at $T_s = 373.15$ K (see Fig. 10.6). The use of the Kelvin scale eliminates all reference to negative values of temperature. They are not required in this scheme, as they are on the Celsius scale.

By defining temperature we have introduced a new dimensional quantity into physics, in addition to length, mass, and time. This proves to be necessary as well as convenient. However, it will be shown in the next chapter that another aspect of heat can be related to the original mechanical units through the energy concept.

With the establishment of the absolute temperature scale, we can now write a simple proportion between the temperature of a sample of gas and the pressure under which it is confined. This expression may take any of the following forms:

For an ideal gas at constant volume,

$$P \propto T \qquad P = K_1 T \qquad \frac{P_1}{P_2} = \frac{T_1}{T_2} \qquad\qquad \textbf{10.1}$$

where K_1 is a constant whose value depends on the kind and amount of gas used and the volume it occupies. P_1, T_1 and P_2, T_2 are any two pairs of observed pressure and temperature values.

To describe completely the condition of a gas at any time, three variables are needed: volume, pressure, and temperature. Several series of experiments can be carried out to find what relation exists, if any, among these quantities. In each of these series one of the three factors is held constant and the relation between the other two is investigated. This was the procedure used (page 126) in finding the experimental connection between force, mass, and acceleration, leading to Newton's second law.

Kelvin. (*British Information Services.*)

In the set of experiments with the constant-volume gas thermometer described on pages 219–221, the volume was held constant, leading to the relation between pressure and temperature embodied in Eq. 10.1. A similar series of experiments can be carried out with a *constant-pressure* gas thermometer, shown schematically in Fig. 10.7. Although the procedures are more difficult than in the previous experiment, the surprising result is that the data for a rarefied gas, when plotted on a volume-temperature graph, again fall on a straight line (Fig. 10.8). When this line is extended downward, it cuts the temperature axis at the very same point ($-273.15°C$) as in the previous experiment.

The constant-pressure gas thermometer.

As before, we use the Kelvin temperature scale, leading to a simple proportion between volume and temperature:

For an ideal gas at constant pressure,

$$V \propto T \qquad V = K_2 T \qquad \frac{V_1}{V_2} = \frac{T_1}{T_2} \qquad\qquad \textbf{10.2}$$

Here K_2 is a quantity that remains constant throughout the experiment. Its numerical value depends on the kind and amount of gas used and on the pressure under which it is confined.

The relations expressed by Eqs. 10.1 and 10.2 were obtained experi-

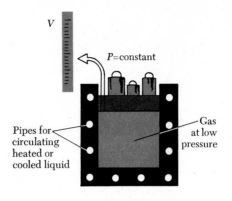

FIGURE 10.7   Schematic representation of a constant-pressure gas thermometer. Pressure may be applied by means of weights placed on the piston.

The relation between pressure and volume of an ideal gas at constant temperature was obtained independently some years later by the French scientist E. Mariotte (1620–1684). In Europe, the names of both Mariotte and Boyle are often attached to this law.

mentally by the French physicist J. L. Gay-Lussac (1778–1850) in 1802, although his countryman J. A. Charles (1746–1823) claimed to have found the same results some years earlier. The two statements are usually referred to collectively as the laws of Charles and Gay-Lussac.

The pair of equations is sufficient to yield a general relation between P, V, and T. However, it is of interest to look at the remaining experimental correlation, the connection between pressure and volume that exists when the temperature of an ideal gas is held constant. Experimentally, this is much easier to obtain than the above relations, since it requires no scheme for measuring temperature but only a provision for holding the temperature constant. As a matter of historical fact, the required relation was found as early as 1660 by Robert Boyle (page 56n) and is usually known by his name. It may be expressed in the following ways:

For an ideal gas at constant temperature,

$$V \propto \frac{1}{P} \qquad V = \frac{K_3}{P} \qquad \frac{V_1}{V_2} = \frac{P_2}{P_1} \qquad \text{10.3}$$

where K_3 remains constant through the entire experiment. Its numerical value depends on the kind and amount of gas used and on the temperature at which it is kept. Notice that the equations express the fact that the volume is *inversely* proportional to the pressure (see Fig. 10.9).

It is now possible to summarize what is known about the behavior of

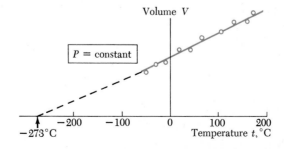

FIGURE 10.8   Volume-temperature curve of a rarefied gas at constant pressure. Again, the intersection with the temperature axis is at −273°C.

TABLE 10.1   TEMPERATURES IN THE UNIVERSE

TEMPERATURE, K

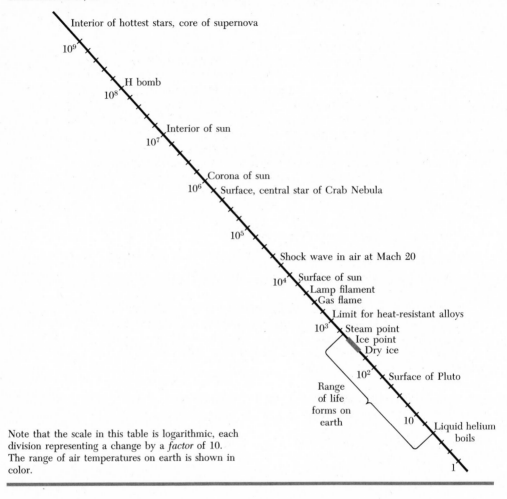

Note that the scale in this table is logarithmic, each division representing a change by a *factor* of 10. The range of air temperatures on earth is shown in color.

an ideal gas by combining Eqs. 10.1 to 10.3. The result is conveniently written in either of these two forms:

$$PV = KT \quad \text{or} \quad \frac{P_1 V_1}{P_2 V_2} = \frac{T_1}{T_2} \qquad \qquad 10.4$$

Practice set*

1. By what factor does the volume of a confined sample of a rarefied gas change if the temperature is held constant while the applied pressure is (a) doubled; (b) reduced to one-third as much as before?
2. By what factor must the pressure be changed in order to double the volume while the absolute temperature is increased by a factor $\frac{3}{2}$?

*Ans.: $\frac{1}{2}$, 3, $\frac{3}{4}$.

The constant K in Eq. 10.4 depends only on the chemical nature and mass of the gas used. It is customary to express it in terms of the chemical *molecular weight* of the gas and the mass of the experimental sample. In chemistry, a *gram molecule* or *mole* of a substance is defined as the amount, in grams, numerically equal to the molecular weight. In the mks system, the **kilogram mole** (k-mol) is defined as the corresponding number of kilograms for each substance. Thus the values are 2.016 kg for hydrogen, 4.003 for helium, 28.02 for nitrogen, etc. These are the masses, in kilograms, of a **standard chemical volume** (22.415 m³ in mks units) of each gas, which is the volume occupied by one kilogram mole of *any* gas at standard temperature and pressure, $t_s = 0°C$, $P_s = 1$ standard atmosphere $= 1.013 \times 10^5$ N/m². See Worked example 10.3.

Chemistry students will be more familiar with the standard volume that corresponds to the use of gram moles, 22.4 liters.

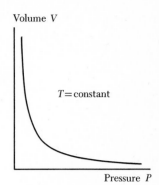

Volume V

$T =$ constant

Pressure P

FIGURE 10.9 For an ideal gas at constant temperature, the pressure and volume are inversely proportional (Boyle's law).

The constant K in Eq. 10.4 can now be replaced by nR, where n is the number of kilogram moles of gas used and R is a constant of proportionality, called the **universal gas constant,** that has the *same value for all gases.* Its numerical value is

$$R = 8.314 \times 10^3 \text{ J/(k-mol)(K)} \qquad \textbf{10.5}$$

Now Eq. 10.4 may be written in the simple form

$$PV = nRT \qquad \textbf{10.6}$$

The relation 10.6 is called the general gas law or **equation of state** of an ideal gas.

No real gas follows this law precisely over a large range of temperature and pressure. The departures become pronounced at low temperature and high pressures. Under such conditions the molecules of a gas are more closely packed, and their finite size and mutual attraction become important disturbing effects.

Worked example 10.1

A chemist collects a sample of chlorine gas occupying a volume of 180 cm³ at a temperature of 17°C. Unfortunately, the pressure gauge on the container is calibrated in English units and reads 20.0 lb/in.² How can he calculate the volume, in cubic centimeters, that the sample of gas would occupy at standard temperature and pressure? In English units $P_s = 14.7$ lb/in.²

Solution   It is convenient here to use the gas law in the proportion form of the second equality in Eq. 10.4. Notice that this equation consists of three ratios of like quantities: a pair each of pressures, volumes, and temperatures. The members of a pair may be measured in any units whatever, as long as they are both expressed in the *same* units. Hence it is no hindrance that the pressure data in the problem are given in English units. These values can be inserted directly in the formula because, if we were to convert to mks units, the *same* conversion factor would have to be applied to each pressure value, and so this factor would cancel out.

To proceed with the solution, we solve Eq. 10.4 for V_2:

$$V_2 = \frac{P_1}{P_2} \frac{T_2}{T_1} V_1$$

One precaution must be observed: The temperatures must be on the Kelvin scale. Thus for T_1 do *not* use 17°C, but 290 K, which is 273 + 17. Substitution of all the numbers yields

$$V_2 = \frac{20.0 \; \cancel{lb/in.^2}}{14.7 \; \cancel{lb/in.^2}} \frac{273 \; K}{290 \; K} \; 180 \; cm^3 \qquad or \qquad V_2 = 230 \; cm^3$$

In some situations, the gas law in the form of Eq. 10.6 leads more directly to a result, as illustrated by the following Worked example. In using this formula, the units are prescribed. Where the mks system is employed, P must be expressed in N/m², V in m³, n in k-mol, R in J/(k-mol)(K), and T in K.

Worked example 10.2   How many kilogram moles of helium gas are needed to inflate a high-altitude balloon whose volume is to be 2,000 m³ at a pressure of 0.10 atm and a temperature of −43°C?

Solution   Direct substitution into Eq. 10.6 gives

$$(0.10)(1.013 \times 10^5)(2 \times 10^3) = n(8.314 \times 10^3)(230)$$

from which $n = 10.6$ k-mol. This result is independent of the kind of gas used. However, since different gases have different molecular weights, the actual mass of gas would be different in each case.

Worked example 10.3   Check the value given for the standard chemical volume on page 225.

Solution   Since this is defined as the volume occupied by 1 mole of an ideal gas at standard conditions, substitute $n = 1$ in Eq. 10.6, together with the other numerical values given on page 225 and solve for V:

$$V = \frac{(1)(8.314 \times 10^3)(273.15)}{1.012 \times 10^5} \qquad or \qquad V = 22.42 \; m^3$$

10.3   THE MOLECULAR MODEL OF A GAS

Beginning with the conjecture of Democritus that all matter consists of discrete particles, the molecular hypothesis has persisted and continually evolved as a model for the structure of matter. In Chap. 3 some observed properties of substances were described in terms of molecular structure. For example, the difference in behavior of gases, liquids, and solids was correlated with the state of aggregation of the molecules in each of these phases. Although a solid represents the most orderly arrangement of molecules in space, a gas is actually a simpler phase of matter because its molecules are, on an average, far apart and so most of the time do not exert appreciable forces on each other.

Can a molecular model yield a *quantitative* representation of the behavior of matter? In particular, can such a model account for the gas laws? It will be shown in this chapter that such a model is able to do this. Further, it is capable of leading to an interpretation of temperature and heat based on mechanical concepts.

The molecular idea was rejected by the followers of Plato and Aristotle but was revived, beginning in the seventeenth century, by a number of

prominent scientists who realized its power in explaining observed phenomena, at least qualitatively. Newton spoke of "... solid, massy, hard,
impenetrable, movable particles ..." and Galileo asserted that "... the
thing that produces heat in us and makes us perceive it ... is a multitude
of minute corpuscles thus and thus figured, moved with such and such
a velocity. ..." The Russian chemist M. W. Lomonosov (1711–1765) also
considered matter to be made up of atoms and molecules, and heat to
be a result of the motion of molecules.

The key feature of these speculations is the idea that the particles are
in motion, leading to a *kinetic-molecular* hypothesis of the nature of matter.
It is precisely this feature that makes the model amenable to treatment by
the laws of mechanics. An important step in this direction was taken by
Daniel Bernoulli (1700–1782), who showed that the pressure exerted by
a gas on the sides of its container could be explained by thinking of a
gas as consisting of "very minute corpuscles ...". practically infinite in
number ... driven hither and thither with a very rapid motion."

Bernoulli was one of more than a dozen eminent scholars and scientists from a single Dutch-Swiss family. One of his accomplishments is the hydrodynamic law basic to airplane flight.

Bernoulli's calculations led to a deduction of Boyle's law (page 223),
as will be explained below. It is interesting to see how many properties
of gases and of the other phases of matter can be explained qualitatively
by the kinetic-theory model of a gas as a swarm of swiftly moving particles,
continually colliding with each other and with the sides of their container.

The relatively great average distance between molecules is in accord
with the observed low density of gases at ordinary temperatures and pressures (of the order of 10^{-3} to 10^{-4} times those of solids and liquids). For
the same reason, gases are readily compressible. The compression of a
gas should be thought of as making less space available to the flying
particles, rather than as the squeezing together of a continuous rubberlike
material.

The association of molecular motion with heat explains many other features of the behavior of matter. The speeding up of the molecules resulting
from the application of heat was described in Chap. 3 as one cause of
the melting of a solid and of the evaporation or boiling away of a liquid.
The vapor thus formed behaves like a gas. Under the application of extreme
heat, all known materials can be converted into gases. At sufficiently high
temperatures the molecules dissociate chemically and even their constituent atoms break up. As far as is known, most of the matter in the universe
is in the form of a gaseous mixture of electrified fragments of atoms, called
a **plasma.**

The term vapor is usually applied to the gaseous phase of a substance when in contact with the liquid phase.

The **diffusion** of gases gives further evidence for the motion of molecules. If a bottle of ammonia is uncorked in a room, the pungent odor of
this gas is soon detected some distance away. The natural explanation is
that the rapidly moving ammonia molecules tend to disperse. In moving
about, each ammonia molecule encounters molecules of the gases of the
air, as well as others of its own kind. Each collision gives it a sudden change
in direction, so that instead of moving straight away from the source the
molecule follows a zigzag path. Diffusion is observed also for liquids (Fig.
10.10) and even for one solid through another. In solids the atoms or
molecules for the most part remain near fixed lattice positions, but very
gradual migration of the atoms of one solid through another has been
verified by radioactive tracer techniques (Chap. 22).

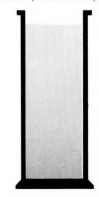

FIGURE 10.10 A layer of colored liquid carefully placed at the bottom of a tall jar of clear water is seen to diffuse upward as time passes.

Brownian motion.

A phenomenon first observed by the British botanist Robert Brown in 1827 provides the nearest thing to direct visual evidence for the motion of molecules. He placed water containing very small solid particles under a microscope and observed that the suspended particles were in continual, haphazard motion. The path of any one particle was seen to be just the kind of jagged succession of short displacements described above for the movement of a gas particle (Fig. 10.11).

The tiny particles used by Brown in his experiment were pollen grains. At first he thought that the grains moved because they were alive but found that the motion continued even after they had been boiled. The same type of motion can be observed for liquid droplets or solid particles suspended in a gas. This continual random motion of small particles is called **brownian motion.**

FIGURE 10.11 Brownian motion.

Before long, a number of scientists saw that the kinetic-molecular hypothesis provides a ready explanation of brownian motion. A solid particle large enough to be seen in a microscope may contain hundreds of billions of molecules, yet when the particle is buffeted by the rapidly moving molecules of the gas or liquid in which it is suspended it reacts visibly. Since there is little chance that the effects of these random impacts will exactly cancel out at any instant, the particle experiences a succession of rebounds in various directions.

The French scientist Jean Perrin (1870–1942) received the Nobel Prize in physics for his investigations of the so-called sedimentation of suspended particles.

The brownian motion of suspended particles prevents them from settling completely out of their state of suspension, but they collect in greater numbers toward the bottom. Gravity produces a similar effect on the molecules of the atmosphere so that, except for local disturbances, the air density decreases continuously as the altitude increases (Fig. 10.12).

Einstein calculated that the average of the square of the displacement of a brownian particle should be proportional to the elapsed time. Perrin,

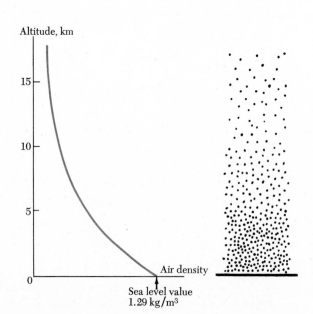

FIGURE 10.12 In a standard atmosphere the density drops by a factor $\frac{1}{2}$ for each 5.5-km increase in altitude.

R. A. Millikan, and others were able to confirm this by experiment, thus giving great support to the kinetic-molecular hypothesis. Observations of brownian motion and of the distribution of suspended particles even led to a determination of the number of molecules in a given quantity of matter and showed how numerous molecules really are. The result indicated that there are about 6×10^{26} molecules in a standard chemical volume (page 225) of *any* gas measured at standard temperature and pressure. It will be shown later that this value is confirmed by other observations of the most diverse kinds.

J. P. Joule (1818–1889), an industrialist whose avocation was science, published his first research paper when he was 20. His great achievement was to show that heat is a form of energy (Chap. 11).

10.4 THE JOULE-BERNOULLI EQUATION

The first kinetic-theory computation of any importance was Bernoulli's deduction of Boyle's law (page 223). More than a century later, Bernoulli's rough calculation was refined by the British physicist Joule, resulting in a quantitative expression for the pressure exerted by a confined gas in terms of certain characteristics of the molecules. It will now be shown how this relationship can be deduced by making the following assumptions:

1. A sample of a confined gas consists of a large number of molecules, all of one kind. This is a reasonable restriction for our purposes, although the results we shall derive can be extended easily to mixtures of different gases.

Joule. (*British Information Services.*)

The word gas derives from the Greek word chaos.

2. The molecules are in constant motion, and this motion is entirely random. The molecules collide with each other and with the (molecules of the) sides of the container.

3. In a rarefied gas the molecules are, on the average, far apart compared with their own size.

4. The molecules do not exert appreciable forces on each other except at their moments of impact, whose duration is negligible compared with the intervals between collisions. Thus the path of any individual molecule can be expected to be a random sequence of short, straight segments, similar to Fig. 10.11.

5. In addition to conservation of linear momentum, the total kinetic energy of the colliding molecules remains, on the average, the same before and after impact.

On the basis of these assumptions we can now calculate an expression for the pressure exerted by the gas, following the method used by Joule. Work through this derivation carefully, making sure that each step is understood before proceeding to the next.

Deduction of the equation for P.

For simplicity, assume that the sample of gas is in a rectangular container whose dimensions are a, b, and c (Fig. 10.13). The directions of the three edges may be taken to be the x, y, and z axes of a coordinate system. The mass of each molecule is m, and the total number of molecules in the box is N.

Consider a single molecule that is about to strike the wall A of the container (Fig. 10.14). Designate by $\mathbf{v}_x$ the component of the velocity of this molecule that is perpendicular to the wall. In general, there will be two other independent velocity components $\mathbf{v}_y$ and $\mathbf{v}_z$, parallel to the plane of the wall A, but they will have no effect on the interaction of the molecule with this wall and need not be further considered.

The effect of a perfectly elastic collision with A, then, is to send the molecule in the minus x direction with a speed numerically the same as it had on approach. This means that the original x component of velocity has been changed from $\mathbf{v}_x$ to $-\mathbf{v}_x$, and so at impact the molecule has experienced a change in linear momentum amounting to

$$\Delta p_x = mv_x - (-mv_x) = 2mv_x \qquad\qquad 10.7$$

With regard to the x component of its motion, the molecule may be thought of as bounding back and forth continually between the opposite walls A and B, always moving with a speed v_x. In reality, a given molecule

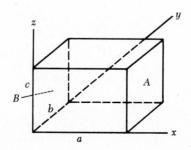

FIGURE 10.13 Box containing a gas.

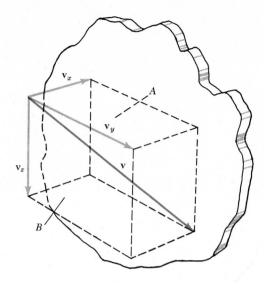

FIGURE 10.14 Velocity components of a molecule about to strike
a wall.

will make a large number of collisions with other gas molecules before going very far, but its momentum will be handed on just as if it had actually traveled the whole length of the box and collided with the opposite wall directly.

The time taken for a single molecule, moving as described, to go from A across to B and back again to A at the constant speed v_x is $t = 2a/v_x$. Then, with reference to its impacts with A, the rate of change of momentum of this molecule will be, using Eq. 10.7,

$$\frac{\Delta p_x}{\Delta t} = \frac{2mv_x}{2a/v_x} = \frac{mv_x{}^2}{a} \qquad\qquad \textbf{10.8}$$

By Newton's second law, the rate of change of linear momentum is equal to the (average) force on the molecule, and the equal and opposite reaction of the molecule on the wall will have a magnitude of

$$f_A = \frac{mv_x{}^2}{a} \qquad\qquad \textbf{10.9}$$

Equation 10.9 gives the magnitude of the average, in time, of the force exerted by a single molecule on one wall of the container. This average, computed for a single molecule, has little physical meaning because it represents merely an arithmetical averaging of sudden, strong impulse forces occurring at relatively long intervals (Fig. 10.15a). Actually, a tremendous number of molecules take part in these impacts, and the succession of blows amounts to essentially a steady force (Fig. 10.15b) whose magnitude will now be calculated.

At any instant, the molecules of a gas have a wide range of speeds and are moving in all conceivable directions. How can we take account of this fact and arrive at a definite conclusion when we must deal with astronomical numbers of molecules? It is precisely because the molecules are so numerous that it becomes possible to handle the problem. No matter what suc-

(a)   (b)

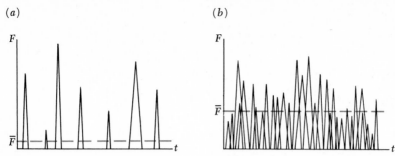

FIGURE 10.15 Average of a succession of impulses. When the blows are numerous, the net effect is a nearly steady force.

cession of impacts each molecule experiences, it continues to obey the laws of mechanics. There is no reason to believe otherwise. But it is obviously impossible, in a practical calculation, to keep track of the mechanics of all the molecules individually. We therefore must resort to considering the *average* behavior of a large number of molecules. The subject that deals with the collective mechanical behavior of large assemblages of objects is called **statistical mechanics.** It proves to be a powerful method of attacking many important physical problems.

We can apply statistical methods to the assemblage of gas molecules. The single molecule that we have been considering was moving back and forth in the x direction. The choice of this direction was perfectly arbitrary, and nothing about the behavior of the gas would be expected to change if the orientation of the box were changed. This amounts to saying that, on the average, the three coordinate directions are on a par with each other. Specifically, if v is the instantaneous speed of an individual molecule, then

$$v^2 = v_x^2 + v_y^2 + v_z^2$$

where v is the magnitude of the resultant velocity of the molecule. Since the motion is truly random, the values of the three terms on the right will be the same: $v_x^2 = v_y^2 = v_z^2$, so that, from the above equation, $v^2 = 3v_x^2$, or $v_x^2 = v^2/3$.

Substituting this value into Eq. 10.9 yields

$$f_A = \frac{mv^2}{3a} \qquad\qquad\qquad \textbf{10.10}$$

Summing up over all N molecules, the total force exerted on A because of molecular impacts is

$$F_A = \frac{m}{3a}\left(v_1^2 + v_2^2 + \cdots + v_N^2\right) \qquad\qquad \textbf{10.11}$$

Next, divide both sides of this relation by bc, the area of wall A, to get the gas *pressure* (force per unit area), which has the same value on any portion of the container:

$$P = \frac{F_A}{bc} = \frac{m}{3V}\left(v_1^2 + v_2^2 + \cdots + v_N^2\right) \qquad\qquad \textbf{10.12}$$

where $V = abc$ is the volume of the container.

We define the quantity

$$\overline{v^2} = \frac{v_1^2 + v_2^2 + \cdots + v_N^2}{N} \qquad\qquad 10.13$$

as the **mean-square speed** of the molecules. The **mean-square speed** is the average value of the squares of all the molecular speeds. The bar over the symbol indicates that it represents an *average* value.

Substituting for the quantity in parentheses in Eq. 10.12, the result is

$$P = \frac{mN\overline{v^2}}{3V} \qquad\qquad 10.14$$

the celebrated **Joule-Bernoulli equation** for the pressure of a gas.

From Eq. 10.14, the dependence of P on the other quantities seems reasonable: The pressure would be expected to increase with the mass m of each molecule as well as with the "crowdedness" of the molecules as measured by N/V. Also, increasing the speed v of the molecules increases both the frequency of collision with the walls and the momentum change involved in each collision, and so the pressure depends on v on two counts, or on v^2. However, only the exact mathematical analysis can supply the factor $\frac{1}{3}$ or show that the averaging of the v's must be done in a certain way.

10.5 SOME CONSEQUENCES OF THE JOULE-BERNOULLI EQUATION

Equation 10.14 represents a relationship between the large-scale quantities P and V and the microscopic quantities m, N, and $\overline{v^2}$. The first two are readily measurable without any regard for the molecular constitution of matter, whereas the latter three are properties of molecules themselves and cannot be observed directly.

It would seem impossible to find the values of the three microscopic quantities in terms of the two large-scale ones, since there is only one equation (10.14) relating them. However, the product mN represents the mass of each molecule multiplied by the total number of molecules, so that it is equivalent to the total mass M of the sample of gas. When this is divided by the volume V of the container, which appears in the denominator of the equation, the result is simply M/V, which is the *mass density* D of the gas. Hence Eq. 10.14 may now be written $P = D\overline{v^2}/3$, or solved for $\overline{v^2}$,

Density was defined in this way on page 54.

$$\overline{v^2} = \frac{3P}{D} \qquad\qquad 10.15$$

The density of the gas is readily measured by weighing, and the pressure is directly measurable by using a suitable gauge. Equation 10.15 is then capable of giving information about the average of the squares of the molecular speeds from values of the directly observable quantities. If we substitute numbers for the case of air at standard conditions, we must set $P = 1.013 \times 10^5$ N/m^2 (page 225) and $D = 1.293$ kg/m^3 (from Table 3.4, page 55). The result is

$$\overline{v_{\text{air}}^2} = 2.35 \times 10^5 \text{ m}^2/\text{sec}^2$$

The $\overline{v^2}$ represents the *average of the squares* of all the individual molecular speeds, and this is not numerically the same as the *square of the average* speed (see Prob. 10.10). Since the molecular speeds enter only to the second power in the Joule-Bernoulli equation, the correct average to take is that of the squares. After carrying out this averaging process, the square root may be taken in order to get a quantity of the nature of speed itself. This value is called the **root-mean-square** speed and usually differs a few percent from the (arithmetical) average speed (see Fig. 10.18). Taking the square root of the number given above yields, for air at standard conditions,

$$v_{\text{rms}} = 485 \text{ m/sec}$$

which is over 1,000 mi/hr. The value for hydrogen, with a density about 14 times smaller than that of air, would be $\sqrt{14}$ times *greater* than the above, according to Eq. 10.15. The result is a speed close to 4,000 mi/hr. It is shown below that such extremely high molecular speeds actually exist and that there is a way of measuring them directly.

At this point we can deduce an important conclusion from a comparison of the Joule-Bernoulli result with the experimental ideal-gas law discussed earlier. Equation 10.14 can be written in the form

$$PV = \frac{mN\overline{v^2}}{3}$$

Rewrite this as

$$PV = \frac{2}{3} N \left(\tfrac{1}{2} m\overline{v^2} \right) = \frac{2}{3} N\overline{K} \qquad\qquad 10.16$$

where $\overline{K} = \tfrac{1}{2} m\overline{v^2}$ will be recognized as the *average translational KE per molecule* of the gas.

Return now to the general gas law (equation of state of an ideal gas) in the form of Eq. 10.6:

$$PV = nRT$$

where n is the number of kilogram moles of gas in the sample and R is the universal gas constant. The result of comparing this with Eq. 10.16 is

$$\overline{K} = \frac{3nR}{2N} T \qquad\qquad 10.17$$

Chemical evidence shows that *equal volumes of all gases, measured at the same temperature and pressure, contain equal numbers of molecules.* This statement is called **Avogadro's law,** after the Italian scientist Amadeo Avogadro (1776–1856). This makes the number of molecules, N_0, in a standard chemical volume of any gas a universal constant, called **Avogadro's number.** Its numerical value is

Root-mean-square speed.

Avogadro's number.

$$N_0 = 6.023 \times 10^{26} \text{ particles/k-mol} \qquad\qquad 10.18$$

Consistent numerical results for N_0 are obtained from a great variety of experiments. Avogadro's number is one of the important constants of physics.

In Eq. 10.17, n/N can now be replaced by $1/N_0$, since N_0 is the number of molecules corresponding to 1 mole. With this substitution, Eq. 10.17 can be written

$$\bar{K} = \frac{3R}{2N_0}\,T \qquad\qquad 10.19$$

where the combination $3R/2N_0$ is a constant. As a result, we can conclude that *the average translational KE of the molecules of a gas is proportional to the absolute temperature.*

This gives a direct mechanical picture of the elusive concept of temperature. At the same time it confirms the idea that heat is a form of energy—the translational KE of the molecules. The only difference between heat energy and the KE of finite objects is that heat is the *random, disorganized* motion of the individual molecules whereas the KE of a body as a whole represents the orderly motion possessed by all its particles. Through friction, impact, or other dissipative processes, organized particle motion may be converted into additional random motion of the molecules.

Why is no energy "lost" in molecular impacts? Why is it that the molecules of a gas do not eventually lose their translational KE and settle to the bottom of the container in the form of a liquid? The answer is that the collision of gas molecules cannot be pictured in the same terms as a collision between billiard balls. The impact of two molecules must not be thought of as the actual bumping together of hard spheres but as a close approach of two intense centers of force which repel each other strongly by means of electric forces when the distance between them becomes small (see Fig. 3.9, page 46). This kind of encounter, visualized in Fig. 10.16, involves no net change in energy, *on an average.* Although molecules having a complex structure may be set into rotation or their atoms may be set into vibration by the effects of a collision, thus reducing their translational KE, this energy is communicated to other molecules in later collisions so that, statistically, the entire collection of molecules remains in a steady state of constant translational KE.

Figure 10.17 is a representation of what happens on a molecular scale when a bullet strikes a wall. The visible, large-scale motion of the bullet as a whole is the forward-directed motion of all its particles, superimposed

FIGURE 10.16 One way of visualizing the collision of two molecules.

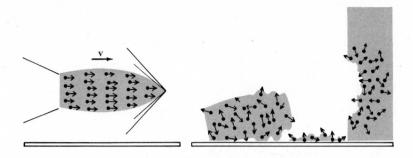

FIGURE 10.17 The common translational motion of the molecules is changed to random motion when the bullet strikes the wall.

on the random heat motions which they possess because of the finite temperature. On impact, the forward motions are changed to additional *random* motion, which is heat energy. For simplicity, the initial heat motion of the molecules is not shown in the figure.

10.6 DISTRIBUTION OF MOLECULAR SPEEDS; MEAN FREE PATH

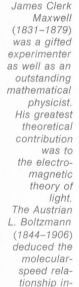

James Clerk Maxwell (1831–1879) was a gifted experimenter as well as an outstanding mathematical physicist. His greatest theoretical contribution was to the electromagnetic theory of light. The Austrian L. Boltzmann (1844–1906) deduced the molecular-speed relationship independently.

In discussing the Joule-Bernoulli equation we made use of the root-mean-square speed of the molecules and of the average KE which depends on this speed. In many situations involving statistics, it is not enough to know average values. In addition, we must have information on how the individual values are *distributed* around the average value. For gas molecules, this means a knowledge of what fraction of the collection is moving in various speed ranges above and below the average.

About a century ago, the British physicist J. C. Maxwell developed the mathematical description of how the molecular speeds are distributed. By assuming that in a random collection of moving gas molecules all energy values are equally probable, he found that the distribution of the speeds at any given temperature is represented by a curve like that of Fig. 10.18. It gives the relative number of molecules in any selected narrow range of speeds. Most of the values cluster around the "hump" of the curve, but all speeds from zero to indefinitely large ones are to be expected.

Ample experimental confirmation of Maxwell's distribution has been obtained by several methods. One of these is shown in Fig. 10.19. A metal is vaporized in a small furnace, and a narrow beam of atoms coming from it passes through slits in two rotating disks. Since the slits are offset from each other, only the atoms in a particular range of speeds pass through

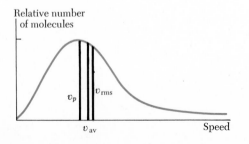

FIGURE 10.18 The Maxwell-Boltzmann distribution curve for a gas at a given temperature.

Maxwell. (*Courtesy of the Granger Collection.*)

The Maxwell distribution function.

at a given rate of turning. By observing the relative thickness of the deposit of metal on the detecting plate for various disk speeds, the distribution of atomic speeds corresponding to the temperature of the furnace can be computed. The agreement with Maxwell's curve is very satisfactory.

The speed corresponding to the crown of the curve is called the *most probable* speed, v_p. On the basis of the Maxwell-Boltzmann calculations the (arithmetical) *average* speed v_{av} is about 13 percent greater and the *root-mean-square* speed v_{rms} about 22 percent greater than v_p. The chance

FIGURE 10.19 At a given rate of turning, only the atoms in a certain range of speeds get through to the detector. The entire arrangement is in a vacuum.

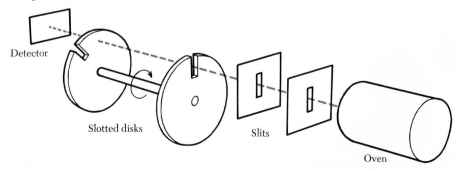

Detector

Slotted disks

Slits

Oven

of finding molecules in a given narrow speed range some distance from v_p becomes quite small, especially on the side of higher speeds. This probability is just over $\frac{1}{2}$ for speeds around $\frac{1}{2}v_p$, but at $2v_p$ the chance drops to about $\frac{1}{5}$. Only one molecule in 10,000 has a speed greater than $3v_p$.

The speeds of all molecules increase with temperature, and there is a Maxwell curve corresponding to each value of T. The curves corresponding to higher temperatures stretch progressively farther out toward large values of v, as Fig. 10.20 shows.

Practice set

1. The area under a Maxwell distribution curve (Fig. 10.18) has a finite value. What does it represent? HINT: Compare pages 152–153.
2. How are the areas under the curves in Fig. 10.20 related? Explain.

Mean free path and molecular size.

In deriving the Joule-Bernoulli equation, no assumption was made about the size of individual gas molecules, and they could well be imagined to be mere points having no extension in space. If this were true, however, no molecule would present any target area to any other, and there would be no collisions among them. In particular, there would be no probability of collision between gas molecules and those making up the wall of the container, so that it would be impossible to confine the gas at all. Obviously, molecules must have a finite cross-section—a fact already mentioned in Chap. 3. A moving molecule will go a finite distance between collisions with other molecules. The average distance traveled is called the **mean free path.** It is denoted by the symbol l.

The magnitude of the mean free path would be expected to depend on how crowded the molecules are (density), how fast they are moving (temperature), and how big their cross-section is (molecular diameter). Analysis shows that the mean free path for the molecules of a gas that follows the Maxwell distribution is

$$l = \frac{RT}{\pi\sqrt{2}\ d^2 N_0 P} \qquad\qquad 10.20$$

where d is the molecular diameter and N_0 is Avogadro's number (page 234).

Effects depending on mean free path.

Several phenomena observed on a laboratory scale for gases can be directly related to the mean free path of the molecules. The *diffusion* of a gas, described on page 227, is one example. The rate of this process must depend on l, which measures how far a given molecule will travel,

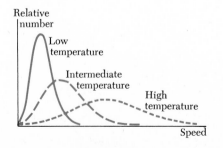

FIGURE 10.20 Maxwell curves for a sample of gas at various temperatures.

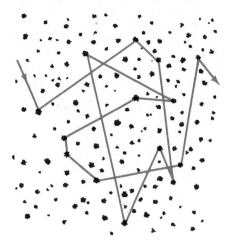

FIGURE 10.21 The successive collisions of a molecule with other molecules of the gas. All the molecules must be considered to move in this way.

on the average, before being deflected by an impact (see Fig. 10.21). It is found that observed rates of diffusion can be predicted from the properties and condition of the molecules.

Another phenomenon involving l is the *conduction of heat* by a gas (Chap. 11). On a molecular scale, this entails the transfer of heat (molecular KE) from a warmer to a cooler part of the gas by the motion of the molecules and hence must depend on l. Again, comparison of the theoretical relations with experiment gives satisfactory agreement.

A third property of fluids, *viscosity* (page 47), concerns the transfer of momentum from one layer of a fluid to an adjoining layer flowing past it. The magnitude of this effect depends on the average distance a molecule travels in passing from one layer to another. Here, too, the theory can be correlated satisfactorily with observations. The fact that all three of these phenomena—diffusion, heat conduction, and viscosity—give consistent values for the mean free path is convincing proof of the correctness of the kinetic-theory assumptions.

TABLE 10.2 MOLECULAR PROPERTIES OF GASES AT STANDARD CONDITIONS

	AIR*	HYDROGEN
Number of particles/m³ (the same for all gases)	2.7×10^{25}	2.7×10^{25}
Diameter of a molecule	3.7×10^{-10} m	2.4×10^{-10} m
Mass of a molecule	4.8×10^{-26} kg	3.3×10^{-26} kg
Mean free path	1.0×10^{-7} m	1.8×10^{-7} m
Number of collisions per second for each molecule	4.9×10^{9}	1.0×10^{10}
Most probable speed	360 m/sec	1,500 m/sec
Average speed	450 m/sec	1,700 m/sec
Root-mean-square speed	485 m/sec	1,840 m/sec

* Average for the mixture of nitrogen and oxygen molecules of which the air largely consists.

10.7 AN EVALUATION OF THE KINETIC THEORY

We have seen how the kinetic-molecular model, founded on simple assumptions about the constitution of matter, has served as a foundation for
a very comprehensive and detailed theoretical structure. This theory succeeded in giving reasonable explanations for a great variety of phenomena.
Starting with macroscopic observations, it went further to yield consistent
values for the properties of the unseen molecules by a variety of experimental methods. By any valid standard, the kinetic theory is one of the
most successful of all scientific generalizations.

The theory did not find general acceptance among scientists, even as
late as the end of the nineteenth century. Men as prominent as Ernst Mach
and the German physical chemist Wilhelm Ostwald (1853–1932) saw no
use for the concept of molecules. Eventually, overwhelming experimental
evidence—especially observation of the brownian motion—secured general
acceptance for the theory.

One of the most important results of this acceptance was recognition
of the fact that newtonian mechanics could be applied to the invisible
molecular world, making thermal phenomena understandable in mechanical
terms. Maxwell's great contribution was in showing that, although the
myriad particles in a gas cannot be followed individually, very definite
statements can be made about the behavior of the whole collection. As
stated by Chalmers (Ref. 10.5):

> To the Victorians, with their fervent belief in natural law, Maxwell's conclusions were
> little short of staggering. He opened their eyes to the possibility that regularity in
> physical phenomena did not necessarily imply the existence of "natural law," but
> might be a mathematical consequence of the operation of pure chance.

We shall find that the admission of the ideas of probability and chance
into physics had a strong influence on the further development of the
subject, especially in connection with quantum theory, which is needed
in order to deal with certain phenomena occurring in the world of very
small dimensions.

Programmed review

Instructions: See page 18.

1. Give an intuitive idea of the meaning of the term *temperature*.

    A measure of the apparent coldness or hotness of a body.   [10.1]

2. What values are assigned to the two fixed points on the Celsius
    temperature scale?

    Ice point, 0°C; steam point, 100°C.   [10.1]

3. In terms of the behavior of an ideal gas, what is meant by the absolute
    zero of temperature?

The temperature to which the gas would have to be lowered in order to make its pressure and volume zero, provided that it remains an ideal gas throughout the process.   [10.2]

4. Express the equation of state of an ideal gas.

$P_1V_1/P_2V_2 = T_1/T_2$, or $PV = nRT$.   [10.2]

5. What is the brownian motion, and what does it show?

The continual trembling motion of small, suspended particles. It is evidence for the motion of the molecules of the surrounding fluid.   [10.3]

6. State the Joule-Bernoulli equation for the pressure exerted by an ideal gas, and explain what each term stands for.

$P = mN\overline{v^2}/3V$, where P is the pressure, m the mass of each molecule, N the number of molecules, $\overline{v^2}$ the average of the squares of their speeds, and V the volume of the gas.   [10.4]

7. State Avogadro's law.

Equal volumes of all gases, at the same temperature and pressure, contain equal numbers of molecules.   [10.5]

8. What is the connection between the KE of gas molecules and the temperature of the gas?

The average KE is proportional to the absolute temperature.   [10.5]

9. What is the basic assumption made by Maxwell in deriving his relation for the distribution of molecular speeds in a gas?

All values of the KE of the molecules are equally probable.   [10.6]

10. Define the mean free path of a gas molecule.

The average distance traveled between collisions with other molecules.
                                                                          [10.6]

For further reading

10.1. *Zemansky, M.* "Temperatures, Very Low and Very High." Read pp. 1–16.
10.2. *Mott-Smith, M.* "Heat and Its Workings." Read chaps. 1–3 on thermometry and thermal expansion.
10.3. *Newman, J. R.* (ed.) "The World of Mathematics," vol. 2. Read pp. 771–777 on Daniel Bernoulli and his work on the kinetic theory.
10.4. *Born, M.* "The Restless Universe." A fascinating, simply written book on classical and modern physics. A novel feature is the set of animated sequences of phenomena, operated by flicking the pages with the thumb. Read chap. 1 at this point.
10.5. *Chalmers, T. W.* "Historic Researches." Read chap. VIII.

Questions and problems

10.1. When a solid block of material is heated, does a cavity inside the block change in volume? If so, does the volume increase or decrease? HINT: Imagine the cavity to be completely filled with the same material. Would you expect either compression of this filler or any voids to form when uniform heating takes place? On this basis, what do you conclude?

10.2. When a solid body is heated, the increase in any linear dimension is given by an empirical expression of the form $\Delta L = \alpha L \, \Delta t$, where L is the initial length, ΔL the increase in length, Δt the rise in temperature, and α the *coefficient of linear expansion* which may be considered constant over a limited temperature range. How much taller is the steel framework of the Empire State Building in summer than in winter? Assume the difference between the two extreme temperatures to be 54 C°. The value of α for steel is 1.3×10^{-5} per C° and the height of the frame is 1,470 ft.

10.3. Is there a definite answer to the question: By what factor does the volume of a sample of gas change when the Celsius temperature is doubled, the pressure remaining constant? Give reasons for your answer.

10.4. Air at $-5°C$ is drawn into a heater, emerging at a temperature of $30°C$. If the pressure is assumed to remain constant, by what factor is the volume of the air increased?

10.5. What mass of air escapes from a room of volume 50 m³ when the temperature is raised from 0 to $22°C$? The density of air at standard conditions is 1.29 kg/m³. HINT: The air pressure in the room may be considered to stay constant.

10.6. If the volume of an automobile tire is assumed to be approximately constant, by what fraction does the pressure in a tire increase when the temperature goes from -5 to $35°C$?

10.7. A tank was filled with air at a pressure of 25 lb/in.² when the temperature was $37°C$. When a valve was opened for a short time the pressure dropped to 21 lb/in.² and the temperature to $17°C$. What percent of the original air was released? HINT: Use Eq. 10.6.

10.8. A gas is confined in a cylinder equipped with a piston. Why does the temperature of the gas drop when the piston is suddenly pulled back? HINT: Compare qualitatively the speed of rebound and the speed of approach of a molecule that strikes the receding piston.

10.9. At what temperature is the rms speed of hydrogen molecules equal to that of nitrogen molecules at $27°C$? A nitrogen molecule is about 14 times as massive as a hydrogen molecule. HINT: Use Eq. 10.19.

10.10. To see the difference between an arithmetical average and a root-mean-square average, consider a set of values 2, 3, 3, 5, 7. Compute (a) their average value; (b) their rms value. Which is greater? Is this in accord with what was stated on page 237 for molecular speeds?

10.11.   Every second, 3×10^{23} helium atoms hit a wall of area 4.0 cm², each striking the wall at an angle of 60° to the normal and with a speed of 700 m/sec. The mass of a helium atom is 6.6×10^{-26} kg. What average pressure is exerted on the wall?

10.12.   Show that the dimensions of both sides of Eq. 10.14 are the same.

10.13.   One k-mol of water has a mass of 18 kg. How far apart, on an average, are the molecules in liquid water? By what factor does this distance increase when the water is changed to a vapor at standard conditions?

10.14.   Suppose a molecule in the air were to "lose" even as little as one-billionth (10^{-9}) of its KE at each collision with its neighbors. Using the value given in Table 10.2 for the number of collisions per second, compute about how long it would take (order of magnitude only) before all the KE of the molecules would be gone.

10.15.   If the evaporation of a liquid involves a greater rate of escape of the faster molecules, how does this affect the average speed of those left behind? Explain, on this basis, the cooling effect produced by evaporation.

10.16.   What meaning, if any, can be ascribed to the expression "the temperature of a molecule." Give reasons for your answer.

10.17.   Show that the average translational KE per molecule of a gas increases by 1.07×10^{-23} joule for each Celsius-degree rise in temperature. Use Eq. 10.19.

10.18.   A sample of gas is at 0°C. To what temperature, in Celsius degrees, must it be raised in order to double the rms speed of the molecules?

10.19.   A box of air at 300 K is moving at a translational speed equal to one-tenth the rms speed of the molecules. How much does the temperature rise when the box is suddenly stopped? Use Eq. 10.17.

10.20.   The speed of escape of a body from the earth is about 11 km/sec (page 208). Compare the value of v_{rms} for hydrogen high in the earth's atmosphere, at a temperature of around 1000 K, and decide if you would expect to find much of this gas there.

10.21.   The speed of sound waves in air at standard conditions is 331 m/sec. How does this compare with v_{rms} for air, given in Table 10.2? Since sound waves are transmitted from molecule to molecule, does it seem reasonable that both speeds are of the same order of magnitude?

10.22.   How long, on the average, would it take an atom of helium from the sun to reach the earth if the atom continued to move at the speed corresponding to the solar surface temperature of 6000 K? (Use Eq. 10.19, page 235). Consult page 237 and decide whether it is necessary to allow for the fact that $v_{\mathrm{av}} < v_{\mathrm{rms}}$.

10.23   Show that Eq. 10.20 can be written

$$l = \frac{1}{\pi \sqrt{2}\, d^2 N_0 q}$$

where q is the number of k-mol of gas per cubic meter.

10.24.   The density of matter in the space between the galaxies is estimated to be of the order of 10^{-27} kg/m^3. Assuming that most of this matter consists of hydrogen molecules of diameter 1.2×10^{-10} m and molecular mass of 2 kg/k-mol, find the mean free path of these molecules. HINT: Use the relation of Prob. 10.23.

10.25.   The best vacuum attainable in the laboratory is about 10^{-18} of standard air pressure. How many air molecules remain, per cubic meter, in such a vacuum? Use the first item in Table 10.2.

10.26.   Using the first and third items in Table 10.2, compute the density of air at standard conditions and compare with the value given in Table 3.4 (page 55).

10.27.   Using the value of the mean free path of "air molecules" given in Table 10.2, compute the average molecular diameter and compare with the value given in the table.

Chapter 11
HEAT AND ENERGY

Long before the development of the energy concept, heat was considered to be some kind of permanent, weightless, invisible fluid, called **caloric,** that could pass from one body to another. This idea seemed to be entirely reasonable and was developed to provide general explanations of a considerable number of heat phenomena. However, in some areas the caloric concept met with difficulties, and as early as the beginning of the eighteenth century, scientists began to speculate on the possibility of interpreting heat as molecular motion. Nearly a century passed before this concept was sufficiently confirmed by experiment to be generally accepted.

Important consequences followed the explanation of the nature of heat. One was the development of practical steam power, leading to the establishment of our industrial society. Another and perhaps more significant result was the extension of the notion of mechanical-energy conservation to include energy of all kinds: the general principle of the conservation of energy, which is taken up in this chapter.

THERMAL EQUILIBRIUM

11.1 QUANTITY OF HEAT; SPECIFIC HEAT

Black taught medicine and chemistry at Glasgow and, later, at Edinburgh. His work on heat was published after his death.

Toward the end of the eighteenth century reliable thermometers were perfected, and this made possible quantitative experiments on heat. One of the foremost investigators of heat was the British scientist Joseph Black (1728–1799). Although he thought of heat in terms of the caloric hypothesis, he succeeded in interpreting fundamental phenomena and in putting the subject on a consistent conceptual basis. He recognized the fact that, if bodies at different temperatures are put in an insulated enclosure, the hotter objects become cooler and the colder ones become warmer until all reach a common equilibrium temperature. By using measured quantities of hot and cold substances and carefully recording the initial and final

Black. (*British Information Services.*)

temperatures, he was able to show the fundamental difference between heat and temperature and to describe a way of measuring heat.

In experiments of this kind, the materials are put into a heat-insulated vessel called a **calorimeter** (see Fig. 11.1). One or more of the objects is usually in the form of a liquid in order to promote heat exchange.

If equal masses of the same material at different temperatures are mixed in a calorimeter, the final (equilibrium) temperature will be found to be halfway between the initial temperatures of the two. This is exactly what would be expected if heat were a fluid whose total amount in the two samples was constant. Whatever is given up by the hot body is taken up by the cold one, and, since equal masses are present, the exchange of a given amount of heat fluid would be expected to have equal effects on the temperatures of the two bodies.

In an unsymmetrical situation, the results are somewhat different, although still understandable in terms of a principle of heat conservation. It is obvious that if a cupful of hot water is poured into a swimming pool filled with cool water, the equilibrium temperature will no longer be midway between the two initial temperatures but will be very close to that of the big bulk of water originally in the pool. When Black performed experiments in which samples of *different* materials were mixed, he found a further complication. For instance, when a hot sample of a metal is put into an equal mass of cool water, the temperature drop of the metal turns out to

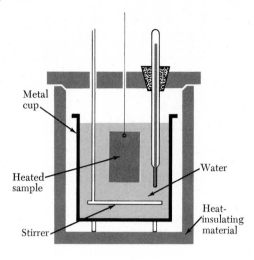

Metal
cup

Heated
sample

Stirrer

Water

Heat-
insulating
material

FIGURE 11.1 Cross-section of a simple mixture calorimeter. Other types, such as steam, combustion, and electrical calorimeters, are used for special purposes.

be of the order of 10 times the temperature rise of the water when equilibrium is reached.

It is possible to correlate all such observations in a straightforward way. Let ΔQ represent the quantity of heat (molecular KE) entering or leaving a body. This quantity may be positive or negative, depending on whether the energy is absorbed or emitted. Then, for any closed system of bodies, it is found that when no further temperature changes occur, the algebraic sum of the ΔQ's is zero:

$$\Delta Q_1 + \Delta Q_2 + \Delta Q_3 + \cdots = 0 \qquad\qquad 11.1$$

where 1, 2, 3, etc., enumerate the bodies that make up the system. It is assumed here that all energy exchanges occurring are purely thermal, that is, they involve only random molecular energy. Further, no phase changes are assumed to occur. Instances where such changes take place will be discussed starting with Sec. 11.3.

Observation shows that each term in Eq. 11.1 has the form

$$\Delta Q = cm\,\Delta t \qquad\qquad 11.2$$

where m is the mass of the body and c is a coefficient characteristic of the material of which the body consists. Over a restricted range of temperature, c may be taken to have a constant value for each substance, especially if the sample is either a liquid or a solid, which is usually the case in calorimetric experiments. It is called the specific heat capacity (or simply *Specific* **specific heat**) of the material. In order to fix numerical values for this *heat.* quantity, water is taken as a standard, with $c = 1$. Then c for any material represents its thermal capacity with respect to that of water, and because of the way c was defined, its numerical value is independent of the system of units used. In this regard, it is similar to *relative density* (page 54). Values of c for some common materials are shown in Table 11.1.

It should be noticed, from Table 11.1, that the value of the specific heat of water is considerably higher than that of other common substances. An important consequence of this fact is the influence of large bodies of water

TABLE 11.1   SPECIFIC HEATS OF VARIOUS SUBSTANCES

MATERIAL	SPECIFIC HEAT, c	TEMPERATURE (°C) OR RANGE WHERE VALUE APPLIES
Solids:		
Aluminum	0.22	0–100
Brass	0.094	15–100
Copper	0.093	0–100
Glass, window	0.20	20
Granite	0.19	12–100
Ice	0.53	−10
Iron, steel	0.11	20
Porcelain	0.26	15–950
Silver	0.056	0–100
Liquids:		
Ethyl alcohol	0.58	0
Glycerin	0.58	15–50
Mercury	0.033	17
Turpentine	0.41	0
Water	1	15

on climate. The high specific heat allows the temperature of sea water to increase only very slowly in the spring and to decline very slowly in the fall. This results in a moderation of the summer-winter temperature differences, particularly for coastal areas. Water may be said to have large thermal inertia.

Water is a good standard for measuring c because its own value of specific heat varies less than 1 percent in the range 0 to 100°C. The specific heats of all materials decrease toward zero as the temperature approaches absolute zero—a fact explained later by the quantum theory.

11.2 HEAT UNITS; SOME CALCULATIONS

Although we accept the interpretation of heat as a form of energy, it proves useful to have a purely thermal measurement of the quantity of heat and to define suitable units for this purpose. The procedure is usually the same in all cases where definition of a unit for a newly introduced quantity is needed: An algebraic relationship that expresses the facts is found, and *unity* is substituted for each quantity. For instance, in defining an absolute unit of force in mechanics, we started with the relation $F = ma$ and substituted $m = 1$ kg and $a = 1$ m/sec², thus defining a unit for F, 1 newton. The same procedure can be followed in any other system of units.

Here, for reasons to be explained later, we do not specify heat quantity itself but *changes* in this entity, ΔQ. Recall that a similar procedure was used in Chap. 9, where changes in GPE rather than absolute values of this quantity were of interest.

In order to define units for ΔQ, we start with Eq. 11.2, taking $c = 1$ (water), $m = 1$, and $\Delta t = 1$. Thus, in the mks system the unit of quantity of heat is 1 kcal. One kilogram calorie, or **kilocalorie** (kcal), is the quantity of heat that enters or leaves one kilogram of water when its temperature changes by one Celsius degree. For very precise work, the degree interval is specified as 14.5 to 15.5°C in order to take into account the slight variation in the specific heat of water. For ordinary purposes these differences can be ignored.

The kilocalorie is the unit used by dietitians in stating the nutritional values of various foods. Food is oxidized (burned) in the process of metabolism in the body. In determining food values in the laboratory, a sample is dried and then undergoes combustion in pure oxygen inside a calorimeter. The heat output is measured, and values such as those given in Table 11.2 are obtained.

Heat units appropriate to two other systems are sometimes encountered and will be mentioned here, although they will not be used in this book. In the cgs system, the unit is based on the gram rather than the kilogram and is called one **gram calorie,** or sometimes one small calorie. One gram calorie (abbreviated cal) is equal to 0.001 kilogram calorie. In the English foot-pound-second system, one **British thermal unit** (abbreviated Btu) is defined as the amount of heat entering or leaving 1 lb-mass of water when its temperature changes by 1 F°. 1 Btu = 0.252 kcal.

With units available for expressing quantity of heat, Eqs. 11.1 and 11.2 can be used to compute what will happen in thermal mixing experiments such as those described on page 246. Two examples are given:

Worked example 11.1

A 50-g block of aluminum is held in steam rising from water boiling at standard conditions. After enough time has passed to make certain that the block is at steam temperature, it is quickly lowered into 150 g of water in a copper cup of mass 80 g and the water is stirred until thermal equilibrium is reached. The cup and water are initially at 20°C. If no exchange of heat with surroundings is assumed, what is the final temperature of the system?

Solution   A convenient way of regarding Eq. 11.1 is to express it:

Heat given up by hot object(s) = heat taken on by cold object(s)

Each term in this equation will be of the form given by Eq. 11.2: $\Delta Q = cm\,\Delta t$. If we call the final equilibrium temperature t_e and take values of the

---

TABLE 11.2   FUEL VALUES OF VARIOUS FOODS

	kcal/g		kcal/g
Apple, raw	0.64	Ice cream, plain	2.10
Bread, white	2.66	Lettuce, leaf	0.20
Butter	7.69	Meat, lean	1.20
Cheese, Swiss	4.43	Milk, whole	0.72
Chocolate	6.31	Oil, olive	9.00
Egg, boiled	2.15	Orange juice	0.43
Gumdrops	3.50	Rice, cooked	1.12

specific heats of the metals from Table 11.1, the equilibrium relation becomes

$$(0.22)(50)(100 - t_e) = (1)(150)(t_e - 20) + (0.093)(80)(t_e - 20)$$

from which

$$t_e = 25.2°C$$

Notice that it was allowable to express all the masses in grams in the calculation. Why?

Worked example 11.2

The method of thermal mixing, as carried out above, is often used to determine the specific heat of a material.

A 150-g piece of lead is heated to 84°C and then immersed in 100 g of water initially at 16°C. If the final temperature is 19°C, find the specific heat of lead.

Solution   Equating the heat given up by the lead to the heat taken on by the water,

$$c(150)(84 - 19) = (1)(100)(19 - 16)$$

where c is the specific heat. The result is

$$c = 0.031$$

THERMAL PROCESSES

11.3 CHANGE OF PHASE; HEATS OF TRANSFORMATION

In the examples described above, the addition or removal of heat from a body was always accompanied by a change in its temperature. However, there are circumstances in which a sample of matter exchanges heat energy with its surroundings, yet keeps its temperature constant. This happens, for example, when the addition or subtraction of heat is accompanied by a change of phase (physical state).

Consider the thermal history of a kilogram of ice in a closed container

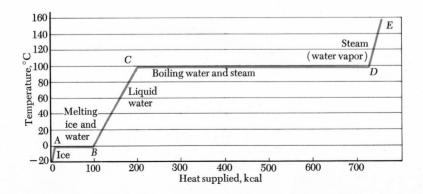

FIGURE 11.2 Thermal history of a sample of ice as it changes to liquid water and finally to steam

to which heat is slowly and continuously supplied. Assume that the sample is initially at a temperature of $-20°C$. When heat is applied, the ice does not immediately start to melt but warms up bodily, as any piece of solid matter would. The specific heat of ice is about 0.5 (Table 11.1), so that to warm the ice to $0°C$, $(0.5)(1)(20) = 10$ kcal of heat must be added. In Fig. 11.2, where the temperature of the sample is plotted against the quantity of heat supplied, the change just described is represented by the line OA.

It is assumed that the system is at standard air pressure throughout the experiment; otherwise the melting would not occur at exactly 0°C.

Now, with further application of heat, a thermometer embedded in the sample continues to read $0°C$, but the ice is observed to be *melting*. The melting proceeds with continued application of heat, but the temperature stays at $0°$ until the entire sample has melted. Measurement shows that, from the time the ice started to melt, about 80 kcal has been added. This stage of the process is indicated by the segment AB on the graph of Fig. 11.2.

With continued application of heat, the temperature of the water formed by melting the ice begins to rise until a quantity of heat amounting to $cm \, \Delta t = (1)(1)(100) = 100$ kcal brings it to $100°C$, as represented by the segment BC. This line has a smaller slope than OA because the specific heat of liquid water is greater than that of ice.

As additional heat is supplied, the water begins to boil. Bubbles of steam (water vapor) form throughout the liquid and rise to the surface, where they escape. All the while, the temperature of the system remains at $100°C$, and this condition continues until the entire liquid has been converted to the vapor phase. The total heat needed to change the kilogram of water at $100°C$ to steam at $100°C$ is determined to be about 540 kcal. The vaporization process is represented in Fig. 11.2 by the line CD. If the system is in a closed vessel, any further heat supplied to it only succeeds in warming ("superheating") the steam—a process represented in Fig. 11.2 by the line DE. This could in principle go on indefinitely. Steam has a specific heat of 0.48, so that this segment has very nearly the same slope as the line OA (see Prob. 11.8).

The experiment described above showed that at two stages heat energy had to be supplied merely to change the phase of the substance without producing any change in temperature. A quantity of heat associated with a change in phase is called a **heat of phase transformation,** and may be represented by the symbol L. The quantity of heat ΔQ associated with changing the phase of a sample of mass m is given by

Heats of phase transformation are also called latent heats.

$$\Delta Q = LM \qquad\qquad 11.3$$

where ΔQ is measured in kcal, m in kg, and L in kcal/kg.

The heat associated with the change from the solid to the liquid phase (melting, or fusion) is called the **heat of fusion.** This quantity was found to have the value $L_F = 80$ kcal/kg for water at standard pressure. The heat associated with the liquid-to-vapor transition is called the **heat of vaporization.** It has the value $L_V = 540$ kcal/kg for water at standard pressure. Values for substances other than water can be found in tables of physical data. Values ordinarily refer to unit mass of the substance and to the normal transition point.

Practice set Explain why Eq. 11.3, unlike Eq. 11.2, does not contain a factor Δt.

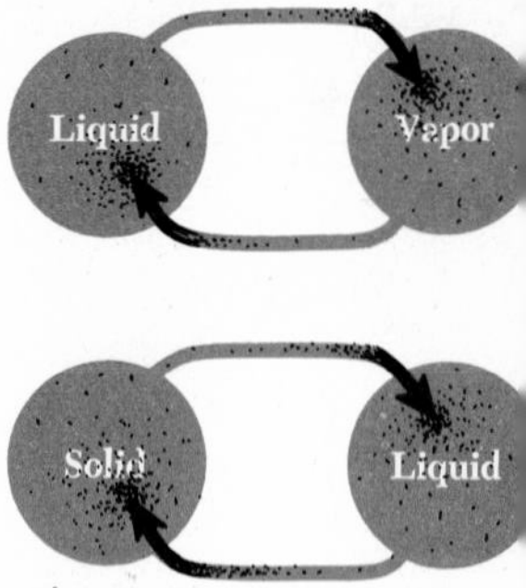

Heats of phase transformation can be interpreted on the basis of the kinetic-molecular model. In a solid, the particles are strongly bound to the lattice arrangement, and work must be done on them to set them free in the transition to the liquid state. This energy is supplied by the heat of fusion. This extra energy does not show up as additional KE, but as additional PE connected with the interparticle forces. As a result, there is no temperature rise connected with the process, since there is no increase in the average particle speed. In a similar way, the heat of vaporization must be supplied to the liquid to change it to a vapor, where the particles are so far apart that their mutual attractions are negligible.

FIGURE 11.3 Cyclical processes involving phase transitions (schematic).

Heats of phase transformation are associated not only with the transitions solid-to-liquid and liquid-to-gas but also with the changes in the reverse direction, liquid-to-solid and gas-to-liquid. The fact that application of heat makes a solid melt implies that heat must be given out when a liquid freezes. Thus, there is a heat of *crystallization,* or *solidification.* Similarly, heat is supplied to make a liquid boil, suggesting that heat is given out when a vapor condenses. This is called the heat of *condensation.*

The two heat quantities applying respectively to the two possible directions of transition between a given pair of phases must be equal in magnitude. Otherwise, it would be possible to obtain unlimited amounts of energy, without doing any work, by taking a sample repeatedly through cycles of phase transitions such as the ones shown schematically in Fig. 11.3. Such a limitless source of energy production without expenditure of work has never been found experimentally and is believed unattainable (see Sec. 11.7).

We shall find, below, that consideration of cyclic processes can be very important in the theory of heat.

In summary: With regard to numerical values,

1. Latent heat of fusion = latent heat of solidification. The value for water, at the ice point, is $L_F = 80$ kcal/kg.
2. Latent heat of vaporization = latent heat of condensation. The value for water, at the steam point, is $L_V = 540$ kcal/kg.

Worked example 11.3 A sample of steam initially at 110°C and a piece of ice originally at $-15°C$ come to the thermal equilibrium as liquid water at a temperature of 40°C at normal atmospheric pressure. Find the ratio of the mass of ice to the mass of steam used.

Solution Use the statement

Heat given up by hot object = heat taken on by cold object with the latent heats of phase changes included. Calling the mass of steam m_S and the mass of ice m_I, we can break down each side of this equality as follows:

1. Heat given up by
   a. cooling of steam to 100°C = $(0.48)(m_S)(10)$ = 5m_S kcal
   b. condensing of steam = $(m_S)(540)$ = 540m_S
   c. cooling of water formed, to 40°C, = $(1)(m_S)(60)$

                                            = 60m_S
                 Total heat given up: 605m_S kcal

2. Heat taken on by
 a. warming of ice to $0°C = (0.53)(m_I)(15)$ $= 8m_I$ kcal
 b. melting of ice $= (m_I)(80)$ $= 80m_I$
 c. warming of water formed, to $40°C, = (1)(m_I)(40)$

$$= \underline{40m_I}$$

Total heat taken on: $128m_I$ kcal

Equating the two quantities,

$$605m_S = 128m_I \quad \text{or} \quad \frac{m_I}{m_S} = 4.7$$

Worked example 11.4   In an experiment to determine the latent heat of fusion of ice, 50 g of ice at $0°C$ is dropped into 200 g of water in a 170-g copper calorimeter cup at $22°C$. When the ice has all melted and thermal equilibrium has been attained, the temperature is observed to be $3°C$. What value of L_F does this yield?

Solution   Equating heat given up by the water and cup to the heat taken on by the ice,

$$[(200 + (0.093)(170)](22 - 3) = 50L_F + (50)(3)$$

from which

$$L_F = 81 \text{ kcal/kg}$$

Notice again that it was allowable to retain all masses in grams when substituting values into the above equation.

11.4 PHASE EQUILIBRIUM

In the preceding section it was implied that when we continue to communicate heat to a solid, it goes first into the liquid state (melts) and eventually into the gaseous state (vaporizes). However, under certain conditions a material can change directly from the solid to the gaseous state, or the reverse. The process whereby a material changes directly from the solid to the gaseous state, or vice versa, is called **sublimation.** An example is the behavior of solid carbon dioxide ("dry ice") which, at ordinary temperatures, does not pass through the liquid state in changing from solid to vapor or vice versa. Another example is naphthalene ("moth flakes").

Corresponding to the process of sublimation is another latent heat, the heat of sublimation, defined in a way similar to the other two latent heats.

The boiling and freezing points of a given substance depend on the pressure to which it is subjected, the boiling point being especially sensitive to pressure changes. The bubbles of vapor that indicate boiling can form in a liquid only when the vapor pressure has become as great as the outside pressure. This explains the fact that on a mountain top, where atmospheric pressure is less than at sea level, water is observed to boil at temperatures considerably less than $100°C$.

The vaporization curve as a phase boundary.

The curve in Fig. 11.4 shows how the boiling temperature of water changes with the applied pressure. For example, at a pressure of 50 atm, the boiling point would be about $265°C$. The vaporization curve has an end point, for water, at $374°C$ and 218 atm. This is called the **critical point.** No substance can exist in the liquid phase at a temperature above its critical temperature, no matter how great a pressure is applied to it. Above that temperature it is always a gas.

The curve in Fig. 11.4 not only expresses the relationship between pressure and boiling point but also represents the dividing line between the two phases, liquid and gas. If the point that represents the state of a given sample lies above and to the left of the curve, the sample is in the liquid phase. If it lies below and to the right, it is in the gaseous phase. A change of phase can be brought about by crossing the vapor-pressure curve. For example, a sample of steam whose condition is represented by point A in Fig. 11.4 could be converted into liquid water by taking it to point B by any path, such as the one shown. As the path crosses the curve, the steam condenses.

Another way to effect the change of phase is to make an "end run" around the critical point (see Fig. 11.4). Beyond this point, the distinction between liquid and gas disappears, and the substance may be thought of as either a liquid or a highly compressed gas. It is best referred to merely as a fluid.

A phase-equilibrium diagram like that of Fig. 11.5 includes the solid phase. Each curve in this figure depicts the relationship between the particular pressures and temperatures at which pairs of phases can be in equilibrium. Each region lying between a pair of these lines is the range where only one phase exists.

All three curves meet at a place called the **triple point.** The **triple point** represents the condition at which all three phases can exist in equilibrium. For water, the coordinates of the triple point are $P = 0.0061$ atm, $t = 0.01°C$.

Unlike the vaporization curve, the fusion curve has no upper terminating point, at least at pressures that can be attained in the laboratory (up to about 100,000 atm). Thus there is apparently no point above which the solid and liquid phases merge.

The temperature 0.01°C, corresponding to 273.16 K, is now the basis for defining the Kelvin temperature scale (see Sec. 11.10).

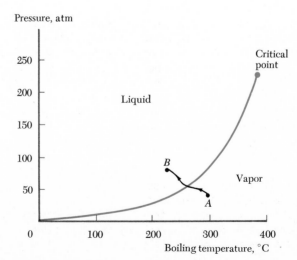

FIGURE 11.4 The vaporization curve of water. It represents a boundary between the liquid and vapor phases. *AB* is an arbitrary transition from the vapor to the liquid phase.

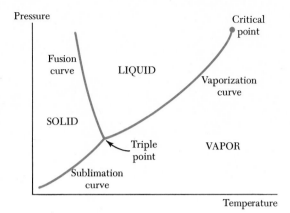

Practice set

1. A sample of ice at $-10°C$ and 0.015 atm is warmed 20 C° at this constant pressure. What is its final phase? Consult Fig. 11.5.
2. The same question when the sample is at a pressure of 0.004 atm.

11.5 HEAT TRANSFER

A transfer of heat takes place between two objects in contact, or two adjoining regions of one body, if they are at different temperatures. If no heat is exchanged with the surroundings, the temperature difference will disappear after a time, and the system will come to thermal equilibrium.

Heat conduction.

An exchange of kinetic energy and momentum between the molecules of two bodies in contact is called **heat conduction.** This exchange is caused by a **temperature gradient.** Temperature gradient is defined as the maximum rate of change of temperature with distance in passing from one place in a material to a neighboring one. Thus if the temperature drops by ΔT in going down the steepest path on a temperature "hill" a distance Δx from a given point P (Fig. 11.6), the temperature gradient at P has a value $\Delta T/\Delta x$. It may be expressed in C°/m.

In the simplest situation, consider an extended slab of solid material of constant thickness (Fig. 11.7), placed in thermal contact with a warmer object on one side and a cooler one on the other. The two objects are assumed to be large heat reservoirs whose temperatures remain constant.

Experiment shows that a constant thermal gradient is set up through the slab and that a constant rate of transfer of heat takes place by conduction. This rate is observed to be proportional to the cross-section area of the slab and to the temperature gradient, so that we may write

$$\frac{\Delta Q}{\Delta t} = kA\,\frac{\Delta T}{\Delta x} \qquad 11.4$$

where ΔQ is the quantity of heat transported in a time Δt, A is the face area of the slab, $\Delta T/\Delta x$ is the temperature gradient across the thickness of the slab, and k is a constant of the material, called its **coefficient of thermal conductivity.** Values for many substances may be found in tables of physical data.

FIGURE 11.6 The temperature gradient is measured by $\Delta T/\Delta x$.

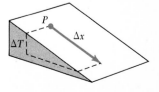

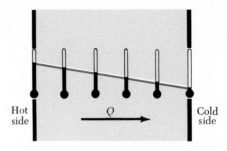

FIGURE 11.7 Thermal conduction. In the steady state, there is a constant thermal gradient through the slab, and heat energy is conducted at a constant rate.

Practice set*

What are the units for k if ΔQ is expressed in kcal, Δt in sec, ΔT in C°, Δx in m, and A in m²?

Metals have the highest values of k, of the order of 100 times those of nonmetals. The presence of free electrons among the atoms of metals accounts for the fact that they are the best conductors of heat as well as the best conductors of electricity. Application of heat to one face of a metal plate causes increased agitation of the atoms. They remain near their lattice positions but transfer some of their additional energy to the electrons, which are free to diffuse toward the cooler side, where they cause increased motion of the atoms. These electrons are responsible also for the conduction of electricity in the metal. In nonmetallic solids, the energy transfer takes place only through the atomic motion handed on along the lattice.

Worked example 11.5

A plate-glass window measures 3 by 5 m and is 0.6 cm thick. If the outer surface is at a temperature of 10°C and the inner one at 16°C, how much heat is conducted through the glass in 1 hr? The value of k for this glass is 2.2 × 10⁻⁴ kcal/sec-m-C° at ordinary temperatures.

Solution In the steady state, the temperature gradient in the glass is constant and equal to the entire temperature difference between the surfaces divided by the thickness. Using Eq. 11.4,

$$\frac{\Delta Q}{\Delta t} = (2.2 \times 10^{-4})(3 \times 5)\frac{16 - 10}{6 \times 10^{-3}} = 3.3 \text{ kcal/sec}$$

The total heat transferred in 1 hr is (3.3)(3600) = 12,000 kcal. This is a surprisingly large quantity of heat—enough to melt 150 kg of ice.

Convection.

Another kind of heat transfer, which is generally more effective than conduction, can take place in fluids. This is called **convection.** Convection is the transport of heat by the actual motion of a portion of heated liquid or gas. The motion is usually caused by buoyancy forces: A heated region of fluid expands, becomes less dense than the surrounding cooler portion and floats upward, while a cooler region sinks through warmer surrounding

Ans.: kcal/sec-m-C°.

fluid. Often the combination of these processes leads to a continuous circulation; this is the nature of ocean currents and the trade winds. The laws describing heat transfer by convection are highly complex and in most cases must be worked out with the help of empirical factors.

Fluids, particularly gases, are very poor heat conductors, largely because there are few molecules available per unit volume for transporting the energy. Thus, if convection is prevented by barriers, gases become especially good heat insulators. The thermal insulating properties of felt, wool, fur, etc., are due to air trapped in the pores and enclosed spaces in such materials.

Radiative heat transfer. Light and other forms of radiation can be a means of transporting heat energy from one body to another. This radiative transport is different from conduction and convection, however, in that it does not involve *direct* transfer of molecular motion. Instead, molecular motion is first converted into energy of radiation. When radiation strikes another body, it may be partly or wholly absorbed, and the absorbed energy is converted once more into molecular motion (Fig. 11.8).

All bodies exchange radiation with their surroundings. If an isolated body receives radiation energy at a greater rate than it sends out, the temperature of the body rises; if the rate of reception is less than the rate of emission, its temperature falls.

Every object at a temperature above absolute zero radiates to some extent, but the rate increases very rapidly with temperature. For thermal radiators of a specified nature, an exact relation between temperature and rate of radiation can be given. Further discussion will be found in Chap. 20. It should be mentioned at this point that attempts to explain the details of thermal radiation led to the development of the quantum theory.

In summary: Conduction, convection, and radiation are processes by which heat transfer can occur. The first two require actual contact, but the third can go on even in empty space, so that thermal energy can be transported from one body to another, with radiation acting as an intermediary.

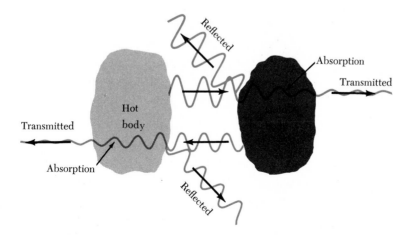

FIGURE 11.8 The radiation striking a given object is partly reflected, partly transmitted, and partly absorbed. The absorbed portion is converted into additional molecular motion (heat). The objects radiate not only to each other (as shown here schematically) but in other directions also.

11.6 EQUIVALENCE OF HEAT AND WORK

In Chap. 10 it was pointed out that, as early as the beginning 1700s, the kinetic theory was able to interpret heat as a form of mechanical energy: the KE of moving molecules. However, this idea was slow in winning acceptance. Meanwhile, the caloric theory seemed to furnish a satisfactory model for explaining, at least qualitatively, such phenomena as thermal expansion, change of phase, heating of a gas by sudden compression, etc. And in the case of thermal mixing, the simple assumption of conservation of the amount of caloric fluid gave quantitative results that agreed with experiment.

During the early years of the study of heat phenomena, the connection between mechanical energy and heat seems to have gone unnoticed although there are many obvious processes that point to it, going all the way back to the kindling of fire by rubbing sticks together. The observations made by Rumford near the end of the eighteenth century showed conclusively that *heat can be produced by doing mechanical work*. While supervising the boring of a gun barrel, he was impressed by the large and apparently unlimited amount of heat generated by the cutting tool and noted that the amount of heat developed increased with the work done by the horse that provided the motive power for the lathe. In another experiment, heat generated by friction caused water to boil.

Rumford concluded from such observations that it was difficult to think of heat as a fluid. He tried to detect the weight of caloric fluid, without definite results. But he succeeded in making rough measurements of the quantity of heat obtained from a given amount of mechanical work and so arrived at the first numerical estimate of the heat-work equivalent.

It required the careful and ingenious experiments of Joule (see biographical note on page 229) to clarify the relationship. By comparing the heat obtained from an electric current with the work done in turning the generator, he arrived at a value for the heat-work equivalent not far from Rumford's rough figure. Moreover, he checked this value by a variety of other experiments extending over about 40 years, using processes such as solid friction, the fluid resistance of liquids, and the compression of air. Regardless of the method used, the "rate of exchange" between heat and work was always found to be the same, within the limits of experimental error. The best present value of what is generally called the **Joule equivalent,** J, is

$$J = 4{,}185.5 \text{ J/kcal} \tag{11.5}$$

The corresponding value in English units is $J = 778.1$ ft-lb/Btu.

The heat equivalent of a moderate amount of mechanical energy is surprisingly small. Conversely, large and obvious amounts of mechanical energy seem to disappear in some processes, yet delicate measurements are often needed to show that heat is generated at all. Numerical calculations will illustrate this.

Count Rumford (Benjamin Thompson; 1753–1814) was one of the most colorful characters in the history of science. He was a politician, adventurer, and engineer as well as a scientist, and was the founder of the Royal Institution of London, which fostered some of the most important scientific research of the nineteenth century. See Ref. 11.2.

The Joule equivalent.

Worked example 11.6

Joule measured the increase in temperature of the water at the bottom of a waterfall. Assuming that all the initial GPE of the water at the top is converted into heat on striking the rocks below, how much higher is the temperature of the water at the base of Niagara Falls (height 50 m) than at the top?

Solution   Consider a mass m kg of water. Referred to the base of the falls, its GPE at the top is $mgh = m(9.8)(50) = 490m$ joules. According to Eq. 11.5, the equivalent quantity of heat is $490m/4186$ kcal. Equating this to $cm\ \Delta t$,

$$\frac{490m}{4186} = (1)(m)(\Delta t) \qquad \text{or} \qquad \Delta t = 0.12\text{C}°$$

Notice that the mass m drops out of the equation, as in previous examples.

Worked example 11.7   How fast must a 1,300-kg automobile travel in order that its KE be equivalent to the energy needed to heat a liter of water (one kilogram) from the ice point to the steam point?

Solution   The KE is $\frac{1}{2}(1,300)v^2$ joules, where v is the speed in meters per second. The required quantity of heat is $(1)(1)(100) = 100$ kcal. Changing the last quantity to mechanical energy by multiplying it by J and equating to the KE:

$$(100)(4,190) = \tfrac{1}{2}(1,300)v^2 \qquad \text{from which} \qquad v = 25.4 \text{ m/sec}$$

This is just under 60 mi/hr.

Because most mechanical speeds are quite small compared with average molecular speeds at ordinary temperatures (see page 234), there is a great numerical disparity between equivalent amounts of mechanical and heat energy.

11.7   CONSERVATION OF ENERGY

Even before Joule's work on the equivalence of heat and mechanical energy, there were indications of the existence of a broader principle concerning the transformations of one form of energy into another. J. R. Mayer (1814–1878), a German physician working in Java, observed that blood drawn from the veins of patients in the tropics was redder than samples taken in the temperate zone. This suggested to him that there must be a lower rate of loss of body heat in equatorial regions. From here he went on to surmise:

Let the quantity of mechanical work performed by an animal in a given time be collected and converted by friction or some other means into heat; add to this the heat generated immediately in the animal body at the same time, we have then the exact quantity of heat corresponding to the chemical processes that have taken place.

Pushing on to the final generalization, Mayer and others concluded in effect that it is impossible to create or destroy energy; any decrease in the amount of one form is necessarily accompanied by an equal total increase in one or more other forms. This is one way of stating what has since become known as the **principle of the conservation of energy.** Mayer himself phrased the idea in other equivalent forms, among them: "Energies are indestructible, convertible entities. Energy once in existence cannot be annihilated, it can only change its form.''

Unlike the limited conservation principle for mechanical energy discussed in Chap. 9, these statements refer to the constancy of the total amount of *all kinds* of energy in a system, including heat, chemical, electrical, and all other forms. Mayer discussed the extension of the principle to chemical, astronomical, and biological processes. However, his ideas were widely opposed, and they did not get recognition for almost 20 years after he first announced them in 1842.

Experiments such as those performed by Joule involve the transformation of work into heat, but the reverse process can be carried out also. Any *heat engine,* such as the steam engine or the internal-combustion engine, is a device for converting heat energy to mechanical work. Similarly, there are devices and processes for converting almost any form of energy into other forms, often in either direction. A few examples involving heat are shown schematically in Fig. 11.9.

Conservation of energy implies the impossibility of constructing a device that would furnish unlimited amounts of energy without input of some form of energy. This statement is given further discussion below. We can apply it here, however, to draw an important conclusion about any energy-conversion process and its reverse: Equivalent amounts of energy are involved in the two processes, and the ''rate of exchange'' between two energy forms is exactly reciprocal. For instance, Eq. 11.5 says that, in the conversion of work into heat, 1 kcal of thermal energy appears for every 4,186 joules of mechanical energy dissipated. Then it must be equally true that, for each kilocalorie of heat that is completely converted to mechanical work, 4,186 joules of mechanical energy must appear—a fact that is not obvious without the conservation principle.

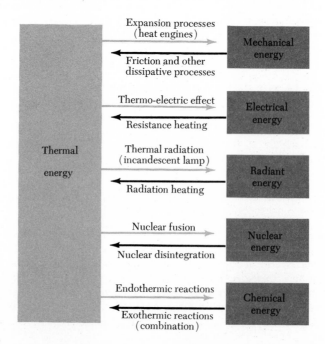

FIGURE 11.9 Thermal energy is mutually convertible with other forms. The various conversions that exist among the nonthermal varieties are not shown here.

See page 583n for a recent confirmation of this principle.

Conservation of energy is without question the most significant and far-reaching generalization in all classical science. The qualification "classical" must be made because the theory of relativity showed, half a century later, that even the conservation-of-energy principle had to be modified. Significantly, it was found possible to preserve this principle by broadening its scope (page 517).

Conservation of energy, like all the important general laws of physics, is not based on mathematical proof but on the concordance of countless observations. There have been one or two notable instances where the principle seemed to fail, but each time it was found to be upheld. As a result, this concept attained even greater stature than before.

THERMODYNAMICS

11.8 THERMODYNAMIC PROCESSES: THE FIRST LAW

A gas can possess energy in addition to the translational KE of its molecules. In the preceding chapter it was mentioned that there may be various forms of potential energy, arising from chemical binding forces between

GLOBAL ENERGY

Man is the only creature who has been able to increase his energy supply far beyond that which his own muscles can provide. Because of its dominant place in modern life, energy is no longer only a scientific or technological topic but a potent social force as well. Our dependence on energy utilization is characterized by two important problems: the extent of our energy resources, and the effect of power production on the environment.

Global-energy statistics are usually expressed in a unit called one Q, which is equal to 10^{18} Btu, or about 1.05×10^{21} joules. Virtually all the energy used on earth can be traced back to radiation received from the sun. Although estimates differ widely, we can assume that about 200 Q/yr is stored by the earth. World energy "consumption" (utilization) amounts to about 0.2 Q/yr at present. This is a small fraction of the above figure, but it would be unrealistic to take comfort in this fact, since the atmosphere is in very delicate thermal balance. Moreover, according to some authorities, the demand could easily reach 1.5 Q in the next 50 years.

About one-third of the world's energy consumption occurs in the United States. It is estimated that about 22 Q in known, recoverable fossil fuels (coal, petroleum, gas) and about 12,500 Q additional in potential reserves are available here. However, some fuels are in quite limited supply, and prompt steps should be taken to regulate their use.

Uranium and thorium in the earth's crust appear to be adequate to furnish nuclear fission fuels for a few billion years, and there is a practically unlimited supply of deuterium in seawater which is available if we find a way to produce energy by controlled nuclear fusion.

Although the reserves of fuel of various kinds seem to be ample for a long time to come, environmental considerations point to the advisability of increased conversion to nuclear power. This, however, brings with it certain difficulties, such as the threat of thermal pollution (see Box, page 219).

For a detailed discussion of the effects of human energy production, see Ref. 11.9.

the atoms and long-range forces between the molecules themselves. Further, if the molecules consist of more than one atom each, the atoms may have rotational or vibrational energy. All forms of energy, taken together, constitute the **internal energy** of a system.

When a system is in any given condition it has certain values of characteristic quantities called **thermodynamic variables.** These quantities may include temperature, pressure, volume, chemical characteristics, magnetic properties, etc. Sometimes it is possible to discover a relation between thermodynamic variables. The relation between thermodynamic variables is called an **equation of state.** An example is the ideal-gas law described in the preceding chapter.

The word state as used here means only a given combination of values of the thermodynamic variables and should not be confused with its use in "change of state" (better, change of phase).

The internal energy U of a system at any given time depends only on its thermodynamic state, that is, on the values of the thermodynamic variables at that time.

The thermodynamic state of a system may change. During such a process the system may absorb or lose heat; it may, at the same time, do work on its surroundings or have work done on it. For example, compression of a gas is usually accompanied by temperature changes: A tire becomes warmed when pumped up, and a gas allowed to expand rapidly is noticeably cooled.

The study of the interrelations of work, heat, and internal energy is called **thermodynamics.** Ordinarily, thermodynamics deals with systems that are at or very near a condition of equilibrium. It is assumed that any changes that take place proceed from one equilibrium state to an adjoining one, so that any change from one state to another can be considered to be a succession of slow, gentle processes. Such changes are called *reversible,* implying that a slight alteration in conditions could make the change proceed in the opposite direction.

Consider a confined sample of a gas that is subjected to changes in its thermodynamic state. On a graph of T versus P (Fig. 11.10) let the gas be initially in the state represented by point 1, corresponding to an absolute temperature T_1 and pressure P_1. Let the state of the sample be changed by directing it over the path a in the diagram, moving its state to point 2, for which $T = T_2$ and $P = P_2$. Assume that, in carrying out this change, an amount of heat ΔQ was supplied to the system and that the system performed an amount of work ΔW on its surroundings.

Now repeat the process, but let the system move over some other path, such as b, in going from point 1 to point 2. Continue in this way, using any other route. It is found that, although ΔQ and ΔW separately depend on the path, their difference $\Delta Q - \Delta W$ has the *same* value, independent of the particular path that is followed. However, the magnitude of this difference *does* depend on the choice of the *end points* 1 and 2.

The situation suggests the familiar one discussed in Chap. 9 in which an object is moved slowly between two points in a gravitational field. In the absence of friction, the work done depends only on the initial and final locations and not at all on the path taken in moving between them. The work done in this case is equal to the difference in the gravitational potential energy at the two places. Similarly, in the thermodynamic case the difference between ΔQ and ΔW is, by conservation of energy, equal to the

FIGURE 11.10 The value of $\Delta U = \Delta Q - \Delta W$ is the same for all paths between the two given states.

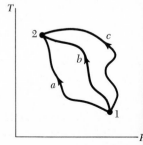

accompanying change in the internal energy U of the system. Thus

$$\Delta U = \Delta Q - \Delta W \qquad\qquad 11.6$$

where all three quantities are expressed in *one* kind of unit, say joules. Stated in this form, the conservation-of-energy principle is known as the **first law of thermodynamics.**

In using Eq. 11.6 it must be remembered that:

1. ΔQ is considered *positive* when heat *enters* the system and *negative* when it *leaves*.
2. ΔW is *positive* when work is done *by* the system and *negative* when work is done *on* the system.

11.9 THE SECOND LAW OF THERMODYNAMICS

The first law of thermodynamics, which states that energy is conserved in all processes, is universally applicable, as we have already seen. However, this law is not able to predict whether any specific process will actually occur. Imagine an arrangement like one that was used by Joule for determining the value of J, in which mechanical work is done in stirring a vessel of water and, as a result, the water becomes slightly warmed. This process is observed to satisfy the conservation of energy: A given amount of mechanical work always converts into an equivalent amount of thermal energy.

The same should be true for the reverse process, but this process is not observed to occur by itself. No one has ever reported seeing a vessel of water spontaneously start swirling while its temperature drops a little, or a stone suddenly leap up to the top of a cliff as it cooled slightly, although such events could perfectly well satisfy the conservation of energy.

It is easy to convert a given amount of mechanical energy completely into heat by a variety of different procedures, but there is no known way of changing a given amount of heat into work, with no heat escaping in the process and no frictional loss.

An independent proposition, supplementing the first law, has been formulated to show what processes can be expected to take place. It is called the **second law of thermodynamics.** This law is subtle, and there are several equivalent ways of stating it. Some of them will be given below without systematic proof, since the details are beyond the scope of this discussion.

The science of thermodynamics developed largely because of practical interest in determining how much work could be obtained from a steam engine. Early in the nineteenth century, the French scientist and engineer N. L. S. Carnot (1796–1832) began to analyze physical processes in which a sample of matter is taken through a *cycle* of changes of its thermodynamic state.

The operation of any kind of heat engine—internal combustion, steam, diesel, etc.—is a cyclic process. In each cycle, heat from some high-temperature source, such as the boiler of a steam engine, is applied to the *working substance,* which is usually a gas. The engine delivers some

Carnot. (*Courtesy of Historische Bildarchiv Handke-Bad Berneck.*)

external work—for instance, by the expansion of the gas against a movable piston. This work is in exchange for heat, but it represents only a fraction of the total heat that was taken from the high-temperature reservoir. The rest is discharged to a lower-temperature body such as the outside air or a steam condenser.

The general nature of such a process is shown schematically in Fig. 11.11. After the completion of each cycle, the working substance is left in the same state as before, so that there is no change in the internal energy U. Thus the work output is equal to the difference between the heat energy taken in at the higher temperature and the heat energy discharged at the lower temperature. The crucial fact here is that, always, *some heat remains unused* and is discarded later in the cycle.

Figure 11.12 should be examined carefully.

Carnot made a study of a hypothetical engine in which an ideal gas is taken reversibly through the cycle shown in Fig. 11.12. He was able to derive a number of important results from this theoretical work. One of the most significant of his conclusions has already been mentioned: It is inherently impossible to construct any device that would convert a quantity of heat energy completely into mechanical energy. This is one way of

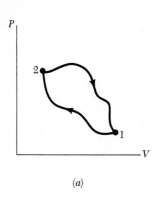

(a)

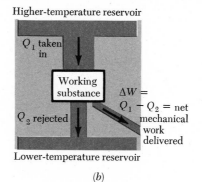

Higher-temperature reservoir

Q_1 taken in

Working substance

$\Delta W = Q_1 - Q_2 =$ net mechanical work delivered

Q_2 rejected

Lower-temperature reservoir

(b)

FIGURE 11.11 (a) Pressure-volume graph of a complete cycle. (b) Energy flow taking place in one cycle (schematic).

stating the second law of thermodynamics. Other useful statements will be given below.

Efficiency of a heat engine.

Carnot also analyzed the efficiencies of heat engines. The efficiency e of a heat engine may be defined as the ratio of the useful work obtained from it to the heat energy put into it. In symbols,

$$e = \frac{\Delta W}{Q_1} \qquad 11.7$$

where Q_1 is the amount of heat taken in at the higher temperature. Since the process is cyclical, ΔU is zero, and Eq. 11.6 (page 263) becomes simply

$$\Delta W = \Delta Q = Q_1 - Q_2$$

where Q_2 is the heat rejected at the lower temperature. Then Eq. 11.7 can be written

$$e = \frac{Q_1 - Q_2}{Q_1} = 1 - \frac{Q_2}{Q_1} \qquad 11.8$$

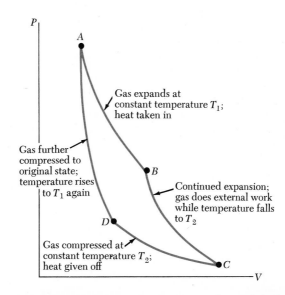

Gas expands at constant temperature T_1; heat taken in

Gas further compressed to original state; temperature rises to T_1 again

Continued expansion; gas does external work while temperature falls to T_2

Gas compressed at constant temperature T_2; heat given off

FIGURE 11.12 The Carnot cycle, represented on a pressure-volume graph. The cycle begins at point A, proceeding in the clockwise direction. All processes are assumed to take place slowly (reversibly).

It is possible to prove that the two amounts of heat exchanged are proportional to the absolute temperatures at which these changes occur:

$$\frac{Q_2}{Q_1} = \frac{T_2}{T_1}$$

and so the expression for efficiency given above becomes

$$e = 1 - \frac{T_2}{T_1} \qquad\qquad 11.9$$

This equation says that the ideal efficiency of a reversible engine depends only on the temperatures of the two heat reservoirs and that the greater the temperature difference, the more efficient the process. The efficiency does not depend on the nature of the working substance.

It can be shown that the statements in the preceding paragraph are true not only for the Carnot cycle but for any reversible cycle operating between the same two temperatures. Moreover, the results can be applied to systems that are not "engines" in the usual sense—to chemical, electrical, magnetic, and other processes. Thus the work represented by ΔW in Eq. 11.6 need not be mechanical but may be electromagnetic, chemical, or of some other form. Such generality makes thermodynamics an extremely useful subject, with a wide range of applicability.

Real engines, which are not reversible because of inevitable dissipation of energy in the mechanism, have lower efficiency than the value given by Eq. 11.9. For example, a steam engine can ordinarily change only a few percent of the heat of its fuel into mechanical work (see Prob. 11.28). Gasoline and diesel engines can be somewhat more efficient largely because the heat is generated within the engine cylinders rather than in a separate boiler.

11.10   THERMODYNAMIC TEMPERATURE SCALE; ALTERNATIVE FORMULATIONS OF THE SECOND LAW

Carnot had obtained all his results on the basis of the caloric theory. One of the changes Kelvin made was to re-interpret them in terms of the emerging concept of heat energy.

The work of Carnot, published in 1824, did not begin to gain general acceptance until it was discussed and improved by Kelvin a quarter of a century later. Kelvin showed that consideration of ideal-engine cycles could lead to a definition of the absolute zero independent of conclusions based on the behavior of gases. In effect, Eq. 11.9 gives the basis for his argument. Imagine a reversible heat engine that could take in heat at some higher temperature, convert *all* this heat to mechanical work, and so have none left to reject at the lower temperature. Such an engine would have an efficiency of 1 (100 percent). According to Eq. 11.9, this would be possible only if T_2 were zero while T_1 remained finite; hence such a cycle serves to fix the absolute zero of temperature. This is, in fact, the basis for the Kelvin scale of temperature, which can be shown to be identical with the absolute scale based on the behavior of an ideal gas.

According to thermodynamic theory, it is fundamentally impossible to reach absolute zero exactly, although temperatures down to about a millionth of a degree have been produced by ingenious experiments. The idea that this point cannot be attained becomes understandable if we realize that the ratio of two absolute temperatures is more fundamental than their difference. This implies that temperature is best measured on a logarithmic scale, such as the one in Table 10.1 (page 224), although linear scales are normally used on practical thermometers. The unattainability of absolute zero is thus connected with the fact that no matter how far a logarithmic scale is extended toward lower temperatures, there is no place on it for $T = 0$.

A statement of the second law of thermodynamics as given by Carnot appears on page 264. It is useful to repeat this statement here, together with two other forms that were expressed by the German physicist Rudolph Clausius (1822–1888) and by Kelvin.

Carnot   It is impossible to construct a device that would convert a quantity of heat energy completely into mechanical energy.

Clausius   It is impossible for any self-operating device to take heat continuously from one body and to deliver it to another that is at a higher temperature.

Kelvin   As a result of natural processes, the energy in our world available for work is continually decreasing.

All these formulations can be shown to be equivalent. An additional statement will be developed in the next section.

11.11 ENTROPY

It has been shown that bulk mechanical energy can be changed into heat completely and easily but that this is not true for the reverse process. Natural processes always tend to move toward a more probable state, and the disorderly molecular motion called heat represents a more likely condition than the orderly motion of matter in bulk. A state of disorder is one of very many possible configurations of the elements that make up a system, whereas the states recognized as orderly are very few.

The natural tendency toward disorganization and randomness is a familiar one in many real-life situations. More specifically, suppose a collection of a large number of consecutively numbered objects is put into a box, which is then thoroughly shaken. When the objects are drawn out at random, the chance that they will turn up in any particular numerical order is extremely small. In the same way, if two different gases are initially put in adjoining chambers of a container and the wall between them is removed, the gases will diffuse through each other, ultimately producing a uniform mixture. It is practically infinitely improbable—although not *entirely* impossible—that at some later time the two gases will again be found completely separated, each in one of the chambers.

The term "entropy" comes from the Greek word for 'transformation.''

In 1854, Clausius defined a mathematical quantity which he named **entropy** to serve as a measure of the degree of disorder of a system. The entropy S is so defined that, for any small, reversible change ΔQ in the heat content of a body, taking place at a temperature T, the change in entropy ΔS is given by

$$\Delta S = \frac{\Delta Q}{T} \qquad\qquad 11.10$$

Using the entropy concept, the second law may be given still another formulation: Changes in nature tend to take place in such a way that the entropy of an isolated system is never observed to decrease. In any actual system there is an increase in entropy; only in an ideal, reversible process can the change be zero. The entropy of a system was shown by Boltzmann (page 236n) to be connected mathematically with its state of disorder.

Worked example 11.8   Find the change in entropy that occurs when 1 kg of water at 61°C is mixed with 0.5 kg of water at 58°C.

Solution   By equating the heat, in kilocalories, given up by the warmer water to that taken on by the cooler,

$$(1)(1)(61 - t) = (1)(0.5)(t - 58)$$

so that the equilibrium temperature is $t = 60°C$. Thus the quantity of heat exchanged is equal in magnitude to 1 kcal: +1 kcal for the heat absorbed by the cooler body and −1 kcal for the warmer, since heat absorbed is considered positive.

Thus the entropy change on the part of the cooler water becomes $\Delta Q / T_C$; this amounts to $+1/(273 + 58) = 1/331$ kcal/K. For the warmer water, the entropy change is $\Delta Q / T_H = -1/(273 + 61) = -1/334$. These figures are approximate, since the temperature changes continuously along with the entropy change. An exact computation would require calculus.

The net change in entropy of the system is then

$$\frac{1}{331} - \frac{1}{334}$$

which is of the order of 10^{-5} kcal/K. This number is positive, showing that the entropy of the system has *increased*. From the preceding discussion, this implies that the disorder has also increased.

Entropy and life processes.

There are ways of deliberately separating mixtures, of course, but they do not violate the second law. Work must be done to carry out such a separation, and the increase of entropy accompanying this work is always greater than the decrease of entropy of the mixture. In the whole process, the total entropy increases. The same is true for processes that occur in living matter. Even the simplest forms of life represent highly ordered structures, and it is extremely improbable that they can be created merely by mixing the needed amounts of chemicals. However, by existing and developing, organisms raise the total entropy of the universe, and this exceeds the decrease of entropy that corresponds to the greater orderliness they succeed in bringing about.

Maxwell's highly original ideas on this subject are described in Ref. 11.8.

The practical usefulness of thermodynamics is extremely far-reaching. It has clarified the whole subject of heat and serves to establish a rational temperature scale. It is indispensable in the design and improvement of heat engines and refrigeration systems. Applied to chemistry, it has led to a better understanding of chemical processes and to the development of many new products.

In addition to such practical achievements, thermodynamics appears to have important implications for philosophy and cosmology as well. An interesting result of the second law is that it prescribes a definite *direction* of flow for time. The laws of physics considered up to this point are all reversible with respect to time, so that if the directions of motion of all the particles of a system are reversed, the same laws hold as before. Thus Newton's laws, conservation of energy, etc., would still be valid. However, the second law of thermodynamics asserts that natural processes always operate in the direction of greater disorder as time passes—that is, of two states of an isolated system, the more disorderly state is the later one.

The conversion of other energy forms into heat always results in an increase in molecular disorder—an increase in entropy—and the fate of all kinds of energy is their ultimate conversion into heat. It is as if all forms of energy other than heat were ripples on a vast thermal sea—ripples that eventually die out.

There is no reason to doubt that the universe as a whole is tending toward a final state in which all matter will be at the same temperature and all forms of energy except perhaps gravitational PE and the energy equivalent of matter (page 49) will have been converted into heat. The German physical chemist Walther Nernst (1864–1941) called this state of maximum entropy, marked by an end to all natural processes, the "thermal death" of the universe. Not all scientists are willing to accept this outlook, and some astronomers have suggested that there may be distant parts of the universe where the second law does not hold. So far, this conjecture has no observational justification.

Entropy, time, and cosmic evolution.

The second law of thermodynamics has been referred to as me's arrow."

Programmed review

Instructions: See page 18.

1.   State the equation that defines quantity of heat.

$\Delta Q = cm \, \Delta t$, where ΔQ is the quantity of heat taken on or given off by a body, Δt is the temperature change, and c is the specific heat of the material.
[11.1]

2.   Define a kilocalorie.

A quantity of heat equal in magnitude to that exchanged by 1 kg of water when its temperature changes by 1 C°.   [11.2]

3.   What is meant by a latent heat?

The quantity of heat involved in changing the state of unit mass of a substance at a specified transition temperature. The change may be vaporization, solidification, sublimation, etc.   [11.3]

4.   Define the triple point of a substance.

> The temperature and pressure at which all three phases of the substance can exist in equilibrium.   [11.4]

5. What is the principal means of heat transfer in (*a*) a solid, (*b*) a fluid, (*c*) a vacuum?

> (*a*) Conduction, (*b*) convection, (*c*) radiation.   [11.5]

6. What is meant by the Joule equivalent?

> The factor giving the numerical relation between a quantity of mechanical energy and the corresponding amount of heat energy.   [11.6]

7. Give a simple statement of the principle of the conservation of energy.

> It is impossible to create or destroy energy; what disappears in one form must reappear in another. The total energy of an isolated system stays constant in time.   [11.7]

8. Define thermodynamics.

> The study of the interrelations of work, heat, and internal energy.   [11.8]

9. What is the first law of thermodynamics?

> A statement of the conservation of energy in terms of heat exchanged, work done, and change in internal energy.   [11.8]

10. Describe the connection between the operation of a Carnot engine and the absolute (Kelvin) scale of temperature.

> The absolute zero can be shown to be the temperature of the cold reservoir of an ideal heat engine when its efficiency becomes 100 percent.   [11.10]

11. What property of a collection of a large number of particles is measured by the entropy of the system?

> Its state of disorder.   [11.11]

12. State the second law of thermodynamics, using the entropy concept.

> Natural changes tend to take place in such a way that the entropy of an isolated system increases.   [11.11]

For further reading

11.1. *Dyson, F. W.* What Is Heat? *Sci. Am.,* September, 1954, p. 58. Includes biographical sketches of Carnot, Rumford, and Joule supplied by I. B. Cohen.
11.2. *Brown, S.* ''Count Rumford: Physicist Extraordinary.'' A fascinating biography of this colorful figure.
11.3. *Einstein, A.,* and *L. Infeld* Conservation of Energy, in A. Beiser (ed.), ''The World of Physics,'' p. 45.
11.4. *MacDonald, D. K. C.* ''Faraday, Maxwell and Kelvin.'' Useful especially at this point for the description of Kelvin's work on heat.

11.5. *Sandfort, J. F.* "Heat Engines: Thermodynamics in Theory and Practice." Particularly interesting for the history and development of heat engines.

11.6. *Gamow, G.* Entropy, in A. Beiser (ed.), "The World of Physics," p. 69. A clear and simple description of the concept.

11.7. *Whitrow, G. J.* "The Natural Philosophy of Time." The relation of thermodynamics to the direction of time is discussed.

11.8. *Ehrenberg, W.* Maxwell's Demon, *Sci. Am.,* November, 1967, p. 103. Can a hypothetical being, capable of dealing with individual molecules, violate the second law?

11.9. *Scientific American,* September, 1971. An entire issue devoted to articles on energy and power.

Questions and problems

11.1. Seven-tenths of a kilocalorie of heat is supplied to a 500-g sample of granite whose temperature thereby rises by 7.4 C°. From these data, what is the value of the specific heat of granite?

11.2. A porcelain cup of mass 200 g contains 150 g of hot coffee (essentially water) at a temperature of 60°C. If the liquid is stirred with a silver spoon of mass 25 g, initially at 20°C, until temperature equilibrium is attained, (*a*) what fraction of the heat energy given up by the cup and coffee goes into the spoon, assuming no exchange of heat with the surroundings? (*b*) What is the equilibrium temperature?

11.3. Show that *c* for a given substance has the same numerical value in Btu/lb-F° as in kcal/kg-C°. Make use of information on page 249 and in Fig. 10.2 on page 218.

11.4. What fraction of his daily food allowance of 2500 kcal does a man consume at dinner if this meal consists of a lean steak weighing 540 g, two slices of white bread of total weight 40 g, a pat of butter of 8 g, and a dish of plain ice cream of 220 g? Use Table 11.2.

11.5. A copper ball of mass 100 g is heated to 100°C and placed in a hollow in a cake of ice at 0°C. What mass of ice will melt?

11.6. Explain how the formation of dew on the leaves of a plant helps to prevent damage to the plant by freezing. HINT: How is the heat of condensation of water involved here?

11.7. A certain quantity of water initially at 0°C is placed on an electric heater and comes to a boil 15 min later. After another hour and twenty minutes under the same conditions, the water has all boiled away. Compute from these data the latent heat of vaporization of water, assuming the water continues to absorb heat at the initial rate.

11.8. Show that the slopes of the ice, water, and steam segments in Fig. 11.2 are proportional to the reciprocals of the specific heats of these phases, as implied on page 251.

11.9.   Look up the phenomenon of *regelation*. Explain why the fusion curve of Fig. 11.5 is very nearly vertical and has a slight backward slope.

11.10.   In what way is steady heat flow like the flow of an incompressible fluid? Is it possible to explain heat conduction satisfactorily on the basis of the caloric theory? Try it.

11.11.   In winter, why is the temperature of the inside wall of a house lower than that of the room, while the outside wall temperature is higher than that of the outdoor air?

11.12.   A cylindrical iron bar of diameter 1.0 cm and length 20 cm has one end immersed in steam and the other packed in melting ice. If it is assumed that no heat is lost through the lateral surface of the bar, how many kilograms of ice is being melted each hour? The value of k for iron is 0.012 kcal/sec-m-C°.

11.13.   A cubical box, 4.0 m on each edge, is made of fiberboard 0.50 cm thick. An electric heater inside the box dissipates 1,000 watts. If the coefficient k of this material is 1.1×10^{-4} kcal/sec-m-C°, what will be the temperature difference between the inner and outer walls of the box in the steady state?

11.14.   Name at least one device that accomplishes directly each of the following energy conversions involving heat: (*a*) chemical to thermal, (*b*) thermal to radiant, (*c*) electrical to thermal, (*d*) nuclear to thermal, (*e*) thermal to chemical. It will be helpful to look at Fig. 11.9 again.

11.15.   An average adult when resting produces about 65 kcal of heat per hour, the so-called basal metabolism rate. To how many watts is this equivalent?

11.16.   An engine is observed to expend energy at the rate of 150 watts while idling. If 20 percent of this energy goes into heat developed in the mechanism, at what rate in kilocalories per minute is heat being dissipated through friction?

11.17.   A floor clock is driven by the descent of two 3.5-kg weights, each of which travels a vertical distance of 1.2 m in 24 hr. If it is assumed that all the original GPE of the weights is transformed into heat, how many kilocalories are produced in the clock mechanism in a day?

11.18.   A car of mass 1,200 kg traveling at 96 km/hr is brought to rest by the brakes. How much heat is generated?

11.19.   How much heat was generated when the sled in Worked example 9.5 (page 206) came to rest after coasting up the hill?

11.20.   Could you throw an ice cube against a brick wall hard enough to melt it? Compute the necessary speed, assuming that all the KE goes into heat on impact.

11.21.   How many kilocalories are needed to bring 1 kg of liquid water from the ice point to the boiling point? If the equivalent amount of work were applied to this mass of water, to what height, in meters, could it be raised?

11.22. Suppose that the temperature of 0.75 m³ of water for a bath is raised from 10 to 45°C. If the required heat energy were completely converted into work, how high could it raise a 70-kg man?

11.23. The atmosphere and the oceans contain enormous amounts of internal energy. What prevents our using these sources to run heat engines and thus getting unlimited amounts of work?

11.24. The lower half of a jar is filled with salt and the upper half with pepper. Next, the mixture is well stirred. (*a*) What is the probability that, at some later time, the two substances will again be completely separated? (*b*) Will stirring half the time in one direction and half in the opposite direction increase the probability of attaining a separation? (*c*) Tiring of the stirring method, the experimenter produces a complete separation by dissolving the salt in water and filtering out the pepper. Does this violate the second law? Explain.

11.25. Using Eq. 11.8, show that any engine that is less than 100 percent efficient cannot utilize all the heat it takes in during a cycle but must reject some.

11.26. Compute the efficiency of a Carnot engine operating between temperatures of 120 and 80°C.

11.27. Calculate the change in entropy when 10 g of ice at 0°C melts. Does this represent an increase or decrease of entropy?

11.28. A certain reversible heat engine takes 10 kcal of heat from a source at 217°C and exhausts part of this heat to a lower-temperature reservoir at 70°C. Calculate (*a*) the efficiency of the engine, (*b*) the work done by the engine in the process, and (*c*) the amount of heat rejected.

Part Three
WAVE PHENOMENA

Chapter 12
VIBRATIONS AND WAVES

Several of the earlier chapters of this book were concerned with the study of special kinds of motion: the motion of a particle with constant speed or with constant acceleration, the motion of a projectile, and the haphazard motion of the myriad molecules of a gas. In every case, we had to adopt a simplified model of the natural phenomenon in question. Vehicles, planets, and molecules were replaced by particles, and dissipative forces were often neglected. In studying the motions of gas molecules, the tremendous number of particles involved made it necessary to resort to statistics, and although it was no longer possible to keep account of the motions of individual particles, we were able to deduce quantitative statements about the average behavior of the whole collection.

In this chapter we shall examine another kind of motion that is not as simple to describe as that of a freely falling stone nor as complicated as that of a gas molecule. It is called **periodic** motion. Periodic motion is any motion that repeats itself at regular intervals of time. Many familiar phenomena are examples: the motion of a violin string, of a planet, of a clock pendulum, of the pistons of an engine, etc. Here also, in order to simplify the discussion, we often idealize the actual situation by replacing the moving object by a particle and by restricting the type of periodic motion to be considered.

PERIODIC VIBRATIONS

12.1 VIBRATIONS; PERIOD AND FREQUENCY

If a particle describing periodic motion moves back and forth repeatedly *over the same path,* it is said to **vibrate** or **oscillate.** In physics, these terms are not confined to the mechanical motion of a particle but are used in a more general way. We refer to periodic oscillations of temperature, pressure, illumination, and electric or magnetic force. In any case, there is a *return to the initial state* after the system goes through a succession of changes.

The complete sequence (or "round trip" in the case of a moving particle) is called one **cycle,** and the time T required to complete each cycle is called the **period** of the motion.

In some cases it is more convenient to specify the **frequency** of the motion. The frequency f is defined as the *reciprocal* of the period:

$$f = \frac{1}{T}$$

12.1

One cycle per second is sometimes called one hertz, after the discoverer of electromagnetic waves (page 475).

Frequency is usually measured in cycles per second, abbreviated cycles/sec. The period would then be expressed in seconds per cycle, the reciprocal of the above. Dimensionally, however, "cycles" expresses a pure number and so has no representation in terms of length, mass, and time. The physical dimension of frequency is the reciprocal of the time dimension.

Many practical examples of vibration occur when a system that is originally in a condition of stable equilibrium is displaced from that position and then left to itself. The position of equilibrium of a particle (page 70) is one in which no net force acts. If the equilibrium is stable, any net force that arises when the particle is displaced from its equilibrium location must be directed back toward that position. It must be of the nature of a **restoring** force (or **return** force), for only under these circumstances would the motion repeat itself.

It is assumed that the glider may be considered a particle, to which the spring forces are applied. This supposition is justified by experience.

Consider a glider on a linear air track (page 125). The glider is connected to two fixed points, one on each side, by means of two identical springs which are under tension (Fig. 12.1). If the glider is slightly displaced from its rest position toward either side, the tension in the spring on that side will be lessened, while the one on the other side will be increased. This leads to a net force tending to bring the glider back to its position of equilibrium. If the glider is released after being displaced, it will oscillate about this position. The greatest distance A that it moves on either side of the equilibrium location is called the **amplitude** of its motion. The amplitude is the maximum value of the displacement x.

In the complete absence of frictional forces, the amplitude would remain constant as the vibrations continued. In practice, however, dissipative forces are always present to some extent, and the motion is a **damped vibration,** in which the amplitude decreases in a regular manner as time passes (compare Fig. 12.2a and b). Thus the glider is observed to come to rest again eventually at the equilibrium position. However, with friction reduced to this degree, it may describe a large number of oscillations before

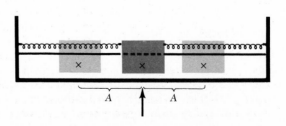

FIGURE 12.1   When the glider i displaced to the right, the net forc acting on it is toward the left, an vice versa. This means that ther is restoring (return) force tha makes the glider oscillate. The am plitude is A.

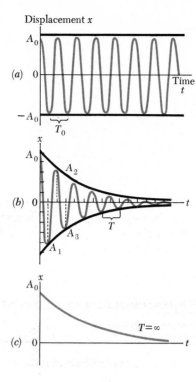

FIGURE 12.2 Comparison of the motion of an oscillatory system when (a) undamped, (b) damped, and (c) overdamped. In the last case, the motion is nonoscillatory and the period is infinite.

the motion becomes unnoticeable. The same kind of behavior is noted for a swinging pendulum, for a tuning fork after it is struck with a mallet, and so on.

In spite of the progressive decrease in amplitude, the period stays constant, but its magnitude is slightly greater than it would be if damping were absent. If the damping exceeds a critical value, the displaced system will not oscillate at all but will slowly settle back toward the equilibrium position (Fig. 12.2c).

12.2 SIMPLE HARMONIC MOTION

There is a special kind of vibratory motion of a particle that is capable of representing a great variety of observed motions quite accurately. It is the type of motion of a particle that moves back and forth along a straight path while subject to a return force whose magnitude is proportional to the displacement. This kind of vibration is called **simple harmonic motion,** abbreviated SHM. Stated algebraically, the condition on the return force is

$$\mathbf{F} = -k\mathbf{x} \qquad\qquad 12.2$$

where $\mathbf{F}$ is the force vector, $\mathbf{x}$ is the corresponding displacement vector of the particle from its equilibrium position, and k is a scalar called the **force constant** of the system. The minus sign indicates that we are dealing

Equation 3.1
(page 44)
was written
without the
minus sign,
since it was
intended to
give only
the magni-
tude of the
force.

with a restoring force: **F** is at all times opposite in direction to **x** and tends to bring the particle back to the origin.

The force law for SHM, stated above, has exactly the same form as the equation expressing Hooke's law for small deformations of elastic bodies. This shows immediately that SHM represents the great variety of motions of elastic bodies when they are slightly displaced from equilibrium.

In addition to elastic vibrations, certain types of nonmechanical oscillation, such as the motion of electrons in a radio circuit, are governed by this same law of force. Even a very complicated force law can often be represented to a good approximation by a *linear* relation like Hooke's law, provided only that the displacements remain very small (see page 46). This means that, in *any* system, very small oscillations of a particle about its position of stable equilibrium may be considered simple harmonic, and that once a description of SHM is developed it should be applicable to many physical phenomena.

In spite of its name, SHM is not as simple as some of the types of motion previously considered. First, it should be noted that for any kind of vibratory motion, the velocity cannot be constant since the particle must have zero velocity as it passes through the turn-around point at either end of its oscillation. Furthermore, Eq. 12.2 shows that the acceleration is also variable. According to this equation, the force acting on the vibrating particle is a function of its displacement x, and so by Newton's second law the acceleration must also depend on x.

Energy
relations
in SHM.

The energy concept is helpful in working out some of the details of SHM. In general, the oscillating particle has both KE and PE. If the mass of the particle is m, the KE is given, as always, by $mv^2/2$ (Sec. 9.2). The PE is the elastic potential energy stored in the spring or in whatever system is furnishing the restoring force. A general expression for this PE can be deduced directly by using an average value of the acting force. In Hooke's law (Eq. 12.2), the magnitude of the restoring force changes linearly with the displacement. The force is zero when the particle passes through the position of equilibrium and attains its maximum value, numerically equal to kA, when the particle is at either end of its path. At any intermediate point, the force F amounts to kx. Because the relation is a linear one, the *average* value of F over a displacement x is just *half* of kx (Fig. 12.3). The elastic PE stored in the spring is equal to the work done in stretching the spring a distance x; this, in turn, is equal to $\bar{F}x = (kx/2)(x)$, or the elastic PE—abbreviated EPE—is given by EPE = $kx^2/2$.

See page
100, where
the same
method was
used for
other pur-
poses.

According to the energy principle, the sum of the KE and the EPE of the particle must remain constant, as long as there are no dissipative or other forces acting (Fig. 12.4). With the expressions given above,

$$\tfrac{1}{2}mv^2 + \tfrac{1}{2}kx^2 = \text{constant} \qquad\qquad \textbf{12.3}$$

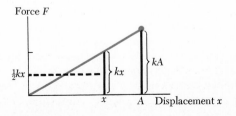

FIGURE 12.3 Because the force increases uniformly with the displacement, the average force is just half the value at the end of the displacement.

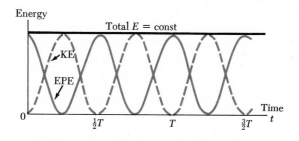

FIGURE 12.4 The energy of the oscillating particle is continually changing between KE and EPE, but the total energy remains constant.

The similarity in form of the two terms on the left is worth noting. Since both v and x enter only as their squares, nothing is changed when either of these variables is replaced by its negative. Physically, this means that the total energy depends only on the momentary location and speed of the vibrating particle and not at all on whether it is at one side or the other of the equilibrium position or in which direction it is moving.

The right-hand side of the equation is constant, and so we can find its numerical value if we know what it is at any particular instant. This is known, for example, when the particle is at either end of its path, for then (1) the speed is zero, making the KE equal to zero, and (2) the displacement is equal to the amplitude A, making the EPE equal to $kA^2/2$. Accordingly, at that time, Eq. 12.3 reads $0 + \frac{1}{2}kA^2 = $ constant, showing that the constant has the value $\frac{1}{2}kA^2$. With this value, Eq. 12.3 becomes $\frac{1}{2}mv^2 + \frac{1}{2}kx^2 = \frac{1}{2}kA^2$. Transposing, and canceling the common factor $\frac{1}{2}$, we get $v^2 = k(A^2 - x^2)/m$, or simply

$$v = \sqrt{\frac{k}{m}(A^2 - x^2)} \qquad\qquad \textbf{12.4}$$

This equation shows how the speed of the oscillating particle varies with its position, given by x.

12.3 COMPUTING THE PERIOD; PHASE RELATIONS

It is possible to derive additional relations between the variables in SHM by using calculus methods. For example, the result for the displacement-time relation is

$$x = A \cos \sqrt{\frac{k}{m}} t \qquad\qquad \textbf{12.5}$$

and its graph has the familiar form of a cosine (or sine) curve (Fig. 12.2a). However, there is also a straightforward geometric way of representing SHM and arriving at the formulas that describe it. Imagine a wheel turning uniformly on a horizontal axle (Fig. 12.5). A short rod projects from the wheel. The sun, directly overhead, casts a shadow of the apparatus on the table. We can now prove that, as the wheel turns, the oscillating motion of the shadow of the rod is strictly simple harmonic.

Figure 12.6 is a diagram of the wheel as seen when looking along the axle. With the rod in the position P, θ represents the angle that the radius drawn to the rod makes with the positive x axis. Point Q is the vertical projection of P on the x axis. If we call the radius of the wheel A and if the distance OQ is designated x, then their relation in the triangle OQP

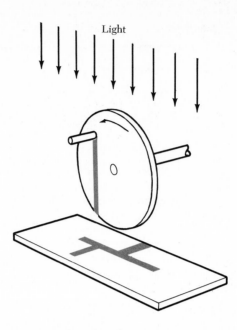

FIGURE 12.5 As the wheel turns uniformly on its axle, the shadow of the projecting rod moves back and forth across the table with simple harmonic vibration.

is merely $x = A \cos \theta$. Further, since the wheel is assumed to turn at a uniform rate, the magnitude of θ is directly proportional to the time measured from position B, and so we can write $\theta = \omega t$, where the Greek letter ω (omega) represents a constant. The above relation may then be written

$$x = A \cos \omega t \qquad\qquad\qquad\qquad \textbf{12.6}$$

This equation is identical with Eq. 12.5 if ω is taken to be equal to $\sqrt{k/m}$. The motion of the projection Q on the diameter of the circle is the same as that of its projection on any horizontal line, and therefore of the shadow of the rod as projected onto the table. In this way it is seen that *the projection, onto a diameter, of a point moving around a circle at constant speed is accurately SHM.* During the time P moves once around the so-called reference circle, Q completes one cycle. The radius of this circle is equal to the amplitude of the SHM.

More information is readily obtained from the reference circle. Draw a vector **V** at the point P (Fig. 12.6) representing the velocity of the rod along the circle. Then the vector **v,** the horizontal component of V, will be the instantaneous velocity of the point Q. From the fact that **V** is perpendicular to A and NM is perpendicular to Ox, the angle at N is the same as θ. Then, in the two similar triangles PNM and POQ,

FIGURE 12.6 Use of the reference circle in analyzing simple harmonic motion.

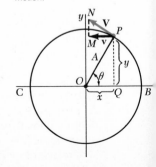

$$\sin \theta = \frac{y}{A} = \frac{v}{V} \qquad \text{or} \qquad v = \frac{V}{A} y$$

With $y = \sqrt{A^2 - x^2}$, this becomes

$$v = \frac{V}{A} \sqrt{A^2 - x^2}$$

This equation has exactly the same form as 12.4. The two become identical if we take

$$\frac{V}{A} = \sqrt{\frac{k}{m}} \qquad\qquad 12.7$$

With the help of Eq. 12.7 we can get a useful expression for the period of the SHM as a function of the given constants of the system. The period T is the same as the time required for the point P to move once around the reference circle. Since the perimeter of this circle is $2\pi A$ and the point P moves at the constant speed V, the time required is

$$T = \frac{2\pi A}{V}$$

Substituting the value of A/V from Eq. 12.7 gives

$$T = 2\pi \sqrt{\frac{m}{k}} \qquad\qquad 12.8$$

which is a general relation for the period of vibration in terms of the mass m of the vibrating particle and the force constant k of the system. The formula for T does not contain A, and so *the period is independent of the amplitude* of the motion. The oscillations are regulated by just two factors: the *inertia* of the particle and the *elasticity* (or its equivalent) of the system that furnishes the return force.

An expression for the *frequency f* of the vibration is given at once by inverting relation 12.8, since f was defined as $1/T$ on page 278:

$$f = \frac{1}{2\pi} \sqrt{\frac{k}{m}} \qquad\qquad 12.9$$

Worked example 12.1

A certain spring (Fig. 12.7) stretches 4.80 cm when pulled with a force of 1.20 newtons.

   a. Compute the force constant of the spring.

   b. What will be the frequency of small vertical oscillations when a 500-g mass hangs from the spring?

Solution   a. According to Eq. 12.2, the value of k is given by F/x, or $k = 1.20/0.048 = 25.0$ N/m. This means, for example, that a force of 25 newtons would be needed to stretch this spring 1 m, provided that such a stretch does not go beyond the proportional limit.

   b. With the value of k known, we can substitute k and m into Eq. 12.9, getting

$$f = \frac{1}{2\pi} \sqrt{\frac{25.0}{0.500}} = 1.13 \text{ cycles/sec}$$

FIGURE 12.7   Weight hanging from a spiral spring.

How were the units for the final result determined? In Eq. 12.9, k is measured in newtons per meter and m in kilograms, and so the fraction under the radical has the units N/kg-m. As defined by Newton's second law, newtons is equivalent to kg-m/sec² (page 128). Then the quantity under the radical has the units kg-m/kg-m sec², or simply 1/sec². Taking the square root yields 1/sec, and we must supply the dimensionless label *cycles* in the numerator to give the result its full physical description.

Phase relations.   We have seen that a particle in SHM is continually undergoing changes in speed and direction. These changes can be followed with the help of the reference circle. It was shown on page 282 that the velocity of the particle at any instant is merely the projection on a diameter of the velocity of the point that moves with constant speed around the reference circle. As seen directly from Fig. 12.6, the *speed* of the particle in SHM is zero at each end of its oscillation and is a maximum when the particle passes through the central (equilibrium) position. The form of the force law shows that the particle has its maximum *acceleration* when at either end of its oscillation, and it has zero acceleration when passing through the center. To summarize these statements:

$v = 0$	when $x = \pm A$	$a = 0$	when $x = 0$
$v = v_{max}$	when $x = 0$	$a = a_{max}$	when $x = \pm A$

The speed and acceleration are "out of step" by a quarter of a cycle, or by 90° as measured around the reference circle. The two quantities are said to differ in **phase** by a quarter of a cycle, or by a phase angle of 90°.

It is often useful to speak of the difference in phase of two particles vibrating with SHM. As long as they have the same period, they remain in a constant phase relation with each other even though they may have different amplitudes of motion. One particle will continue to *lead* the other in phase by a constant amount. For example, it will arrive at each maximum a certain fraction of a period sooner than the other. In the same way, the second particle is said to have a phase *lag* of this same amount with respect to the first (Fig. 12.8). In the figure, the phase difference is 30°.

12.4 THE SIMPLE PENDULUM

It has already been noted that the vibrations of any system approximate SHM, provided that the amplitude is kept small enough. Further, the return force need not be furnished by an elastic body but may have its origin

FIGURE 12.8 One of the vibratory motions (dashed curve) leads the other (solid curve) by a phase angle of 30°, or one-twelfth of a period. The two vibrations shown here also have different amplitudes, but they must have the same period if a constant phase relation is to exist between them.

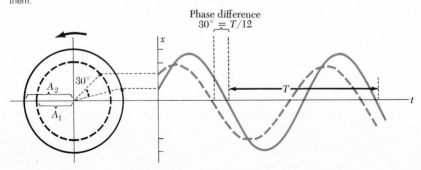

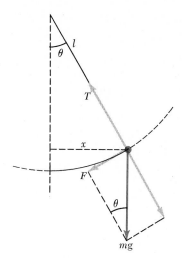

FIGURE 12.9 Analysis of the motion of a simple pendulum. If the displacement is small, it may be measured along the chord rather than along the arc of the circle.

in a mechanism of some other kind. An important and interesting example of SHM is the oscillation of an ideal, or simple, pendulum. As defined on page 203, a **simple pendulum** is a particle suspended from a fixed point by a weightless thread and allowed to swing in a vertical plane. It will be shown that for small oscillations of the pendulum bob, the return force (supplied by gravity) is approximately proportional to the displacement, so that the resulting motion is very nearly SHM.

Figure 12.9 shows a simple pendulum consisting of a particle of mass m suspended by a thread of length l. The particle has been drawn aside from its equilibrium position and released. At the instant shown in the diagram, the pendulum is moving back toward the equilibrium position, and the angle of the thread with the vertical is θ. Two forces act on the particle: its weight mg and the tension T in the thread. Choose a pair of perpendicular axes, one along the circular path of the particle and the other along the thread, and resolve the forces into components along these axes.

The forces along the thread, T and $mg \cos \theta$, combine to provide the centripetal force that holds the moving particle in its circle, while the component of the weight along the tangent, $mg \sin \theta$, acts as a return force tending to bring the particle back toward the center.

The return force is thus proportional to $\sin \theta$, while the displacement of the particle is measured along the arc of the circle and so is proportional to the angle θ itself. We see that the condition for SHM is *not* fulfilled exactly here. It is possible to solve this case of a simple pendulum swinging through an arbitrary angle, but the mathematics is complicated. However, note that, if the angle of swing is kept small, the arc and the half chord x are approximately the same in length (see Fig. 12.9). From the large triangle,

$$x = l \sin \theta$$

and from the vector triangle,

$$F = -mg \sin \theta$$

where the minus sign indicates that F is in such a direction as to *decrease* θ. When sin θ is eliminated from the last two equations the result is

$$F = - \frac{mg}{l} x \qquad\qquad \textbf{12.10}$$

Since mg/l is a constant, this relation has the form of Eq. 12.2, with $k = mg/l$. This shows that for a sufficiently small maximum angle of swing the motion may be considered simple harmonic.

The period of swing may now be calculated by substituting the above value of k into Eq. 12.8. This gives $T = 2\pi \sqrt{m/(mg/l)}$, or

$$T = 2\pi \sqrt{\frac{l}{g}} \qquad\qquad \textbf{12.11}$$

This approximate formula for the period is found to give numerical results that differ very little from the exact solution. For example, if θ does not exceed 30°, the error is less than $\frac{1}{2}$ percent.

Practice set*   What is the effect on the period of small swings of a simple pendulum if (*a*) its length is increased by a factor of 4; (*b*) the mass of the bob is doubled; and (*c*) the pendulum is taken to the moon, where g has about one-sixth the value on earth?

Period is independent of amplitude and of mass.

It is significant that the period formula does not contain either the mass of the particle or the amplitude of swing. Galileo is said to have observed the swinging of a hanging lamp and, timing the swings with his pulse, he found that the period remained constant although the amplitude decreased as time passed. It is precisely this constancy that makes a pendulum suitable for controlling a clock. With driving energy supplied by springs or by descending weights, the clockwork imparts regular pushes to the pendulum, but because of imperfections of the mechanism these impulses are not all of equal strength. Nevertheless, even with such variations, the period remains almost perfectly constant as long as the amplitude is kept small.

Huygens (page 313n) is generally credited with invention of the pendulum clock (1656), although Galileo had designed such a mechanism years earlier.

The absence of the mass of the particle from the period equation is a more fundamental matter. In deducing Eq. 12.11, the factor m was canceled in the numerator and denominator of the fraction under the square-root sign. In making this cancellation we were tacitly assuming that the factor m in Eq. 12.8 and the factor m in Eq. 12.9 represent the same quantity. Actually, the former is the *inertial* mass of the particle whereas the latter is the *gravitational* mass (Sec. 8.7, page 186). The fact that Eq. 12.11 predicts the period with extreme accuracy in every observed instance is definite evidence that the same number m can be used to represent both the inertial and the gravitational effects of a given piece of matter.

If the length and period of pendulum are carefully determined, Eq. 12.11 may be used to find an accurate value of g at any locality. This method, first used by Huygens, was a far better and more accurate way of measuring g than any direct experiment on free fall. Special pendulums are used in

---

*Ans.: T is doubled; no effect; T increases about 2.5 times.

geophysics to locate underground deposits of ore or petroleum by the very slight variations in the value of g caused by their presence.

Any real pendulum, such as one used in a clock, is called a **physical pendulum** because the mass is distributed along its length rather than concentrated in a single particle as in the simple pendulum. The mechanics of the physical pendulum will not be analyzed here.

12.5 FORCED OSCILLATIONS AND RESONANCE

Instead of being allowed to vibrate freely after an initial displacement, a system may be driven by continued application of force from the outside. Whenever a system is made to vibrate by a periodic force, the resulting motion is called **forced oscillation.** Examples are the vibration of a factory structure caused by the running of heavy machinery or the motion of electrons in a circuit connected to an alternating-current generator.

Forced oscillations take place with the frequency (or period) of the driving force rather than with the natural frequency of the system. The amplitude of the response depends on how the driving frequency is related to the natural frequency. If these frequencies are nearly the same, even a very weak driving force can, in time, feed enough energy into the system to give it a large amplitude of motion. This condition is called **resonance.** A heavy pendulum can be made ultimately to swing with a large amplitude by giving it small taps in the tempo of its natural frequency. The taps must be applied in the same direction as the instantaneous motion of the pendulum. Any *submultiple* ($\frac{1}{2}$, $\frac{1}{3}$, $\frac{1}{4}$, etc.) of the natural frequency will also produce resonance.

If the impulses are not close to one of these frequencies, or if they are not periodic at all, the system merely moves erratically without attaining any considerable amplitude of swing because there is no fixed phase relation between the applied force and the motion of the system. Some of the impulses may actually take energy away from the system as they meet it "head on."

Besides depending on how close the driving frequency is to the natural frequency, the amplitude of response of the forced vibrations of a system

In November 1940 the Tacoma Narrows Bridge in the state of Washington was destroyed by wind-generated resonance. (*Courtesy of Prof. F. B. Farquaharson, University of Washington.*)

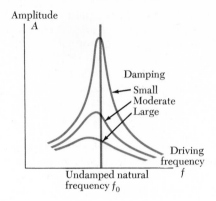

FIGURE 12.10 The response of a harmonic oscillator driven by a periodic force. The amplitude is a maximum when the driving frequency equals the natural frequency and falls off, unsymmetrically, on both sides. For small damping the resonance curve is sharp and has a high peak value.

also depends on the strength of the damping (page 278). Figure 12.10 shows resonance curves of an oscillator with weak, moderate, and strong damping. In each case, the amplitude of vibration is a maximum for a particular driving frequency (the resonant frequency), falling off on both sides of this value. The response is greater and the resonance curve sharper for smaller damping. The resonance frequency is always lower than the natural frequency but approaches it as the damping is reduced. Resonance considerations are of interest in a great variety of mechanical, electrical, and molecular phenomena.

WAVE PROPAGATION

12.6 NATURE OF WAVE MOTION

So far, in this book, several kinds of motion have been described and examined: motion with constant velocity, motion with constant acceleration, and periodic motion. In every case, for simplicity, it was the motion of an individual particle that was studied. However, there is a large and varied category of natural phenomena in which a moving object interacts with all the particles of a surrounding medium. A medium is a substance or region of space in which forces act or other effects are produced. One important kind of effect is called **wave motion.** Wave motion involves the transfer of energy from the moving body, called the **source,** to the medium with which it communicates.

A familiar and highly visual form of wave motion is represented by the set of circular ripples that spread out over the surface of still water after a stone has been dropped into it. The ripples are circles of disturbance that are observed to move outward from the point of impact with constant speed, crests alternating with troughs. A floating chip encountered by the waves is set into oscillation, showing that energy has been transferred from the source to the medium. The chip is seen to oscillate around its original position but it does not move forward progressively with the passing waves.

Leonardo da Vinci (1452–1519), the versatile Italian genius, watched waves moving across a windswept field of wheat and realized that the onward motion of the grain was an illusion, that each stalk merely bent aside slightly as the wave passed. In any wave motion, it is only the *form* of the disturbance that moves along at the characteristic wave speed as one part after another of the medium receives energy and passes it on. Contrasted with other conceivable means of transferring energy from one place to another, such as the throwing of stones, *waves are able to transfer energy without the transport of matter.*

There are other types of waves besides those traveling on the surface of water, on grain, or on a flag rippling in the breeze. A sudden mechanical disturbance in the air, such as an explosion, compresses the adjoining parts of the medium. When this layer of air expands again, it compresses the surrounding parts. In this way, a wave of compression is propagated away from the source in all directions. Sound waves, described in more detail below, are compressional waves.

Still other kinds of waves are torsional (twisting) waves associated with earthquakes, waves of temperature variation, waves traveling along a crystal lattice, electrical surges in power lines, etc. It will be shown later that a wave model is required for understanding the propagation of light and similar radiations. Unlike the wave types already mentioned, these waves are nonmechanical and do not require the presence of matter for their propagation; they are able to travel in empty space. Light and other such radiations can be described as electromagnetic waves (Chap. 19), and the inclusion of light in the category of waves is ample reason for examining the nature of wave motion in general.

In the example of a stone dropped into a pond, the sequence of disturbances at the place where the stone enters is duplicated at increasingly later times for parts of the water surface that are farther and farther from the source. At any instant, all points that lie on any circle drawn with its center at the source are in the same phase of vibration. If the water surface does not have the same character at all points (for instance, if there are patches of oil on it), the particles having the same phase will not lie on circles but on curves of some other kind (Fig. 12.11). The curves connecting particles of a medium that are in the same phase of vibration are called **wavefronts.**

Two ways of examining a wave are useful: We can study what happens at a given place in the medium as time passes, or we can travel along with the wave and look at its profile. For the example of ripples on water,

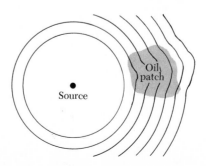

FIGURE 12.11 Nonuniform regions of the medium cause distortions of the wavefronts, as shown here for circular ripples encountering a patch of oil on the surface of the water.

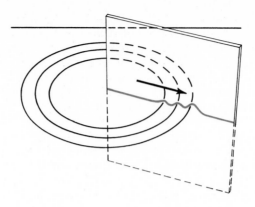

FIGURE 12.12 The profile of the set of ripples can be seen against a vertical plane placed in the water and extending outward from the source.

Fig. 12.12 represents the wave profile, a sort of snapshot of a vertical section of the water surface taken at any instant. The displacement of any point on this contour will become smaller as the wave travels away from the source because the energy "spreads thinner" as the wavefront increases in extent. However, under ordinary conditions the *shape* of the wave will be otherwise preserved as it goes onward.

To see what happens at a particular place as the wave passes through, we must use a graph of vertical displacement of the water surface as a function of time. Because of the constant speed of motion of the disturbance, this graph will have the same shape as the wave profile. Moreover, apart from the continuing decrease of amplitude, the displacement-time curves recorded at more and more distant points will all have the same form (Fig. 12.13).

12.7 THE WAVE EQUATION

In the above example, the sudden entry of the stone into the water gave rise to a limited number of oscillations. After the disturbance passed, the water surface was again quiet. A wave disturbance of short duration is called a **pulse** or, particularly if very sudden and intense, a **shock wave.** (See pages 305–306.) To start or maintain a wave motion, there must be some kind of connection, or coupling, between the source and the surrounding medium.

Now suppose that, instead of dropping a stone into water, a stick that dips into the surface is moved up and down at regular intervals of time. The result is a *continuous wave train.* Circular ripples begin to move away from the source, new ones are periodically produced behind them, and

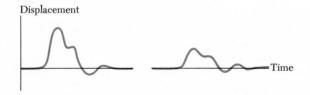

FIGURE 12.13 Although the displacements decrease with time, the form of the disturbance is preserved.

soon the whole surface is covered with ripples. Thus a continuous wave train, as opposed to a single pulse, results when the source is an oscillating body. If the oscillations of the source are periodic, the wave train will consist of a regular repetition of a given form. The wave profile shows that, at any instant, a given phase recurs at equal intervals along it. The distance between any two successive particles of the medium that are in the same phase is called the **wavelength.** It is denoted by the Greek letter λ (lambda).

Each time the source makes one complete cycle, one new wave is produced and incorporated into the wave train, which continues to move outward with the wave speed V characteristic of the medium. The value of this speed depends only on the properties of the medium and not on the form of the particular wave or waves under consideration, as long as the displacements are not too great. If the period of vibration of the source is T, the disturbance moving at the constant speed V will cover a distance λ in this interval of time, so that $\lambda = VT$ or, in terms of the frequency $f(=1/T)$,

$$V = f\lambda \qquad\qquad \text{12.12}$$

This simple wave equation holds for any type of continuous wave coming from a periodic source and traveling in a uniform medium. The relation results from the way the three quantities were defined.

For some kinds of wave motion there is a slight dependence of wave speed on frequency. This occurs for water waves and, more importantly, for light waves moving in transparent media. The latter phenomenon will be described in detail later. For the present discussion, V and f will be assumed to be independent of each other.

Worked example 12.2   A broadcasting station sends out radio waves whose frequency is 710 kc (kilocycles, or thousands of cycles per second).

a. If radio (electromagnetic) waves travel at a speed of 3×10^8 m/sec, find the wavelength.

b. How many waves are there in the wave train extending from the broadcasting station to a receiving set located 20 km away?

Solution   *a.* Equation 12.12, solved for the wavelength, is $\lambda = V/f$. Substitution of the given numbers yields $\lambda = (3 \times 10^8)(7.1 \times 10^5) = 420$ m, or slightly more than $\frac{1}{3}$ mi.

b. At any instant, the number of these waves in a train 20 km long is $(20 \times 10^3)/420$, or about 48 waves.

12.8   TYPES OF WAVES; DIRECTION OF PARTICLE MOTION

Ripples on the surface of a pond may be thought of as waves progressing in two dimensions. A simpler case is that of waves in a rope, which are waves in one dimension. Think of a long, heavy rope with one end tied to a fixed point, the other end held in the hand, and the rope pulled fairly taut in the horizontal direction. A long, coiled spring or even a garden hose can be used in place of the rope.

Transverse waves.   If the hand is now moved suddenly upward, a pulse will travel down the rope. The pulse in this instance is a hump, and it moves along the rope at constant speed. The picture shows the progress of such a pulse on a long spring. The individual coils do not move along the spring as the pulse reaches them; rather, one coil after another is lifted by the passing hump, falling back again as the pulse moves on. This pulse is an example of a transverse wave. A **transverse wave** is one in which the particles of the medium move *crosswise* to the direction of wave travel. Later, it will be shown that light waves are transverse, although electric and magnetic forces rather than mechanical motions are involved.

Compressional waves.

Transverse waves may be set up in a long steel rod by striking it with a *sidewise* blow from a hammer. However, another kind of wave pulse, traveling with a different speed, can be sent along the rod by hitting the end with a *lengthwise* blow. This gives rise to **compressional** (or **longitudinal**) **waves** in the steel. In a compressional wave, the displacements of the particles of the medium are parallel to the direction in which the waves advance. Sound waves are an important example of compressional waves and will be discussed more fully below.

Water waves are composite.

Waves on the surface of water represent a composite situation. The particles move in small circular or elliptical paths as the wave passes, and so the oscillations have both transverse and longitudinal components. Figure 12.14 shows the profile of a simple surface wave on water. The rather steep crests are separated by relatively long, gentle troughs. Notice that the particle motion is in the same direction as the wave motion at the crests and in the opposite direction at the troughs. When sea waves approach a beach the oscillatory motion is changed into the translational motion of large masses of water. The reason is that the sloping bottom retards the motion of the lower parts of the wave, making the crests spill over, or "break."

A compressional wave can be started in air by a mechanical disturbance, as mentioned briefly above. If the disturbance is of limited duration (an explosion, for example), it gives rise to a pulse, similar to dropping a stone into water or jerking the end of a rope. In order to produce a continuous train of compressional waves in the air, the single disturbance must be replaced by a source that oscillates repeatedly.

Look up the origins of the words subsonic, infrasonic, and ultrasonic.

If the frequency of vibration of the source is between about 20 and 20,000 cycles/sec, the waves are able to produce the sensation of sound in the human ear. Compressional waves that have frequencies below this range are known to occur in the atmosphere. They are called **infrasonic** waves. Frequencies above the audible range are called **ultrasonic.** Values above 20 billion cycles/sec have been attained by using electrically driven sources.

Sound waves are longitudinal.

Every sustained sound can be traced to a mechanically vibrating object such as a violin string, tuning fork, or the air inside a pipe or other enclosure. For purposes of discussion, let the source be a small balloon connected to a simple reciprocating pump (Fig. 12.15). If a little air is alternately moved into and out of the balloon, the surface of the balloon

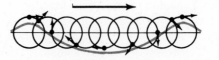

FIGURE 12.14 Profile of a surface wave on a liquid. As the wave passes, each particle in the surface describes a small, closed path—in this case, a circle. The particle motion is directly forward at a crest and directly backward at a trough.

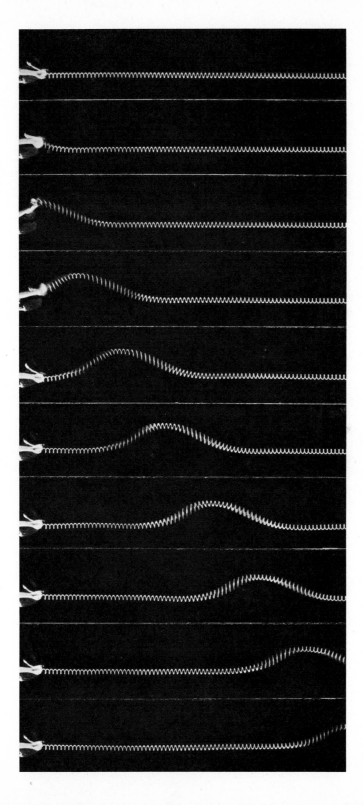

A transverse wave pulse traveling along a coiled spring. (*From PSSC Physics, D.C. Heath & Co., 1965.*)

pulsates and a train of compressional waves spreads out through the surrounding air. If the properties of the air are the same in all directions, the wavefronts are spheres with the balloon as their common center.

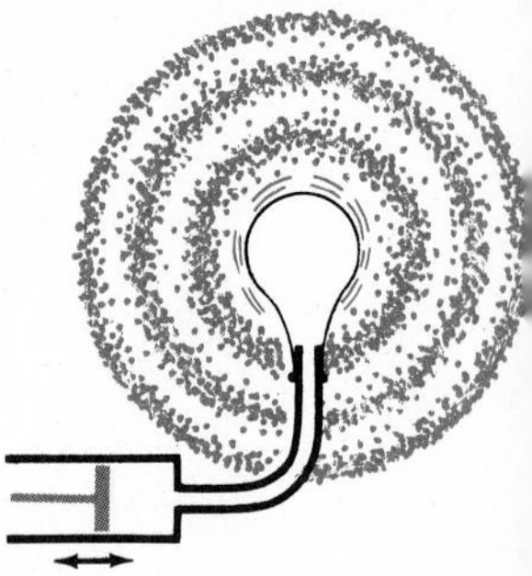

The wave train consists of alternate regions of compression and expansion, and at all points the air molecules move back and forth short distances along the direction of advance of the waves. Thus these waves are purely longitudinal (page 292). This is to be expected, since air approximates an ideal fluid and, unlike a rope or an iron bar, cannot transmit sidewise (shear) forces. The motion of the air particles in the compressions is *forward,* in the direction of advance of the wave, and is *backward* in the expansions. Figure 12.16*a* shows one way of depicting a compressional wave, similar to the scheme used in Fig. 12.15. A more useful and precise representation is a graph showing how either the pressure or the particle displacement in the medium varies with time or with position along the wave train. Figure 12.16*b* shows, for instance, the way the pressure varies along the wave train represented in the upper part of the figure. In the simple example shown, where the source vibrates with SHM, the curve is a *sine wave* (the graph of the sine function). In any case, it must be remembered that this is merely a graph of the pressure variation and that the direction of the particle motion is *not* sidewise but entirely longitudinal. The pressure values represent a condition of the medium at each point and have no directional property.

Underwater sound waves should not be confused with the surface waves previously discussed.

Longitudinal waves can propagate in any substance in any physical state, since all materials are compressible to some extent. For instance, sound waves are often transmitted along a building structure in this way. Underwater compressional waves produced by a ship's propellers can be detected some distance away by a suitable receiver. One kind of earthquake wave is longitudinal.

Factors determining wave speed.

The speed of transmission of a mechanical wave of any kind is determined entirely by the elasticity and the inertia of the medium, and the formula for the speed in any given instance can be deduced by using the laws of mechanics. When this is done for compressional waves in a medium such as air, the result is

$$V = \sqrt{\frac{B}{D}} \qquad\qquad 12.13$$

where V is the wave speed, B a measure of resistance to compression of the medium, and D its mass density. This result was first derived by

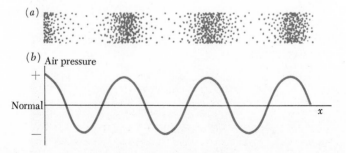

FIGURE 12.15   A pulsating balloon generates spherical compressional waves in the surrounding air.

(*a*)

(*b*) Air pressure

+

Normal

—

x

FIGURE 12.16   Two ways of representing a compressional wave in air. In sound waves of moderate loudness, the pressure amplitude is only of the order of 0.0001 of normal air pressure.

TABLE 12.1 SPEED OF SOUND

MEDIUM	TEMPERATURE °C	SPEED, m/sec	ft/sec
Air	0	331.4	1,087
Hydrogen	0	1,270	4,160
Water	20	1,460	4,790
Aluminum	20	5,100	16,700
Brass	20	3,500	11,500
Glass	0	5,500	18,000
Granite	0	3,950	14,500
Iron	0	5,100	16,700

Newton, who used it to calculate a numerical value of V. Later, measurements of the speed of sound in air showed Newton's value to be too small by about $\frac{1}{6}$. The difficulty was not resolved until almost a century afterward when the French scientist P. S. Laplace (1749–1827) found that Newton had erred by using the value of B corresponding to constant temperature conditions. This is not exactly the case when each part of the medium is expanding and contracting as the waves traverse it, and substitution of the proper elastic constant gave good agreement with the experimental values.

For the case of transverse waves on a rope, the analysis yields the following expression for wave speed:

$$V = \sqrt{\frac{T}{d}} \qquad \text{12.14}$$

where T is the tension force with which the rope is stretched and d is its mass per unit length. Here again, as for a vibrating particle, the determining factors are an elastic constant (T) and an inertial constant (d). The equation is valid also for transverse waves in a stretched musical string or wire, where the stretching force is normally so great that there is no complicating effect due to the transverse rigidity (shear) of the medium.

For surface waves on liquids, the speed is governed by two kinds of elastic factors. One, which dominates for short waves, is the surface-tension constant. This is a measure of the tendency of the liquid surface to resist stretching (page 47). The other, which dominates for long waves, is not an elastic quantity as such but is gravity. It furnishes a restoring force here, just as it does for the vibration of a pendulum (page 285). The inertial factor for both short and long liquid surface waves is the mass density of the liquid. The explicit formulas for wave speed can be found in books on advanced mechanics.

12.9 THE SUPERPOSITION PRINCIPLE; BEATS

Previous sections of this chapter discussed the passage of a single pulse or of a given set of waves through media of various kinds. We now examine what happens when more than one set of waves traverse the same medium.

What effect, if any, do two intersecting sets of waves have on each other? How can the resulting disturbance of the medium at any point be found?

It is easy to create a situation of this kind by dropping two stones into water at different places and then observing what happens when the two sets of circular ripples cross. Each set passes through the other without affecting its shape, direction, or speed. Each set moves across the surface as though the other were not present. The same kind of behavior is observed for two pulses moving in opposite directions in a rope (Fig. 12.17); each retains its identity after passing through the other. The pictures show that, during the time the two disturbances are on the same portion of rope, the composite disturbance there is found by adding vectorially the two individual displacements at each point. This is equivalent to saying that the two disturbances are completely *independent;* each produces its own effect on the medium without any influence on the other.

Experience shows that this is true also for sound waves. For example, the air in a room can convey the sounds of several independent conversations with no effect of one wave motion on another. The actual disturbance of the air at any point in the room is the combination of the separate disturbances and may be quite complicated.

A general statement of this behavior is called the **principle of super-position.** The principle of superposition states that, when two or more disturbances are present in a given part of the medium at the same time, each proceeds independently of the other. This principle holds for *any* kind of mechanical wave, provided that the displacements are not very large. It is no longer valid, for instance, for intersecting shock waves (page 305). Superposition holds for electromagnetic waves, including light, but the intense beams generated by lasers (page 491p) are able to violate the principle of superposition.

Complex waves; sound quality.

One consequence of the superposition principle is that any periodic wave form, no matter how complicated, can be reproduced by superposing simple harmonic waves having suitably chosen amplitudes and frequencies that are integral multiples of some lowest value. The wave of lowest frequency is called the **fundamental mode** of vibration, and all the frequencies collectively are called **harmonics.**

The French mathematician J. B. Fourier (1768–1830) showed how to calculate the relative amplitudes of the harmonic components. This can also be done experimentally by feeding a complex wave into a mechanical or electrical analyzing device. When sound waves from various sources are introduced, different distributions of amplitude values are found. Examples are shown in Fig. 12.18. These patterns are recognized by the ear as differences in the **quality** of the tones produced by different musical sources. Thus the characteristic of a tone that musicians call quality (or *timbre*) is determined largely by the relative amplitudes of the harmonics.

As a further example of superposition, consider two sets of continuous waves having slightly *different frequencies* moving in the same direction. According to the superposition principle, the disturbance at any point in the medium is, at any instant, the resultant of the two individual disturbances. Suppose the two wave trains to be momentarily in the same phase at a given place. Then, because of their difference in frequency, there will be a progressive change of phase of one set with respect to the other.

FIGURE 12.17 Oppositely directed wave pulses traversing a rope illustrate the principle of superposition. The pulses retain their form after passing through each other.

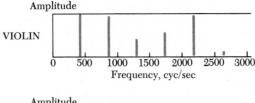

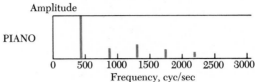

FIGURE 12.18 Frequency analysis for the tone "A," whose fundamental frequency is 440 cycles/sec, as produced typically by the violin (above) and by the piano (below). Only the first few overtones are shown, the remaining ones being relatively weak. Tones in other parts of the range of these instruments may yield an amplitude distribution somewhat different from the ones shown.

The resultant displacement at any instant can be found by vector addition of the two separate displacements. The variation of the resultant in time has the form of a wave whose frequency is the average of the individual frequencies and whose amplitude rises and falls at regular intervals.

Figure 12.19 shows the result for two wave trains of slightly different frequencies but equal amplitudes. The periodic changes in amplitude of the combination are called **beats,** and the number of beats per second is equal to the difference in the two original frequencies. For sound, the beats manifest themselves as a periodic fluctuation in loudness when two sources of different frequency are sounded at the same time. For example, the throbbing noise from a twin-motored airplane is a beat effect produced when the two propellers are turning at slightly different rates. The phenomenon of acoustic beats is complicated by certain subjective aspects. Despite much study, it remains imperfectly understood.

Practice set*   Two sources of waves having frequencies of 199 and 202 cycles/sec, respectively, produce beats. If both sets of waves have amplitude 1 at the place of observation, find (a) the number of beats per second, (b) the maximum amplitude of the composite wave, and (c) its minimum amplitude.

12.10   THE DOPPLER EFFECT

If the frequency of vibration of a source of waves is f, an observer receives these waves at the rate of f each second as long as he and the source are not moving relative to each other. However, when source and observer are in relative motion, the received frequency differs from the source frequency. This phenomenon is called the **Doppler effect,** after the Austrian physicist C. J. Doppler (1803–1853) who first discussed it. Whether the object that moves is the source, the observer, or both, the result is that the received frequency is *greater* than that of the source for relative *approach* and *less* for relative *recession.*

Pitch designates the place of a tone in the musical scale.

Many common observations make us aware of the Doppler effect for sound waves, where the phenomenon manifests itself as an apparent change in the **pitch** of a tone. This occurs in many familiar situations. For

* Ans.: 3; 2; 0.

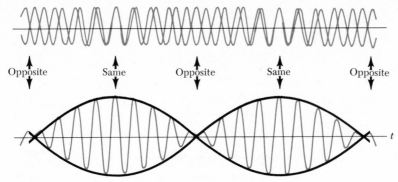

FIGURE 12.19 The two wave trains shown separately in the upper part of the figure have slightly different frequencies. As a result, they alternately get into and out of phase, producing beats. The composite wave is shown in the lower diagram, and the periodic change in amplitude is evident.

Perceptually, the pitch of a simple tone is determined almost entirely by the frequency.

example, if the horn of an approaching car is sounding, the pitch drops abruptly as the car passes. The pitch can be verified to be higher than normal while the car is approaching and lower while receding. Usually there are accompanying changes in the *loudness* of the sound, due to changing distance, but these are of no concern here.

The amount of the change in perceived frequency can be computed. This leads to two formulas that differ slightly for the two cases (moving observer or moving source) represented in Fig. 12.20. However, both relations reduce to the approximate formula

$$\frac{\Delta f}{f} = \frac{v}{c}$$

12.15

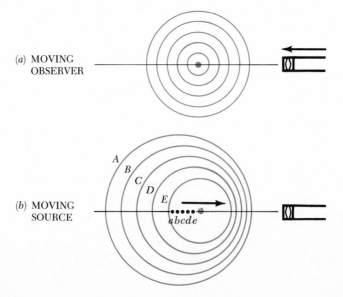

(a) MOVING OBSERVER

(b) MOVING SOURCE

FIGURE 12.20 The Doppler effect. In (a) the source is at rest and the observer is in motion; in (b) the source is in motion and the observer is at rest. In both cases the two are approaching each other, and the observer receives a greater number of waves per unit time than the normal frequency of the source. The opposite is true for motion of recession.

if the relative speed v of source and observer is small compared with the wave speed c. Here f is the actual frequency of the source and Δf is the observed change in frequency. This change represents an *increase* in frequency if source and observer are approaching each other, and a *decrease* if they are receding from each other.

Worked example 12.3 The tone of a whistle has a frequency of 400 cycles/sec. What pitch is perceived by an observer approaching this whistle at a speed of 44 ft/sec? The speed of sound waves in air can be taken to be 1,100 ft/sec.

Solution Substitution into Eq. 12.15 gives $\Delta f = (44)(400)/1100$, or 16 cycles/sec. According to the exact formulas (not given here), the results are:

1. Source stationary, observer approaching: $f = 416.0$ cycles/sec. This is identical with the result obtained by using Eq. 12.15.
2. Observer stationary, source approaching: $f = 416.7$ cycles/sec, which is slightly different from the above.

Intermediate values are found if both source and observer are in motion with respect to the surrounding air.

Doppler effect in relativity. Consider any case where a Doppler effect is observed in a material medium. If the wave speed and the source frequency are known and if the observed frequency is measured, it is then possible to determine whether it is the source or the observer that is moving. This is done by finding out whether the value of the observed frequency is given accurately by the equation for one or for the other situation pictured in Fig. 12.20. Although the theory of relativity denies the possibility of determining the "absolute" motion of a body, there is no contradiction because the exact equations for the Doppler effect are deduced on the assumption that there is a material medium, such as the air, to which all motions can be referred. However, there is no such frame of reference for light and other electromagnetic radiations, as the Michelson-Morley experiment showed (Sec. 5.3, page 92).

In an equation for the Doppler effect for light in the theory of relativity, only the relative velocity of source and observer enters. The resulting expression, for source and observer approaching each other in the line joining the two, is

$$\frac{\Delta f}{f} = \sqrt{\frac{1 + v/c}{1 - v/c}} - 1 \qquad \textbf{12.16}$$

where v is the relative speed of the pair and c is the speed of light in a vacuum. It can be shown that this equation reduces to Eq. 12.15 when v/c is very small.

The Mössbauer effect. A technique devised in 1958 by the German physicist R. Mössbauer uses the Doppler effect to measure extremely small mechanical speeds. The method employs very short electromagnetic waves (gamma rays) emitted by certain kinds of atoms embedded in a crystal. Other atoms of the same kind located in another body can absorb these waves by a resonance effect (Sec. 12.5, page 287). The resonance is so sharp that a frequency change of as little as 1 part in 10^{12} destroys the response completely. If a change of this kind is produced by a Doppler effect, a relative motion of source and receiver of only about a millionth of a millimeter per second can be determined. The Mössbauer effect has also made possible the detection of a minute change in frequency of radiation due to changes in the location of the source in a gravitational field, as predicted by the general theory of relativity.

Programmed review

Instructions: See page 18.

1.   What is meant by vibration, or oscillation?

> Regular repetition of a process, especially of the motion of a particle over a given path.   [12.1]

2.   Define *frequency* and *period* of an oscillatory process.

> Frequency: number of complete cycles of events described in unit time. Period: the reciprocal of frequency—the time for each cycle.   [12.1]

3.   Define simple harmonic motion.

> Vibration of a particle acted upon by a return force that is proportional to the displacement.   [12.2]

4.   Give a general expression for the period of a particle in SHM.

> $T = 2\pi \sqrt{m/k}$, where m is the mass of the particle and k is the constant in the force equation $F = -kx$.   [12.3]

5.   What does the period formula become for a simple pendulum making small vibrations?

> $T = 2\pi \sqrt{l/g}$, where l is the length of the pendulum and g is the acceleration due to gravity.   [12.4]

6.   What is resonance?

> The vigorous response of a system when driven by a period force having a frequency approximating the natural vibration frequency of the system.   [12.5]

7.   Explain what is meant by wave motion.

> The propagation of a disturbance through a medium at a characteristic speed whose value is determined by the properties of the medium.   [12.6]

8.   Define wavelength.

> The distance λ between any point in a wave train and the next point that is in the same *phase*, or state of vibration. In a uniform medium it is related to the wave speed V and the frequency f by the wave equation $V = f\lambda$. [12.7]

9. Explain why sound waves are classed as longitudinal waves.

The particle displacements take place in the direction in which the waves advance.   [12.8]

10. What is meant by the principle of superposition?

If several disturbances pass through a given place in a medium, each produces its own effect, independent of the others, provided that the displacements are small.   [12.9]

11. Describe the Doppler effect.

The apparent change in frequency of a train of waves resulting from relative motion of the source and the observer.   [12.10]

12. Give the approximate formula for the apparent change of frequency in the Doppler effect.

$\Delta f / f = v/c.$ [12.10]

For further reading

12.1. *Pierce, J. R.,* and *E. E. David, Jr.* "Man's World of Sound."
12.2. *Benade, A. H.* "Horns, Strings and Harmony." This and Ref. 12.1 are two simple presentations of various aspects of sound and music.
12.3. *Wood, A.* "The Physics of Music." A more detailed and advanced book.
12.4. *Beranek, L. L.* Acoustics, *Phys. Today,* November, 1969. An article on the varied applications of sound waves.
12.5. *Henry, G. E.* Ultrasonics, *Sci. Am.,* May, 1954. A good description of the production and effects of ultrasonic waves.

Questions and problems

Use 1,100 ft/sec or 340 m/sec for the speed of sound in air.

12.1. In Fig. 12.4 (page 281), at what point in its path—middle, either end, or other location—was the particle at $t = 0$? At $t = \frac{1}{4}T$? Give reasons for your answers.

12.2. A particle of mass m, vibrating with SHM, has a period T and an amplitude A. What is its speed when at a distance x from the equilibrium position? HINT: Refer to Sec. 12.3.

12.3. A particle describing SHM has a mass of 200 g, a period of 2.0 sec, and an amplitude of 20 cm. What is the magnitude of its acceleration (*a*) as it passes through the equilibrium position and (*b*) at either end of its path?

12.4. When a rock is hung from a certain spiral spring of negligible mass, the spring stretches 4 cm. If the rock is pulled down slightly from its equilibrium position and released, with what frequency does it oscillate? Explain why the mass of the rock does not have to be given.

12.5. A metal ball of mass 250 g oscillates with a frequency of 2.5 cycles/sec in the up-and-down direction when suspended from a certain spring. When the ball is hanging at rest, how much of a downward pull is needed to stretch the spring an additional 1 cm?

12.6. A simple pendulum of length 3.27 m is observed to take 363 sec to complete 100 cycles. Find the value of g at the place where the observations were made.

12.7. The period of a simple pendulum is observed to be 3.6 sec. What will its period be if it is taken to the moon, where the acceleration due to gravity is one-sixth its value on earth?

12.8. At a certain place on a phonograph record the speed of the groove relative to the needle is 1.00 m/sec. If the frequency of the tone produced at that time is 440 cycles/sec, compute the distance between the crests of the wavy indentations in the groove.

12.9. Compute the wavelength in air of (a) sound waves of lowest audible frequency; (b) sound waves of highest audible frequency; (c) ultrasonic waves of the highest frequency so far produced (2.5×10^{10} cycles/sec). Express all results in feet.

12.10. A long, heavy rope hangs from a hook in the ceiling, the lower end being free. A transverse pulse is started by hitting the rope near its lower end with a stick. As the pulse moves up the rope, will its speed remain constant? HINT: Examine Eq. 12.14.

12.11. A sounding device on a ship sends out an underwater "beep" and receives the echo from the sea bottom after 1.6 sec. If the speed of sound in seawater is 4,900 ft/sec, how deep is the water at that location?

12.12. Two whistles, of frequencies 1,000 and 1,015 cycles/sec, respectively, are sounded simultaneously. (a) How many beats are heard each second? (b) How far apart in the air are the places of minimum wave intensity?

12.13. A man standing beside a railroad track measures a drop of 13.6 cycles/sec in the pitch of the whistle of a train as it passes him. If the actual frequency of the whistle is 110 cycles/sec, how fast, in feet per second, was the train going?

12.14. The Doppler effect can be observed with microwaves (short radio waves) by reflecting a continuous train of such waves from a moving object. This is the principle of the electronic devices used by police to measure the speed of approaching cars. The returned waves are superposed on the direct waves and the beat frequency is measured. Its value is approximately $2(v/c)f$, where v is the speed of the car, c is the speed of the waves, and f is the frequency of the microwaves used (compare Eq. 12.15). On a certain occasion, when waves of length 0.12 m were used, a beat frequency of 400 cycles/sec was measured. Find the speed of the approaching car.

Chapter 13
FURTHER PROPERTIES OF WAVES

Wave motion was introduced in Chap. 12 as an effect produced by the vibration of a source. Examples were drawn from a variety of physical situations, and some that were selected from the field of sound proved to be interesting in their own right. In this chapter the discussion of general wave properties continues, and since certain of these attributes are of interest primarily in connection with light waves, the illustrations tend more in that direction. Finally, in Chap. 14, some wave properties will be treated in their purely optical manifestations.

13.1 WAVE INTENSITY

An important characteristic of waves, mentioned on page 288, is their ability to hand on energy from one particle of the medium to the next. The flow of energy at any place can be determined by absorbing all the energy delivered to a receiver in a given time and measuring it by converting it into another form such as heat or electric energy. The intensity I of a wave is defined as the time rate of transport of energy per unit area S of the wavefront (see Fig. 13.1). Thus the intensity is given by

$$I = \frac{W/t}{S}$$

or since W/t is the *power* L (page 209), this can be written

$$I = \frac{L}{S} \qquad\qquad \textbf{13.1}$$

If mks units are used, L is measured in watts and S in square meters, so that wave intensity has the units watts per square meter in this system.

For sound waves, the numerical values are ordinarily very small. Although the sound intensity measured near the mouth of a person talking in a moderate voice is only of the order of 10^{-5} watt/m², a normal human ear is able to detect sounds that are about a million times less intense than this and can still respond without pain to intensities a million times greater (see following section).

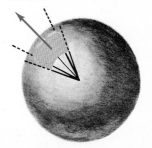

FIGURE 13.1 The wave intensity is measured by the rate at which energy streams out through each unit area.

The intensity of a simple harmonic wave is given by

$$I = 2\pi^2 A^2 f^2 DV \qquad\qquad \textbf{13.2}$$

where f is the wave frequency, D is the mass density of the medium, and V is the wave speed. The quantity A is the amplitude of motion of the particles of the medium at the place where the intensity is measured; it is *not* the amplitude of vibration of the source. However, the value of A must depend in some way on the distance from the source.

Effect of distance from source.

Consider the particular case of a small source of waves (referred to as a **point source**) immersed in a uniform, nondissipative medium of indefinite extent. The wavefronts are spherical surfaces, and as the wave moves outward the energy passing through the surface of a given sphere will be spread out over larger spheres at later times. Since the surface area of a sphere is proportional to the square of its radius, *the intensity of a wave at any* point *is inversely proportional to the square of the distance from the source.* This statement is valid only for an isolated point source in an ideal medium, but not for horns, lamps, or radio antennas, which do not radiate uniformly in all directions. If the dimensions of the source are not small compared with the distances considered, the source must in effect be divided into an infinite number of small sections whose contributions are added to get the total effect at the point of reception. Such a summation process usually requires the use of integral calculus.

13.2 INTENSITY AND LOUDNESS OF SOUNDS; SHOCK WAVES

The loudness of a sound, as judged by the ear, is found to depend not only on the intensity of the waves as defined above but also to some extent on the pitch and wave form. In addition, physiological and psychological factors play an important part.

The ear does not respond in direct proportion to the stimulus but at a more moderate rate, roughly proportional to the logarithm of the intensity. Experiments show that a normal ear judges a sound to be *twice* as loud as another of the same frequency when the wave intensity of the second is *ten* times that of the first. The reference level is taken to be 10^{-12} watt/m^2, which is about the minimum intensity the average ear can detect at a frequency of 1,000 cycles/sec. The intensity level of a given sound is specified by giving the exponent to which 10 must be raised in order to equal the intensity ratio of this sound to the standard minimum value. This exponent is expressed in a unit called a **bel,** after Alexander Graham Bell who patented the telephone in 1876.

A unit of more convenient size, the **decibel** (dB), is equal to 0.1 bel. The intensity level L, in decibels, is given by

$$L = 10 \log \frac{I}{I_0} \qquad\qquad \textbf{13.3}$$

where I is the intensity of the sound in question and I_0 is the standard minimum intensity. Thus if one sound is 1 dB louder than another, its physical intensity is 1.26 ($=10^{1/10}$) times as great. The intensity doubles for an increase of about 3 dB. On this basis, the rustling of leaves has a rating of about 10 dB; ordinary conversation, 65 dB; street traffic, 70 to 80 dB; etc. Levels above 85 dB are considered harmful, and the sensation becomes painful when a level of about 120 dB is reached.

The displacement amplitudes and the changes in pressure for ordinary sound waves in air are surprisingly small. The human hearing mechanism is so sensitive that, in a wave of minimum audibility, the pressure departs from normal by only about 2×10^{-10} atm, and the displacement amplitude is 10^{-11} m—less than the diameter of the air molecules. Even for sounds approaching the level of pain, the pressure variations are less than 10^{-3} atm and the amplitudes less than a hundredth of a millimeter. It must be noted also that the minute air displacements in a sound wave are superimposed on the random thermal motions of the air molecules. Nevertheless, these displacements succeed in coming through the thermal background to convey their information to the ear.

A mechanical **shock wave** may be thought of as a compressional wave of large amplitude. An example is a pulse produced in the surrounding air by a violent mechanical impulse such as an explosion. It is characterized by a steep rise in pressure whose magnitude may be several times as great as normal air pressure, followed by a more extended region where the pressure is somewhat below normal. The principle of superposition does not hold for shock waves, since Hooke's law is not applicable for large displacements. One result is that a shock wave travels with a speed considerably greater than that of compressional waves of ordinary intensity.

Rockets and jet airplanes can maintain **supersonic** speeds, that is, speeds greater than that of sound. As a result, a shock front moves along with the source, producing a ''sonic boom'' as it sweeps across an observer.

To see how a shock front develops, look back at Fig. 12.20*b* (page 298), which shows the wavefronts produced by a source whose speed is *less* than that of sound. The effect is to crowd the waves in the forward direction, but they still move out in advance of the source. If, however, the source moves faster than sound, the wave pattern changes to resemble Fig. 13.2. Spherical sound waves originate along the path of the moving object and form a conical surface that constitutes a shock front. The angle at the vertex of the cone is determined by the ratio of the speed of the source to the speed of sound, a quantity called the **Mach number.** For example, a guided missile moving with twice the speed of sound is rated ''Mach 2.'' Pulses have been produced in experimental shock tubes at Mach 200, giving rise to temperatures approaching 10^6 K.

The familiar V-shaped bow wave produced by a ship moving at a speed greater than the wave speed corresponds to the above kind of conical shock wave. In 1934 the Russian physicist P. A. Čerenkov discovered a similar phenomenon with light waves emitted by fast electrons moving through a transparent medium. The speed of light in such materials is considerably less than the speed of light in a vacuum, c. Charged particles can be given speeds close to c, so that the condition for the generation of a conical wavefront is fulfilled. The bluish Čerenkov radiation is used in nuclear research to detect the passage of high-energy particles.

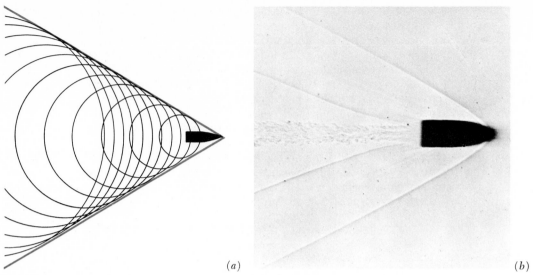

FIGURE 13.2   (a) Shock front accompanying a body moving through a medium at Mach 2. (b) Shock waves produced by the nose and tail of a bullet. The photograph was made by a shadow technique. (*Olin.*)

13.3   REFLECTION OF A PULSE

The motion of a transverse pulse along a stretched rope or spring was described on page 291. It is of interest to examine what happens when the pulse, traveling at constant speed, reaches the far end of the rope.

Suppose this end is tied to a hook fastened to a wall (Fig. 13.3). It is observed that, when the original pulse reaches the wall, another pulse immediately starts back in the opposite direction with the same speed. If the original pulse is a hump (upward disturbance), the returning pulse is observed to be a hollow (downward disturbance) of the same shape: If one of these profiles is turned upside down as well as reversed right and left, it looks exactly like the other. In the same way, a hollow sent down the rope returns as a hump having the same form.

This behavior of a pulse on a rope is a specific example of what happens to any wave under similar conditions. In general, whenever a wave meets a discontinuity in the medium, a returning wave is produced. The process is called **reflection.** The original wave is the **incident wave,** and the one that returns is the **reflected wave.** The terms are perhaps most familiar in connection with the behavior of light waves on striking a mirror, but they apply to any type of wave encountering a sudden change in the properties of the medium.

Effect of end conditions.

In the experiment with the rope, why is a hump that strikes the fixed support reflected in the form of a hollow? When the hump-shaped pulse reaches the ring it tends to move the entire wall suddenly upward, which is obviously impossible. Instead, the reaction snaps the rope downward, sending a hollow back along the rope. This pulse is similarly reflected when it returns to the experimenter's hand, and it may travel back and forth several times before its energy is completely dissipated.

FIGURE 13.3   When a pulse reflected from the fixed end, it returns with its displacement reversed.

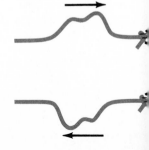

Suppose that, instead of the distant end of the rope being fixed in position, a loop at the far end is slipped over a smooth post so that this end can slide freely up and down while still keeping the rope under tension (Fig. 13.4). A pulse that now arrives at the far end is no longer confronted by an object of infinitely great inertia, such as the wall, and there is nothing to oppose the transverse motion. The loop end of the rope accelerates and overshoots the original disturbance, giving rise to a reflected pulse that lies on the *same* side of the rope as the incident one. Again the incident and reflected pulses have the same shape but are reversed right and left.

For transverse waves on a rope, there is a great difference between reflection from a *fixed* end and reflection from a *free* end. This difference may be summarized compactly:

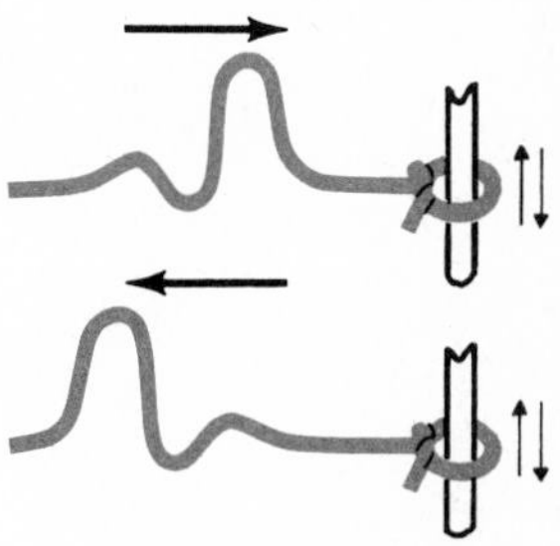

Fixed end: Reflected pulse has its direction of disturbance *reversed.*
Free end: Reflected pulse has the direction of disturbance *unchanged.*

It will be shown later that a similar description holds for the reflection of waves of any kind.

FIGURE 13.4 When a pulse is reflected from a free end, it returns with the direction of its displacement unchanged.

13.4 CONTINUOUS WAVES IN A ROPE; STANDING WAVES

In place of applying a single snap at one end of a stretched rope, suppose the end is shaken up and down periodically. Two continuous wave trains now travel along the rope, the first set going toward the distant end and a reflected set coming back. At any given instant, the position of any particle of the rope is determined by the principle of superposition (Sec. 12.9) as the vector sum of the displacements caused by the two wave trains. Each

NOISE POLLUTION

Noise usually consists of a random mixture of sounds extending over a wide range of frequencies. Any unwanted sound may be classed as noise. Public health officials recognize the dangers of the high sound levels to which people are subjected and refer to this condition as a form of environmental pollution.

More than one percent of the United States population suffers from partial hearing loss induced by noise. Industrial noise is responsible for most of this impairment; transportation noises are probably the second most important contributor.

In recent years, electronic sound reproducers have added another factor. Measurements of the sound levels produced by "rock" bands showed that they often exceeded the risk criteria used by industries. No data are available on hearing loss of listeners, but tests made on musicians showed that about 25 percent of them experienced temporary hearing loss. It is known that permanent damage results when this type of exposure is continued over a span of years.

Psychological studies show that continued exposure to high-level noise causes tension and irritability. In industry, this leads to reduction in the efficiency and accuracy of the workers. According to a report by the World Health Organization, the total annual monetary loss from inefficiency and accidents caused by industrial noise exceeds 4 billion dollars.

Consult Ref. 13.1 for an interesting and comprehensive discussion.

set of waves passes through the other set without affecting the form of either.

As already stated, the character of the reflected set depends on conditions at the far end of the rope. Suppose this end is fixed in position. The end that is held in the hand is, very nearly, a fixed point, too: The hand need only be moved with very small amplitude in order to maintain the waves. Hence both ends of the rope may be considered fixed in position.

If the rope is now shaken with a very low frequency, the intermediate parts of the rope are observed to undergo an irregular motion that is difficult to follow visually. If the frequency of the applied force is gradually increased, the appearance of progressive motion along the rope vanishes when a definite frequency of shaking is reached. At that time, all parts of the rope vibrate as a single arch that swings periodically up and down across the rest position (Fig. 13.5a). This condition of vibration is an example of a **standing wave,** as contrasted with the traveling waves that move along the rope in both directions. The standing wave is the result of superposition of the two traveling waves of the same amplitude when they have a particular frequency, f_1.

If the frequency of shaking is now gradually increased, the rope reverts to an irregular, erratic motion until the frequency reaches a value of exactly $2f_1$, when a standing wave again appears. This time the configuration has two arches, each occupying half the rope, and the particle motions in one arch are exactly opposite in direction to those in the other (Fig. 13.5b). The midpoint of the rope is observed to remain fixed at all times even though it is not constrained in any way by an outside agency. It is merely a place where the two traveling waves combine to produce zero displacement at all times, just as they do at the two ends.

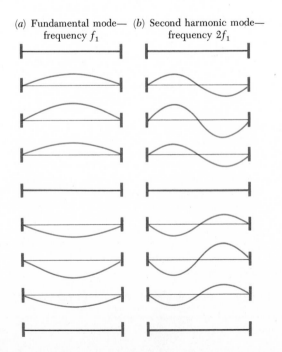

(a) Fundamental mode— frequency f_1 (b) Second harmonic mode— frequency $2f_1$

FIGURE 13.5 Configurations of rope vibrating (a) in a single loop (fundamental mode) and (b) in two loops (second harmonic mode). Each sequence, from top to bottom, shows the rope at intervals of one-eighth of a period.

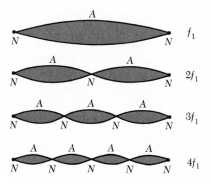

FIGURE 13.6 Appearance of a rope or string vibrating in each of its first four normal modes. In the fundamental mode the string forms a single loop, in the next it forms two loops, etc. The possible frequencies of a string are harmonic. Review, at this point, the definitions of fundamental and harmonic. (page 296).

Verify these observations by experimenting with a length of heavy rope or, better, a long spiral spring.

When the experiment is continued, it is found that for vibration rates exactly 3, 4, 5, . . . times f_1 the rope again breaks into steady patterns, pictured in Fig. 13.6 as blurred "time exposures." Thus the rope is able to divide into 1, 2, 3, . . . equal segments or loops, corresponding to definite standing-wave configurations. The points where the rope remains motionless are called **nodes,** and the places halfway between the nodes, where the amplitude is a maximum, are **antinodes.** All particles in any one loop vibrate in the same phase, while neighboring loops are in opposite phase. The distance between two adjacent nodes or two adjacent antinodes is *one-half* the wavelength of the traveling waves.

Unlike traveling waves, standing waves do not involve any net transport of energy along or through the medium (Fig. 13.7). The energy possessed by the medium remains localized in each loop and merely alternates between kinetic energy and elastic potential energy. It is possible to consider the two traveling wave trains as carrying equal amounts of energy in opposite directions, thus combining to give zero energy flow.

Musical strings.

The vibration of a stretched rope, discussed above, is of precisely the same nature as that of a wire or string in a musical instrument such as the violin, harp, piano, etc. However, unlike the shaken rope, a musical string is not driven by an applied force of specific frequency but determines its own configurations of vibration (**normal modes**), some of which are

FIGURE 13.7 The standing-wave pattern is the vector sum of the direct and reflected wave trains. Represented here at intervals of a quarter period are all three sets of waves. In the mode of vibration shown, the rope is divided into four loops.

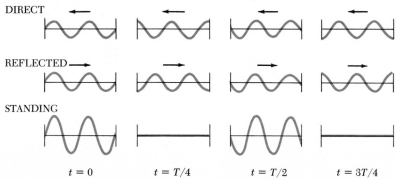

DIRECT

REFLECTED

STANDING

$t = 0$ $t = T/4$ $t = T/2$ $t = 3T/4$

*Be careful
to avoid
confusing
the terms
mode and
node.*
illustrated in Fig. 13.6. The applied stimulus may be considered to be a complex disturbance which, according to Fourier's principle (page 296), consists of a combination of a great many simple vibrations of different frequencies. The string absorbs energy from the frequencies adjoining the normal ones, building up this particular set at the expense of all the others. Thus *the response of the string is a resonance phenomenon* (Sec. 12.5, page 287). A single-particle system such as a simple pendulum has only one resonance frequency, whereas a string has an infinite set of resonances, or normal modes.

The actual condition of vibration of the string depends on how it is bowed, plucked, or hit. In general, it will vibrate with some combination of several of its normal modes. Each mode corresponds to one of the partial tones given off by the string, and the relative prominence of the various partials determines the quality of the tone produced (page 296).

How can the frequency of each normal mode of a string be computed? The distance between two adjoining nodes or antinodes is $\frac{1}{2}\lambda$, and so if n is the number of loops into which the string resolves,

$$\lambda = \frac{L}{\frac{1}{2}n} = \frac{2L}{n}$$

Then, according to the wave equation, $f = V/\lambda$, or

$$f = \frac{nV}{2L} \qquad\qquad 13.4$$

Putting $n = 1, 2, 3, \ldots$ gives the frequencies of the various normal modes.

Worked example 13.1   A piano string 1.5 m long, mass per unit length of 10 g/m, is stretched under a tension of 400 newtons. With what frequency is it vibrating when there are three loops on the string?

Solution According to Eq. 12.14, page 295, the speed of transverse waves in a string is given by $V = \sqrt{T/d}$. Using this value in Eq. 13.4, we get

$$f = \frac{n}{2L}\sqrt{\frac{T}{d}} \qquad\qquad 13.5$$

for the frequency. Substituting the given numbers and remembering to change d from g/m to kg/m, the result is

$$f = \frac{3}{2(1.5)}\sqrt{\frac{400}{0.01}} = 200 \text{ cycles/sec}$$

13.5 STANDING WAVES IN AIR COLUMNS

In the preceding section it was shown that a stretched string can vibrate with a definite set of frequencies that correspond to the normal modes of vibration associated with standing waves. In the same way, standing compressional waves in a long, narrow column of air produce tones of definite frequencies in wind instruments such as the flute, trumpet, pipe organ, etc.

Compressional waves moving down an air column in a pipe are reflected from the far end, and the superposition of the direct and reflected waves can establish a standing-wave pattern, just as for transverse waves in a string. Both ends of a string on a musical instrument are fixed in position, or very nearly so. In air-column instruments such as a whistle or flue-type organ pipe, the end into which air is blown is very nearly free, while the far end may be either fixed or free with respect to the air vibrations. If the far end is fixed, we speak of a "closed" pipe; if free, an "open" pipe (Fig. 13.8).

n most reed instruments (clarinet, saxophone, etc.) the mouthpiece s effectively a closed end.

A compressional pulse moving down a closed pipe and meeting the fixed wall at the far end rebounds as a compression. The reversal in direction of motion of the air molecules produces a change to the opposite phase, just as in the case of a transverse wave reflected from the fixed end of a string. However, if a compression is sent down an *open* pipe, there is nothing to restrict the motion of the air molecules at the far end, and they move freely into the surrounding air when the compression arrives. The same thing happens when a hole on the side of the pipe is uncovered, as in playing a note on the flute. The hole in effect becomes the open end of the pipe.

The inertia of the escaping molecules causes them to overshoot and, instead of merely relapsing to normal pressure, the air at the open end experiences a slight decrease of pressure. This partial vacuum (rarefaction) makes the air molecules just inside the tube move toward the open end, causing the neighboring molecules to move in a similar way. The result is that the rarefaction continues to travel back up the tube. Since the air molecules move forward in a compression and backward in a rarefaction (page 294), the direction of the displacements is the same for the reflected waves as for the direct waves. Thus there is no change of phase in the process of reflection from the open end. The situation is analogous to that of a transverse wave reflected from the free end of a rope (page 307).

In a practi- al case the antinode may lie just beyond the pen end of the pipe.

The above information can be summarized by saying that *at a closed end there is a node of displacement,* and *at (or near) an open end there is an antinode of displacement.* This indicates at once the kind of standing-wave pattern that can be expected in each kind of pipe: A simple closed pipe has an antinode at the blown end and a node at the closed end; an open pipe has an antinode at each end. Figure 13.8, which should be studied carefully, shows the configuration of the first three partial tones for open and closed pipes. As a result of the end conditions, the frequencies of the partials of the open pipe are all *consecutive* integral multiples of the fundamental, whereas for the closed pipe only the *odd* multiples occur.

FIGURE 13.8 The modes of vibration of open and closed pipes are determined by the conditions at the ends: There must be a node (N) at each closed end and an antinode (A) at each open end. The black lines show the possible standing-wave configurations that result. For the open pipe, the frequencies of the partials are 1, 2, 3, . . . times the fundamental frequency f_o, whereas for the closed pipe they are 1, 3, 5, . . . times the fundamental frequency f_c.

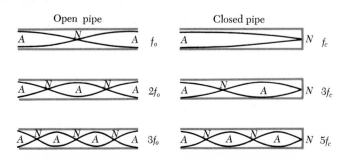

Standing waves are set up and maintained in a simple air-column source such as the flute or a flue organ pipe (Fig. 13.9) when a thin stream of air is blown against the sharp edge of an opening at one end of the pipe. This air stream has a highly irregular motion which can be looked upon as a mixture of a great many frequencies. If the pipe were absent, the air jet itself would not emit musical tones but only a noisy hiss—a mixture of a large number of unrelated frequencies. Each one corresponds to one of the possible normal modes. However, with the pipe in place, the intensity of vibration builds up at the frequencies that correspond to the possible normal modes, and all others die out. The process is exactly the same as that for the vibrating string.

This discussion of standing waves has been confined to one-dimensional media such as strings and narrow columns of air. A further example of standing waves in one dimension is the transverse *bending* vibrations of bars and rods, represented by musical sources such as the xylophone, chimes, etc. Two-dimensional standing waves can be produced on the surface of a liquid, on a drumhead, or on the diaphragm of a loudspeaker. In such cases the nodes are lines rather than points. Three-dimensional standing waves may be maintained in confined fluids and in suspended liquid drops. In general, frequencies of the higher modes of two- and three-dimensional vibrating bodies are not *harmonic* (integral multiples of the fundamental), as is true for strings and pipes. The details can become very complicated and will not be elaborated here.

FIGURE 13.9 Open and close flue organ pipes, shown in cross section.

Worked example 13.2 A certain open pipe has an effective length of 70 cm. Find the frequency of (*a*) its fundamental tone; (*b*) its third partial. Take the speed of sound in air under the prevailing conditions to be 350 m/sec.

Solution   The distance between two adjacent nodes or antinodes in a pipe is $\frac{1}{2}\lambda$, just as for a string (page 309).

a. For the fundamental, according to the top drawing (left) in Fig. 13.8, this distance is equal to the pipe length. Hence the wavelength of the standing waves in this example will be $\lambda = 70 \times 2 = 140$ cm $= 1.40$ m. The corresponding frequency is $f = V/\lambda = 350/1.40 = 250$ cycles/sec.

b. The third partial has a frequency three times that of the fundamental, or 750 cycles/sec (see Fig. 13.8).

13.6   HUYGENS' CONSTRUCTION

Experience shows that in ordinary media such as air or water a disturbance emanating from a point source moves outward as a sphere of constantly increasing radius. This is because the wave speed is the same at all points and in all directions. A medium that has the same properties at all points is said to be **homogeneous;** one having the same properties in whatever direction they are measured is **isotropic.** In what follows, a medium that is both homogeneous and isotropic will be characterized as **uniform.** A crude, one-dimensional analog is provided by the cutting edge of a saw: It is homogeneous (on a large scale) but not isotropic.

Spherical waves of the kind described were produced by the pulsating balloon illustrated in Fig. 12.15 (page 294). At a place far from the source, a limited area of a spherical wave is almost undistinguishable from a plane. A disturbance having plane wavefronts is called a **plane wave.** Light waves received on earth from a distant star or from the sun are essentially plane. It is possible to change the curvature of wavefronts from nearby sources of radiation and make them plane by using suitable optical devices such as lenses and mirrors (Chap. 15).

Instead of drawing the waves themselves, it often proves simpler to trace lines along which the disturbance travels. A line giving the direction of advance of the waves at any point is called a **ray.** A ray may be drawn at any place through which the waves pass. Unlike the waves themselves, rays have no physical reality but are simply construction lines showing the direction in which a set of waves is moving, just as stream lines may be traced to show the direction of travel of the particles of a fluid.

The rays corresponding to a set of expanding spherical waves are straight lines drawn outward from the source in all directions (Fig. 13.10). Any number of such rays may be drawn. The rays associated with plane waves are parallel lines, all normal to the wavefronts. In an isotropic medium the rays are always perpendicular to the wavefronts at any point, regardless of what form the wavefronts have.

A perfectly uniform medium is an ideal seldom approached in practical situations where waves often encounter irregularities, discontinuities, and boundaries. A means of tracing the progress of a wave under such general circumstances is provided by a construction announced by Huygens in 1678. He pointed out that, since the particles of the medium are set into vibration when a wave passes, these particles in turn may be thought of as new sources of waves. Huygens' construction may be described in this way: *Each point on an advancing wavefront acts as a source from which secondary waves spread, and succeeding wavefronts are determined by the superposition of these secondary waves.* At any given place, the secondary waves ("Huygens wavelets") have a speed equal to the local wave speed in the medium.

To see how this applies in a particular case, consider the propagation of expanding spherical waves in a uniform medium, as already described. In Fig. 13.11, W_1 represents such a wave at any instant. Take any points on W_1, such as $A, B, C, \ldots$, to be new sources of waves. These wavelets start out simultaneously and, after a short time Δt, have a radius $V \Delta t$, where V is the wave speed. The surface that envelopes (touches) all these small spheres is W_2, which is a sphere whose center is also at S, and this is the new wavefront. A short time later, wavelets from points on W_2 similarly indicate a new spherical surface W_3 to be the wavefront, and so on.

It may seem that another spherical envelope could be drawn on the opposite side of the series of wavelets coming from W_1, leading to a new wavefront propagating back toward the source, as indicated by the dashed curve in Fig. 13.11. However, this is not observed to happen. It can be proved that superposition (page 296) exactly cancels all parts of the Huygens wavelets except the advancing rim, and so the waves are propagated only in the forward direction.

Huygens' construction is useful in a variety of cases of wave propagation. An example is the spreading of ripples when they encounter a patch of

The word "ray" comes from the Latin radius. One of its meanings is the spoke of a wheel."

Christian Huygens (1629–1695), Dutch physicist, astronomer, and mathematician, made many important contributions to optics and other branches of physics.

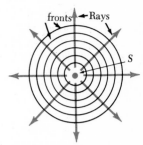

FIGURE 13.10 Spherical waves and some rays associated with them.

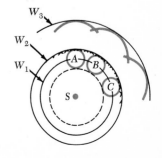

FIGURE 13.11 Propagation of a spherical wave according to Huygens' construction.

oil on the surface of water (page 289). The resulting distortion of the circular ripples can be traced in detail if we know the wave speed on the oily part of the surface and the contour of the patch. All the wavelets originating within this area have a radius differing from the ones on clean water, and if they are drawn accordingly the progress of the wavefront can be correctly traced.

**Practice
set**   Using the general procedure described in this section for spherical waves in a uniform medium, show how Huygens' construction can be used to trace the progress of plane waves in a uniform medium. Make a sketch comparable to Fig. 13.11.

13.7 LAW OF REFLECTION

Huygens' construction may also be used to predict the contour of a wave after reflection from a barrier. A specific example is the case of expanding spherical waves reflected from a plane surface. This includes the familiar optical situation in which light from a point source is reflected from a flat mirror.

In Fig. 13.12, let W_1 be an expanding spherical wavefront at the instant it first touches the plane reflecting barrier MM' at the point O. This point immediately becomes the source of a Huygens wavelet, which can spread only in the medium from which W_1 has already come. Other pairs of points on MM', such A and A', B and B', etc., are reached by W_1 at successively later times, and these points become, in turn, sources of new wavelets, as shown in Fig. 12.13. At a given instant, the whole set of wavelets has the surface W_2 as an envelope. As long as the medium is uniform, this

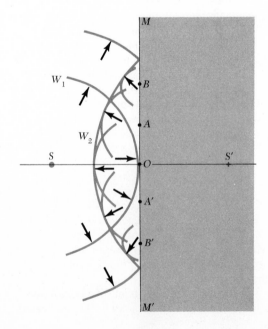

FIGURE 13.12 Huygens' construction for the reflection of spherical waves from a plane barrier. The wavefronts after reflection appear to come from the point S'

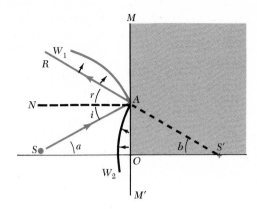

FIGURE 13.13 The law of reflection.

surface is again part of a sphere, but it is turned opposite from W_1. Its center S' is just as far behind MM' as the source S is in front of it, and S and S' lie on the same perpendicular to MM'.

The reflection of spherical waves from a plane surface can be described much more simply by using rays rather than the wavefronts themselves. Figure 13.13 corresponds to a portion of Fig. 13.12, with an incident ray SA and the corresponding reflected ray AR added. Since the wavefront W_2 is a sphere with its center at S', AR will be an extension of $S'A$. Also, as stated in the preceding paragraph, $S'O = SO$. Thus the right triangles ASO and $AS'O$ are similar, so that angles a and b are equal. Draw AN perpendicular to MM' at A. AN is also called the **normal** to the reflecting surface at the point A. Since the normal is parallel to SS', $r = b$ and $i = a$. Consequently,

$$r = i \qquad\qquad\qquad \textbf{13.6}$$

Angle i is the angle between the incident ray and the outward-drawn normal to the reflecting surface; it is called the **angle of incidence.** Angle r is the angle between the reflected ray and the normal and is called the **angle of reflection.** The two angles lie on opposite sides of the normal. Moreover, all three lines—the incident ray, the reflected ray, and the normal—lie in a single plane called the **plane of incidence** (Fig. 13.14).

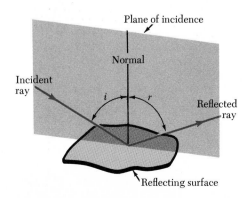

FIGURE 13.14 The incident ray, the reflected ray, and the normal to the reflecting surface all lie in one plane, the plane of incidence.

These facts were known to the ancients and constitute the celebrated **law of reflection:** *When waves are reflected from a surface, the angles of incidence and reflection are equal. The incident and reflected rays lie on opposite sides of the normal to the reflecting surface, and all three lines are in the same plane.*

Notice that i and r are also the respective angles between the wavefronts and the reflecting surface.

It may seem more direct to measure the inclinations of the two rays with respect to the reflecting surface itself rather than with the normal, but the latter procedure is preferred because the normal provides the simplest way of specifying the plane of incidence. The law of reflection is valid not only for a plane reflector but also for a *limited* portion of a surface of any shape. The normal is well defined at every point of a surface, so that the course of a ray reflected at a given place can always be determined, no matter how irregular the surface may be (Fig. 13.15).

Whether Huygens' construction or the simpler ray construction is used, a significant result is found for the reflection of spherical waves from a spherical surface. The reflected waves are also approximately spherical, but in general their curvature is changed on reflection. The consequences are especially important for light waves and radar waves, where reflection from curved surfaces is put to practical use in a number of ways.

Practice set*

1. If the angle of incidence of a ray is increased from 15° to 40°, by how many degrees is the angle between the incident and reflected rays increased?
2. What is the largest possible value, in degrees, that the angle of incidence can have?
3. In Fig. 13.13, find by measurement the value of either i or r corresponding to W_1 and its reflection W_2.

13.8   REFRACTION OF WAVES: SNELL'S LAW

Waves advancing in a uniform medium are observed to travel straight ahead, which means that the rays are straight lines. Departures from uniformity can cause the rays to deviate. A change in direction of the rays, resulting from nonuniformity of the medium, is called **refraction.**

Refraction is observed for waves of all kinds. Many unusual phenomena result from the refraction of sound waves and light waves passing through

*Ans.: 50°; 90°; 29°.

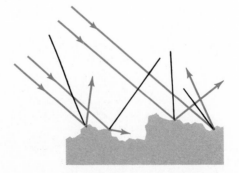

FIGURE 13.15   Parallel rays, corresponding to plane waves, are reflected in many different directions by an irregular surface. The direction of each reflected ray is readily found, however, by using the law of reflection, since there is a definite normal (shown here in black) at each point of the surface.

Refraction is based on the Latin refractus: broken up.

air that is not at the same temperature throughout. However, refraction effects are most prominent when the waves cross a sharp boundary between two media, as when light goes from air into water or glass. Such phenomena were observed in ancient times, but a quantitative description of refraction was not attained until the seventeenth century. The correct relation is generally connected with the name of the Dutch astronomer and mathematician W. Snell (1591–1626). Even with an empirical law of refraction available, the wave model of this phenomenon did not prevail until much later. An account of this controversy will be given in the next chapter. Meanwhile, we proceed with the wave interpretation.

Consider a train of plane waves incident obliquely on a plane boundary that separates two uniform but different media (Fig. 13.16). A set of reflected plane waves travels back into the first medium, as already described. In addition, a train of plane waves is observed in the second medium. These are the refracted waves, which arise when the incident waves cross the boundary plane. Their direction differs from that of the incident waves except in the special case of perpendicular incidence.

In some crystals the refracted ray does not lie in the incident plane.

As in the case of reflection, refraction at the boundary between two media can be discussed conveniently by using rays rather than wavefronts. Like the waves themselves, the rays undergo an abrupt change in direction on crossing the interface, but the refracted ray lies in the plane of incidence (see Fig. 13.16). The angle R between the refracted ray and the normal is called the **angle of refraction.**

The position of the refracted ray is given by Snell's law, the **law of refraction:** The refracted ray lies in the plane of incidence, and the angle of refraction is given by

$$n_1 \sin i = n_2 \sin R \qquad\qquad 13.7$$

where n_1 and n_2 are constants characteristic of the initial medium and the refracting medium, respectively.

The symmetry of form shown by Eq. 13.7 indicates that either medium may be considered the refracting one. This is a particular instance of a general proposition about waves, called the **principle of reversibility.** The principle of reversibility states that, if the direction of travel of the waves is reversed, they will retrace their former path exactly, regardless of any reflections or refractions they may have experienced. This is illustrated in

Reversibility of rays.

Fig. 13.18, where a ray from the point source S strikes a reflecting surface at M. The reflected ray (constructed according to the law of reflection)

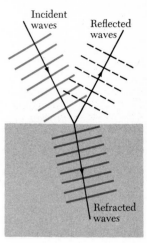

FIGURE 13.16 Waves striking the boundary between two media are partially reflected and partially refracted.

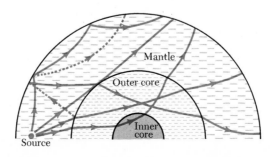

Source

FIGURE 13.17 Reflection and refraction of earthquake waves. The solid rays represent compressional waves; the dashed rays indicate the paths of transverse waves formed by reflection. The tracing of such disturbances has yielded much information on the internal structure of the earth.

then enters a second medium at N. Let the refracted ray be NP, traced according to the law of refraction. If we now imagine that P is a point source and we trace a ray from it along the line PN, we find that this ray follows exactly along NM after emerging into the upper medium and is reflected from the barrier in the direction MS. Thus by reversing their direction of travel, the rays were made to retrace their former path exactly.

Huygens' construction can account for Snell's law and at the same time provide an interpretation of the constant n_1 and n_2. Figure 13.19 shows a limited portion of a train of plane waves incident on the interface of two uniform media. The incident wavefronts are shown at intervals of time equal to the wave period T, so that these fronts are spaced at a distance equal to the wavelength, λ_1. According to Huygens, each of these wavefronts may be thought of as the envelope of wavelets of radius λ coming from points on the preceding wavefront.

Assume that the speed of the waves in the refracting medium is *less* than that in the initial medium, or $V_2 < V_1$. The wavefront AA' is about to enter the lower medium at A, so that after a time T the wavelet from A will have a radius of $\lambda_2 = V_2 T$, while wavelets originating at places higher on AA' have the larger radius $\lambda_1 = V_1 T$. As a result, a wavefront that has moved part of the way into the second medium has a broken appearance, such as BUB' or CWC', etc. After a given wavefront has passed entirely through the interface, it propagates onward as an unbroken plane wave except that the wavelength is reduced in the ratio V_2/V_1 as compared with the value in the initial medium:

$$\frac{\lambda_2}{\lambda_1} = \frac{V_2}{V_1} \qquad\qquad \textbf{13.8}$$

Compare Fig. 13.16.   Figure 13.20 is similar to 13.19 but shows only the wavefront about to enter the second medium and then completing its passage through the interface at some time t later. From geometry, the angle $A'AZ$ is equal to the angle of incidence i, and the angle AZY is equal to the angle of refraction R. Then, in the upper right triangle, $V_1 t = s \sin i$, and in the

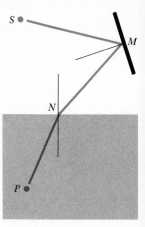

FIGURE 13.18 The principle of reversibility. If the direction of travel is reversed along any ray segment, the disturbance will retrace its former path.

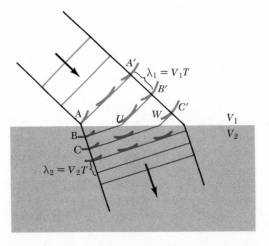

FIGURE 13.19 Plane waves entering a medium in which the wave speed is smaller. Huygens' construction shows that the refracted waves are again plane, but they are traveling in a direction closer to the normal and the wavelength is reduced in the ratio V_2/V_1.

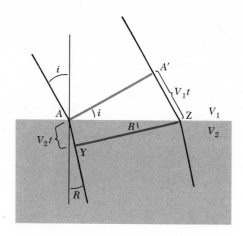

FIGURE 13.20 Snell's law deduced from Huygens' construction.

lower right triangle $V_2 t = s \sin R$. Division of one equation by the other yields

$$\frac{\sin i}{\sin R} = \frac{V_1}{V_2} \qquad \text{or} \qquad V_2 \sin i = V_1 \sin R \qquad \qquad \textbf{13.9}$$

Comparison with Eq. 13.7 shows that the constants n_1 and n_2, characteristic of the two media, may be taken to be inversely proportional to the respective wave speeds. Thus Snell's law of refraction states that the sines of the angles between the rays and the normal are proportional to the wave speeds in the two media. It will be shown in the next chapter that this proposition played a decisive part in establishing the wave theory of light.

Worked example 13.3 Plane waves of wavelength 16 cm enter a second medium where the wave speed is three-fourths as great as in the original medium. If the angle of incidence is 45°, find (a) the angle of refraction and (b) the wavelength in the second medium.

Solution   a. According to the law of refraction as expressed by Eq. 13.9, we have

$$\frac{\sin 45°}{\sin R} = \frac{4}{3} \qquad \sin R = \frac{3(0.707)}{4} \qquad \text{or} \qquad R = 32°$$

b. Substituting into Eq. 13.8 gives $\lambda_2 = 3(16)/4 = 12$ cm for the wavelength in the slower medium.

Table 13.1 presents values of the **index of refraction** $n = V_1/V_2$ for a few transparent substances of optical interest. Notice that n, being the ratio of two speeds, is a pure number. The numerical values generally increase with the density of the material but not in any simple way except for a gas, where the index is found to be proportional to the density. For a vacuum, $n = 1$.

Worked example 13.4 A ray of yellow light strikes one of the plane facets of a cut diamond at an angle of incidence of 40°. How much is the ray deviated on entering the diamond?

TABLE 13.1   INDEX OF REFRACTION OF SOME TRANSPARENT SUBSTANCES

MEDIUM	INDEX
Water	1.33
Optical glass, ordinary	1.52–1.67
Air (1 atm and 20°C)	1.0003
Quartz	1.54
Rock salt (NaCl)	1.54
Fluorite (CaF$_2$)	1.43
Carbon bisulfide	1.63
Ethyl alcohol	1.36
Ice	1.31
Diamond	2.42

The values are for the refraction of the yellow light from a sodium-vapor source.

Solution   The index of refraction of the first medium, air, may be taken to be 1.00, and Table 13.1 gives 2.42 as the value of n for diamond. According to Eq. 13.7, $n_1 \sin i = n_2 \sin R$; substituting the numbers,

$$\sin R = \frac{1.00}{2.42} \sin 40 = 0.266$$

from which $R = 15°$, approximately. The change in direction of the ray then amounts to $40 - 15 = 25°$.

13.9   INTERFERENCE OF WAVES

The principle of superposition was introduced on page 296. It leads to results of great interest under special circumstances. Superposition applies to waves of any kind—ripples on water, sound waves, light waves—as long as the disturbances are small.

As a consequence of superposition, two or more wave trains may partly or wholly annul one another at certain times and places, combining to produce an increased effect at other times and places. These effects are given the name **interference.** Although the term is much used, it is not accurate because it suggests a mutual disturbing effect of one wave train on another, whereas it is their complete *independence* that makes so-called interference effects possible.

These effects were described in Secs. 13.4 and 13.5.

Perhaps the simplest example of interference is the superposition of the direct and reflected transverse waves on a rope or the compressional waves in the air in a pipe. It was shown that in general the resulting configuration of the medium is fleeting and irregular but that stationary patterns result for certain values of the frequency, corresponding to the normal modes.

Interference effects in two dimensions may be produced by ripples on the surface of water. If two sticks are dipped in and out of the water in a random way, the resulting configuration of the surface changes erratically and no regularity is evident. However, if the two sources have the same frequency and operate continuously, the pattern on the surface becomes a steady one. Figure 13.21 shows the appearance of the water surface

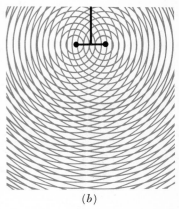

FIGURE 13.21   (a) A continuously vibrating pair of rigidly coupled point sources produces this steady pattern of ripples. (b) Diagram showing a plan view of the wave pattern.

when two round objects, mounted on a crossbar and dipping into the surface, are moved slightly up and down at constant frequency. The entire surface becomes covered with rows of waves extending radially outward from the source. Separating these areas are narrow paths of undisturbed water.

In Fig. 13.21b, the two sets of concentric circles represent the crests of waves from the source pair. When two crests meet, they combine to produce a crest that is twice as high as either one. When two troughs meet, they produce one that is twice as deep as a single trough. However, when a crest from one source meets a trough from the other, they mutually cancel and the result is undisturbed water at that place (Fig. 13.22).

The resultant lanes of maximum and minimum disturbance can be seen by holding the page in a horizontal position just below eye level and looking

FIGURE 13.22   Vertical section of the water surface, showing the resultant disturbance at the meeting place of (a) two crests, (b) two troughs, (c) a crest and a trough.

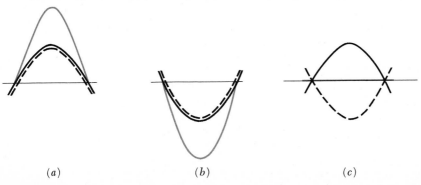

at Fig. 13.21*b* with the line of sight just grazing the page. The curving lanes of *destructive* interference are nodal *lines.* They correspond to the nodal *points* of standing waves in a rope or air column. Between the nodal lines there are regions of *constructive* interference corresponding to antinodes in the one-dimensional examples.

In order to have steady, observable interference patterns, the two sources of waves must have the same frequency and a *constant phase relation* with each other. For this reason, the spheres dipping into water were connected by a rigid bar which kept the frequency the same and the phase difference constant (zero). Two point sources of the same frequency whose wave trains stand in a constant phase relation are called **coherent sources.** Two radio or radar antennas connected to the same circuit serve as an example. However, in the production of sound or light waves, there is ordinarily no way of locking the individual sources in step. Thus the sound waves emitted by a violin string change phase abruptly and unpredictably each time the string slips along the bow, producing a succession of wave trains of limited lengths and random relative phase. If two instruments sound the same note with the same intensity, the disturbance in the air at the listener's ear is sometimes much greater than that of one instrument and at other times much less, depending on the momentary phase relationship of the two sets of waves. What is heard is the average effect, which amounts to a steady sound of twice the intensity of either source.

Similarly, an ordinary light source is a collection of a tremendous number of independently radiating atoms or molecules, and the total radiation output is not coherent. Nevertheless, optical interference experiments using ordinary light sources can be carried out by taking two beams from the same source and reuniting them at the place where the interference effects can be observed. Since each limited wave train from the source contributes to both beams, constancy of phase is realized at all times, and a stationary interference pattern can be observed. This highly significant class of optical phenomena will be discussed in more detail in Chap. 15.

Practical sources of coherent light and electromagnetic waves have become available since 1960 with the invention of **lasers** and **masers.** In such devices, the radiations from individual centers are released in the same phase by a wave that traverses the material of the source. Lasers and masers are becoming increasingly important in physical research and in technological and biological applications (see Refs. 13.7 and 13.8).

There is no ''destruction'' of energy in any interference experiment, for it can be proved that the total energy of the pattern is the same as that furnished by both sources. All that happens as a result of interference is that the energy is redistributed in space, the disturbance being increased at some places and diminished at others.

13.10 DIFFRACTION OF WAVES

In the previous discussion, there have been frequent references to a highly evident and plausible feature of wave motion: The rays associated with waves traveling in a uniform medium are straight. Such phenomena as the casting of shadows, the appearance of shafts of sunlight coming through a rift in the clouds, and our inability to see around corners suggest that this straight-line travel is characteristic of light waves. On the other hand, water waves that strike an obstacle or pass through an aperture are usually observed to bend into the "shadow" region quite markedly (Fig. 13.23). Sound waves, too, obviously deflect around barriers. A listener indoors does not have to be in line with an open window in order to hear noises coming in from the street. Departure from straight-line propagation when waves pass an obstacle or go through an aperture is called **diffraction.** It can be observed for light as well as for other kinds of waves, but the effect is ordinarily small and may escape notice in most optical situations. Nevertheless, the observation of diffraction effects gave important evidence for concluding that light has wave properties. Details will be brought out in Chap. 15.

The word "diffraction" comes from the Latin diffractus: broken in pieces.

Under what circumstances are diffraction effects noticed? The photographs show plane water waves passing through an opening in a barrier. The shorter waves are seen to undergo less sidewise bending (diffraction) than the longer ones. Similarly, sea waves passing an extended obstacle such as a breakwater cast a fairly definite shadow, but on passing around a slender pole projecting from the water, the two parts of each wave rejoin smoothly in a very short distance and the influence of the narrow obstacle is no longer noticeable at places beyond. *Observations of this nature, with waves of all kinds, show that diffraction effects are prominent whenever the aperture or obstacle is comparable in size with the wavelength, whereas the effects are insignificant when the aperture or obstacle is wide compared with the wavelength.* The principle of superposition (interference), taken together with Huygens' construction, is able to explain diffraction phenomena in detail. The solution of such problems in a general case presents great mathematical difficulties, but it is possible to understand qualitatively how diffraction effects arise.

For a wavefront that spreads freely in a medium without encountering any barriers or openings, the Huygens wavelets need not be considered. The progress of the disturbance can be traced adequately by drawing rays. If, however, a part of the wavefront is blocked by an obstacle so that some of the Huygens wavelets are held back, the results are different. If they had not been blocked, these wavelets would have combined with the others to ensure the straight-line propagation of the wave in the region beyond the barrier. For instance, Fig. 13.23*a* represents plane wavefronts passing

Diffraction of straight water waves by an aperture. Of the three wavelengths shown, the bending is strongest for the one most nearly comparable with the width of the opening.

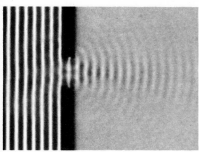 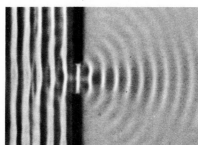 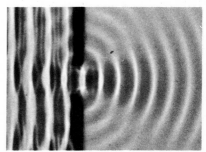

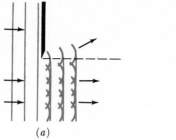
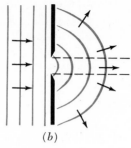

(a)   (b)

FIGURE 13.23 Plane waves encountering (a) the edge of an obstacle and (b) a narrow slit. Diffraction effects make the waves bend into the region of the geometric shadow.

the edge of an obstacle. Several Huygens wavelets are shown. The ones passing freely into the space beyond the edge interfere destructively at all points except in the forward direction, as described earlier, and combine to produce the plane wavefronts that travel on beyond the barrier. However, the wavelet originating at the edge has no companion above it to annul its sidewise contribution; as a result some of the disturbance spreads into the shadow region beyond the ray drawn to the edge of the obstacle. Thus we conclude that *diffraction effects are the result of limiting the extent of the wavefront.*

The intensity of each Huygens wavelet would be expected to decrease very rapidly with the distance it has traveled from its point of origin. Calculation shows that only the contributions of the wavelets in the immediate neighborhood are important. Thus diffraction effects are observed only very near the edge, while at greater distances straight-line propagation still holds. Figure 13.23b represents the diffraction of plane waves by an aperture whose width is comparable with the wavelength. If there were no diffraction, the disturbance on the right would be confined to the region between the rays (dashed lines) drawn through the edges of the hole. In reality, because diffraction cannot be neglected, a disturbance exists almost everywhere on the right, and the aperture acts in some respects like a new source of waves. However, if the aperture were a real point source, the strength of the disturbance would be the same in all directions. Actually, the disturbance on any wavefront is a maximum in the forward direction, falling off rapidly on each side, as suggested in the diagram by the decreasing thickness of shading of the wavefronts.

Programmed review

Instructions: See page 18.

1. Define wave intensity.

   The rate of transport of energy through unit area of the wavefront.   [13.1]

2. What is meant by supersonic speed?

   Speed greater than that of sound waves (Mach number greater than 1).

   [13.2]

3. Describe standing waves.

    Stationary vibration patterns produced by superposition of two wave trains of the same frequency and amplitude traveling in opposite directions.   [13.4]

4. In a closed pipe in which a standing-wave pattern exists, what condition holds at the closed end?

    There is a node of air displacement at that point.   [13.5]

5. Describe Huygens' construction.

    Each point on a wavefront is a source of secondary waves. Acting together, the latter determine succeeding wavefronts.   [13.6]

6. What does the law of reflection assert?

    Angles of incidence and reflection are equal; the incident and reflected rays and the normal to the surface all lie in one plane.   [13.7]

7. Define refraction.

    A change in direction of waves (or rays) resulting from nonuniformity of the medium through which they pass.   [13.8]

8. State Snell's law in analytical form, using the wave speeds in the two media.

    $V_2 \sin i = V_1 \sin R.$ [13.8]

9. What is meant by the reversibility of rays?

    If the direction of wave travel is reversed, the waves retrace their former path exactly.   [13.8]

10. What is interference of waves?

    The changes of amplitude, at various times and places, caused by superposition of two or more sets of waves.   [13.9]

11. When are two sets of waves said to be coherent?

    When they have (a) the same frequency and (b) a constant phase relation.   [13.9]

12. What is diffraction, and under what conditions is it prominent?

    Departure from straight-line propagation when waves pass·through an aperture or around an obstacle. The effects become prominent when the width of the aperture or obstacle is decreased to a size comparable with the wavelength.   [13.10]

For further reading

13.1. *Berland, T.* "The Fight for Quiet."
13.2. *McChesney, M.* Shock Waves and High Temperatures, *Sci. Am.,* February, 1963.

13.3.   *Van Bergeijk, W., J. R. Pierce,* and *E. E. David, Jr.* "Waves and the Ear." How the human ear responds to sounds.

13.4.   *Jeans, J. H.* "Science and Music." This and the following two references are agreeable presentations of the physical and physiological basis of music.

13.5.   *Backus, J.* "The Acoustical Foundations of Music."

13.6.   *Wood, A.* "The Physics of Music."

13.7.   *Schawlow, A. L.* Optical Masers, *Sci. Am.,* June, 1961, p. 52.

13.8.   *Schawlow, A. L.* Lasers and Coherent Light, *Sci. Am.,* September, 1967, p. 197.

Questions and problems

13.1.   How would (*a*) the amplitude and (*b*) the intensity of surface water waves from a small source be expected to vary with the distance from the source? Explain, giving quantitative answers.

13.2.   The opening of the human ear is effectively a circle 8 mm in diameter. How long would one have to listen, using both ears, to a sound at the lower limit of audibility in order to take in 1 erg (10^{-7} joule) of sound energy?

13.3.   How long would a radio announcer have to speak into his microphone at a sound level of 40 dB in order to deliver 1 joule of energy? Take the effective area of the microphone to be 4 cm².

13.4.   Check the dimensional consistency of Eq. 13.2, page 304.

13.5.   Account for the pattern of concentric circular corrugations seen on the surface of a cup of coffee on a train or airplane.

13.6.   One end of a rope 18 ft long is tied to a hook in the wall. With what frequency must the free end of the rope be shaken in order to make it break into exactly three loops if the speed of transverse waves in the rope is 30 ft/sec?

13.7.   A musical string 40 cm long has a mass of 1.0 g. When vibrating in its fundamental mode it produces sound waves in the air whose wavelength is 1.0 m. Find the tension, in newtons, with which the string is stretched.

13.8.   Show how the mass density of a sample of metal, available in the form of a uniform wire, can be determined by an acoustic method.

13.9.   In Fig. 13.8, what is the ratio of f_c to f_0, assuming the two pipes to be of the same length? Justify your answer.

13.10. Find the length, in inches, of the shortest closed tube that will reinforce the tone of a vibrating tuning fork of frequency 495 cycles/sec held at the open end of the tube.

13.11. Using the fact that, for standing waves, the distance between adjoining nodes or antinodes is half a wavelength, show that the partial frequencies of a closed pipe are odd multiples of the fundamental, as stated on page 311.

13.12. What set of partials can be expected from an air column that is closed at both ends? This kind of pipe would not be a practical musical source, since very little energy could reach the surrounding air.

13.13. How tall a mirror, mounted flat against a wall, is needed in order that a man 6 ft tall may just be able to see himself head to foot when he stands erect before it? How must the mirror be placed? Does it make any difference how far away the man stands? Draw a ray diagram of the situation as seen from one side.

13.14. A ray of light, fixed in position, strikes a plane mirror obliquely. Prove that, if the mirror is rotated through an angle θ around an axis perpendicular to the plane of incidence, the reflected ray will turn through an angle 2θ.

13.15. Waves are incident at an angle of 60° on the surface of separation of two media. If the wave speed in the first medium is one and one-half times that in the second, find the angle between the refracted ray and the ray reflected from the interface.

13.16. The index of refraction for light passing from air into glass is 1.50, and from air into water 1.33. Prove that in passing from water into glass the effective index is 1.13.

13.17. Light waves go from water into a block of glass whose index of refraction is 1.60. At what angle of incidence will the refracted ray be perpendicular to the ray reflected from the interface?

13.18. *Total internal reflection* concerns light rays in a material medium approaching the plane boundary between this medium and one of *smaller* index of refraction. As the angle of incidence is increased (Fig. 13.24), the refracted ray becomes more nearly horizontal until, at C, it lies along the surface ($R = 90°$). The corresponding angle of incidence is called the critical angle, i_c. Any ray incident at a greater angle will not emerge but will undergo total internal reflection. Starting from Eq. 13.7, with n_1 and n_2 representing the respective indices of refraction, show that the critical angle is about 49° for light approaching the surface of water from within. For an important application of total reflection, see "Fiber Optics" in Ref. 15.4 (page 364).

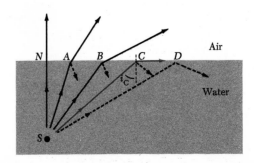

FIGURE 13.24 Total internal reflection.

Chapter 14
WAVES AND RAYS

The search for a satisfactory model for light and other *electromagnetic radiations* constitutes an important part of the effort to understand the physical world. Radiation implies energy transfer between separated places, and this is perhaps the circumstance that must be accounted for above all others. Energy transport can occur only by means of projected particles or by means of waves—there are no other possibilities. The former involves the transfer of matter, whereas the latter does not.

Some of the ancients thought of light as a stream of high-speed particles shot out from luminous objects. Plato and others believed that an emanation from the eye made objects visible when they were struck by it, whereas Aristotle preferred to think of light as something nonmaterial occurring in the space between the eye and the object.

Aristotle's interpretation did not take on a more definite form until the seventeenth century, when Hooke and Huygens independently suggested that light might be a wave motion. Huygens conceived of light as a longitudinal wave in an all-pervading "ether" which was assumed to fill all of space, including the interior of matter. Newton speculated about the two ideas—light as a stream of particles or as a train of waves—but he tended to favor the particle model, which was called the *corpuscular theory.* Newton's great authority and prestige kept this notion alive for nearly a century after his time, in spite of the fact that certain aspects of the behavior of light are impossible to explain on the corpuscular basis but have a ready explanation in terms of waves.

In what follows, the main phenomena of light and radiation will be examined in an effort to decide which of the two models is more successful in accounting for the observations. Later, in Chap. 20, we shall find that neither model by itself is adequate for all known optical phenomena.

14.1 REFLECTION AND REFRACTION OF LIGHT

As already mentioned, it is a matter of common experience that light travels in straight lines in a uniform medium as long as no obstacles or apertures are present but that it can change its direction abruptly if it is reflected from surfaces. The wave model has been shown to account for these phenomena in a completely satisfactory way, yielding a law of reflection which is readily verified for light. However, a particle model can do equally well because the straight-line travel of light is just what is expected of a stream of independent particles, and reflection is accounted for by the perfectly elastic and frictionless collision of such particles with the surface of a material body.

In Fig. 14.1, **p** represents the linear momentum of a corpuscle that is about to strike the reflecting surface MM' at an oblique angle. This vector can be resolved into a component $\mathbf{p}_x$ parallel to the surface and a component $\mathbf{p}_y$ normal to the surface. If the impact is frictionless and ideally elastic, the parallel component will be the same after collision, whereas the normal one will be reversed in direction but unchanged in magnitude. Thus the resultant linear momentum **p'** after impact is also unchanged in magnitude, but its direction is now different from that of **p.** From the triangles it is evident that **p** and **p'** make the same angle with the normal; also, all the vectors lie in the same plane. Thus the law of reflection follows readily from the particle model. The wave and the particle models appear to be equally valid for the phenomena of reflection and straight-line propagation.

A more decisive conclusion is reached when comparing the relative merits of the two models to explain the refraction of light. The theory of refraction of waves discussed in Sec. 13.8 (page 316) showed that when waves pass from one medium into another where the wave speed is lower, the rays are deviated *toward* the normal. By the principle of reversibility, the deviation is *away* from the normal if the second medium is the one with the higher wave speed. Geometric considerations based on the wave idea led to the quantitative law of refraction expressed by Eq. 13.9 (page 319).

Is it possible to explain refraction by means of the corpuscular theory? Newton realized that additional assumptions would have to be made about the corpuscles in order to account for refraction. In particular, why are some reflected at the interface but others are transmitted? He was forced *Corpuscular* to assume that the state of a corpuscle could change periodically, taking *theory of* on "fits of easy transmission" alternating with "fits of easy reflection." Thus *refraction.* a given particle is either transmitted or reflected, depending on its state when it encounters the surface. Although such an assumption seems bizarre as well as arbitrary, we shall follow the corpuscular model through to see where it leads.

Figure 14.2 represents a light corpuscle approaching the surface of separation MM' obliquely. The momentum vector **p** may be resolved into components $\mathbf{p}_x$ along the surface and $\mathbf{p}_y$ along the normal.

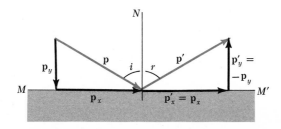

FIGURE 14.1 Conservation of linear momentum applied to a light corpuscle striking a plane reflector is able to account for the law of reflection.

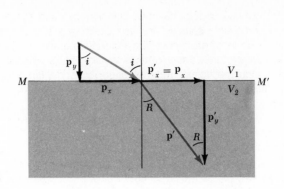

FIGURE 14.2 If light consists of corpuscles, their momentum—and hence their speed—should increase on entering a denser medium. Measurement shows that the opposite is true: The speed of light is less in the lower medium.

Assume that the lower medium attracts a corpuscle more strongly than the upper one does, which can be expected if the lower medium is the denser one. Then, provided it is in the proper "state," a corpuscle acquires additional momentum in the downward direction when it gets very close to the interface. Since only the momentum in the direction of the normal is affected, the result is that the downward component is suddenly increased as the particle passes through the surface, while the component along the surface is unchanged.

In the upper vector triangle, $p_x/p = \sin i$, and in the lower vector triangle $p_x/p' = \sin R$. Divide the first expression by the second: $\sin i/\sin R = p'/p$. If m is the mass of the corpuscle and V_1 and V_2 represent the speeds of light in the upper and lower medium, respectively, we can substitute mV_1 for p and mV_2 for p', and the preceding equation becomes, after canceling the factor m,

$$\frac{\sin i}{\sin R} = \frac{V_2}{V_1}$$

14.1

This relation has the same form as Eq. 13.9, except that the positions of V_1 and V_2 are reversed. What does this imply? Equation 13.9 was deduced from Huygens' construction on the assumption that the wave speed is smaller in the lower medium ($V_2 < V_1$). This assumption was necessary in order to conform with the observed fact that $i > R$ when light passes, for example, from air into water. On the other hand, to account for the deviation of the light on the corpuscular model, it had to be assumed that the speed of light is *greater* in the lower medium ($V_2 > V_1$). The two models evidently lead to a contradiction in accounting for refraction, a dilemma that can be resolved only by measuring the speed of light in the two media.

Methods of measuring the speed of light are described in Sec. 14.2, below.

In Newton's time, methods for making such measurements did not exist. It was not until nearly two centuries later that Foucault (page 123) was able to measure the speed of light in a long tube of water. He found that the speed in water was only about three-fourths the value of the speed in air. Thus *the phenomenon of refraction supports the argument in favor of the wave model.* It must be remembered that this statement simply compares the wave theory with the specific particle model conceived by Newton. It leaves open the possibility that some other model, based on the particle idea, may apply in connection with certain light phenomena. Historically, this is what happened (Chap. 20).

Index of refraction.

The speed of light in any material medium is less than the speed c in free space (vacuum). The refractive ability of a substance is specified by describing the passage of light from free space into the substance in question. On page 319 the index of refraction of a material was defined as the ratio of the speed of light in free space to its speed in the material. In the statement of Snell's law in Eq. 13.7 (page 317), n_1 and n_2 represent the indices of refraction of the two media. This can be seen by replacing n_1 by c/V_1 and n_2 by c/V_2, according to the above definition. Then the equation reduces to $V_2 \sin i = V_1 \sin R$, which is exactly Eq. 13.9 (page 319).

There is another way in which the phenomenon of refraction supports the wave theory of light. Maxwell (page 236n) developed a comprehensive mathematical theory of light and similar radiations, based on the idea of electromagnetic waves propagated in free space with the speed c. According to Maxwell's theory, these waves consist of a combination of electric and magnetic forces of periodically changing strength. Thus, in an electromagnetic wave there is no medium oscillating in a mechanical sense, as there is in water waves or sound waves. Maxwell's theory will be discussed later in more detail.

In interpreting refraction, the electromagnetic theory asserts that a light wave, on entering a material medium, continues to travel with speed c in the space between the atoms or molecules. As the wave passes through, its electric force sets the electrons and the nuclei of the atoms into oscillation, which causes the particles to send out new electromagnetic waves. Since the original wave sweeps across some of the particles later than others, the secondary waves do not all have the same phase. A complicated mathematical calculation shows that all the secondary waves, together with the incident wave, add up to produce a new wave that travels with a speed less than c. This new speed turns out to be the same as the measured speed of light in the substance in question. Thus the apparently simple fact of observation that light "slows down" on entering glass has actually a quite intricate explanation—one that adds further support to the wave model. Reflection can be explained in a similar way as the superposition of waves reemitted by atoms in the surface of the material.

14.2 MEASUREMENT OF THE SPEED OF LIGHT

The earliest indication that the speed of light is finite and measurable was obtained in 1675 by the Danish astronomer Olaf Römer (1644–1710). He observed that one of Jupiter's moons appeared to come out from behind the planet at successively later times when the earth was receding from Jupiter and at correspondingly earlier times when the earth was approaching. The true period of motion of the satellite is constant, and the apparent changes are due to a kind of Doppler effect resulting from relative motion of source and observer. Römer interpreted the total time delay to be the time taken by light to cross the earth's orbit. From the figures then available, the speed of light was determined to be about 220,000 km/sec, of the same order of size as the best present value of

$$c = 299{,}792 \text{ km/sec} = 186{,}282 \text{ mi/sec}$$

The first successful determination of c carried out entirely on the surface of the earth was made by the French experimenter H. L. Fizeau (1819–

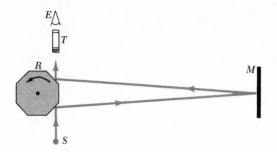

FIGURE 14.3 Scheme of the Michelson method for measuring the speed of light.

1896) in 1849. A modification of his method was used by Foucault in 1862 and later by other workers. This approach attained its utmost refinement in the measurements made by Michelson (page 93n) between 1924 and 1931. His arrangement is shown schematically in Fig. 14.3. Light from an intense source is reflected from one face of an octagonal mirror R rotating at high speed. The light goes to a distant mirror M, located in this instance about 22 mi away, and on its return is reflected from another face of the eight-sided mirror to the observing telescope T.

If R is stationary and in the position shown in the diagram, the light will be seen in the telescope. If the mirror R is now set into rotation, the light will not be seen in the field of view until the octagon is spinning at just the right speed to bring one face into an exact reflecting position in the time taken by light to reach the distant mirror and return.

Measurements were also made by reflecting the light back and forth in a partly evacuated pipe about a mile long. The final values for c were very close to the number given above.

One of the more recent methods uses an electronic shutter in place of a mechanically rotated mirror, and the much higher rate of interruption of the beam makes it possible to carry out the experiment within the laboratory. In another method, radar waves are tuned to resonance in a cavity, much like the production of standing waves in a pipe. The results confirm the fact that these electromagnetic waves have the same speed as light waves.

A new laser method permits the independent measurement of frequency and wavelength of such a source of light. Then c can be computed directly from the wave equation $c = f\lambda$. Determinations to nearly one part in a hundred million are expected.

References 14.1 and 14.2 will be found of interest.

14.3 LENSES AND THEIR FUNCTION

There are indications that crude lenses were made and used as far back as 25 centuries ago. In modern times, lenses are an essential part of many optical devices such as microscopes, projectors, eyeglasses, cameras, telescopes, range finders, etc.

The purpose of a lens is to change the curvature of wavefronts by means of refraction. In practice, an optical lens consists of a portion of some transparent material bounded by two curved surfaces. The surfaces are

CONVERGING

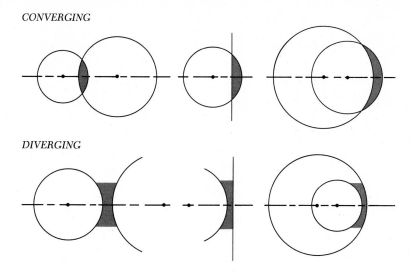

DIVERGING

FIGURE 14.4 The surfaces of ordinary lenses are portions of spheres.

If one surface is plane, the axis is the line through the center of curvature of the other surface and perpendicular to the plane surface.

usually spherical because this form is comparatively easy to attain by grinding and polishing. In particular, one surface of a lens may be plane. Cross-sections of typical lens shapes are shown in Fig. 14.4. Lenses that are thicker at the center than at the edge are called **converging lenses;** those thinner at the center are called **diverging lenses.** The line joining the centers of the two spheres that determine the surfaces is called the **principal axis** of the lens. It is the line of symmetry of the lens itself.

What happens when plane waves of light, advancing along the axis, strike a typical converging lens (Fig. 14.5a)? Because the speed of light is less in glass than in air, the part of the wavefront passing through the thicker central region of the lens is retarded more than the other parts, the retardation diminishing smoothly toward the edge. As a result, the emerging

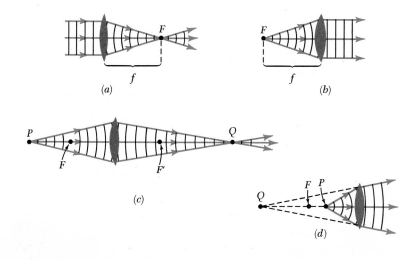

FIGURE 14.5 Image formation by a converging lens for point sources of light. The waves as well as the rays are shown.

The above
statement
about the
speed holds
for all
practical
lens mate-
rials (see
Table 13.1,
page 320).

wavefront is concave. If the glass surfaces are spherical, the wavefront is nearly spherical, closing down approximately to a point called the **principal focus** of the converging lens. After passing through this point, the waves again expand as they continue onward. If plane waves pass through the lens in the reverse direction, they are observed to come to a focus at a point lying the same distance on the other side of the lens. Thus there are *two* principal foci, one on each side of the lens, and they are at equal distances f from the lens. f is called the **focal length** of the lens. The principal focus on either side of a converging lens is the point where waves moving along the axis and coming from a very distant point source converge after passing through the lens. It may also be described as the point where the corresponding rays cross.

This discussion of lens optics will be restricted to **thin lenses,** where the central thickness is small compared with the focal length—say not more than one-tenth as large. In this case it makes little difference whether distances such as focal lengths are measured from the center or from any other part of the lens. The behavior of thick lenses is more complex and will not be treated here.

If a point source is placed at the principal focus of a lens (Fig. 14.5*b*), waves diverging from the source strike the lens and emerge as plane waves, obviously the reverse of the preceding situation. On the other hand, if the source is at a point P (Fig. 14.5*c*) located *beyond* the principal focus, the lens reverses the curvature of the wavefronts and they close down to a point Q that lies beyond the other focus. To an observer located a considerable distance to the right, the waves appear to originate at Q rather than at P. Q may be called the **image** of P.

If the point source is placed closer to the lens than its principal focus (Fig. 14.5*d*), the lens reduces the curvature of the wavefronts somewhat but cannot reverse it or even make the waves plane. To an observer on the right-hand side of the lens, the wavefronts appear to have come from some point Q that is more distant from the lens than their actual source P. The point Q is appropriately called a **virtual focus** and may be designated a virtual image of P.

With a diverging lens, waves originating at any point, no matter what its distance from the lens, have their divergence increased on passing through the lens, as seen in Fig. 14.6. If parallel rays strike the lens, the point from which the light *appears* to diverge after passing through the lens is the (virtual) principal focus on the side of the lens from which the light comes.

FIGURE 14.6 Action of a diverging lens.

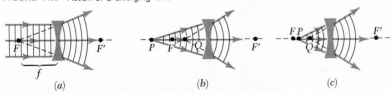

(a)   (b)   (c)

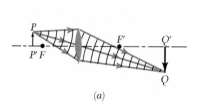

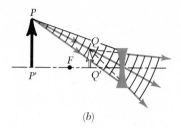

(a)   (b)

FIGURE 14.7 Imaging of off-axis points.

14.4 IMAGE FORMATION BY LENSES

The main function of any optical system is to produce images, usually of objects of finite size. In optics, an **object** is any body that acts as a source of light, whether it is self-luminous or merely reflects or scatters light from other sources. An image is an optical counterpart (replica) of the object. If the incoming light does not go along the lens axis, the image points no longer fall on this axis but are displaced sidewise (Fig. 14.7).

Consider a finite object placed before a lens. The object as well as the image may be represented by a heavy line segment, with an arrowhead to distinguish one end from the other, as shown in Fig. 14.7. The image may be found by regarding each point of the object either as a source of spherical waves or as a point from which rays diverge.

On page 334 it was stated that, for simplicity, the discussion would be confined to thin lenses. A further restriction is now introduced: Assume that all rays make small angles with the lens axis. Then it is found that,

If the aperture of a lens is not small compared with the focal length, various image defects arise. The picture shows narrow light beams passing through a lens. Notice that the rays near the rim of the lens focus closer in than those near the axis, a defect called spherical aberration. (*Klinger Scientific Apparatus Corp.*)

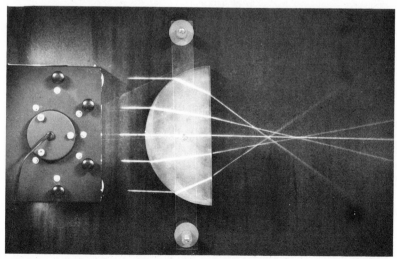

to a good approximation, if the object lies entirely in one plane, so does the image. The two planes are called the **object plane** and the **image plane,** respectively. For almost all practical situations, the two planes are perpendicular to the lens axis.

In the special case where the object is infinitely far away, the image plane is called the **focal plane** of the lens. In a camera, it is the plane in which the film must be placed in order to get a sharp image of a distant object.

When waves from points on the object *converge* to points on the image after passing through the lens, the image can be received on a screen. This is called a **real image.** The image on a motion picture screen is real. Diffuse reflection of light allows the image to be seen from any location in front of the screen. Examples of the formation of real images are shown in Figs. 14.5a and c and 14.7a.

If the waves *diverge* after passing through a lens, an image cannot be formed on a screen directly. In this case, however, there exists an image from which the waves *appear* to diverge. It is called a **virtual image.** A virtual image cannot be caught on a screen but can be seen with the eye by looking through the lens. In using a microscope or telescope (see below), the eye views virtual images. Examples of virtual-image formation are shown in Figs. 14.5d, 14.6, and 14.7b.

In many of the preceding figures, waves as well as rays were drawn, but it is usually simpler to deal only with rays. The optics of light rays, called **geometric optics,** can represent most of the features of image formation, as will be shown below. On the other hand, geometric optics ignores effects ascribable to the wave nature of light, such as diffraction, interference, and polarization, described later (Chap. 15). In many applications, these effects are small enough to disregard and in such cases geometric optics is a valuable simplification.

Geometric optics.

By tracing the course of certain rays in an optical system, the image point corresponding to each object point can be determined uniquely. An image point is given by the intersection of rays. In order to find the image of any given object, it is necessary to know only the focal length of the lens. The focal length f can be determined in several ways. For instance, for a converging lens, f can be found experimentally by moving a card back and forth beyond the lens until the image of a very distant object is seen in sharp focus on the card. The distance between card and lens is then equal to f.

The detailed course of any ray can be traced through an optical system if the contour of each refracting surface and the index of refraction of each medium in the system are known. Figure 14.8 shows two examples of a ray passing from air into a piece of glass, then through a second surface and emerging again into air. In Fig. 14.8a, the piece of glass is a plane-parallel plate. If the refractive index of the glass and the angle of incidence i of the incoming ray are known, the angle of refraction R can be calculated by using Snell's law, as in Worked example 13.3, page 319. Because the sides of the plate are parallel, the other angles are as shown, and the emerging ray is parallel to the incoming one but displaced sidewise with respect to it.

In Fig. 14.8b, the small portion of the glass surface at the point of entry of the ray may be considered plane. Then, using Snell's law, the angle

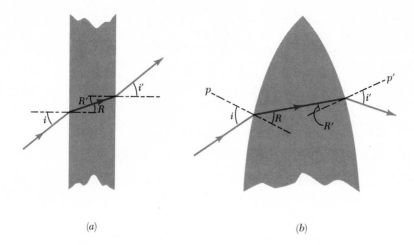

FIGURE 14.8 Effect on a single ray of (a) a plane-parallel plate and (b) a lens-shaped body.

R can be computed as before. Next, the angle R' between the internal ray and the normal p' must be measured on the diagram and finally i' must be computed.

Use of ray construc-tions to find images.

Although this kind of point-by-point tracing of rays can be carried out, there is a much simpler way to determine images in thin-lens optics by using certain special rays whose course can be predicted. An example is shown in Fig. 14.9a where the ray PC comes from the tip of the arrow that represents the object and heads toward the center of the converging lens. This ray continues onward with no change in direction because it has in effect gone through a plane-parallel plate (see Fig. 14.8a, above). The small sidewise displacement of this ray is negligible since the lens is thin.

Now trace a ray PA coming from the tip of the arrow and traveling parallel to the lens axis. What will its path be after going through the lens? In Fig. 14.5a (page 333) it was pointed out that all rays entering parallel to the principal axis pass through the principal focus F' after emerging. As long as the object is farther from the lens than F (as it is in the figure), the

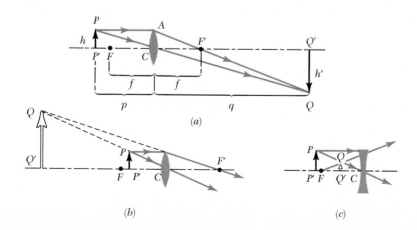

FIGURE 14.9 Graphical tracing of rays.

two transmitted rays cross at some point Q, the image point of the tip of the arrow. Image points corresponding to all other object points are found in their proper order in the image plane that passes through Q, which makes it possible to construct the complete image at once. An inverted, real image of the arrow can be seen on a card held in the plane QQ'. Since Fig. 14.9a represents *any* section of the system made by a plane passing through the lens axis, any object that extends out of the plane of the diagram will have an image that is reversed side to side as well as up and down.

The formation of an image by a diverging lens is shown in Fig. 14.9c. As already mentioned, this type of lens always forms virtual images. The image QQ', shown as a white arrow, is erect and smaller than the object. Each point of it, such as Q, is determined by the intersection of an actual ray (solid line) and a virtual, back-projected ray (dashed line).

Practice set

Is it possible to interchange object and image in Fig. 14.9a? Make a ray diagram, starting with QQ' as the object. Do you get PP' as the image? Answer the same questions for Fig. 14.9b. To what circumstance do you ascribe the difference?

14.5 THE SIMPLE MICROSCOPE; LATERAL MAGNIFICATION

Figure 14.9b shows how to determine the image graphically when an object is placed closer to a converging lens than the principal focus. As shown in Fig. 14.5d (page 333), a virtual image is obtained in this instance because the ray through the lens center and the one passing through F' diverge after passing through the lens but *appear* to come from some point Q located by projecting them backward (dashed lines) until they cross. This point is the virtual image of the tip of the arrow, and the entire image is represented by the white arrow. It may be viewed by looking into the lens from the right.

The eye perceives the direction of a ray that enters it but is unaware of the previous course of the ray. In the present situation it sees an enlarged, erect image (virtual) QQ'. When a lens is used in this way it acts as a magnifying glass called a **simple microscope.** The object to be examined is brought just within the focal point, and the eye is as close behind the lens as convenient in order to have a wide field of view. The optical system of the eye, acting like a converging lens, forms a final real image on the retina. Because a normal eye can see details most distinctly at a distance of about 25 cm, the magnifier is usually adjusted so that the virtual image falls at this distance from the eye.

The **lateral magnification** of a lens or optical system is defined as the ratio of the height of the image to the height of the object. For a simple microscope, the object is essentially at the focal distance f from the lens. Inspection of the two similar triangles $QQ'C$ and $PP'C$ in Fig. 14.9b shows that the lateral magnification M_s of the simple microscope is given approximately by

$$M_s = \frac{25}{f_{(cm)}}$$

14.2

In effect, the magnifier enables the user to bring the object close to the eye and yet observe it comfortably. Because of inherent image distortions produced by a single lens, simple microscopes with magnifications greater than 2 or 3 are not practical. With combinations of lenses, this can be increased to about 20. Whenever high optical magnification is needed, the compound microscope (page 341) is used.

14.6 LENS FORMULAS

The geometry of ray tracing may be translated into algebraic form so that the details of image formation may be computed rather than constructed graphically. This kind of conversion was already made in arriving at Eq. 14.2 for the lateral magnification of a simple microscope. However, in every case of image formation discussed above, it can be seen from the similar triangles $PP'C$ and $QQ'C$ in the various figures that the lateral magnification M is given by

$$\frac{\textit{Height of image}}{\textit{Height of object}} = \frac{\textit{distance of image from lens}}{\textit{distance of object from lens}}$$

If h and h' are used to represent the respective heights and p, q to represent the corresponding distances from the lens, the preceding relation for the lateral magnification may be written

$$M = \frac{h'}{h} = \frac{q}{p} \qquad\qquad \textbf{14.3}$$

Further, the character and location of the image can be determined as follows: In Fig. 14.9a, the triangles ACF' and $QQ'F'$ are similar, so that

$$\frac{h'}{h} = \frac{q - f}{f}$$

If we make use of Eq. 14.3, this relation becomes

$$\frac{q}{p} = \frac{q - f}{f}$$

Cross-multiply, divide each term by pqf, and transpose to get

$$\frac{1}{p} + \frac{1}{q} = \frac{1}{f} \qquad\qquad \textbf{14.4}$$

This equation can be shown to hold not only for the situation shown in Fig. 14.9a but for any instance of image formation by a thin lens, either converging or diverging, provided that the following conventions are observed:

1. Take f to be positive for a converging lens and negative for a diverging lens.
2. The standard arrangement, from left to right in the diagram, is object, lens, image. q is positive for a real image and negative for a virtual image. A real image lies to the right of the lens, a virtual image to the left.

Worked example 14.1 The lens system of a certain portrait camera is equivalent to a thin converging lens of focal length 12 in. How far behind the lens must the film be located to receive a sharp image of a person seated 5 ft from the lens? How large is the image in comparison with the object?

Solution   a. The object distance p is 60 in. Substitution into the lens formula (Eq. 14.4) gives

$$\frac{1}{60} + \frac{1}{q} = \frac{1}{12} \qquad \frac{1}{q} = \frac{1}{12} - \frac{1}{60}$$

In finding the value of q, it is essential to avoid the common mistake of inverting separately each term in the equation. Both fractions on the right must first be brought over a common denominator. In fact, this is the best way to avoid inaccuracies caused by rounding off decimal values. After the right side of the equation has been converted to a single fraction, its terms may be inverted and then divided to yield a numerical result in decimal form.

Proceeding with the solution,

$$\frac{1}{q} = \frac{60 - 12}{(60)(12)} \qquad q = \frac{(60)(12)}{60 - 12} = 15 \text{ in.}$$

b. From Eq. 14.3, $M = \frac{15}{60} = \frac{1}{4}$; the image is one-fourth as large as the object.

Worked example 14.2 When an object is placed 18 cm from a certain thin lens, a virtual image is formed 8 cm from the lens. What is the nature of the lens, and what is its focal length?

Solution   According to the above convention, the fact that the image is virtual means that q is negative. From the lens equation,

$$\frac{1}{18} - \frac{1}{8} = \frac{1}{f} \qquad f = \frac{8 \times 18}{8 - 18} = -14.4 \text{ cm}$$

The negative sign resulting for f shows that the lens is diverging.

14.7   OPTICAL INSTRUMENTS

An optical instrument is a system composed of various optical elements such as mirrors, lenses, prisms, plates, etc. It is possible to trace rays, step by step, through such a system. For example, in a sequence of two or more lenses the image that would be formed by the first lens in the absence of the others may be considered the object for the second lens. The image formed by the second lens can then be treated as the object for the third, and so on.

Of the many important optical instruments now in use, attention here will be confined to the compound microscope and the astronomical telescope. Both the telescope and the microscope have basically the same function, which is to magnify objects so that details can be more readily seen. The microscope examines accessible objects whose linear dimensions are small; the telescope magnifies objects that appear small because of their great distance.

In instruments intended for direct viewing, the human eye is the final element. The element just in front of the eye is the equivalent of a single lens used as a simple microscope. It serves to examine an image formed by preceding parts of the system and is called the eyepiece, or **ocular.**

Compound microscope. When very high optical magnification of small objects is required, the **compound microscope** is used. The first instrument of this kind was built by Hooke (page 44n) in 1665. It consists of two converging lens systems: an **objective** of very short focal length and an ocular of moderate focal length. The object to be examined is placed just beyond the principal focus of the objective, producing a somewhat enlarged real image within the tube of the instrument. This image is then examined with the ocular, which serves as a simple magnifier. Figure 14.10 is a ray diagram of the instrument, each of the two systems being represented by an equivalent single thin lens. The first image, QQ', is real. It is not caught on a screen but merely forms in space. The light goes on to meet the ocular, which is moved closer until QQ' lies just within the principal focus F_2. The final image RR' is virtual, enlarged, and reversed with respect to the object. It is possible to prove that, with the instrument adjusted to put the final image at a comfortable viewing distance of 25 cm, the lateral magnification, M_c, is approximately

$$M_c = \frac{25q}{f_1 f_2} \qquad \text{14.5}$$

where all distances are measured in centimeters. In practice, the largest magnifications attained approach 4,000. Much greater magnification may be reached by means of the electron microscope (page 57).

The refracting telescope appears to have been invented in 1608 by the Dutch optician Hans Lippershey. Like the compound microscope, it consists of an objective lens system and an ocular, except that the telescope uses an objective of very long focal length. This objective forms a real image of a distant object within the tube (Fig. 14.11a). The ocular, used as a simple magnifier, then forms an inverted virtual image.

Increasing the magnification of a direct-viewing optical device increases the size of the image formed on the retina of the eye. For the microscope, the lateral magnification (page 338) is a convenient measure, but for the inaccessible objects examined by telescopes, **angular magnification** is

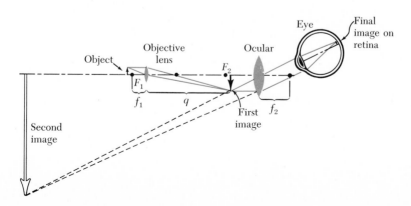

FIGURE 14.10 Ray diagram for the compound microscope.

more suitable. The **angular magnification** is defined as the ratio of the angle subtended at the eye by the final image formed by the instrument to the angle subtended by the object when viewed without the instrument. In Fig. 14.11a, these angles are θ_2 and θ_1, respectively. With the telescope adjusted to put the final image at infinity, the angular magnification M_θ is approximately

$$M_\theta = \frac{f_1}{f_2} \qquad\qquad 14.6$$

This relation suggests that M_θ may be made as large as desired by choosing a very long-focus objective and a very short-focus ocular. The Yerkes Observatory telescope has an objective whose focal length is about 760 in. If it were used with an ocular of focal length 0.2 in. a magnification of nearly 4,000 would result. Other factors, however, drastically limit the values employed in practice, and magnifications exceeding 1,500 to 2,000 are seldom used in ground-based astronomy.

Newton was apparently the first person to construct a **reflecting telescope** in which a concave mirror replaces the objective lens system. One such arrangement is shown in Fig. 14.11b. A reflector has many advantages over a refractor, including better image quality, greater mechanical stability, and lower cost. The optical principle is exactly the same as that of the

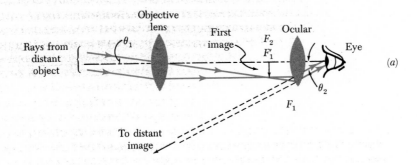

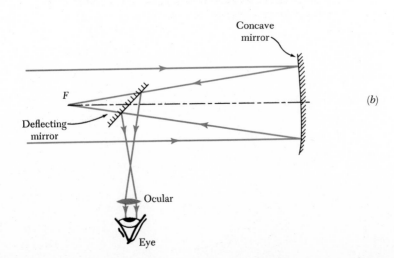

FIGURE 14.11 Ray diagrams for astronomical telescopes.

refractor except that the first image is formed by reflection from the concave mirror.

Curved mirrors—convex as well as concave—can form the same kinds of images that lenses can. A study of the ray diagrams shows that a concave mirror serves the same purposes as a converging lens, and a convex mirror corresponds to a diverging lens; Eqs. 14.3 and 14.4 apply to curved mirrors as well as to lenses. For a spherical mirror whose aperture is small compared with its radius of curvature, f is half the radius of curvature.

A feature of the telescope that is often of greatest importance in astronomy is its light-gathering power; this is one of the main reasons for making telescopes with objectives of large diameter. The brightness of the image formed in a telescope depends on the amount of light collected by the objective. This quantity, in turn, is proportional to the area of the objective. Since the area of a circle is proportional to the square of its diameter, the amount of light energy collected (from a given object located at a given distance) by the objective of the 200-in. telescope on Mount Palomar is about $(200/0.2)^2 = 1,000,000$ times as much as with the unaided eye (pupil diameter 0.2 in.). Thus, objects that are far too faint to be seen with the eye alone are visible through a large telescope. At present, most of the larger astronomical telescopes are reflectors. The largest, located in the Caucasus, has a mirror 6 m (236 in.) in diameter.

Programmed review

Instructions: See page 18.

1. Are the wave theory and the corpuscular theory equally capable of giving a satisfactory explanation of the reflection and refraction of light?

    Both theories explain reflection equally well, but the corpuscular theory leads to a contradiction in attempting to explain refraction.   [14.1]

2. In general terms, what is the function of a lens?

    To change the curvature of wavefronts by means of refraction.   [14.3]

3. Parallel rays of light, moving along the axis of a thin lens, either converge at or appear to diverge from a point on the axis after going through the lens. What is such a point called?

    A principal focus. There are two such points, one on each side of the lens and equally distant from it.   [14.3]

4. What name is given to the distance from a lens to either principal focus?

    The focal length.   [14.3]

5. Under what circumstances does the image formed by a thin lens lie very nearly in one plane?

    When all the rays make small angles with the lens axis.   [14.4]

6. Define *real image.*

    An image that can be cast on a screen.   [14.4]

7. What is a *simple microscope?*

    A single lens used as a magnifier.   [14.5]

8. What is meant by the lateral magnification of an optical system?

    The ratio of the height (or any lateral dimension) of the image to the corresponding dimension of the object.   [14.5]

9. State the image equation of a thin lens.

    $1/p + 1/q = 1/f$. For this equation to be valid in all situations, certain algebraic sign conventions must be observed (see text).   [14.6]

10. Name the two essential optical elements that form a refracting astronomical telescope.

    A long-focus objective lens (or lens system) and a short-focus ocular. [14.7]

For further reading

14.1. *Jaffe, B.* "Michelson and the Speed of Light." An interesting and well-written account of his life and work.

14.2. *Froome, K., and L. Essen* "The Velocity of Light and Radio Waves." On a somewhat technical level but includes various newer methods.

14.3. *Taylor, L. W.* "Physics, the Pioneer Science." See especially the treatment of lens defects in chap. 33.

14.4. *Bragg, W.* "The Universe of Light." The eye and vision are described in chap. 2.

Questions and problems

14.1. Does a lens have to be as tall as the object in order to form a complete image of it? (Think of the relative sizes of a camera lens and a person to be photographed.) Refer to Prob. 13.13 (page 327) where the corresponding question was raised concerning a plane mirror.

14.2. In what way, if any, do the kind, size, and brightness of the image formed by a fragment of a broken lens differ from the same characteristics of the image formed by the intact lens? Mask down a converging lens, using adhesive tape, and examine the images it forms.

14.3. What would be the optical properties of a double convex lens made of a material with a *smaller* index of refraction than the surrounding medium, such as a convex lens-shaped cavity in a block of glass?

14.4. The two surfaces of a thin converging lens have different radii of curvature. The lens forms a real image of a small object. In what way, if any, would you expect the location of the image to change when the lens is turned to face the opposite direction?

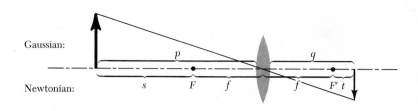

Gaussian:

Newtonian:

p q

s F f f F' t

FIGURE 14.12 Comparison of gaussian and newtonian notations. The following sign conventions must be used with the newtonian form: (1) If the object is to the left of the first principal focus, s is positive; if to the right, s is negative. (2) A positive t indicates that the image falls to the right of the second principal focus; negative t indicates that it is to the left.

14.5. An erect image formed by a certain lens is one-fourth as tall as the object and is located 16 in. from the lens. Determine the focal length and character of the lens.

14.6. Show that when a thin converging lens forms a real image the distance between object and image is never less than $4f$.

14.7. Find the position of the image formed by a diverging lens of $f = -20$ cm if the object is 40 cm from the lens.

14.8. Derive the *newtonian* form of the thin-lens equation, $st = f^2$, where s and t are, respectively, the distances of object and image measured from the principal focus rather than from the lens (Fig. 14.12). Start with the gaussian equation (Eq. 14.4), replacing p by $s + f$ and q by $t + f$. Notice that the newtonian form is algebraically simpler than the gaussian.

14.9. A diverging lens of $f = -10$ cm forms a virtual image of an object 1.0 cm high. If the object is 30 cm from the lens, compute the position of the image and its height.

14.10. A lantern slide 3.40 cm high is projected onto a screen 5.00 m away, using a projection lens of $f = 12.5$ cm. How tall must the screen be in order to accommodate the entire image? HINT: The illuminated slide acts as the object. Use Eq. 14.3.

14.11. An object and a screen are placed 100 cm apart. (*a*) At what *two* points between them can a converging lens of focal length 9 cm be placed to form a sharp image on the screen? (*b*) Compute the lateral magnification in each case.

14.12. Fill in the remainder of the following table. Making ray diagrams similar to those in Fig. 14.9 will be helpful.

	LOCATION OF OBJECT	LOCATION OF IMAGE	CHARACTER OF IMAGE	LATERAL MAGNIFICATION
Converging lens	Beyond F	Beyond F''	Real, inverted	$M > 1$ if $p < 2f$, $m < 1$ if $p > 2f$
	Within F			
Diverging lens	Beyond F			
	Within F			

14.13. The first astronomical telescope, devised by Galileo, was really an opera glass and is sometimes called a galilean telescope. Look up the optical description of this instrument.

14.14. A telescope is to be made of two thin, converging lenses of focal length 10 cm and 1.0 cm. (a) How far apart must the lenses be placed if the object and the final virtual image are very far away? (b) What is the value of the angular magnification?

14.15. A refracting telescope has an objective of focal length 90.0 cm and an ocular of focal length 3.00 cm. (a) How far apart must the lenses be placed to focus on the moon? (b) After setting for the moon, how must the ocular be moved to focus on a tree 20.0 m away? In both cases, assume the final image to be placed at infinity.

Chapter 15
WAVE OPTICS

15.1 DISPERSION OF LIGHT; NEWTON'S EXPERIMENTS

Up to this point in the discussion of light, one of its most striking characteristics has not been mentioned: color. The ancients were aware of the brilliant colors produced when sunlight passes through transparent gems and crystals. As early as the fourteenth century, it was recognized that the rainbow was caused by refraction and reflection in suspended raindrops. When Newton set out to build an astronomical telescope he was troubled by the objectionable fringes of color surrounding the images. This led him to begin a series of investigations that ultimately led to an understanding of the nature of color and initiated the important and useful branch of optics called **spectroscopy,** the analysis of light.

The crucial experiments were performed when Newton was 24 and a student at Cambridge. He held a triangular glass prism in the path of a narrow beam of sunlight that entered a darkened room through a hole in the window blind. The refracted beam, striking the opposite wall, produced an elongated colored patch of light, red at one end and violet at the other (Fig. 15.1). Between these limits, other colors could be seen—orange, yellow, green, and blue—one color merging imperceptibly into the next. Newton called the spread of colors a **spectrum,** from the Latin word for *appearance* or *form.*

Newton suspected that what he observed was the result of a spreading, or **dispersion,** of colors that were present in the beam of sunlight and that the prism did not *produce* the colors but merely *separated* a mixture that was already there. To test this idea, he placed an identical prism behind

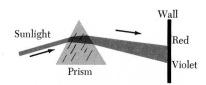

FIGURE 15.1 Newton's fundamental experiment on the dispersion of sunlight (white light) by a prism. Each color present in the original beam is refracted at a slightly different angle to reveal the spectrum.

the first one, but reversed in position and found that the rays were reunited in a colorless, round patch of light. Any kind of light which, like sunlight or the light from any high-temperature source, appears to be essentially colorless is called **white light.**

In a further experiment, a spectrum was cast on a screen having a small hole in it (Fig. 15.2*b*), to see whether a single color passing through the hole could be dispersed further by the second prism. Nothing of this kind occurred. Green light, for instance, coming through the aperture was deviated (refracted) by the second prism but the transmitted light was still of the same green color as the original.

Newton further demonstrated that the various colors composing white light do not in any way react with each other but simply contribute to the resulting mixture. He showed this by holding back a single color—for example, red—by inserting a narrow barrier in the spectrum and allowing the remaining light to go through a second prism. No red was seen in the resulting spectrum.

Read Newton's account of his prism experiments (his first scientific publication) in Ref. 15.1.

From these experiments and a number of refinements and extensions of them, Newton concluded that *white light consists of a mixture of colors,* each color being refracted by a different amount in a transparent material. This means that the index of refraction of a given substance is really slightly different for each color—least for red and progressively greater for orange, yellow, green, blue, violet. It follows, according to the wave theory, that the various colors travel with different speeds in matter, red having the greatest speed and violet the least. The difference in speed is small. In ordinary glass, the speed of red light is only about 1 percent greater than that of violet.

The adherents to the corpuscular theory attempted to explain dispersion by assuming that each spectral color consists of corpuscles of a given size, an assumption that could not be checked operationally. It was established later that what the eye perceives as the *color* of a particular part of the spectrum is physically related to the *wavelength* of that radiation. The length of the waves is greatest for light at the extreme red end of the spectrum, amounting to about 7×10^{-7} m and decreasing progressively through the sequence of colors to about 4×10^{-7} m at the extreme violet. Thus, although Newton did not use the term wavelength, what he found

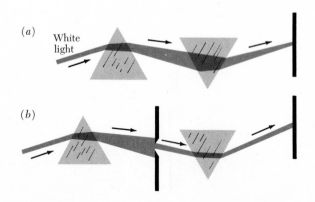

(a) White light

(b)

FIGURE 15.2 Two additional dispersion experiments performed by Newton: (a) The separation produced by the first prism is nullified by the second, and a white spot is observed on the screen. (b) Light of a single color, coming through the hole, is not further dispersed by a second prism.

is essentially that white light consists of a mixture of a great many different wavelengths. In general, dispersion is the separation of a mixture of waves according to some property, such as wavelength.

In a vacuum there is no dispersion of light waves whatever; all colors (or wavelengths) travel with the same speed. This is proved observationally by the fact that when the moon or a planet passes in front of a distant star, cutting off its light from our view, the disappearance is sudden and shows no color effects. If the various components of the light of the star traveled with different speeds in empty space, this would not be the case, and the star would appear to change color as it disappears. However, radio waves coming from pulsars are slightly dispersed by the tenuous plasma in space, the shorter waves arriving slightly earlier than the longer.

Sound waves in air ordinarily show no dispersion, as proved by the fact that musical sounds of different pitch coming from a distance arrive exactly in their proper time sequence. On the other hand, waves on the surface of water do exhibit dispersion because their speed is found to depend noticeably on wavelength.

15.2 INTERFERENCE OF LIGHT; YOUNG'S EXPERIMENT

Thomas Young 1773–1829), English physician, scientist, linguist, and archaeologist, was among the most versatile and accomplished geniuses in the history of science. Read Ref. 15.2.

One of the strongest and most convincing pieces of evidence in favor of the wave model of light was supplied by Thomas Young in 1803. His experiment demonstrated definitely that light showed the property of wave interference, as described in Sec. 13.9 (page 320), whereas a corpuscular model could not account for the observations.

Young's experimental arrangement is the optical counterpart of the water-wave demonstration, described on page 321, in which two sets of waves start from two coupled point sources placed a short distance apart. Figure 15.3 is a schematic view of Young's experiment. A source of light is placed behind a narrow slit S_1. In another screen just beyond are two more narrow slits S_2 and S_3, which are very close together, and both are parallel to S_1. In Young's original arrangement, the apertures were pinholes, but the effects are much easier to observe if slits are used. Also, the visibility is increased if the source of light is of a single wavelength (**monochromatic** light). This will be assumed to be the case in the following discussion.

If light traveled in straight lines, the illumination to be expected on the screen W would consist merely of two slitlike bright lines, one lying in the line S_1S_2 and the other in the line S_1S_3. However, what Young observed was completely different: The pattern on the screen W consisted of a number

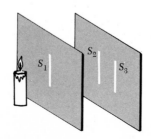
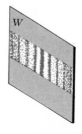

FIGURE 15.3 In Young's experiment a series of alternate light and dark bands appears on the screen W when illuminated by light coming through the pair of slits S_2 and S_3 which in turn receive light coming through S_1 from the source.

of diffuse, alternate bright and dark bands parallel to the slits. If either slit S_2 or S_3 was covered, the pattern of bands disappeared and the illumination of the screen became faint and quite uniform in appearance. Thus, when both slits were open it was evidently possible for light added to light to produce darkness at certain places on the screen. This suggested to Young an explanation on the basis of the destructive interference of waves.

Each of the narrow slits becomes the source of a set of Huygens wavelets whose subsequent interference (superposition) produces the pattern of interference fringes observed on the screen. The interference effects exist throughout the space beyond the double slit, just as in the corresponding ripple pattern in Fig. 13.21 (page 321). The purpose of the screen W is merely to make the pattern visible in the form of a set of fringes.

Figure 15.4 represents a plan view of the arrangement. The wavelets starting from S_2 and S_3 have a fixed phase relation; they are coherent. In particular, if the distances S_1S_2 and S_1S_3 are exactly equal, the two sets of waves start out in the *same* phase. The illumination on W at any point P depends on the relative phase in which the two sets of waves arrive. If they arrive in the same phase, P will be a place of maximum brightness; if they arrive in opposite phase, P will be dark. Arrival in any other relative phase will produce some intermediate level of illumination.

If, as in Young's experiment, the beams always travel in the same medium, the phase relation on arrival at P depends only on the *path lengths* of the two sets of waves. If the path difference is a whole number of wavelengths, they arrive in the same phase; if it is an odd number of half wavelengths, they arrive in opposite phase. To summarize:

PLACE ON SCREEN	RELATIVE PHASE OF ARRIVAL	PATH DIFFERENCE
Center of bright fringe	Same phase	0, 1, 2, 3 . . . wavelengths
Center of dark fringe	Opposite phase	$\frac{1}{2}, \frac{3}{2}, \frac{5}{2}, \ldots$ wavelengths

From Fig. 15.4 it is possible to calculate the positions of the bright and dark fringes. Corresponding to any point of arrival P, there is a path difference $S_2P - S_3P$. In practice, the separation d of the slits is always quite small compared with the distance F to the screen, and so a simple approximate value of the path difference can be obtained by dropping a perpendicular S_2N from one of the slits to the light path from the other and taking

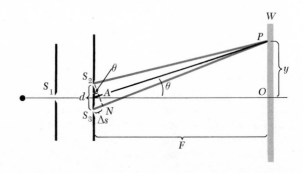

FIGURE 15.4 Young's experiment, shown schematically. For clarity, the sizes of d and y have been made relatively large.

the segment Δs to be the difference in path. From the construction of the diagram, the right triangles S_3NS_2 and AOP are similar. Equate the values of $\sin \theta$ in the two triangles:

$$\sin \theta = \frac{\Delta s}{d} = \frac{y}{AP} \qquad\qquad \textbf{15.1}$$

or, since θ is always kept small, AP can be replaced by F, the two being always approximately the same length. Then the above equation becomes

$$\Delta s = \frac{yd}{F} \qquad\qquad \textbf{15.2}$$

If, in this relation, Δs is set equal to 0, λ, 2λ, 3λ, etc., we get the positions of the centers of the successive bright fringes. If Δs is set equal to $\frac{1}{2}\lambda$, $\frac{3}{2}\lambda$, $\frac{5}{2}\lambda$, etc., we get the positions of the centers of the dark fringes. Thus the wavelength of the light can be computed by measuring the slit separation, the distance to the screen, and the distance from the center of the fringe pattern to the middle of any fringe. Young performed such calculations and found that the values obtained were in rough agreement with those from other optical experiments.

Worked example 15.1

In a Young experiment, the two slits S_2 and S_3 are 0.15 mm apart, and the interference pattern is received on a screen 75 cm away. If the fourth bright fringe on either side is found at a distance of 1.1 cm from the central fringe, what is the wavelength of the light used?

Solution   The central bright fringe corresponds to $\Delta s = 0$ in Eq. 15.2, and the fourth bright fringe to $\Delta s = 4\lambda$. We use the latter value and substitute the numbers into the equation, expressing all distances in centimeters:

$$\lambda = \frac{(1.1)(0.015)}{(4)(75)} \qquad \text{or} \qquad \lambda = 5.5 \times 10^{-5} \text{ cm}$$

Light of this wavelength lies near the middle of the visible spectrum, in the yellow-green region.

A. J. Ångström (1814–1874) was a Swedish spectroscopist.

Spectroscopists often use a special unit for the measurement of wavelengths in and near the visible range. It is called one **angstrom** (abbreviated Å) and is defined by

$$1 \text{ Å} = 10^{-10} \text{ m}$$

Thus the limits of the visible spectrum are about 7000 Å for the extreme red and 4000 Å for the extreme violet.

15.3 INTERFERENCE IN THIN FILMS

In his experiment Young obtained two coherent sources by sending light from a single source through two apertures. Some workers raised the objection that the observed effects were due to the passage of the light through small holes rather than to actual interference. Objections of this kind are not valid because interference can be produced by a variety of experimental arrangements that make use of broad beams of light. In many of these experiments the two coherent sources are obtained, in effect, by passing light into a *thin film* of matter.

More than a century and a half before Young's experiment, Newton made certain observations that clearly showed the phenomenon of interference. However, he seems to have stopped just short of recognizing this fact explicitly as well as its bearing on the concept of the nature of light. In Newton's experiment, a very slightly convex piece of glass is placed on a flat piece. When the combination is viewed from above (Fig. 15.5), a system of concentric circular rings, alternately bright and dark, is seen.

The explanation on a wave basis is straightforward. In the figure, *SABCE* is a ray that reaches the eye after reflection from the curved surface, and *SFGBHE* is a ray passing through *B* after reflection from the flat surface. Although these two rays start from the source in the same phase, they will, in general, arrive at the eye in a different phase. One reason is that the path of the second ray is longer than that of the first. Also, for the portions of the paths that lie inside the glass, the wavelength is shortened by a factor n, as expressed by Eq. 13.8 (page 318), and this introduces a further phase shift. Further, when light waves are reflected at the surface of a medium of greater index of refraction, as they are at *G*, a phase change of 180° takes place. This is the exact counterpart of what happens when a mechanical wave is reflected from a fixed wall, as described in Sec. 13.3 (page 306).

In general, if the thickness of the "air film" in the Newton ring experiment is such that the two rays emerge from the system exactly in the same phase, this place on the glass will appear bright. It will be dark if they emerge

FIGURE 15.5 Newton's rings constitute an interference system formed by light reflected from the two surfaces of a thin film of air. A complementary set of circular fringes can be seen in the light that passes through the apparatus, but the light coming direct from the source makes it less prominent. (*Photo from Bausch and Lomb.*)

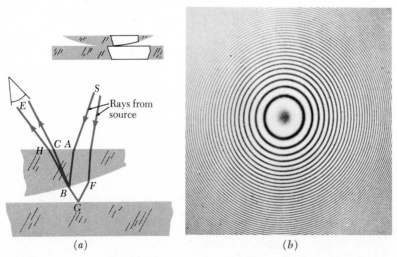

(*a*)   (*b*)

exactly in opposite phase. Outward from the center, there are places where there is alternately constructive and destructive interference. Since the whole arrangement is symmetric around a vertical line through the center, a set of concentric circular fringes is observed. The two rays that lie along the central axis have identical optical paths, since the air film has zero thickness there, but the 180° phase shift described above makes them interfere destructively, which accounts for the darkness of the central spot.

If the curvature of the upper glass is known, the thickness of the air film at any distance from the center can be computed. It is then possible to calculate the wavelength of the light used by measuring the diameters of the circular fringes. Young actually used some of Newton's data to compute wavelengths in this way. Newton himself evidently came very close to doing this because he recorded the observation that the ring pattern was larger for red light than for blue. In addition, he tried to theorize about the phenomenon in terms of some kind of periodicity inherent in a beam of light (see Ref. 15.1). Certainly this situation is one of the most remarkable examples of a "near miss" in the entire history of scientific discovery.

A familiar example of interference from a thin film is the rich coloration of a soap bubble or of a film of oil on a wet pavement. As in the case of Newton's rings, the color seen at any point is a composite of all the colors in the incident light (usually white) that are *not* destroyed by interference. The colors are most vivid when the film is between about 3×10^{-5} and 7×10^{-5} cm thick. If the film is thinner, the number of different wavelengths destroyed is small and the remaining mixture is not very brightly colored. If the film is thicker, the wavelength regions removed or weakened by interference are narrow and well spaced through the spectrum so that the remaining mixture appears essentially white.

A carpenter tests the flatness of a board by placing the edge of a ruler in contact with it in various orientations. This crude method would naturally be inadequate for testing the flatness of the surface of a glass prism, mirror, or plate which often must be made truly plane to within a fraction of a wavelength of light for use in accurate optical work. During the process of grinding and polishing such a piece, it is repeatedly tested for flatness by laying it on another plate that is known to be flat and inspecting the interference pattern produced. If the interference bands are irregular, the piece requires further polishing.

In a compound-lens system such as a photographic objective, light reflected from the several glass surfaces impairs the efficiency of the unit in two ways: (1) Less light is transmitted, weakening the image, and (2) the reflections from curved surfaces produce troublesome secondary images. It has been found possible to coat the surfaces with a thin layer of transparent material and thus reduce the reflected light by destructive interference. By using a coating of the proper thickness and index of refraction, waves of a given length are reflected from its two surfaces, emerging in opposite phase and destroying each other. The wave energy is then thrown back into the lens instead of being reflected off the surface. Where a compound lens consists of several units, coating the surfaces may increase the image brightness by 25 percent or more.

15.4   INTERFEROMETERS

The term **interferometer** is applied to a number of devices that use interference phenomena for purposes of measuring such quantities as the wavelength of a certain kind of light, a distance (such as the thickness of a film or the diameter of a star), the index of refraction of a sample of matter, etc. One of the most useful forms of interferometer, devised by Michelson, is shown schematically in plan in Fig. 15.6.

An instrument of this kind was used by Michelson and Morley in their ''ether drift'' experiment (page 93).

Light from a source falls on the thinly silvered surface of a plane, parallel-sided glass plate P (Fig. 15.6). Part of the light is reflected to the mirror M, which sends it back once more through P and onward into the observing telescope. The other part goes to the mirror N, which returns it to P, from which it is also directed into the telescope. In principle, the arrangement amounts to reflecting coherent beams of light from the two surfaces of a layer of air whose thickness is the difference between the optical paths PM and PN. In this case, the film thickness is not necessarily small, as in the interference experiments already described. Nevertheless, path differences comparable to the wavelength can be attained by small changes in the inclination of the beam, producing a pattern of interference fringes in the telescope.

One of the mirrors, M, is mounted on guides and may be moved back and forth along the line MP by means of a very accurate screw. As M is moved slowly in one direction, the entire interference pattern moves across the field of view. If M is moved a distance equal to just *half* a wavelength of the light being used, the length of path of the beam reflected from this mirror changes by exactly one wavelength and the phase relation with the beam going to N will be the same as before. In this process, the interference pattern is moved along by the distance between adjoining bands. The wavelength of the light can be calculated by counting the number of fringes passing when the screw is advanced a given distance. When the procedure is reversed, the standard meter bar can be measured in wavelengths of a specified kind of light, as described on page 24. This wavelength then becomes the true standard for length measurements of all kinds.

The use of light of known wavelength in the interferometer makes it possible to determine displacements of the movable mirror with great accuracy. In this way the Michelson interferometer can be used to measure displacements of the order of a millionth of an inch.

15.5   DIFFRACTION OF LIGHT

The phenomenon of the diffraction of waves was described in Sec. 13.10, and it was pointed out that these effects become prominent when waves meet an aperture or obstacle whose size is comparable with the wavelength. We have seen that some of the optical phenomena discussed above can be accounted for on a wave basis, and they indicate that the wavelengths of visible light are very small. Hence if a wave model proves suitable for light, diffraction effects should be expected whenever light passes through small holes or around narrow objects.

Francesco Grimaldi (1618–1663), Italian mathematician and scientist,

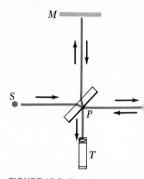

FIGURE 15.6 The Michelson interferometer (schematic). The instrument can be adjusted to show system of parallel interference fringes in the field of view of the telescope.

discovered the diffraction of light when he examined a spot of light formed by sunlight coming through a pinhole. He observed that the bright area was smaller than would be expected if light traveled in straight lines and that the edges were marked by colored rings. Although Newton repeated and refined Grimaldi's tests, he did not draw any conclusion from them as to the wave aspect of light. Huygens, too, was aware of Grimaldi's observations but did not arrive at any explanation.

Diffraction remained a neglected phenomenon for nearly a century after Grimaldi's discovery, when it was brought to prominence again by Young and by the French engineer and scientist A. J. Fresnel (1788–1827). The latter made extensive calculations based on Huygen's construction in order to account for diffraction effects, and the results began to convince many scientists of the correctness of the wave theory. Fresnel was able to explain in detail the diffraction band patterns observed when light passes through a narrow slit or a circular opening. At the same time, the French theorist S. D. Poisson calculated that a bright spot should be found at the center of the shadow cast by a round object. When experiment showed such a spot (see photograph at right), the status of the wave theory was greatly enhanced.

Figure 15.7 shows the appearance (greatly magnified) of the edge of the shadow of a straight obstacle, made by light from a very small source. The band pattern is visible only because a small, concentrated source was used. With an extended source, each point produces such a diffraction pattern, but the various patterns overlap and the bands are not seen. The picture on the next page shows the remarkable diffraction pattern produced by placing an opaque object in the light from a point source.

Resolving power.   In Sec. 13.10 it was shown that diffraction effects can be attributed to limiting the extent of the wavefronts and that diffraction patterns are the

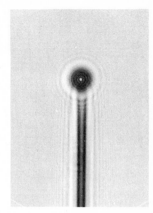

Shadow of a steel ball. There is a central bright spot, as predicted by the wave theory. Diffraction bands are also seen around the edges of the supporting rod. (*From Richards et al., "Modern University Physics," Addison-Wesley, 1960.*)

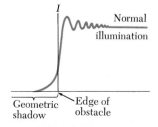

FIGURE 15.7 Appearance of the edge of the shadow of a straight obstacle. A little light is diffracted into the shadow region, and a set of fringes is seen immediately outside the edge of the shadow. (*Photo from Bausch and Lomb.*)

Shadow of a star-shaped washer Diffraction bands are seen inside the hole and around the outer edge. (*Bausch and Lomb.*)

result of superposition (interference) of Huygens wavelets originating on unobstructed portions of the wavefronts.

When a point source of light is viewed through a small circular opening, what is seen is no longer a point of light but a diffraction pattern consisting of a disk surrounded by concentric rings. The edge of the lens or mirror of a telescope or a microscope is a circular aperture. Each point in the object viewed will produce a ring pattern, and the entire image will be made up of such ring systems, overlapping. As a result, the instrument has limited **resolving power,** and it cannot reproduce fine detail, no matter how optically perfect the lenses or mirrors themselves may be.

The two components of a double star may not be resolved at all by a telescope if their diffraction patterns overlap too much. In order to separate them, a telescope with an objective lens or mirror of larger diameter must be used. This, together with increased light-gathering power, is the main reason for constructing large-aperture astronomical telescopes.

In a microscope, the resolving power of the optical system limits the amount of detail that can be seen. The resolving power, besides increasing with the aperture, increases also if light of shorter wavelength is used. The ultraviolet microscope, equipped with lenses made of quartz which transmits these short waves, has high enough resolution to allow the use of magnifications twice as great as for an instrument using visible light. The image cannot be viewed directly but must be photographed or thrown on a fluorescent screen.

15.6 DIFFRACTION BY A SLIT

A simpler diffraction arrangement than the ones described above is shown in Fig. 15.8, where plane waves of light are incident normally on an opaque screen S with slit AB. The slit runs perpendicular to the plane of the

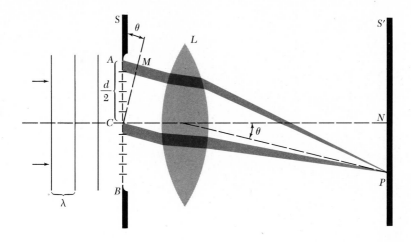

FIGURE 15.8 Diffraction of plane waves by a slit. Superposition of the Huygens wavelets from various parts of the wave surface produces a band pattern on the screen.

drawing, and its width d is assumed to be not too great compared with the wavelength λ of the incoming light. A suitable lens L placed just beyond the slit concentrates the light in an image on the screen S'. What is observed is a diffraction pattern of alternating bright and dark bands. If it were not for diffraction, this image would be only a bright line parallel to the slit, and the rest of the screen S' would be dark.

Some of the features of the diffraction pattern can be computed quite readily. Imagine that the wavefront at the slit is divided into a large number of narrow strips parallel to the edges of the slit. Consider the two sets of Huygens wavelets that originate (1) in the topmost strip and (2) in the strip just below the middle. In particular, consider the portions of the wavelets that travel at an angle θ with the center line CN. These two wave trains are shown in color in Fig. 15.8.

Both sets of waves are in phase as they start out from the plane of the slit, but the upper one travels a longer path than the lower in reaching the screen. The path difference can be found, as in Young's experiment (see Fig. 15.4, page 350) by dropping a perpendicular from C to the upper beam. The distance $AM = (d/2) \sin \theta$ represents this path difference. The illumination at any point P on the screen depends on the phase relations of the light waves arriving there from the various strips into which the incoming wavefront was divided.

At the central point N, waves from all the strips arrive in the same phase, and this place on the screen is bright. Going outward from the center, the path difference of the two beams increases, and at the place where this difference becomes equal to $\frac{1}{2}\lambda$, destructive interference results. Figure 15.8 shows that the path difference for the strips just below the two under consideration will also be $\frac{1}{2}\lambda$. Proceeding in this way, it is seen that the waves from each strip in the upper half of the slit are canceled by those from the corresponding strip in the lower half. The result is that no light reaches the screen at a point P lying at an angular distance θ from the center line, provided that $(d/2) \sin \theta = \frac{1}{2}\lambda$, or

$$\sin \theta = \frac{\lambda}{d} \qquad\qquad\qquad \textbf{15.3}$$

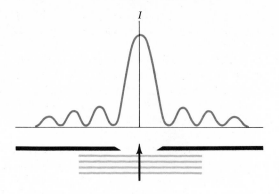

FIGURE 15.9 Diffraction pattern produced by a narrow slit, showing the intensity distribution. For clarity the secondary maxima have been drawn much taller than they really are.

This relation shows that narrowing the slit makes the diffraction pattern broader, in agreement with observation.

By the same kind of reasoning, it is seen that additional dark places are found for values of $\sin \theta$ that are 2, 3, 4, . . . times the above value. Between these places the illumination on the screen reaches a maximum, as shown by Fig. 15.9. Notice that most of the light goes "straight through" the slit to form the central peak and that the other places of maximum illumination fall off rapidly in intensity in an outward direction from the center of the pattern. Calculation of the relative strengths of the maxima or the distribution of light around each one requires a more detailed analysis, which will not be presented here.

Practice set

1. Show from Eq. 15.3 that the central bright band in the diffraction pattern of a single slit has about twice the angular width as the other bright bands.
2. If a slit could be made for which $d = \lambda$, describe the appearance of the screen. See Eq. 15.3.

15.7 THE DIFFRACTION GRATING

In Young's experiment, the fringe pattern is rather flat, and consequently accurate measurement of the spacing between fringes is difficult. Theory shows that by adding a large number of identical, equally spaced slits to the two in Young's arrangement the sharpness of the peaks can be greatly increased. Each slit produces a diffracted beam, and the superposition of all these beams results in the pattern observed. Hence the action of the whole set may be looked upon as a combination of diffraction and interference.

An apparatus of this kind was used extensively for measuring optical wavelengths by the German optician Joseph Fraunhofer (1787–1826). It is called a **diffraction grating.** The first gratings were made by winding wires back and forth around the threads of two parallel machine screws, and they ranged between about 40 and 600 openings to each inch of width. Now it is possible to make highly precise gratings with an instrument that draws a fine diamond point repeatedly across a glass plate that is moved onward the desired distance after each ruling. The slits are the untouched strips between the scratches. The latter act as effective barriers because they transmit light diffusely. Alternatively, a **reflection grating** can be made by ruling the surface of a metal mirror.

The ruling of a good grating is an extremely delicate operation requiring elaborate precautions to avoid vibration, temperature changes, or other disturbances that would spoil the uniformity of the spacing. High-quality gratings several inches wide with 15,000 to 50,000 lines per inch are in use, and a precision grating has been successfully ruled with 274,000 lines per inch. It is used for the diffraction of x rays.

The maxima in the diffraction pattern of a grating are in the same locations as those for the double slit but are much sharper, as mentioned earlier. In fact, they are mere lines of light called **spectrum lines.** Figure 15.10 represents the simplest arrangement of a diffraction grating, of which only a few slits are shown. Plane waves of light of a single wavelength are incident perpendicularly on the grating, and the diffracted light is brought to a focus on the screen by means of a converging lens or its equivalent. The Huygens wavelets shown in the figure come from *corresponding* points of the several slits, and the rays that are drawn indicate the contributions of these wavelets in a direction making an angle θ with the normal to the grating. Precisely as in the case of a pair of slits (Young's arrangement), phase agreement occurs in those directions for which

$$\sin \theta_n = \frac{n\lambda}{d} \qquad\qquad \textbf{15.4}$$

where λ is the wavelength of the light and n takes the values 0, 1, 2, 3, ... Thus, with light of a single wavelength, there will be an array of maxima (spectrum lines) along the screen. If the arrangement is such that the angles are small, the spacing will be very nearly uniform, since $\sin \theta$ is very nearly proportional to θ itself when θ is small. Each line corresponds to a particular value of n, which is called the **order** of the spectrum. The zero-order line is at the center of the screen, and there will be a first-order line on either side, a second order beyond each of these, etc. However, the number of orders produced is limited, since the maximum value $\sin \theta$ can have is 1.

If a *mixture* of several wavelengths enters the grating, a *set of lines* is produced in each order, rather than a single line. There is a line for each

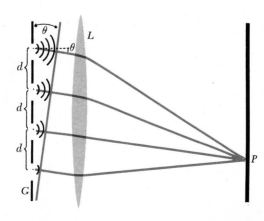

FIGURE 15.10 Schematic representation of the action of a diffraction grating. Wavelets coming from corresponding places in each opening give rise to a set of sharp images (spectrum lines) on the screen.

wavelength present in the incoming light, and each line falls at a different position on the screen. This is implied by Eq. 15.4: With given values for d and n, sin θ increases with λ. Thus the grating produces *dispersion*, spreading out the components of the mixture according to their wavelengths. This makes the grating an indispensable device in spectroscopy. For several reasons it is to be preferred to a prism for producing spectra. In the prism, the deviations that result for various wavelengths depend on the properties of the prism material in a complicated way. In the grating, however, the angles of deviation are related to the wavelengths, the order of the spectrum, and the grating space in the simple and direct way represented by Eq. 15.4. Thus, by using this equation, the wavelengths can be calculated with what has become known as "spectroscopic accuracy." Under favorable conditions, wavelengths can be measured with a grating to perhaps 1 part in 10 million. Interferometer methods, such as the one on which the new definition of the standard meter is based, can reduce the error by two more orders of magnitude.

1. When n is put equal to zero in Eq. 15.4 sin θ becomes zero, quite independently of the value of λ. This being so, what is the color of the zero-order spectrum of white light as formed by a grating?
2. In the grating spectra of white light in the first and higher orders, which color falls farthest from the center of the screen—red or violet?

The characteristic yellow color of the sodium-vapor lamps used in highway lighting is due mainly to a strong pair of lines in the sodium spectrum having wavelengths of 5890 and 5896 Å. The spectrum formed by a grating having 10,000 lines per centimeter of width is focused on a photographic film 2.00 m away. How far apart will the images of the sodium lines fall in the third-order spectrum?

According to Eq. 15.4, the third-order image of λ = 5890 Å falls at an angle to the grating normal (GN in Fig. 15.11) given by sin θ_3 = (3)(5.89 × 10^{-5})/(1/500) = 0.0884, making θ a little over 5°. For an angle this small, it makes little difference numerically whether we use sin θ, tan θ, or θ itself (in radians) for the left side of the equation.

If f is the focal length of the lens used to form the images and x is the distance from N to the image in question, we may to a good approximation replace sin θ by x/f, since the lens is usually placed close behind the grating. Further, since Eq. 15.4 is linear in λ, the equation may be written

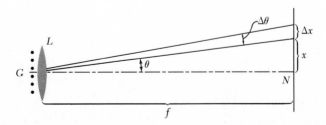

Arrangement of plane diffraction grating for the formation of spectra.

in terms of the linear separation Δx of the two images and the wavelength difference $\Delta\lambda$ (which is 6 Å here):

$$\frac{\Delta x}{f} = \frac{n\,\Delta\lambda}{d} \qquad \text{or} \qquad \Delta x = \frac{nf\,\Delta\lambda}{d}$$

Substitution of the numbers yields

$$x = \frac{(3)(200)(6 \times 10^{-8})}{(1/500)} = 0.018 \text{ cm}$$

15.8 POLARIZATION OF LIGHT

The wave concept explains in detail so many aspects of optical phenomena that there can be no doubt that light is propagated in the form of waves. But what kinds of waves are they? Huygens and his followers assumed that light waves, like sound waves, were *longitudinal,* and there is nothing in the wave theory of refraction, diffraction, and interference that contradicts this. However, the discovery of an entirely different aspect of the behavior of light, called **polarization,** led ultimately to the conclusion that light waves must be *transverse.* This idea, advanced independently by Fresnel and by Young, met with widespread objections because the scientists of that time could not conceive of any kind of waves other than mechanical ones, in which the high speed of light would require the medium (the *ether,* referred to on page 93) to be as rigid as steel yet so tenuous that it would not hinder the motions of the planets. This objection was to be met, later, by Maxwell's electromagnetic theory of light.

What is implied by the statement that transverse waves can be polarized? Recall that in an ordinary light source the observed output of light is the totality of countless independent events—the random contributions of individual atoms or molecules. Assume that, in each act of radiation, a microscopic source of this kind sends out a train of transverse waves traveling in a given direction. The vibration vector in such a wave train lies in a plane containing the ray, and because these occurrences are completely random, there are vibration planes in all possible orientations around the direction of propagation, as shown schematically in Fig. 15.12a.

There are ways of suppressing the vibrations in every orientation except a selected one. The remaining vibration, confined to just one transverse direction, is said to be **plane-polarized.** A crude mechanical analog of such

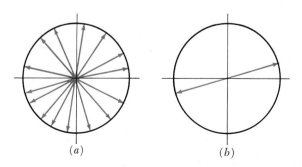

(a) $\qquad$ (b)

FIGURE 15.12 The structure (diagrammatic) of a beam of light. In ordinary, unpolarized light the vibration vectors lie in random directions around the line of propagation as in (a). If the vibrations are confined to just one direction, the light is plane-polarized, as shown in (b).

an operation is represented by a rope passing through a slotlike box (Fig. 15.13). If one end of the stretched rope is shaken in a plane making an arbitrary angle with the flat sides of the box, the only part of the vibration that gets through is the component in the plane of the slot. Thus, if the shaking takes place in a plane parallel to the slot, the entire vibration is transmitted. If the shaking takes place in the crosswise plane, nothing is transmitted. Now suppose that the rope is replaced by a spiral spring. *Longitudinal* waves generated in the spring pass through the box regardless of the orientation of the slot. The same is true for a stream of particles shot through the slot.

Long before the time of Young and Fresnel, optical phenomena corresponding to the rope-and-slot demonstration were observed by Newton, Huygens, and others. It was found that the vibrations in a beam of light could be "combed out" by passage through crystals of certain minerals such as calcite. Other ways of polarizing ordinary light were also discovered: reflection from a nonmetallic mirror at a particular angle of incidence; the scattering of sunlight by molecules of the air (a process that is responsible for the blue color of the sky). The many phenomena connected with the polarization of light can be explained satisfactorily by the assumption that light waves are transverse. Further details will not be given here but can be found in Ref. 15.6.

Practice set

1. Ordinary light becomes polarized to some extent when reflected obliquely from a smooth surface. Suppose you are looking at the glare of wet pavement through a disk of polarizing material. What would you expect to see if you rotate the disk slowly in front of your eye?

2. How could a disk of polarizing material be used to show that direct sunlight is unpolarized but that the light of the sky (sunlight scattered by the air) is partially polarized?

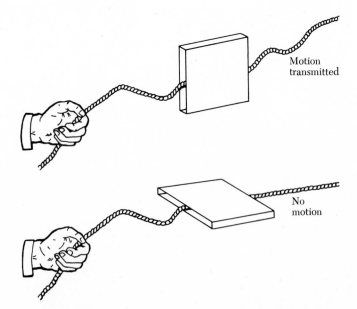

Motion transmitted

No motion

FIGURE 15.13 Waves on the rope pass freely through the box when the slot is lined up with the plane of vibration but are suppressed when the slot is turned in the crosswise direction. This behavior is a good analog of what happens when plane-polarized light strikes certain crystals.

15.9 LIMITATIONS OF THE WAVE MODEL

In the preceding sections, optical phenomena such as reflection, refraction, dispersion, interference, diffraction, and polarization were examined. An important goal of this study was to find a satisfactory physical model for light and similar radiations. The impossibility of devising a consistent and straightforward corpuscular explanation of optical phenomena showed that a wave model is required. Such a model is able to account for the varied and numerous observations quantitatively and in great detail.

It was Maxwell who removed certain fundamental difficulties of the assumption that light waves were of a mechanical nature. Through his electromagnetic theory, developed about 1860, he showed that light is essentially an electrical phenomenon propagated in free space rather than in some hypothetical medium. The Michelson-Morley experiment (page 93) confirmed the suspicion that no such medium exists, and Einstein's relativity, in harmony with the null result of that experiment, dispensed with the use of such a medium altogether.

All conceptual models must be based on a finite amount of experience, and there is always the possibility that a particular model may fail when confronted with further observations. That is what happened with the wave model of light. The phenomena on which the wave idea was tested in the preceding discussion all deal with light in transit, where the theory is eminently successful. However, it was subsequently found that the wave model is unable to explain phenomena involving the interaction of radiation with individual atoms or elementary particles such as electrons. In order to account for such phenomena, an entirely different model (the quantum or photon model) had to be devised. Chapter 19 will show that both the wave and photon models are needed, each with its own range of applicability. A unified theory of the nature of light does not exist at this time except, perhaps, in the highly abstract mathematical description provided by the quantum theory.

Programmed review

Instructions: See page 18.

1. Define dispersion.

    The separation of a mixture of waves according to some property, such as wavelength.   [15.1]

2. Describe the spectrum of white light formed by a prism.

    A continuous band of merging colors: red, orange, yellow, green, blue, and violet.   [15.1]

3. At a certain place on a screen, two portions of a beam of monochromatic light coming originally from the same source meet to produce an interference maximum. If the light traveled entirely in air, what can be said about the path difference of the two beams?

    It is equal to some integer multiple of the wavelength.   [15.2]

4. Describe what is meant by thin-film interference.

    Interference effects produced by superposition of the two beams formed by reflecting a single beam from two surfaces that are close together.   [15.3]

5. Explain why the central spot in the Newton's rings interference pattern is black.

    Reflection from the lower glass surface causes a phase shift of 180°. [15.3]

6. What does the resolving power of an optical instrument measure?

    Its ability to separate, in the image, object points that are close together, thus revealing fine detail.   [15.5]

7. The formula for the positions of the light maxima produced by a pair of slits is the same as that of grating consisting of a very large number of carefully ruled slits. What, then, is the justification for taking the trouble to rule many-lined gratings?

    Increasing the effective number of slits makes the maxima sharper.   [15.7]

8. Give the formula for the angular position of the spectrum lines formed by a grating, expressed as a function of the wavelength, the grating space, and the order of the spectrum.

    This is stated by Eq. 15.4 as $\sin \theta_n = n\lambda / d$.   [15.7]

9. What is a plane-polarized beam of light?

    One in which the transverse vibrations are confined to a single plane that contains the direction of propagation.   [15.8]

10. Describe the present status of the theory of the nature of light.

    A unified theory capable of accounting for all known radiation phenomena in a visualizable form does not exist.   [15.9]

For further reading

15.1. *Taylor, L. W.* ''Physics, the Pioneer Science.'' See pp. 474–477 for Newton's description of his prism experiments. Newton's rings are discussed on pp. 506–510. Polarization of light and applications are described in chap. 38.
15.2. *Wood, A.,* and *F. Oldham* ''Thomas Young, Natural Philosopher.'' A biography.
15.3. *Bragg, W.* ''The Universe of Light.'' Read chaps. 3 and 4 on color and chap. 6 on polarization. Still one of the best, easily read expositions of these topics.
15.4. *Tolansky, S.* ''Revolution in Optics.'' Describes recent applications of optical methods to many fields.
15.5. *Schawlow, A. L.* Lasers and Coherent Light, *Sci. Am.*, September, 1967, p. 197. This reference was cited also in Chap. 13.
15.6. *Shurcliff, W.,* and *S. Ballard* ''Polarized Light.'' See chap. 10 for applications.

Questions and problems

15.1.   Account for the colors seen when a cut diamond is viewed at various angles.

15.2.   Explain the fluttering of a TV picture that occurs when a low-flying airplane passes overhead. HINT: The waves reach the TV antenna via two different paths in such a case.

15.3.   In Young's experiment, the purpose of slit S_1 in Fig. 15.4 is to provide the equivalent of a point source of light. What would happen to the interference pattern if the screen containing S_1 were removed?

15.4.   Is it possible to perform Young's experiment by replacing the single source and slit S_1 in Fig. 15.4 (a) by two lamps, one placed behind each of the slits S_2 and S_3; (b) by a laser? Explain.

15.5.   Describe the appearance of the screen in Young's experiment when white light is used. What is the color of the central bright band; of the others?

15.6.   The two slits in a Young's experiment are 0.0200 cm apart. The bright fringes, viewed on a screen 80 cm away, are 0.240 cm apart. What is the wavelength, in angstroms, of the light used?

15.7.   What is the essential difference between the production of color when (a) a soap bubble and (b) a glass prism are illuminated by sunlight? Explain in detail.

15.8.   The iridescent color of an opal or of the inside surface of an oyster shell is not caused by pigmentation of the material. How, then, can it be explained?

15.9.   Two flat, rectangular glass plates are in contact along one edge and are wedged apart at the opposite edge by a strip of metal foil 5×10^{-4} cm thick. When illuminated broadside by light of a certain wavelength, the width of the pair of plates is seen to contain exactly 20 dark fringes. Find the wavelength, in angstroms, of the light used. HINT: In going from one dark band to the next the difference in thickness of the air film must be one-half wavelength.

15.10. In a pioneering process of color photography devised by G. Lippmann in 1891, light is allowed to fall on a photographic emulsion backed by a mirror. This produces a set of standing waves for each color present in the incident light. After development, the emulsion contains parallel strata half a wavelength apart. Then, when viewed at the proper angle, the plate reflects the corresponding color strongly. In a given trial, light of wavelength 4800 Å was used. Find (a) the wavelength inside the emulsion and (b) the distance between adjoining strata in the sensitive layer. The index of refraction of the gelatin is 1.52.

15.11. Using light of a certain wavelength in a Michelson interferometer, it is found that 2,500 complete interference bands pass across the field of view when the movable mirror is advanced a distance of 0.080 cm. Find the wavelength of the light used.

15.12. Light of wavelength 5000 Å is used in a Michelson interferometer. Calculate the number of interference fringes that pass across the field of view when the movable mirror is backed away a distance of 0.200 mm.

15.13. A slit 0.20 mm wide is illuminated by parallel light whose wavelength is 5000 Å. Diffraction bands are observed on a screen 50 cm away. How far to either side of the center of the pattern does the middle of the fourth dark band fall?

15.14. A grating having 5,000 lines per centimeter forms a spectrum on a screen 1.00 m away. How long a piece of photographic film is needed to record the entire visible spectrum (4000 to 7000 Å) in the first order?

15.15. A grating forms spectra of the entire visible range, 4000 to 7000 Å. Using Eq. 15.4 show that the violet end of the third-order spectrum overlaps the red end of the second-order spectrum. This difficulty can be avoided by admitting a more limited range of wavelengths to the grating at any one time.

15.16. What wavelength in the fourth-order spectrum of a grating falls at the same position as $\lambda = 6000$ Å of the overlapping third-order spectrum? How could these two radiations be distinguished if viewed directly by the eye?

15.17. Of the following types of waves, which can be polarized: blue light, radio (electromagnetic) waves, sound waves? Give reasons for your answer.

15.18. The intensity of a beam of light can be controlled by passing it in succession through a pair of polarizing units, analogous to a pair of slotted boxes in Fig. 15.13. The intensity of the transmitted light is regulated by rotating the second polarizer with respect to the first and is given by $I = I_0 \cos^2 \theta$, where I_0 is the intensity of the beam after it has come through the first polarizer and θ is the angle through which the second element has been turned from its position of maximum transmission. Through what angle must it be turned in order to reduce the intensity to (a) half its original value; (b) one-fourth?

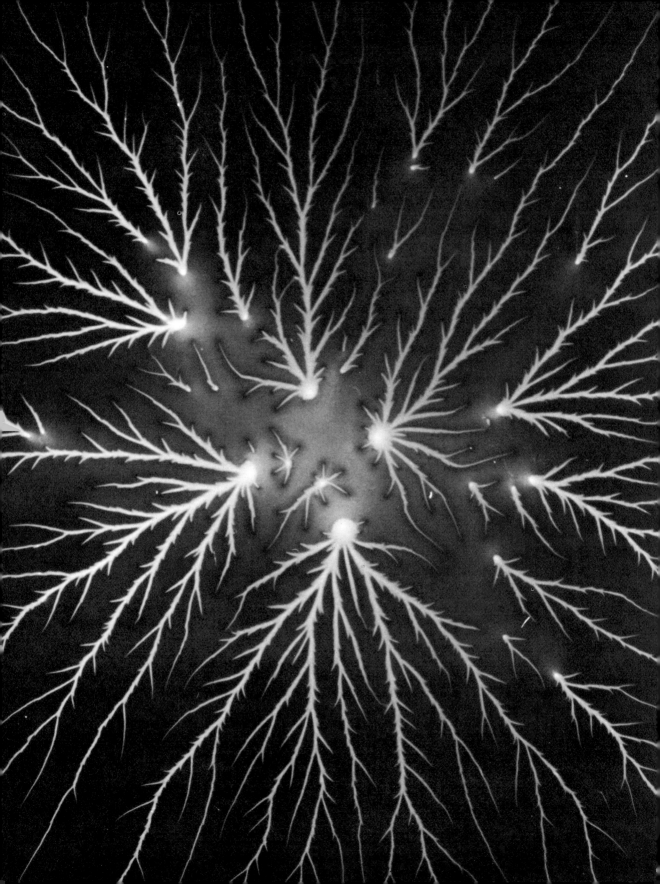

Part four
ELECTRICITY

Chapter 16
ELECTROSTATICS

Few branches of science have applications with such far-reaching consequences as those of electricity. It is hardly necessary to enumerate the ways in which electrical devices and processes have influenced the evolution of our modern way of life. The development of electrotechnics, beginning only about a century ago, has brought about tremendous innovations and changes in transportation, communication, manufacturing, lighting, and many other technical fields, and the social, political, and philosophical consequences of these developments are of supreme importance to all people, not only in the industrialized countries but in the developing ones as well.

Over 3 billion kilowatt-hours of electric energy is generated daily in the United States.

In examining the subject of electricity in this and the following chapters, the main concern will continue to be with matters of concept and principle rather than with applications or technical devices. It will become apparent that an exposition of the fundamental ideas of electricity is basic to the understanding of many aspects of pure science, including the nature of matter and radiation.

16.1 EARLY DISCOVERIES; ELECTRIC CHARGE

The primitive observations that gave rise to the science of electricity appear to have dealt with obscure and trivial phenomena. It is not possible to determine when electrical manifestations were first observed and recorded, but it is known that some of them, which we now recognize as electrical, were noted by the ancient Greeks. The philosopher Thales (640–546 B.C.) is said to have observed that a piece of amber, after it had been rubbed, would attract and pick up small bits of straw, wood, or feathers. This behavior was sometimes confused with the attraction of natural magnets for bits of iron.

The word "electric" derives from the Greek elektron, meaning amber.

Apart from this, little of importance was discovered until 22 centuries later when Gilbert (page 179n) investigated these matters more thoroughly. He found that many other materials besides amber have the power to attract light objects, and he was able to arrange substances into two classes which he called "electrics" and "nonelectrics." Later, these categories were designated as insulators and conductors, respectively.

It is now known that, when *any* two substances are rubbed together, each acquires attractive power. The bodies are said to become *electrified,* or to acquire electric *charge.* Observations of electrification by rubbing are common, particularly in dry weather. When a comb is run through the hair, it makes crackling noises, and a rubbed sheet of paper clings to the desk. All electrical experiments of this kind are most successful when everything is quite dry, for reasons to be explained later.

A convenient indicator of electrification (or of the presence of electric charges) can be made by covering a small ball of some light material such as balsa wood or foamed plastic with metal foil and hanging it from a support by means of a silk thread. With this crude **electroscope,** a number of fundamental observations can be made:

1. If any previously electrified object (such as a plastic rod that has been rubbed with fur or wool cloth) is brought within an inch or two of the hanging ball, the ball will move toward the charged body and take a displaced position. In this new equilibrium position, the thread lies along the resultant of the attracting force and the weight of the ball (Fig. 16.1*a*). The sidewise displacement of the ball, which is a measure of the force of attraction, is often comparable with the length of the thread, which means that the electric force is of the same order of magnitude as the weight of the ball.
2. If, intentionally or otherwise, the ball makes contact with the charged rod, the force immediately changes to one of *repulsion* and the ball is driven away if the rod is again brought near (Fig. 16.1*b*).
3. With the ball in the last condition above, a glass rod that has been rubbed with silk is brought near. The ball is strongly attracted by the glass rod (Fig. 16.1*c*).
4. The three observations described are unaffected if the plastic and glass rods are interchanged: The uncharged ball is attracted by the glass rod, then promptly repelled after touching it, and subsequently shows attraction for the plastic rod.

Dual nature of electrification.

Experiments of this kind demonstrate that electric charges are of two varieties that show opposite forms of behavior. The observations clearly point to the *dual* nature of electrification. In each case, when the ball

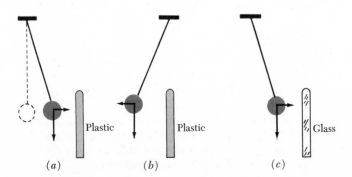

(a)   *(b)*   *(c)*

FIGURE 16.1 Basic experiment with charged bodies. The initially uncharged ball is attracted by the rubbed plastic rod, but after contact it is repelled. If a rubbed glass rod is brought near, the ball is attracted.

touched an electrified rod, it presumably acquired electric charge of the kind already on the rod. Thus, purely as a means of summarizing the results, we can make the following statements:

1. Objects carrying *unlike* charges *attract* each other.
2. Objects carrying *like* charges *repel* each other.
3. An *uncharged* body is attracted by a charged body of either kind.

By the beginning of the eighteenth century, electrification was looked upon as a property of the bodies involved. However, as time passed there was a tendency, suggested by the transfer of charge from one object to another, to associate the phenomenon with some kind of fluid. There seemed to be a vogue in that period for the invention of fluid models for various purposes. For instance, chemists proposed a fluid called *phlogiston* to explain combustion and similar phenomena, and the physicists conceived of heat as *caloric* fluid (page 245).

The dual character of electricity was first established by the experiments of Stephen Gray (1696–1736) and those of Charles Dufay (1698–1739). Gray succeeded in transferring charge from one body to another, sometimes through rods or strings of considerable length. Dufay built upon these experiments and went on to show that the nature of the force between charged objects, attraction or repulsion, depended on the *combination* of charges they possessed, which was essentially equivalent to rules 1 and 2 stated above. Dufay spoke of two kinds of electrification: "vitreous" (the kind produced on glass) and "resinous" (found on amber or similar substances), and these later came to be considered two kinds of fluid.

In addition to his well-known talents in politics and in business, Franklin had unusual abilities as a scientist and engineer. Among his inventions are the lightning rod, bifocal eyeglasses, and watertight compartments for ships.

Dufay's experiments attracted the attention of Benjamin Franklin (1706–1790), who promptly began a long series of electrical experiments of his own. Franklin preferred to think of electricity as a single fluid, which was carried in a normal amount by every unelectrified object. He assumed that an object could gain extra fluid and become *positively* electrified or lose some fluid and become *negatively* electrified. This terminology is consistent with the fact that opposite kinds of charge tend to cancel each other when put onto the same body. The type of charge found on a glass rod after rubbing it with silk was arbitrarily termed positive ($+$) by Franklin, and the kind present on a hard-rubber rod after rubbing it with fur was called negative ($-$).

Whether we think of electrical phenomena along the lines of Dufay's two-fluid theory or Franklin's one-fluid theory is of little consequence. The important point is the dual character of electrification.

16.2 THE ELECTRICAL STRUCTURE OF MATTER; CHARGE CONSERVATION

The concept of the atomic nature of matter was introduced and developed in earlier chapters of this book. Prior to the present century, atoms could be treated as fundamental, indivisible entities, but now it is known that they have a complex internal structure. It is to this structure that we must look for the source of electric charge, and for this reason a brief description of the internal composition of an atom will now be given, although presentation of the detailed evidence must be left to later chapters.

It is now known that electric charge does not occur in the form of a continuous fluid but in definite, discrete amounts. Anticipating a term to be used extensively later, we may say that electric charge is *quantized*. This idea was presented in Chap. 3 in connection with the masses of atoms: Any physical object is made up of an integral number of atoms, never a fraction. The quantization of electric charge is even simpler. Although there are hundreds of different varieties of atoms, there is only one basic amount of electricity, called the **electron charge.** It is the smallest quantity of electric charge that can exist by itself. No smaller charge has ever been detected (see Sec. 16.6).

Electric charge never occurs independently of matter but is always associated with certain subatomic particles. The basic unit (the electron charge) occurs in two forms, one positive and one negative. They are designated $+e$ and $-e$, respectively. In atoms, the charge $+e$ is associated with a subatomic particle called the **proton,** and the charge $-e$ is associated with the **electron.** Although protons and electrons have equal (but opposite) amounts of charge, their masses differ greatly: A proton is nearly 2,000 times as massive as an electron. There is a third kind of particle that is part of the structure of the atom. It is called the **neutron** and has no resultant electric charge. The mass of the neutron is very slightly greater than that of the proton.

The equality of the amounts of electric charge on the various particles has been verified to 1 part in 10^{17}.

The internal "geography" of an atom may be described in general in this way: A group of the heavier particles (protons and neutrons) cluster together to form a compact entity called the **nucleus** of the atom. Arranged around this central nucleus there are electrons equal in number to the number of protons in the nucleus, so that in an atom in its **ground** (normal) **state** the total positive charge is equal to the total negative charge, and the atom as a whole is electrically neutral. For example, the nucleus of an ordinary atom of iron is composed of 26 protons and 30 neutrons. Disposed about this nucleus are 26 electrons.

In an atom having several electrons, some are closer to the nucleus than others. The innermost electrons are, in general, held more tightly by electrical attraction to the nucleus, and in some cases one or more outer electrons are relatively easy to remove. The chemical behavior of an element is determined by the arrangement of the electrons in the atoms that compose it and by the ease with which its electrons can be removed.

If two materials are brought into good contact, some electrons will generally be transferred from one material to the other. This happens when charges result from rubbing. For instance, glass gives up some electrons when rubbed with silk. The glass then has a deficiency of negative electron charges and is said to be positively charged. If the silk is neutral at the start, it accepts these electrons, gaining an amount of negative charge equal to the positive charge on the glass. These facts can be verified by experiment. It is possible to transfer detectable amounts of charge simply by pressing two dissimilar materials together, the main purpose of the rubbing being to promote good contact over a considerable area of surface of the two objects.

It is not correct to say that charges are "produced" or "generated." What occurs is a very slight reapportioning of the existing mixture of the two kinds of charge. Whenever a charge of either kind is given to an object, an equal and opposite charge must appear somewhere else. This is a fact of experience, known long before the discovery of protons and electrons. It was first stated by Franklin and is called the principle of **conservation of electric charge:** *The total net charge of an isolated system remains constant.* This conservation law has an importance that ranks it with two other propositions of this type that have already been examined: conservation of momentum and conservation of energy.

Conservation of charge.

One way of formulating the conservation of charge is to say that electric charge cannot be created or destroyed. Processes are now known in which the carriers of charge—protons, electrons, and still other charge-bearing particles not yet mentioned here—themselves are created or destroyed. Nevertheless, charge conservation still holds under these circumstances, for it is found that such events always involve *pairs* of oppositely charged particles.

Modern theory has been able to localize electric charge and relate it to certain aspects of the behavior of elementary particles of matter. It may even be said that, because it is responsible for chemical binding forces, electric charge is what holds the material world together. Nevertheless, the concept of charge remains an elusive one. Bertrand Russell (Ref. 16.5) expresses this idea as follows:

Some readers may expect me at this stage to tell them what electricity "really is." The fact is that I have already said what it is. It is not a thing like St. Paul's Cathedral; *it is a way in which things behave.* When we have told how things behave when they are electrified, and under what circumstances they are electrified, we have told all there is to tell. When I say that an electron has a certain amount of negative electricity, I mean merely that it behaves in a certain way. Electricity is not like red paint, a substance which can be put on to the electron and taken off again, it is merely a convenient name for certain physical laws.

It is evident that Russell is pointing out the necessity of defining charge in an operational manner.

16.3 CONDUCTORS AND INSULATORS

In experimenting with electric charges it soon becomes evident that certain substances such as metals allow charges to move readily along or through them, whereas other materials such as silk or glass do not. The two classes of substances are called, respectively, **conductors** and **insulators**—a distinction first drawn by Gray. The difference is one of degree only. The best conductors are metals, followed by carbon and certain minerals. The best insulators are glass, amber, mica, etc. Intermediate in this property are **semiconductors,** such as the metals germanium and silicon, which now find wide application in transistors and other "solid-state" electronic devices.

Alternative names for an insulator are nonconductor and dielectric.

A charge applied anywhere to a metal object will spread over the entire surface. If the object is not supported on some insulating material, the charge flows off to the earth, which is a conductor. On the other hand, a charge applied to or brought forth on an insulator remains localized. Thus a charge left on one end of a glass rod after rubbing it with silk remains on the rod even though the other end is held in the experimenter's hand.

A charge can be given to a metal ball by rubbing, provided that the ball is held by an insulating handle. If the ball is held directly in the hand, no net charge remains on it because the human body is a moderately good conductor, particularly when the skin is moist.

Solids are generally conductors because there are always some electrons that are very loosely bound or even temporarily detached from atoms. These **free electrons**—usually about one per atom in metals—form a cloud that permeates the interior of the substance. The free electrons share in the general thermal motion just as though they were the particles of an ordinary gas. In solid metallic conductors it is always the electrons that move; the positive charges in the atomic nuclei remain fixed in the structure of the metal. A movement of large numbers of free electrons in a conductor constitutes an **electric current.** Electrons can also be carriers of charge across free space such as the vacuum in electron tubes used in radio and television circuits.

Electric currents are described in detail in later chapters.

Conduction can also take place by means of **ions.** An ion is an atom or group of atoms carrying a net charge resulting from the detachment or attachment of electrons. The atoms that furnish the free electrons in solid conductors cannot move away; they remain in place as positive ions. When solids melt or dissolve, however, the crystal structure breaks down and the ions may be free to move. Their motion constitutes electric currents in such materials. For example, when sodium chloride is dissolved in water, each molecule breaks into two parts. One of these parts is a sodium ion carrying a charge $+e$ as a result of coming away with one electron missing. The other part is a chloride ion with charge $-e$ due to the attachment of this electron. Similarly, gases can be made conductors by converting the normally neutral atoms or molecules into positive and negative ions for this purpose, as in a neon tube, sodium-vapor lamp, etc.

When a charged conductor is **grounded** by connecting it by a wire to the earth or some other large conductor, it loses essentially all its charge. If the conductor was originally negative, grounding permits the excess electrons to flow from it to the earth through the medium of the free electrons in the connecting wire; if it was originally positive, electrons are attracted from the earth in sufficient number to neutralize its charge. The earth acts merely as a tremendous receiver or contributor of electrons, its own electrical condition being virtually unaffected by such transfer.

16.4 COULOMB'S LAW

The early experiments in electrostatics were understandably of a qualitative nature, but further progress pointed to the need for a more quantitative grasp of the phenomena. A significant advance in this direction was made by the French scientist and engineer Charles Coulomb (1725–1806). It had been suggested earlier by such workers as Daniel Bernoulli, Joseph Priestley, and Henry Cavendish—probably by analogy with Newton's law of gravitation—that the mutual attractive force of charged bodies should vary inversely as the square of the distance between them. Coulomb built a torsion balance (Fig. 16.2), similar to the one used by Cavendish, and carefully verified that the inverse-square law holds for electrostatic forces.

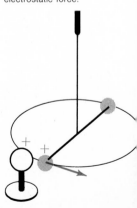

FIGURE 16.2 Schematic diagram of Coulomb's torsion balance. Electrostatic force between the charges makes the suspended system rotate, twisting the fiber until equilibrium is attained. The angle turned is a measure of the electrostatic force.

Since that time, numerous tests of the accuracy of the exponent 2 have been made. Most of these experiments for investigating the distance factor do not depend on the direct measurement of force but on other consequences of electrostatic interaction. A recent test (1971) shows that the exponent is equal to 2 to within 6×10^{-16}.

Besides distance, there is another factor affecting the magnitude of the force between two charged objects. This factor is called the **quantity of charge** on each. Coulomb, again by analogy with gravitation, made the conjecture that the force is proportional to the product of the two charges. But how can the experimenter measure out known amounts of charge and apply them to the two objects?

Coulomb saw, from the assumed form of the force law, that it was not necessary to deal with previously measured amounts of charge but it would be sufficient to be able to *change* the net charge on one or both of the electrified balls *by a known factor.* One obvious way of reducing any charge on a ball to exactly half its value is simply to touch the charged ball with an identical but neutral ball which is then removed. By symmetry, the original charge must divide equally between the two balls when they make contact, leaving the first ball with just half its original amount afterward. This operation may be repeated, leaving the first ball with $\frac{1}{4}$, $\frac{1}{8}$, $\frac{1}{16}$, . . . the initial quantity. By such manipulations, Coulomb was able to verify that the electrostatic force is proportional to the two quantities of charge.

Another factor affecting the magnitude of the force between charged bodies is the nature of the surrounding medium. The force between given charges a given distance apart is greatest when they are in a vacuum, becoming less if a material medium intervenes.

The experimental facts discussed above may be gathered and expressed in algebraic form by what has become known as **Coulomb's law:**

$$F = \frac{kQ_1Q_2}{r^2}$$

16.1

where F is the magnitude of the force that one charged object exerts on the other, Q_1 and Q_2 are the magnitudes of the two charges, r is the distance between the two objects, and k is the constant of proportionality whose numerical value depends on the nature of the medium and the units used for the several quantities. Coulomb's law, like the law of gravitation which it resembles in form, is valid only for *point charges* (point masses in the gravitational case). The sizes of the charged bodies are assumed to be small compared with the distances between them; otherwise, the computation of the forces requires complex summing operations by calculus methods, as in the corresponding gravitational examples.

With the analytical form determined, Coulomb's law can then be used to *define* units for measuring Q. Originally cgs units were used for F and r, and the value of k for empty space was taken to be 1. The resulting unit of charge was called one *electrostatic unit.* Other systems have been defined and are still in use. However, a scheme of electrical units based on the mks system seems best suited to practical needs and is the one used in this book.

The mks unit of charge quantity is defined not through the use of Coulomb's law but from a definition of electric current (Chap. 17). This unit of quantity is called one **coulomb,** abbreviated C. For the present, its value will be given in terms of the electron charge:

$$1\ C = 6.242 \times 10^{18}\ e \quad \text{or} \quad e = 1.602 \times 10^{-19}\ C$$

The order of size of the coulomb makes it well suited to electrical engineering but somewhat large for electrostatic problems. For example, the charge on a briskly rubbed rod may amount to only about a ten-millionth of a coulomb, whereas electrons equivalent to about half a coulomb pass through the filament of a 60-watt lamp each second. The charge on a storm cloud may be of the order of several coulombs.

Since the units for F, r, and now Q are defined in the mks system, the units for the constant k are fixed, but the numerical value must be determined by experiment. The value for free space (vacuum) is

$$k_0 = 8.987 \times 10^9\ \text{N-m}^2/\text{C}^2 \qquad\qquad\qquad 16.2$$

For rough computations, this may be rounded off to 9×10^9 in problems involving charges in vacuum or in air. The value of k for air is only 0.06 percent less than that of k_0.

Worked example 16.1

How does the magnitude of the electrostatic attraction of a proton for an electron compare with their mutual gravitational attraction?

Solution   The electrostatic force in free space is given by Coulomb's law as $F_E = k_0 e^2/r^2$, and the gravitational attraction is given by Newton's law as $F_G = G m_p m_e/r^2$, where r is the distance apart and m_p and m_e are the masses of proton and electron, respectively. The ratio of these forces is, by division, $F_E/F_G = k_0 e^2/G m_p m_e$. The distance factor has canceled out: The result is independent of the distance between the charges.

With $G = 6.67 \times 10^{-11}$ N-m^2/kg^2, $m_p = 1.67 \times 10^{-27}$ kg, $k_0 = 8.99 \times 10^9$ N-m^2/C^2, $m_e = 9.11 \times 10^{-31}$ kg, and $e = 1.60 \times 10^{-19}$ C, the result is $F_E/F_G = 2.37 \times 10^{39}$. Hence, at any given distance, the electrostatic force is enormously greater than the gravitational force.

Michael Faraday (1791–1867) was an English chemist and physicist and probably the greatest experimenter of his time. His contributions to the study of electricity are among the most fundamental in the entire subject.

16.5 THE ELECTROSTATIC FIELD

It was shown in Chap. 8 that an essential feature of Newton's concept of gravitation is the ability of bodies to exert gravitational forces across intervening space. This idea was always difficult to accept, and with the development of electrostatics and the investigation of magnetism, science was confronted with additional examples of apparent "action at a distance." It remained without explanation until early in the nineteenth century when Faraday proposed a model called the field concept. He pictured the interaction of magnets or charged bodies to reside in some way in the space surrounding these objects.

The space in the neighborhood of charged bodies, where electrostatic forces are detectable is called an **electrostatic field.** It is thought of as a condition produced in space by the presence of charged bodies. If another charged object is introduced, the force it experiences is ascribed to the field itself rather than to any direct action of the field-producing

Faraday. (*Courtesy of the Granger Collection.*)

charges. Thus the field is considered an intermediary for the transmission of electrostatic forces.

The presence of an electric field can be detected by bringing in a *test body* such as a small, charged ball suspended by a silk thread. Wherever it is placed, the test body is observed to experience a force of given magnitude and direction. In principle, the entire field may be mapped by drawing an appropriate force vector at each point (Fig. 16.3). In order to standardize the procedure, the test body used is a small object carrying a *positive* charge. The **electric field intensity** (or field strength) **E** at any point is defined as the resultant force per unit charge exerted on a positive test charge placed there. In symbols,

Operational description of an electric field.

$$\mathbf{E} = \frac{\mathbf{F}}{Q} \qquad\qquad 16.3$$

where **F** is the force experienced by the test charge of amount Q. In the mks system, the units of E are newtons per coulomb.

The act of bringing the test charge to the point in question must not in itself appreciably alter either direction or strength of the existing field, for a measurement is of doubtful value if the act of making it changes what is being measured. A field-strength determination can be made by measuring the force on a very small test charge of, say, $+1 \times 10^{-6}$ coulomb. Then the measured force can be multiplied by 10^{6} to get the field strength,

FIGURE 16.3 At each point in an electrostatic field a test charge experiences a force having a definite direction and magnitude. This is an example of a vector field.

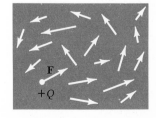

since Coulomb's law states that the force exerted on the test body by each bit of the field-producing charges is proportional to the amount of charge on the test body. In general, the force $\mathbf{F}$ experienced by a small charge Q at a place where the field intensity vector is $\mathbf{E}$ is given by Eq. 16.3 as

$$\mathbf{F} = Q\mathbf{E} \qquad\qquad\qquad 16.4$$

In addition to electrostatic fields, other examples of vector fields exist. The region near a massive body such as the earth was referred to as a gravitational field in Chap. 8. Any material body can serve as a test body for measuring the intensity of such a field, and the force experienced by it is called its weight. But the weight is expressed by $\mathbf{w} = m\mathbf{g}$, where m is the mass and $\mathbf{g}$ is the acceleration due to gravity. Comparison with Eq. 16.4 suggests that $\mathbf{g}$ represents the intensity of the gravitational field. Notice that the analog of Q (quantity of charge) is m ("quantity of matter").

Lines of force.

Faraday was not satisfied to think of electric and magnetic fields as mere mathematical abstractions and was led to formulate a more intuitive description. More than 200 years earlier, Gilbert had observed that iron filings align themselves in the direction of magnetic forces, and many experimenters had noted a similar behavior for small fibers in the neighborhood of electrified bodies. Faraday found it useful to think of "lines of electric force" permeating the field and even imagined them to be filaments having physical properties: tension in the direction of their lengths as well as a tendency to push sidewise on their neighbors. Although such attributes cannot be assumed to exist literally, the idea of field lines is mathematically sound. An electric field line (line of force) may be traced operationally by placing a small, positively charged body at any place in the field and moving it at every instant in the direction of the resultant electric force acting on it. The path of this test body will be a field line.

There is evidence that the French philosopher René Descartes (1596–1650) had earlier suggested the concept of lines of force.

In general, the field lines are curved, and the field vector $\mathbf{E}$ at any point is tangent to this curve. In the simple case of the field around an isolated point charge or charged conducting sphere, the field lines are straight, as shown in Fig. 16.4. Figure 16.4*a* represents a plane section of the field around a positive charge. Since a positive test charge placed anywhere in this region would tend to move radially outward from the central body, the field lines are straight lines running outward in all directions. Similarly, in Fig. 16.4*b*, the field lines around a central negative charge converge radially from all directions. Field lines may be imagined to originate on positive charges and terminate on negative charges.

Figure 16.5*a* shows the field produced by a pair of equal charges of opposite kind. The lines arising on the positive charge curve around and

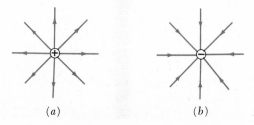

(a) (b)

FIGURE 16.4 Field lines arour a positive charge and around negative charge.

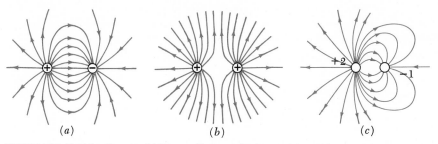

FIGURE 16.5 Field-line diagrams: (*a*) Two equal but opposite charges; (*b*) two equal, like charges; (*c*) two unequal charges of unlike kind.

into the negative charge. Think of the field as filled with a highly viscous liquid. If a positive test particle is released in the field at any point, it will follow the line of force that goes through that point and eventually reach the fixed negative charge. It will *not* move straight toward the negative charge nor straight away from the positive charge but always in some intermediate direction, that of the field vector at each place.

The field strength E at any place in the neighborhood of a single point charge (or small charged sphere) is readily determined by substituting the

Electric fields made visible as patterns of grass seed floating in an insulating liquid. In (*a*) the rods carry equal and opposite charges; in (*b*) the two charges are identical. Compare these patterns with Fig. 16.5, (*a*) and (*b*).

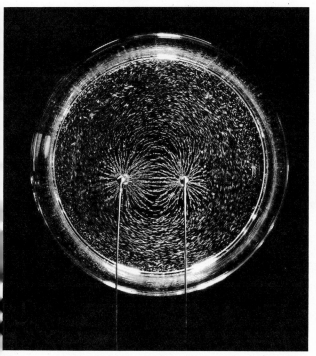

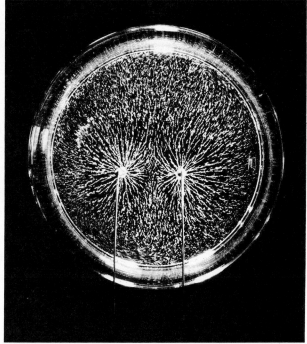

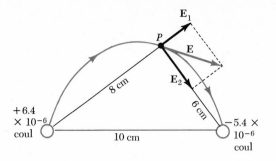

FIGURE 16.6   Calculating the field vector in the neighborhood of two point charges.

value of F from Coulomb's law (Eq. 16.1) into Eq. 16.3. If Q_1 is the field-producing charge, which is now called simply Q, and Q_2 is the unit test charge, the result is

$$E = \frac{kQ}{r^2} \qquad\qquad\qquad \textbf{16.5}$$

Worked example 16.2

Two small conducting spheres carrying charges of $+6.4 \times 10^{-6}$ and -5.4×10^{-6} coulomb, respectively, are 10 cm apart in air. Determine the field vector at a point 8.0 cm from the positive charge and 6.0 from the negative.

Solution   Calculate the field strength at the point P (Fig. 16.6) due to each of the fixed charges alone and then find the resultant. From Eq. 16.5, $E_1 = k_0(6.4 \times 10^{-6})/(0.08)^2 = 10^{-3}k_0$; $E_2 = k_0(5.4 \times 10^{-6})/(0.06)^2 = 1.5 \times 10^{-3}k_0$. Both E's are expressed in newtons per coulomb. Notice that k_0 has been carried along as a symbol; the numerical value need not be inserted until later.

Since the triangles in Fig. 16.6 are right triangles, the resultant field strength is given by $E = \sqrt{E_1{}^2 + E_2{}^2} = \sqrt{3.25 \times 10^{-6}k_0{}^2}$. Inserting the value $k_0 = 9 \times 10^9$ (page 378) and taking the square root, the result is $E = 1.6 \times 10^7$ N/C. The vectors have the directions shown in the figure. If desired, the exact position of **E** can be computed from the information given above (see Prob. 16.11).

The method used in this worked example is valid in principle for determining the field due to *any* distribution of charges. At each point a vector **E** must be computed for each bit of field-producing charge, after which the resultant of all these vectors must be found. For all but very simple charge distributions the details may become quite troublesome.

FIGURE 16.7   The field intensity at any point may be represented by the number of lines drawn per unit of cross-section area.

16.6   UNIFORM FIELD; THE MILLIKAN EXPERIMENT

It has been pointed out that the direction of the field vector at any place is that of the tangent to the corresponding field line, and it would seem that an unlimited number of lines could legitimately be drawn when mapping any field. However, by limiting the number, the field lines may be used to indicate not only the direction of the field at any point, but the intensity

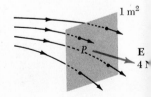

can be shown by the spacing of the lines. In a field-line diagram, *the number of lines drawn through each unit area of a surface placed normal to the field is made proportional to the field strength at that point.* Thus, in Fig. 16.7, a field of intensity of 4 N/C at the point P is represented by drawing four lines through the 1-m² area. The field drawings in Figs. 16.4 and 16.5 are seen to be consistent with this representation: The lines are closer together near the field-producing charges and farther apart at greater distances away.

A field whose lines are everywhere parallel and equally spaced is called a **uniform field.** This means that the field vectors at all points are identical. A uniform field may be produced by setting up two plane, parallel conducting plates carrying equal and opposite charges, with a uniform insulating medium filling the space between (Fig. 16.8). Except near the edges, the field between the plates is uniform. Such fields find practical application in electrical devices of many kinds.

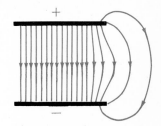

FIGURE 16.8 In a uniform electric field, the field lines are parallel and equally spaced. Except near the edges of the plates, the field vector is the same everywhere.

Robert Millikan 1868–1953), American physicist, was awarded a Nobel Prize for fundamental experiments on the photoelectric effect. He was also known for his investigation of cosmic rays.

A uniform electric field is the basic element in the method used by Millikan beginning about 1909 to measure the electron charge e. Droplets of oil from a spray (Fig. 16.9) are allowed to enter the space between a pair of parallel metal plates, where they are strongly lighted from one side and are observed in a low-powered microscope.

There is air at normal pressure in the space between the plates, and a drop of the size used—about a thousandth of a millimeter in diameter—reaches its terminal speed almost immediately, then continues to fall at a constant rate of the order of 1 mm/sec. The terminal speed of a sphere falling in a fluid is known to be related to the mass of the sphere in a given way, and so the mass of the drop can be determined by clocking it as it moves between reference marks in the field of view.

Now, by connecting a high-potential battery to the metal plates, a uniform electric field is set up in the space between them. As soon as the field is switched on, the drop changes its speed of fall abruptly. It may even start to rise. The reason is that the spraying process usually leaves such a drop with an electric charge and, depending on the sign of this charge and the direction of the field between the plates, the drop experiences an electrostatic force, either up or down. A valuable feature of Millikan's method is the possibility of altering the charge on a drop by allowing it to pick up an occasional ion from the air. Abundant ions may be produced for this purpose by the action of x rays on the air between the plates. A given drop can often be used for a whole series of observations, once its mass has been determined from the rate of fall in the absence of the electric field.

Calculations based on the observations make it possible to compute the charge on the drop in each case, and the remarkable result is that these charges are without exception exact *integral multiples* of the electron charge e.

FIGURE 16.9 Scheme of the Millikan oil-drop experiment for determining the electron charge.

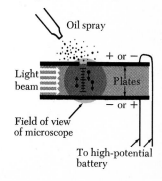

16.7 ELECTROSTATIC INDUCTION

As soon as an electric field is applied to a conductor, any freely movable carriers of charge in it are set in motion. To the extent that they are unrestrained, positive charges move in the direction of the field, and nega-

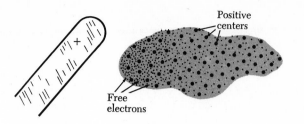

FIGURE 16.10 Induction effects a metallic conductor. The charge rod makes free electrons in th conductor gather in excess at on end.

tive charges move in the opposite direction. This brings about a separation of charges in an originally uncharged, isolated conductor.

In a metal, there are movable charge carriers of only one kind—the free electrons. Application of an external field makes them gather in excess near one side of the body, leaving a deficiency of electrons—a positive charge—near the opposite side (Fig. 16.10). This separation of charges gives rise to a field *inside* the body, and this field is opposite in direction to the external one. The charges distribute themselves to make the resultant field inside the conductor zero.

The above description holds not only for a solid conductor but for a hollow one as well. An external field cannot penetrate the interior of a metal enclosure, and this fact is used in protecting delicate electrical apparatus from disturbances caused by outside fields. Merely surrounding the device with a wire mesh is often sufficient. Faraday constructed a large box covered with sheet metal, placed it on insulating blocks, and connected a powerful electrostatic generator to the metal coating. He described his experiences in this way:

I went into the cube and lived in it, and using lighted candles, electrometers and all other tests of electrical states, I could not find the least influence upon them . . . though all the time the outside of the cube was powerfully charged, and large sparks and brushes were darting off from every part of its outer surface.

Polarization of a dielectric. When a *nonconducting* (insulating) body is placed in an electric field, slight shifting or alignment of the charges making up the molecules can take place, although there are no free electrons or other readily movable charges in such a material. If the normal charge distribution of the molecules is symmetrical, they are said to be **nonpolar** (the oxygen molecule

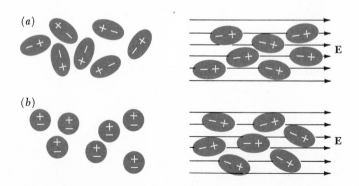

FIGURE 16.11 Polarization e fects in dielectrics. In (a), applica tion of the field makes the pola molecules line up in the genera direction of the field. In (b), initial nonpolar molecules become pola ized in the field direction. Therm motions prevent the alignment fro becoming complete.

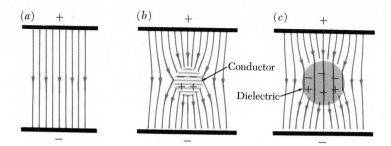

FIGURE 16.12 Effect of introducing into a uniform field (*a*) a conducting sphere shown in (*b*) and an insulating sphere as in (*c*). Inside the dielectric the field lines are reduced in number, but inside the conductor they are absent altogether. The diagrams show that the external field is distorted.

is an example). If it is not symmetrical, the molecules are **polar** (the water molecule is an example). When an external field is applied, the polar molecules, usually oriented at random, are lined up to some extent. The nonpolar molecules become polarized in the direction of the field and tend to remain lined up in this direction (Fig. 16.11). In either case, the net effect is to leave opposite ends of the body oppositely charged, while the interior has equal amounts of the two kinds of charge and so is electrically neutral. The result is again a distortion of the external field near the body. In an insulator, the internal field is weaker than the external one (Fig. 16.12) but is not zero, as it would be within a conductor.

The appearance of opposite charges on the ends of an isolated body placed in an electric field is called **electrostatic induction.** The existence of these charges may be demonstrated directly by placing a charged rod near one side of a pair of insulated, uncharged metal conductors that touch each other (Fig. 16.13*a*). If the conductors are then separated while the charged rod remains nearby, each conductor is left with a resultant charge. The object on the right side is found to have the same kind of charge as the inducing rod, and the one on the left has the opposite kind.

Electrostatic induction explains the attraction of uncharged bodies for charged ones of either sign (page 372). The field near a charged body is usually far from homogeneous, its strength decreasing with distance from the body. For example, the induced negative charge on the near end of the object in Fig. 16.14 is in a slightly stronger field than the more remote positive induced charge. Although the body as a whole is neutral, the net effect is a force urging it toward the charged rod. The result is the same when the inducing charge is negative and does not depend on whether the neutral body is a conductor or an insulator. It should be noted that in a *uniform* field such forces are absent, as shown by the symmetry of the field-line patterns in Fig. 16.12*b* and *c*.

Any excess charge placed on or within an initially neutral conducting body goes entirely to the surface of the body, where it distributes itself in such a way that each bit of charge is in static equilibrium. Qualitatively,

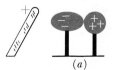

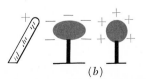

FIGURE 16.13 Separating the two parts of the conducting system while the charged rod is nearby leaves each conductor with a resultant charge.

(a)   *(b)*

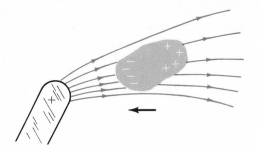

FIGURE 16.14 Explaining the attraction of an uncharged object for a charged one of either sign.

this is what would be expected, since mutual repulsion makes each charge carrier tend to get as far as possible from the others. The absence of interior charge can be verified experimentally to a high order of accuracy, and it can be proved that this is a confirmation of the inverse-square law (see page 377).

Faraday's "ice-pail experiment" was an extension of his earlier experiment with the shielded box (page 384).

In 1843, Faraday investigated quantitatively the charge obtained by induction in what has become known as his "ice-pail experiment." In this experiment a metal ball carrying, say, a positive charge is introduced into an insulated, uncharged hollow conductor through a small opening (Fig. 16.15). The field of the charged ball induces negative charge on the inside wall of the hollow conductor, leaving a positive charge on the outside. An electroscope (page 372) connected to the outside surface immediately indicates this positive charge, which remains constant even if the charged ball is moved around inside the cavity or even if it touches the inner wall. When the ball is removed it is found to be uncharged, but the electroscope still indicates the same positive charge as before. What conclusions can be drawn?

Virtually all the field lines originating on the ball when it is in the cavity terminate on the inside surface of the container, suggesting that the total charge there is equal to the charge on the ball. Since the container was

FIGURE 16.15   The Faraday ice-pail experiment (schematic).

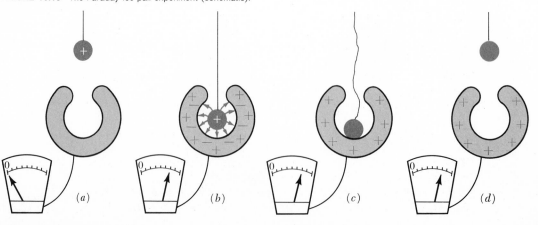

(a)   (b)   (c)   (d)

originally neutral, the appearance of a negative charge on its inner surface must leave an equal positive charge on the outer surface. But when the ball touched the inner surface, the electroscope indication did not change, showing that there was no change in the charge on the exterior.

The ice-pail experiment shows that (1) the total amount of induced charge (of either sign) is equal to the amount of the inducing charge and (2) extra charge applied to a conductor goes entirely to the outside surface, leaving no net charge within. In connection with result 1, it has already been remarked that field lines may be assumed to originate on positive charges and terminate on negative. The lines going outward from an isolated positive-charged ball terminate on surrounding objects such as the earth or the walls of the laboratory. Wherever such lines end, charges of opposite kind are induced, just as in the ice-pail experiment. Result 2, as we have seen, can be interpreted as a confirmation of the inverse-square factor in Coulomb's law.

16.8 ELECTRIC POTENTIAL

The study of mechanics in an earlier part of this book showed that Newton's laws constitute, in principle, an adequate basis for investigating any problem in classical mechanics. It was brought out, however, that a reformulation of these laws by introducing the energy concept often provides a simpler and more direct approach. A similar development took place in the study of electricity, where the energy concept again proved its value. This is plausible because electrical effects are usually detected and measured by the mechanical forces associated with them. Further, the similarity in form between the law of gravitation and Coulomb's law has already been noted.

Up to this point, electrostatic fields have been described by using the concept of field intensity. This is analogous to using the force concept in solving problems in mechanics. Now we introduce into the discussion of electricity a concept called **electrostatic potential,** which is based on energy considerations. This offers a simpler way of describing the phenomena than does the field concept.

The potential may be described intuitively as a measure of the tendency of a charged body to move from one place to another in an electrostatic field. This is not meant to be a strictly operational definition but such a definition will be developed below. If a positive test charge is released at any point P_1 in an electrostatic field, any other point P_2 to which the field forces may move it is said to be at a lower potential than P_1. In particular, P_1 and P_2 may be two points on a wire or other conductor, in which case the *difference in potential* between the two points may be considered a measure of the agency that maintains an electric current in that part of the conductor.

There is a close similarity between electrostatic potential and the gravitational potential energy of an elevated weight. If a small stone is released near a large body such as the earth, the gravitational field makes the stone fall. On any part of its path, the stone moves from a position where it had greater gravitational potential energy to one where it has less. If the stone falls freely, the difference in GPE appears as an equivalent amount of KE.

The work needed to lift the stone slowly from a lower to a higher position is equal to the difference in its GPE in the two locations.

It was also shown in Chap. 9 that at all points the same distance from the center of the earth (that is, at the same level), a stone has the same GPE. The work done in raising a weight from one of these level surfaces to any other depends only on the location of the two surfaces with respect to the center of the earth. The work is determined solely by the *vertical* distance between the level surfaces in which the initial and final locations lie (page 201).

In most problems involving either electric or gravitational potential, absolute values are not of interest, and only a knowledge of the *difference* of potential between the two positions is needed. Sea level has already been mentioned (Chap. 9) as a convenient level surface from which to measure GPE. For electricity, when a reference level is required, the earth as a whole may conveniently be taken to have zero potential. A body is at a *positive* electrostatic potential if, when it is grounded, electrons *flow up to it* from the earth and at a *negative* potential if electrons *flow from it* to the earth when a ground connection is made.

In the preceding sections of this chapter the *force per unit charge* was chosen as the measure of the strength of an electric field at any point. Similarly, in changing to the energy approach, we find it convenient to define a measure based on the electric *energy per unit charge* of a test body at a given place in the field. This quantity has been referred to descriptively above as the electrostatic potential.

The potential at any point in an electric field is defined as the electric PE of a positive test charge located at that point, divided by the amount of the charge. In symbols, if V represents the potential, Q the quantity of charge, and W its electric PE, then

$$V = \frac{W}{Q}$$

16.6

In the mks system, W is expressed in joules and Q in coulombs, so that V has the units *joules per coulomb*. A potential of 1 joule per coulomb is called 1 **volt,** after the Italian scientist Alessandro Volta (1745–1827).

Notice that, whereas the field strength **E** is a vector, the potential V is a scalar and so can be dealt with more easily. Thus, if several sources contribute to the electric field at any point, the field vector due to the presence of each must be found and the vectors combine into a single resultant. By contrast, to find the potential at that point we need only add a set of *numbers,* attaching plus signs to the contributions from positive field-producing charges and minus signs to those from negative charges. By using the potential rather than the field intensity, we have replaced a vector field by a scalar field which can provide a great deal of information in a simpler way.

There is a direct relation between the potential and the field intensity, which will now be developed. In Fig. 16.16, a body carrying a charge $+Q$ is moving along any path MN under the influence of an electric field. **E** is the field vector at any point P, and it is tangent to the field line through that point. The force acting on the charge $+Q$ is given by Eq. 16.4 (page

Besides his experiments in electrostatics, Volta invented the chemical electric battery. See Ref. 16.2.

Potential and field strength.

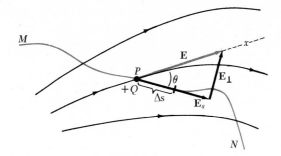

380) as $F = QE$. Now consider a very small displacement Δs of the charged body along the path MN. According to Eq. 9.6 (page 197), the work done by the field in this displacement is given by

$$\Delta W = F \, \Delta s \cos \theta = QE \, \Delta s \cos \theta$$

But $E \cos \theta$ is E_s, the component of **E** in the direction of Δs, so that the above equation becomes

$$\Delta W = QE_s \, \Delta s$$

A body moving in the direction of the field *loses* PE, as in the case of a stone falling in the earth's gravitational field, and a minus sign must be used in Eq. 16.6. If this equation is written for small changes in V and W, it becomes

$$\Delta V = -\frac{\Delta W}{Q}$$

Finally, the result of combining this with the previous equation is

$$E_s = -\frac{\Delta V}{\Delta s} \qquad\qquad \textbf{16.7}$$

The quantity $\Delta V/\Delta s$ represents the rate at which the potential changes in moving a short distance in the direction s. It is called the **potential gradient** in that direction. Equation 16.7 asserts that at any point in an electric field the component of the field vector in any direction is equal to the negative of the potential gradient in that direction. This means that the components of **E** at any point can be computed, provided that the values of V in the neighborhood of that point are known.

Compare the concept of temperature gradient, introduced on page 255.

If the mks system is used, the right side of Eq. 16.7 must have the units *volt per meter*. However, from its definition, E is measured in *newtons per coulomb*. These two combinations of units are equivalent, for

$$\frac{\text{volt}}{\text{m}} = \frac{\text{joules/coulomb}}{\text{m}} = \frac{\text{joules}}{\text{m-coulomb}} = \frac{\text{newton-m}}{\text{m-coulomb}} = \frac{\text{newtons}}{\text{coulomb}}$$

Thus, electric field strengths may be expressed either as *difference of potential per unit distance* or as *force per unit charge*. In air, a disruptive breakdown (spark, lightning) is apt to occur when the potential gradient approaches 3×10^6 volts/m.

Practice set*

In moving 1 mm eastward from a certain point in an electric field, the potential increases by 0.04 volt.

1. Is the east-west component of the field directed eastward or westward?
2. How strong is it?
3. What average force (newtons) acts on a 2×10^{-6} coulomb charge located on this 1-mm segment?

16.9 THE POTENTIAL FIELD

In general, the calculation of V in the neighborhood of an arbitrary distribution of charge requires calculus methods. It is indicated here how such a calculation can be carried out for the simple case of a single point charge. The result holds also for a uniformly charged conducting sphere, since the field is radial for all points outside such a sphere (Fig. 16.4, page 380) and therefore is identical with that of a point charge located at the center.

Consider two points P and P' at distances r and r' from a positive point charge Q (Fig. 16.17). We wish to find the difference in potential between these two locations by computing the work done by the field on a test charge in going from P to P'. To do this, think of the distance PP' divided into a large number n of very short segments $\Delta r_1, \Delta r_2, \Delta r_3, \ldots, \Delta r_n$. Using Eq. 16.7, the difference of potential across the first segment is $\Delta v_1 = -E_1 \Delta r_1$ or, substituting from Eq. 16.5, $\Delta V_1 = -kQ \Delta r_1/r_1{}^2$. Similar expressions hold for ΔV_2, ΔV_3, etc., and the total difference in potential between P and P' is the sum of these expressions:

$$V_{PP'} = -kQ\left(\frac{\Delta r_1}{r_1{}^2} + \frac{\Delta r_2}{r_2{}^2} + \cdots + \frac{\Delta r_n}{r_n{}^2}\right)$$

The sum of the quantities in parentheses approaches a finite limit as n becomes very large, but the value must be found by calculus methods. The result is $1/r' - 1/r$; hence the preceding equation reduces to

$$V_{PP'} = -kQ\left(\frac{1}{r'} - \frac{1}{r}\right) = kQ\left(\frac{1}{r} - \frac{1}{r'}\right) \qquad \text{16.8}$$

Next, suppose that P' is at an infinite distance from Q. This makes r' infinite, and the second fraction in parentheses becomes zero. Then, according to the above equation, the potential difference between the point P and a point infinitely far away becomes

$$V_P = \frac{kQ}{r} \qquad \text{16.9}$$

Notice that the reduction of Eq. 16.8 to the form 16.9 amounts to taking the zero of potential to be at an infinitely distant point rather than at the

Ans.: west; 40 volts/m; 8×10^{-5} newton.

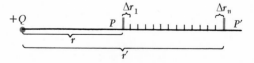

FIGURE 16.17 Calculating potential difference for the field produced by a point charge.

surface of the earth. The former proves to be a convenient reference point for problems of this kind. Thus V_P represents the amount of work done in moving a unit positive charge from an infinite distance to the point P, the work being done against the force of repulsion between this test charge and Q.

As mentioned above, the central charge may be considered to be on the surface of a conducting sphere rather than at a single point. If the radius of this sphere is a, the potential at the surface is given by Eq. 16.9 as

$$V_a = \frac{kQ}{a} \qquad \qquad \textbf{16.10}$$

All portions of the surface of this sphere are at the same potential; in fact, all points on the surface of *any* conductor must be at the same potential. This follows from the fact that, if a conductor is in electrostatic equilibrium, there is no flow of charge along it and so there can be no potential difference between any two points on it.

Whether it forms the actual boundary of a conductor or not, any surface drawn through all the points in a field that are at the same potential is called an **equipotential** surface. The equipotential surfaces of the field of a point charge (or small, charged conducting sphere) are themselves spheres around the common center. They correspond to the level surfaces in the gravitational field of the earth. At every point, the lines of force are perpendicular to the level surfaces. This follows because no work is done in moving a test charge from one place to another on the same equipotential surface, which means that the field vector has no component along the surface and therefore must be perpendicular to it. Examine Fig. 16.18.

The way in which charge distributes itself over the surface of a conductor is determined by the surface contours. Since each bit of charge in effect tends to get as far as possible from every other bit, the charge will concentrate at places where the surface is more strongly curved, and especially where there are projections or sharp points. Accordingly, the field lines leaving the surface at such points will be more concentrated than elsewhere, and the equipotential surfaces in the neighborhood will be closer together (Fig. 16.19). The truth of the last statement can be seen from Eq. 16.7: Moving outward from the surface of the conductor along a field

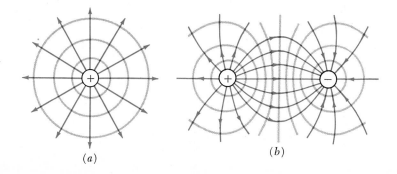

(a)

(b)

FIGURE 16.18 Equipotential surfaces and field lines (a) for an isolated, positive-charged sphere and (b) for two equal, oppositely charged spheres. The field lines are shown in color, and the equipotentials in gray.

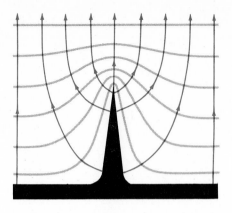

FIGURE 16.19 Equipotentials (gray lines) and field lines (color) near a sharp point on a charged conductor.

line, the potential changes more rapidly with distance at places where the field is stronger.

If the field strength at the surface is great enough, ionization of the surrounding air may take place, resulting in the discharge of the conductor. Such **corona discharge** often produces a visible glow or, if the field is great enough, a spark results. The lightning rod, invented by Franklin, uses the discharging effect of a pointed conductor connected to the ground to dispose of the charge on a cloud, preventing dangerous flashover to a structure.

Worked example 16.3

Two flat metal plates are placed horizontally 7 mm apart in air. The upper one is charged to a potential of 35 volts and the lower one is grounded.

a. Describe the form of the equipotential surfaces in the space between the plates.

b. Compute the field intensity there and give its direction.

c. Find the amount of work done in moving a small test charge of -2×10^{-6} coulomb a distance of 10 cm in a horizontal plane midway between the plates.

d. Find the distance of the 25-volt equipotential surface above the lower plate.

Solution   *a.* Except at the edges, the field is uniform and its lines are vertical (Fig. 16.12*a*, page 385). The equipotential surfaces, which are everywhere perpendicular to these lines, are horizontal planes.

b. The potential increases uniformly by 35 volts over a distance of 7 mm. By Eq. 16.7, the magnitude of the field intensity is $35/(7 \times 10^{-3}) = 5{,}000$ volts/m. The direction is downward, the direction of decreasing potential.

c. A horizontal path lies entirely in a single equipotential surface, and so no work is done in such a displacement.

d. Since the potential increases by 5 volts for each millimeter of distance above the grounded plate, the 25-volt surface will be located at a height of 5 mm.

16.10 KE ACQUIRED BY A CHARGED BODY IN A FIELD

A stone allowed to fall freely in a gravitational field acquires kinetic energy in exchange for an equivalent amount of gravitational potential energy (Chap. 9). The same is true for a charged body that falls freely in an electrostatic field. Suppose a small object carrying a charge of magnitude Q moves freely, under the influence of a field, between two places whose difference of potential is ΔV. If Q is positive, the electric force is in the direction of the field; if it is negative, the force is in the opposite direction. In either case, according to the energy principle, the KE of the body will change by an amount ΔK such that

$$\Delta K = Q \, \Delta V \qquad\qquad \textbf{16.11}$$

where ΔK may be measured in joules, Q in coulombs, and ΔV in volts.

The above method of imparting energy to charged particles is used in many atomic and nuclear experiments. The charged bodies used for such purposes are primary particles such as electrons, protons, ions, etc., produced in large numbers and then allowed to fall in a vacuum through potential differences amounting in some cases to tens of millions of volts. Since the particles carry charges that are multiples of the electron charge e, it is convenient to define a special unit for expressing the energy by using e rather than the coulomb as a measure of the charge. The new unit is called one **electron volt,** abbreviated eV. One electron volt is equivalent to the work done by the electric field when any object carrying a charge of amount e moves through a difference of potential of one volt.

$$1\,\text{eV} = (e)(1\,\text{volt}) = (1.60 \times 10^{-19}\,\text{coulomb})(1\,\text{volt}) = 1.60 \times 10^{-19}\,\text{C-volt}$$

According to page 389, ''coulomb-volts'' are equivalent to newton-meters, or joules, so that

$$1\ \text{eV} = 1.60 \times 10^{-19}\ \text{joule}$$

Large multiples of the electron volt are used in nuclear physics:

GeV is the abbreviation for giga-electron volt (see Table A.2, page 631).

10^3 eV = 1 kilo-electron volt (keV)
10^6 eV = 1 mega-electron volt (MeV)
10^9 eV = 1 billion electron volts (GeV)

These abbreviations are read by pronouncing each of the letters in turn.

The KE imparted to a freely movable charged body by a field depends only on the amount of its charge and the difference of potential through which it falls, as expressed by Eq. 16.11. The KE does not depend on the mass of the body, but the speed acquired does. Thus, if $\frac{1}{2}mv^2$ is put for KE in Eq. 16.11, it becomes

$$\tfrac{1}{2}m(v_2{}^2 - v_1{}^2) = Q \, \Delta V$$

where v_1 and v_2 are the speeds at the two ends of the path over which the potential difference ΔV was applied. If the particle starts from rest, its final speed in the field is given by $\frac{1}{2}mv^2 = Q \, \Delta V$, or

$$v = \sqrt{\frac{2Q \, \Delta V}{m}} \qquad\qquad \textbf{16.12}$$

This relation is strictly valid only for low speeds, where the relativistic mass increase is negligible. Some of the particles used in nuclear physics are accelerated to speeds very close to c, and in such cases the relativity formulas must be used.

Practice set* A body carrying a charge $+4e$ falls freely through a potential difference of 80 volts.

1. Express its increase in KE, in electron volts.
2. The same for a body of twice the mass and half the charge.
3. The same for the original body carrying a charge of $-4e$.

16.11 ELECTROSTATIC GENERATORS

Continuously operating devices for the separation of electric charges by friction were invented as early as the seventeenth century. "Influence machines" operating by electrostatic induction were built toward the end of the eighteenth century, and electrostatic generators of this kind are still demonstrated in physics classes and science museums today.

A modern form of such devices was developed by the American physicist R. Van de Graaff (1901–1967). It makes use of the principle of the ice-pail experiment (page 386): Charges applied to the inside of a hollow conductor move entirely to the outer surface, no matter how high the potential of this conductor may be. Figure 16.20 shows a cross-section of the Van de Graaff generator. A large hollow metal shell stands on a tubular insulating support column. A fabric belt runs between a motor-driven pulley at the bottom of the column and a roller inside the shell. Charge supplied by a source at the base leaks off a set of pointed rods, sprays onto the belt, and is carried up into the shell. There, another set of points allows the charge to leak off and pass to the outer surface of the shell. As charge

*Ans.: 320; 160; 320.

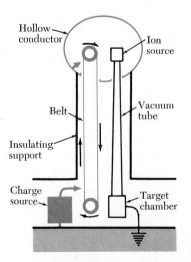

FIGURE 16.20 Schematic diagram of a Van de Graaff electrostatic generator. Ions or other charged particles accelerated inside the vacuum tube strike the target, where they can produce nuclear changes.

accumulates on the shell, its potential rises to a value limited only by leakage along the supporting column and through the surrounding air.

If the generator is to be used to accelerate atomic particles, a long vacuum tube running from the shell to the ground is installed. The particles are liberated in a source at the upper end and fall through the existing potential difference toward a grounded target at the lower end of the vacuum tube. In this way it is possible, in a single operation, to impart energies of the order of 10 MeV to all varieties of atomic particles, producing an intense beam of particles having extremely constant KE.

The tandem Van de Graaff accelerator is a more recent modification of the instrument in which the high potential difference is used to accelerate particles twice by changing the sign of the charge on the particles half way through the process (see Ref. 16.6).

Another product of such an installation is an x-ray beam capable of penetrating several inches of steel.

16.12 CAPACITANCE; DIELECTRIC COEFFICIENT

How much charge can be put on a conductor? In principle, there is no limit as long as the conductor is adequately insulated from its surroundings. Putting charge on a conductor is not like pouring water into a tank; it is more like pumping in air. A tank has a definite capacity for water but not for air. If the tank is strong enough, more air can be forced into it by increasing the applied pressure.

Imagine an isolated conductor which is to be charged by bringing up small bits of charge, one after another, from an infinite distance. The process becomes increasingly difficult with each operation, since each succeeding bit must be brought in against the repulsion of the charge already on the conductor. Another way of saying this is that the potential of the conductor increases as the amount of charge on it builds up.

It is found by experiment that the increase in potential is directly proportional to the increase in the amount of charge. The ratio of the quantity of charge on the conductor to the potential to which this charge has raised it is constant for a given conductor. This ratio is called the **capacitance** of the conductor. In symbols,

$$C = \frac{Q}{V} \qquad\qquad \textbf{16.13}$$

where C is the capacitance, Q the quantity of charge, and V the potential. The value of C for an isolated conductor depends only on its size and shape and on the nature of the surrounding medium. It is independent of the amount of charge present.

In mks units, C is measured in coulombs per volt (see Eq. 16.13). This combination of units has been given a special name, the **farad.** Thus a conductor is said to have a capacitance of 1 farad if the application of 1 coulomb of charge raises its potential by 1 volt. In practice, the farad proves to be inconveniently large in size, and submultiples are frequently used (see Table A.2, page 631).

1 microfarad (μf) = 10^{-6} farad
1 picofarad (pf) = 10^{-12} farad

Some of the circuit elements used in electrical communications devices have capacitances of the order of only a few picofarads.

For an isolated spherical conductor of radius a, the potential is given by Eq. 16.10 as $V_a = kQ/a$. On substituting this value into Eq. 16.13, the capacitance of the sphere is seen to be

$$C_a = \frac{a}{k} \qquad\qquad \textbf{16.14}$$

Capacitors.

Up to this point, the discussion of capacitance has dealt only with single, isolated conductors. In reality there can never be such a situation, for the field lines from a charged conductor M (Fig. 16.21) must end somewhere. The discussion on page 390 assumed the existence of a conductor an infinite distance away whose potential is taken to be zero. Suppose that this distant conductor N is now brought close to the original conductor, being kept at ground potential all the while. The charge on M will then induce a charge of opposite sign on N. The resultant charge on the latter (which it gets via the ground connection) will be equal in magnitude to that on M, provided that all the lines originating on M end on N. It is now easier to bring up additional positive charge to M, thus increasing the capacitance of M, because of the induced negative charge on the grounded conductor N.

Until recently the device now termed a capacitor was often called a condenser, for historical reasons.

A pair of conductors in close proximity, carrying equal and opposite charges, constitutes a **capacitor.** In order to increase the inductive effect—and hence the capacitance of the combination—the two conductors usually have the form of metal sheets or plates of large area, placed close together. A capacitor may be given its charge by joining the plates to the terminals of a battery or by connecting it into a suitable circuit. In Franklin's kite experiment he succeeded in charging a capacitor (''Leyden jar'') by connecting one plate to the conducting kite string, conclusively identifying lightning as an electrical phenomenon (see Ref. 16.4). Capacitors are indispensable elements in technical applications of electricity. They serve to reduce voltage variations, generate electromagnetic waves, provide time delays in circuits, store electromagnetic energy, etc.

Spherical capacitor.

The increase in capacitance resulting from the presence of a second, grounded conductor can be illustrated by calculating the capacitance of a system consisting of two concentric conducting spherical shells separated by an insulating medium, the outer shell being grounded (Fig. 16.22).

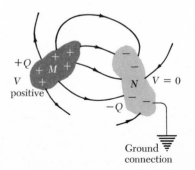

FIGURE 16.21 The combination of an insulated charged conductor and a nearby grounded conductor constitutes a capacitor.

If a charge, say $+Q$, is put on the inner sphere, electrons will flow onto the outer sphere through the ground connection until a total amount $-Q$ has accumulated on the outer sphere. The field between the spheres is then the same as it would be in the absence of the outer sphere. In fact, it is the same as it would be if the charge $+Q$ were concentrated at the center.

According to Eq. 16.10, the potential of the inner sphere is given by $V_a = kQ/a$ and that of the outer sphere by $V_b = kQ/b$. The difference in potential between the two conductors is

$$\Delta V = kQ\left(\frac{1}{a} - \frac{1}{b}\right) = kQ\frac{b - a}{ab} \qquad \text{16.15}$$

Then, from Eq. 16.13, it follows that the capacitance of this system is

$$C_\odot = \frac{Q}{\Delta V} = \frac{1}{k}\frac{ab}{b - a} = \frac{ab}{kd} \qquad \text{16.16}$$

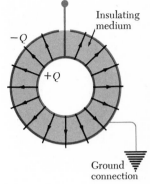

FIGURE 16.22 A spherical capacitor (schematic).

where d is the distance of separation of the two spheres. It should be noted that here again C is independent of Q, which has canceled. Further, Eq. 16.16 shows that the capacitance has been increased by the presence of the grounded outer sphere. This can be seen by dividing Eq. 16.16 by Eq. 16.14, with the result $C_\odot/C_a = b/d$. Since b can be made large compared with d, $C_\odot$ can be very much greater than C_a.

In practice, a form of device more common than the spherical capacitor is the **parallel-plate capacitor.** As its name implies, it consists of two parallel conducting plates separated by a relatively thin layer of insulating material or a vacuum. If one plate is charged and the other grounded, an equal and opposite charge is induced on the other plate, just as in the case of the spherical capacitor. Here, however, the field between the plates, except for slight departures at the edges, is uniform.

Parallel-plate capacitor.

An expression for the capacitance of a parallel-plate capacitor can be derived from Eq. 16.16 by using the fact that a limited portion of a large sphere is a good approximation to a plane. If a and b are both large in comparison with their difference $d = b - a$, then ab in Eq. 16.16 can be replaced by its approximate equal, a^2, and so the equation becomes $C_\odot = a^2/kd$. But the area of a sphere is given by $S = 4\pi a^2$. Eliminating a^2 between the last two expressions leaves $C_\odot = S/4\pi kd$. Because of the symmetry of the system, the value of C for any limited portion of the spheres is given by replacing S by the area A of the portion in question. Therefore, the capacitance of a plane-parallel capacitor is

$$C_\parallel = \frac{A}{4\pi kd} \qquad \text{16.17}$$

where A is the area of either plate and d the separation of the two, which is assumed to be small compared with the plate dimensions.

Effect of introducing a dielectric.

Figure 16.23 shows, schematically, a charged parallel-plate capacitor with an electroscope connected across the plates for the purpose of indicating the difference of potential between them. If a slab of insulating material such as glass or plastic is inserted between the plates, the potential difference is observed to fall. If the slab is removed, the electroscope shows that the potential difference returns to its original value.

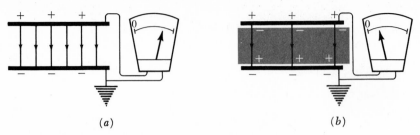

(a) (b)

FIGURE 16.23 Insertion of a slab of insulating material between the plates of a charged capacitor lowers the potential difference.

These observations can be understood on the basis of polarization effects in dielectric materials, discussed in Sec. 16.7. The charged plates induce charge of opposite sign on the adjoining faces of the dielectric slab, and these surface charges effectively weaken the field between the plates and so reduce the potential difference. The capacitance of the system is always given by $C = Q/V$, where Q is the actual charge on either plate and V is the difference in potential. Since Q is not altered by the insertion of the dielectric, a reduction of V means a consequent increase in C. If C_0 is the capacitance with a vacuum between the plates and C_D is the capacitance when the space is filled with a material dielectric, then the **dielectric coefficient** of the insulating material is defined as $K = C_D/C_0$.

For any system, the expression for the capacitance always has the factor k in the denominator, where k is the constant of proportionality in Coulomb's law. In all situations where material dielectrics are present, k replaces the vacuum constant k_0. Then we may write

$$K = \frac{k_0}{k} \qquad\qquad \textbf{16.18}$$

and $K = 1$ for a vacuum. The values for all material dielectrics are greater than 1, as indicated in Table 16.1. For practical purposes, K for air may be taken as 1. Notice that K is a pure number.

For convenience in computations, K can now be introduced into some previous relations where k appeared. From Eq. 16.18,

$$\frac{1}{k} = \frac{K}{k_0}$$

TABLE 16.1   DIELECTRIC COEFFICIENTS FOR SOME INSULATING MATERIALS

SUBSTANCE	K	SUBSTANCE	K
Vacuum	1.	Ethyl alcohol (0°C)	28.4
Air (1 atm)	1.0006	Glycerin (15°C)	56
Glass	5–10	Petroleum	2
Polystyrene	2.6	Water (pure)	80
Porcelain	6.5	Barium titanate	1,200

or, with the numerical value $k_0 = 9 \times 10^9$ N-m²/C² from page 378, this becomes

$$\frac{1}{k} = K(1.11 \times 10^{-10}) \text{ C}^2/\text{N-m}^2 \qquad\qquad \textbf{16.19}$$

With this value substituted into Eq. 16.16 it takes the form

$$C_\odot = \frac{abK}{d}(1.11 \times 10^{-10}) \text{ farad} \qquad\qquad \textbf{16.20}$$

and Eq. 16.17 becomes

$$C_\parallel = \frac{AK}{4\pi d}(1.11 \times 10^{-10}) \text{ farad} \qquad\qquad \textbf{16.21}$$

Worked example 16.4   Each plate of a parallel-plate capacitor has an area of 0.1 m², and the distance of separation is 0.2 mm. The space between the plates is filled with an insulating material of dielectric coefficient $K = 3$. One plate of the device is grounded, and a charge of $+6 \times 10^{-7}$ coulomb is placed on the other.

1. a. What is the capacitance?
   b. What difference of potential exists between the plates?
2. The ground connection is now removed and the dielectric material is withdrawn from between the plates.
   a. Find the charge on each plate.
   b. Find the potential difference between them.

Solution   1. a. By Eq. 16.21,

$$C = \frac{AK(1.11 \times 10^{-10})}{4\pi d} = \frac{(0.1)(3)(1.11 \times 10^{-10})}{4(3.14)(2 \times 10^{-4})} = 1.32 \times 10^{-8} \text{ farad}$$

b. $\Delta V = \dfrac{Q}{C} = \dfrac{6 \times 10^{-7}}{1.32 \times 10^{-8}} = 45.5$ volts

2. a. Q remains the same as before, 6×10^{-7} coulomb.
   b. Since C is reduced by a factor of 3 when the dielectric is removed, ΔV will become three times as great as before, or 136.5 volts.

Programmed review

Instructions: See page 18.

1. State the principle of the conservation of charge.

   The net charge of an isolated system remains constant.   [16.2]

2. Write Coulomb's law in algebraic form, and state what each symbol represents.

   $F = kQ_1Q_2/r^2$, where Q_1, Q_2 are two point charges, r is the distance between them, and k is a proportionality constant.   [16.4]

3. Define, in words, the intensity of an electric field at any given point.

   The resultant force, per unit charge, exerted on a small, positive test charge placed there.   [16.5]

4. Besides yielding a value for the electron charge, what did the Millikan oil-drop experiment show about the properties of electric charge?

   That all charge is quantized; every charge is an integral multiple of the fundamental electron charge.   [16.6]

5. What is the name given to the temporary separation of charge produced on an insulated body when a charged body is brought near it?

   Electrostatic induction.   [16.7]

6. Give an exact definition of the electrostatic potential at any point in a field, and name the mks unit for measuring it.

   The electric PE of a positive test charge located at the point in question, divided by the amount of its charge. The unit is 1 volt.   [16.8]

7. What is meant by saying that the KE of a particle is 1 eV?

   That it has a KE equal to that acquired by a body bearing a charge of magnitude e after moving freely through a potential difference of 1 volt.   [16.10]

8. Define the capacitance of a conductor or system of conductors.

   The constant ratio of charge to potential: $C = Q/V$.   [16.12]

9. State a formula for the capacitance of an isolated charged sphere of radius a surrounded by a medium whose Coulomb-law constant is k.

   $C_a = a/k$.   [16.12]

10. What is meant by the dielectric coefficient of an insulating substance?

    The ratio of the capacitance of any given capacitor with the material filling the space between the plates to the capacitance when there is a vacuum between them: $K = C_D/C_0$.   [16.12]

For further reading

16.1. *Holton, G., and D. H. D. Roller* "Foundations of Modern Physical Science." Read chaps. 26 and 27.
16.2. *Taylor, L. W.* "Physics, the Pioneer Science." Read chap. 40. This and the preceding reference are two excellent treatments of the historical evolution of electrostatics.
16.3. *Magie, W. F.* "A Source Book in Physics." Read pp. 400–420. Extracts from the writings of Coulomb and Franklin.
16.4. *Cohen, I. B.* (ed.) "Benjamin Franklin's Experiments."
16.5. *Russell, B.* "The ABC of Atoms."

16.6.   *Rose, P. H.,* and *A. B. Wittkower*   Tandem Van de Graaff Accelerators, *Sci. Am.,* August, 1970.

16.7.   *Moore, A. D.*   Electrostatics, *Sci. Am.,* March, 1972. Many interesting applications of electric fields.

Questions and problems

Use the value 9.0×10^9 N-m^2/C^2 for the Coulomb-law constant for a vacuum.

16.1.   An experimenter rubs a fountain pen on his coat sleeve and then touches the pen to an insulated metal ball. If a charge of -3×10^{-12} coulomb is hereby given to the ball, about how many electrons were transferred to the ball?

16.2.   Two equal, small conducting spheres carrying charges of $+1.0 \times 10^{-7}$ coulomb and -2.4×10^{-7} coulomb are brought into contact and then placed 3.5 cm apart. With what force do they then act on each other? Is the force one of attraction or repulsion?

16.3.   Four identical conducting spheres, A, B, C, and D, are suspended by silk threads. Ball A is touched by a rubbed glass rod, and then each of the other three is brought into contact with A and removed, one after the other. When A and B are then placed with their centers 5 cm apart, they are found to repel each other with a force of 1.6×10^{-6} newton. How much charge did the rod originally deliver to A?

16.4.   Check the units of k in Coulomb's law, as given in Eq. 16.2.

16.5.   There are about 8×10^{22} free electrons in each cubic centimeter of copper. To how many coulombs is this equivalent? If this much charge were placed on the earth and also on the moon, what would be the magnitude of the force of mutual repulsion? The earth-moon distance is 3.85×10^8 m.

16.6.   Compute the force of attraction between the proton and the electron in a hydrogen atom. The two particles are 5.3×10^{-11} m apart.

16.7.   List the ways in which gravitational attraction and the Coulomb force are similar; also the ways in which they differ.

16.8.   Find the strength and direction of the electric field at a location 10 cm due north of a small spherical object carrying a charge of 3×10^{-6} coulomb. What force (direction and magnitude) would act on an electron placed there?

16.9.   Two small charged bodies, carrying $+1.0 \times 10^{-8}$ and $+4.0 \times 10^{-8}$ coulomb, respectively, are placed 6 cm apart. Find the location of a point between them where the electric field strength is zero.

16.10. A small metal sphere bearing a charge of $+2 \times 10^{-9}$ coulomb is placed 20 cm from a similar sphere carrying -1×10^{-9} coulomb. At what point on the line joining the objects will the field intensity be zero?

16.11. In Worked example 16.2, find the angle between **E** and the line joining the two charges.

16.12. In conformity with the convention of drawing field lines to represent the intensity of an electrostatic field, as described on page 382, how many lines must be assumed to originate on a spherical conductor having a charge $+Q$? HINT: Consider that the lines will be uniformly distributed over the surface of a sphere of radius r concentric with the conductor.

16.13. In the Millikan experiment, why do we not try to observe the motion of individual electrons in the field rather than that of electrons attached to liquid drops?

16.14. An alpha particle is an entity having a mass 6.68×10^{-27} kg and a charge of $+2e$. How strong an electric field is needed to balance it against gravity?

16.15. The plates in a Millikan oil-drop apparatus are 1.0 cm apart, and a potential difference of 100 volts is applied, the upper plate being positive. An oil drop of mass 6.5×10^{-16} kg, carrying a charge of $5e$, is introduced. (a) What is the magnitude of the electrostatic force acting on it? (b) In what direction does it move? (c) How must the potential difference be changed to balance the drop against gravity?

16.16. The system of two conductors shown in Fig. 16.13 (page 385) has electric PE after the two are separated and the charged rod is removed. This follows from the fact that the conductors, if free to move, would acquire KE under their mutual attraction. What is the source of this energy? Explain in detail.

16.17. An electron is released from rest in an electric field. Will it start to move to points of higher or lower potential? Explain.

16.18. A lightning stroke may transport a total charge of 45 coulombs in a flash between the earth and a cloud that is at a potential of 1 million volts. How much electric energy is dissipated in the process? Assume the potential of the cloud to remain constant at the above value.

16.19. The potential difference between two points in an electric field is 45 volts. If 9.0 joules of work must be done to move a small charged object between these locations, how much charge does it carry?

16.20. Point charges of $+1.0 \times 10^{-9}$ coulomb are located at two vertices of an equilateral triangle whose sides are 20 cm long. How much work must be done to bring a test charge of $+1.0 \times 10^{-9}$ coulomb up to the third corner of the triangle from an infinite distance away?

16.21. At the surface of the earth there is a potential gradient of average magnitude of about 100 volts/m, directed downward. The charge on the earth causing this field is largely the result of lightning discharges. Find the sign and magnitude of this charge. The radius of the earth is 6.37×10^6 m.

16.22. The maximum potential gradient that an insulating material can support without disruptive breakdown is called its **dielectric strength.** For air, this is about 3×10^6 volts/m (page 389). What is the maximum charge that can be put on a metal sphere of radius 30 cm, surrounded by air? What is the potential of the charged sphere?

16.23. An electron having a KE of 225 eV is projected directly into a parallel-plate capacitor through a small hole in the positive plate (Fig. 16.24). Describe the subsequent motion of the electron, assumed to be in a vacuum. Does the particle reach the negative plate? If not, how close does it get?

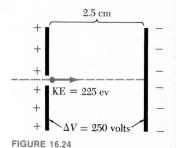

FIGURE 16.24

16.24. A proton falls freely from rest through a potential difference of 200 volts in a vacuum tube. (a) Express its final KE in joules; (b) find its final speed. The mass of a proton is 1.67×10^{-27} kg.

16.25. The KE of a particle may be specified in electron volts even if the particle does not acquire its energy by being accelerated in an electric field, for the electron volt is merely a special kind of energy unit. Compute the KE, in electron volts, of the thermal motion of an "air molecule." The mass may be taken as 4.8×10^{-26} kg and v_{rms} as 485 m/sec (page 234).

16.26. How much additional charge must be put on an isolated system whose capacitance is 60 pf in order to raise its potential by 6.0 volts?

16.27. Calculate the capacitance of a metal sphere the size of the moon ($r = 1.74 \times 10^6$ m).

16.28. A metal sphere of radius a carries a charge Q. It is then joined momentarily by a wire to an uncharged, metal sphere of radius b. Both bodies are on insulating supports and are far apart. What is the final charge on the second sphere? HINT: While connected, how do the potentials of the two spheres compare?

16.29. An isolated spherical conductor 50 cm in diameter is in air. By what factor does its capacitance increase when it is surrounded by a grounded spherical metal shell of inner diameter 51 cm, placed concentric with it, the space between the shells being filled with glycerin? See Table 16.1.

16.30. A compact form of capacitor used in communications work consists of two long strips of metal foil separated by an insulating strip of treated paper. The combination is rolled into a small cylinder. If each foil is 6 cm wide and 18 m long and the paper separator is 0.010 cm thick and of dielectric coefficient 2.2, what is the capacitance of the unit in μf?

Chapter 17
ELECTRIC CURRENTS AND CIRCUITS

The phenomena described in the preceding chapter were classed as *electrostatic*, since they represent the behavior of charges at rest. It was seen, however, that such phenomena appear only when some separation of a normally neutral mixture of positive and negative charges is brought about. All separation processes involve the motion of one or more kinds of electrified particles, and new and highly important phenomena are associated with such motions. These effects were first clearly recognized as originating in moving charges by the brilliant French physicist and mathematician A. M. Ampère (1775–1836), who gave this branch of physics the name **electrodynamics.** The development of the subject is taken up in this chapter and the following ones.

17.1 ELECTRIC CURRENTS

The displacement of charge in an insulated conductor that is placed in an electric field is a momentary one; it stops as soon as the internal field has been reduced to zero (page 384). At the same time, the potential difference that existed at first between the ends of the conductor becomes zero. There are, however, agencies capable of *maintaining* a field, and hence a potential difference, in a conductor. In addition, they supply movable charge carriers at one place on the conductor and remove them at another place. They must, in fact, operate very much like a circulating pump in a hydraulic system. The pump maintains a difference of pressure between the ends of a pipeline, thereby forcing water through the whole system, including the pump itself. The flow established in this way is called a hydraulic current. In electricity, the flow of charge carriers constitutes an **electric current,** and its complete, continuous path is called an electric **circuit.**

   The analogy between an electric current and the flow of a fluid is a close one and is helpful in keeping in mind certain electrodynamic phenomena. The circuit element that corresponds to a water pump may be a battery, generator, or other device whose detailed construction and operation need not concern us now. For the present, we recognize that such a device has two terminals, between which it maintains a difference of potential, analogous to the difference of pressure that a circulating pump maintains between its inlet and outlet ports.

When a steady flow is established in a hydraulic system, there is no accumulation or depletion of liquid at any point. Similarly, in a conductor in which a current exists, there is no piling up or thinning out of charge carriers, so that an initially uncharged current-bearing conductor remains uncharged. In a metal, for example, the free electrons begin to drift through the conductor as soon as the battery is connected. Because of the intense Coulomb repulsion between electrons when they are close together, the electron cloud acts like an incompressible liquid. The instant that a current is established anywhere in a simple circuit, current exists in all parts of the circuit. The battery or its equivalent does not "produce" electricity any more than the circulating pump manufactures water; it serves only to move charged particles already present in the system.

These particles may be of various kinds, depending on the nature of the conductor. In metals they are (negative) electrons; in conducting liquids and gases they may be ions of both signs. Positive carriers have a net movement in the direction of the field, that is, in the direction of *decreasing* potential. Negative carriers move opposite to the field direction, or in the direction of *increasing* potential. In *its external effects, a stream of negative charge carriers is entirely equivalent to a stream of positive carriers moving in the opposite direction.* The **conventional direction** of a current is that in which *positive* carriers move or would move if free to do so; it is the direction from higher toward lower potential. Thus it must be remembered that the negative ions in fluids and the free electrons in metals move counter to the conventional direction of current.

The free electrons in a conductor continually participate in the random thermal motion characteristic of all particles of matter (Chap. 10), and their average speed at ordinary temperatures is of the order of 10^6 m/sec. If the conductor is connected to a battery, the resulting electric field super-imposes a very small drift on the enormously greater thermal motions.

The electrons do not continue to pick up speed indefinitely under the influence of the field. Instead, because of their frequent collisions with the atoms of the metal, they settle down to what amounts to a constant, slow progress toward the high-potential end of the conductor with a net drift speed of the order of only a millimeter per second (see Worked example 17.1, below). Thus the electrons do not move like freely falling stones in a gravitational field but more like objects that have reached terminal speed in falling through a viscous fluid. This exceedingly deliberate process, in which tremendous numbers of electrons participate, constitutes the current in a solid conductor: a slow oozing of the electron cloud through the interatomic spaces. The current in a wire is not something that always moves with lightning-like speed, as may seem to be the case. What really happens when the switch is closed in a simple circuit like the one in Fig. 17.1 is that a *signal* travels along the wire at high speed, which may be as great as the speed of light c in special cases. Such a signal is analogous to the compressional pulse accompanying the starting of the pump in a

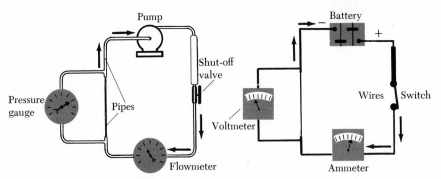

FIGURE 17.1   Liquid and electric circuits, showing their analogous elements. The flow of an incompressible liquid in a pipe system is similar in many ways to the flow of current in an electric wire circuit.

hydraulic circuit. However, neither the electrons nor the drops of water move at high speed in these systems.

Quantitatively, the current I may be defined as follows: The current is the net amount of charge passing through any cross-section of a conductor per unit of time. Thus, if Q is the quantity of charge transported in a time interval t,

$$I = \frac{Q}{t}$$ **17.1**

For currents that vary in time, how would you set up a definition of instantaneous current?

In the mks system, Q is expressed in coulombs and t in seconds, so that I has the units coulombs per second. One coulomb per second is called one **ampere,** abbreviated A. Because of their thermal motion alone, the electrons passing through any cross-section of a copper wire in *one* direction would constitute a current of about 100 billion amperes if this were not balanced, on an average, by an equal transfer in the opposite direction. The current in an ordinary lamp is about 0.5 ampere, whereas the feeble currents used in telephone circuits may be only a ten-millionth as large, a condition in which the thermal motions of the electrons are put off balance by only about 1 part in 10^{18}.

In a simple circuit consisting of a single, continuous path with no side branches, the value of the current is the same at all cross-sections, for the reasons given on page 405. Although electric current need not be considered a vector, a complete specification must include a statement of its direction around the circuit (or through a given conductor).

The current can be expressed in terms of the properties of the moving electrons as follows: Fig. 17.2 represents a portion of a cylindrical metallic conductor (wire). Under the influence of the electric field within the metal, the free electrons are moving with an average drift speed v_D in a direction opposite to that of the field E. In a time interval t, all the free electrons contained in a cylinder of length $v_D t$ will pass through a given cross-section. If the cross-section area is A and the average number of free electrons per unit volume of the conductor is n, the total number in the above cylinder will be n multiplied by the volume of the cylinder, or $n v_D t A$. Since each electron transports a charge of amount e, the total charge carried through

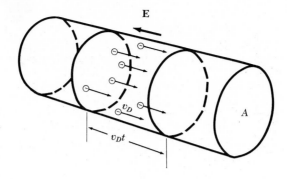

FIGURE 17.2 Motion of electrons in a metallic conductor.

A in the time interval t is $Q = env_D tA$. The corresponding value of the current in the conductor is, from Eq. 17.1,

$$I = env_D A \qquad\qquad\qquad 17.2$$

Worked example 17.1

Estimate the drift speed of the free electrons in a copper wire of cross-section area 1 mm² carrying a current of 3 amperes.

Solution   The number of free electrons in copper may be taken to be about 8×10^{28} per m³. Use $e = 1.6 \times 10^{-19}$ coulomb. Here $A = 10^{-6}$ m². When Eq. 17.1 is solved for v_D and the numbers substituted, we get

$$v_D = \frac{3}{(1.6 \times 10^{-19})(8 \times 10^{28})(10^{-6})}$$

$$= 0.23 \times 10^{-3} \text{ m/sec, or } 0.23 \text{ mm/sec}$$

Thus, if the power is supplied by a station a mile away, the individual electrons take an average of almost 4 hr to make a complete trip around the circuit.

17.2 RESISTANCE OF A CONDUCTOR; OHM'S LAW

As indicated in Fig. 17.1, a simple electric circuit carrying a steady current consists essentially of (1) a battery or similar element that can maintain a constant difference of potential across its terminals and (2) a wire or succession of wires or other conductors joined to these terminals to provide a continuous path for the charge carriers. A useful addition is a simple switch for completing or interrupting the circuit. Suitable instruments are available for measuring the magnitude of the current in the circuit or the potential difference between any two points. These are called **ammeters** and **voltmeters,** respectively, and their operating principle and construction will be described in the next chapter.

Figure 17.3 is an electrician's diagram of a simple circuit showing the conventional symbol for each part. B is a battery consisting, in this case, of three cells. The long stroke represents the positive terminal of each cell and the short, thick line the negative. The zigzag line R is the conducting element of main interest, and the straight lines represent heavy connecting wires. The ammeter A is connected directly into the circuit; the voltmeter

FIGURE 17.3 Conventional representation of a simple electric circuit. The straight, heavy lines are connectors having negligible resistance.

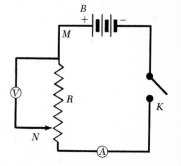

V is in a side circuit (or **shunt** circuit), its terminals being connected between the two points M and N whose difference of potential is to be determined. The arrowhead on the end of one of the voltmeter connectors indicates that this end is movable. It may be joined to any desired point on the conductor, and in some arrangements it is an actual sliding contact.

When the switch K is closed, a steady current arises in the circuit and the meters show constant readings. The ammeter may be placed in the circuit at any point, and changing its location will not have any effect on its reading because *the current is the same at all points in a simple circuit.*

What determines the magnitude of the current in such a circuit? This question was answered early in the nineteenth century by the careful experiments of a German physicist, G. S. Ohm (1789–1854). He found that for a wire inserted as the element R (Fig. 17.3) the current is directly proportional to the cross-section area of the wire and inversely proportional to its length, and that it depends also on the kind of metal of which the wire is made. The best conductors were found to be silver, copper, and gold.

Ohm saw the parallel between his results for the conduction of electricity and the experimental facts concerning the conduction of heat in solids (see page 255). He realized, too, that metals that conduct heat most readily are also the best electrical conductors, which circumstance was found, much later, to be no mere accident but a consequence of the fact that free electrons are responsible for both kinds of conduction in metals.

Using a given wire, Ohm tried the effect of altering the potential difference applied to its ends. This could be done, for example, by changing the number of cells in the battery. He discovered that *the current is strictly proportional to the applied potential difference.* Expressed in symbols, this could be written $I = kV$, where k is some characteristic of a given conductor that may be called its **conductance.** It is more common, however, to consider that the wire limits the magnitude of the current, and so the reciprocal of k, called the **resistance** R of the conductor, is the quantity usually specified. Then the general result of Ohm's experiments may be stated as

$$I = \frac{V}{R} \qquad\qquad 17.3$$

This is the renowned **Ohm's law** for an electrical conductor. In words, it asserts that *the magnitude of the current in a conductor is directly proportional to the applied potential difference and inversely proportional to the resistance.* Two limitations on this statement must be noted: (1) External conditions, such as the temperature of the conductor, must be maintained constant and (2) the law is strictly valid only for ordinary metallic conductors. It is not applicable, except over limited ranges, to the conduction of electricity in gases or liquids or to the electron current in a vacuum (see Fig. 17.4), where the mechanism of the process is quite different.

Following the usual procedure, the fundamental equation may be used to define units for the new quantity R. If I is measured in amperes and V in volts, R is measured in **ohms.** The resistance of a conductor is 1 ohm if the current is 1 ampere when the potential difference between the ends of the conductor is 1 volt. For steady currents in an ordinary metallic

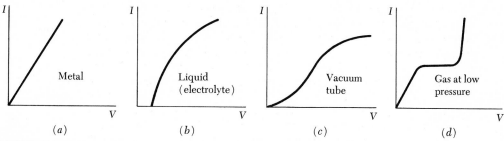

FIGURE 17.4 Characteristic current-potential curves for various kinds of conductors. Only in (a) is the relation linear, and hence the resistance constant, over a considerable range.

For important examples see pages 426 and 427.

conductor, the value of R is the same whether the current is in one direction or the other. There are circuit elements and devices, called rectifiers, for which this is not true.

There is an enormous range of resistance values among ordinary solid materials. A 1-m cube of one of the best metallic conductors has a resistance across opposite faces of the order of 10^{-8} ohm, semiconductors range from about 10^{-5} to 10^7 ohms, and the best insulators have resistances of the order of 10^{17} ohms.

Factors affecting resistance.

The resistance of a given conductor is dependent to some extent on external conditions, as indicated above. Changes in applied pressure, the presence or absence of magnetic fields or illumination, and other circumstances are known to alter the resistance for some materials. All conductors change in resistance with change in temperature; for most conductors the resistance increases slightly with temperature. Once the rate of rise has been measured for a given kind of wire, the procedure may be reversed and temperatures determined by measuring the change in resistance of a coil of such wire. This constitutes a **resistance thermometer.** The most direct method for determining the resistance of a conductor is to apply a definite potential difference to its ends, measure the resulting current by means of an ammeter, and compute R from Ohm's law. Portable instruments called ohmmeters read directly the resistance when they are connected between two points on a conductor. If such an arrangement is combined with the coil of a resistance thermometer, the instrument scale can be marked to read the temperature directly.

Although the resistance of most metals decreases regularly with decreasing temperature, certain substances show an unusual behavior at very low temperatures within about 20° of absolute zero. At a temperature characteristic of each material the resistance drops abruptly to zero: the material is said to become **superconducting.** A current started in a superconducting metal ring continues to flow for months with undiminished strength, although there is no battery or other source of potential difference in the circuit. This behavior may be pictured as a sudden drop in the resistance to motion experienced by the conduction electrons, so that their inertia makes them continue without slowing down. A satisfactory explanation of superconductivity requires the use of the quantum theory (see Ref. 17.2).

Mercury, tin, and lead are examples of metals that become superconducting.

The free-electron model can account for the form of Ohm's law in a

Classical electron model of conduction in metals. simple way. In a metal, each electron, because of its random motion, collides frequently with the relatively fixed atoms, as described on page 405. The electron gas is initially in thermal equilibrium with those atoms. When an electric field is applied, each electron is given an additional *drift velocity*. It may be assumed that whenever a collision with an atom occurs, the electron loses whatever drift speed it had acquired since its last collision. It then starts to drift anew under the influence of the electric field and continues in this way.

Between collisions with atoms, each electron moves with constant acceleration under the influence of the electric field within the wire. According to Eq. 16.7 (page 389), the magnitude E of this field is given by $E = V/L$, where V is the applied potential difference and L is the length of the wire (Fig. 17.5). The force acting on the electron is $F = eE = eV/L$, and its direction is opposite to that of **E** because of its negative charge. Newton's second law then gives the acceleration as $a = F/m$, or

$$a = \frac{eV}{Lm} \qquad\qquad 17.4$$

The average time between collisions of a free electron with the atoms of the solid depends on the net translational speed and on the nature of the crystal lattice. If the mean free path of such an electron is l and its average speed is v, this time is given by

$$t = \frac{l}{v} \qquad\qquad 17.5$$

In this interval the electron moves with constant acceleration and acquires a final drift speed of amount at. The average drift speed v_D is half this amount, or

$$v_D = \tfrac{1}{2}at \qquad\qquad 17.6a$$

Substitution of the values of a and t from Eqs. 17.4 and 17.5 yields

$$v_D = \frac{eVl}{2Lmv} \qquad\qquad 17.6b$$

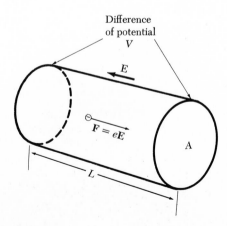

FIGURE 17.5   The classical c duction of Ohm's law on the fre electron model.

An expression for the current is now obtained by substituting this result into Eq. 17.2:

$$I = \frac{e^2 n l A}{2 m v L} V \qquad \qquad 17.7$$

The fraction on the right has a constant value at a constant temperature. If we call the reciprocal of this fraction R, we have

$$R = \frac{2 m v L}{e^2 n l A} \qquad \qquad 17.8$$

and Eq. 17.7 can be written $I = V/R$, which is Ohm's law. Moreover, from Eq. 17.8 it can be seen that R has the observed dependence on the length L and section area A of the conductor (page 408). It also appears to have the expected dependence on temperature because v, which increases with temperature, occurs in the numerator of the expression. Nevertheless, on closer examination it was found that this simple theory of metallic conduction had to be modified by quantum-theory considerations in order to get good agreement with experiment.

It must be emphasized that Ohm's law is subject to certain limitations, as mentioned on page 409, and does not have the universal applicability of Newton's second law or the law of charge conservation. Ohm's law describes the conduction of electricity by certain materials where the ratio of applied potential difference to current is independent of either V or I. Generalizations of Ohm's law have been developed for handling problems involving complicated networks of conductors and circuits in which the current strength varies with time. It is worth noting that Ohm's law, together with its extensions, placed the whole subject of electric circuitry on a firm, exact basis and opened the way for the remarkable electrotechnical developments of recent times.

Worked example 17.2

A steady potential difference of 2.0 volts is applied to the ends of a copper wire 10 m long, of cross-section area 5×10^{-7} m².

  a. What is the value of the electric field within the metal?

  b. Compute the force on a free electron.

  c. Find the resistance of the wire, using the fact that the **resistivity** of copper, the resistance between opposite faces of a 1-m cube, is 1.7×10^{-8} ohm-m.

  d. What is the value of the current in the wire?

Solution a. From Eq. 16.7, page 389, $E = 2/10 = 0.20$ volt/m.

  b. The force is $F = eE$, or $F = (1.6 \times 10^{-19})(0.20) = 3.2 \times 10^{-20}$ newton.

  c. According to page 408, the resistance is proportional to L/A, where L is the length and A the cross-section area. The constant of proportionality is just the resistivity defined above, for a 1-m cube has unit length and unit cross-section area in mks units. With the numbers given here, we have

$$F = \frac{(1.7 \times 10^{-8})(10)}{5 \times 10^{-7}} = 0.34 \text{ ohm}$$

  d. Ohm's law yields $I = 2.0/0.34 = 5.9$ ampere.

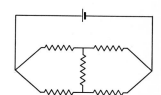

FIGURE 17.6 Example of a circuit that cannot be solved by successive applications of Ohm's law.

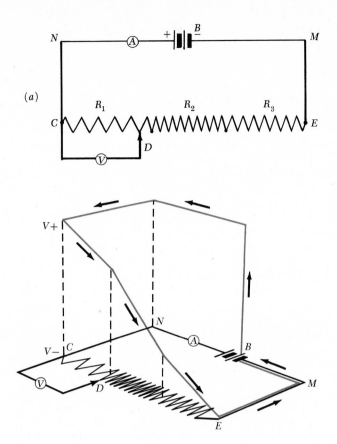

FIGURE 17.7 Part (b) is an oblique view of the circuit diagrammed in (a), with the potential at each point vertically upward. The potential falls at a different rate in each resistor because the values of the resistance per unit length differ.

17.3 SERIES AND PARALLEL CONNECTIONS

The simple circuit in Fig. 17.7a, where a number of conductors are connected **in series** (end to end), is an extension of the one in Fig. 17.3. In this type of joining, the path of the charges is a single loop, with no side branches or shunts. Although the voltmeter and its connectors constitute a bypass, its resistance is usually very high compared with that of the other circuit elements and the current in it is negligible compared with that of the main circuit. The earlier discussion (page 405) brought out the fact that the current is the same at all points in a series circuit.

Suppose one terminal of the voltmeter is connected at C and the other at some point D on the wire R_1. The point C is at the same potential as the positive terminal of the battery, since both the ammeter and the connecting wires have very low resistance, and this makes C in effect merely an extension of the battery terminal itself. On the other hand, D will be at a lower potential than C, there being a steady fall of potential along R_1. This is analogous to the constant decrease of pressure observed at successive points along a pipe carrying a steady current of fluid, or the constant decline of temperature along a rod that is conducting heat from a hot body to a cold one.

The existence of a fall of potential along a current-carrying conductor follows directly from Ohm's law, since if Eq. 17.3 is solved for V it reads

$$V = IR \qquad\qquad\qquad 17.9$$

Thus, because the current exists, there must be an "IR drop" of potential across any conductor whose resistance is R. If R for any conductor is negligibly small, as it is for the thick connecting wires in a circuit, then the above relation shows that V must be small, whatever the value of I may be. Therefore, as concluded above, the point C is at the same potential as the battery terminal to which it is connected.

The voltmeter reads the difference in potential existing between C and D. If the contact point D is moved farther and farther to the right along R_1, the meter reading increases, since the potential of C remains fixed but the potential of the other side of the voltmeter becomes progressively lower. The graph of Fig. 17.7b represents the way in which the potential of the right-hand voltmeter terminal varies as D is moved to the right along the resistor. All potentials are here arbitrarily referred to that of the negative terminal of the battery, the point in the circuit that has a lower potential than any other. If R_2 has a higher resistance per unit length than R_1, the graph line for the former will be steeper, as shown. For R_3, which has a smaller resistance per unit length, the line is less steep. The moving contact of the voltmeter has now arrived at E, which is at the same potential as the negative terminal of the battery. The battery itself may be looked upon as an agency that maintains the abrupt lift in potential, V_B, which brings conditions back to those at the positive terminal.

Conductors connected in series.

A circuit may contain a number of resistors, connected in certain ways. In many cases, Ohm's law makes it possible to reduce complicated groups of resistors effectively to a single equivalent unit whose resistance value can be computed from those of the individual parts. As an example, consider a number of resistors joined in series, as in Fig. 17.7a. What is their combined resistance? The total drop in potential across a number of conductors in series is the sum of the separate drops, so that, if V_1, V_2, V_3, etc., are the individual values, the total drop is

$$V = V_1 + V_2 + V_3 + \cdots \qquad\qquad 17.10$$

where the dots indicate that there may be any number of resistors in the chain. If R represents the equivalent resistance of the whole set, then Ohm's law applied to the entire group is, according to the form 17.9, $V = IR$. Similarly, written for each resistor separately, $V_1 = IR_1$, $V_2 = IR_2$, $V_3 = IR_3$, etc.

Note that the current is the same in all the resistors.

Substituting these values in Eq. 17.10, $IR = IR_1 + IR_2 + IR_3 \cdots$, or

$$R = R_1 + R_2 + R_3 + \cdots \qquad\qquad 17.11$$

The equivalent resistance of a number of conductors in series is the sum of the individual resistances.

Conductors in parallel.

In order to interrupt the current in any one of a set of conductors connected in series, the current must be stopped in all of them. Electrical appliances, however, should be independently operable, so that switching one of them on or off does not affect the others. For this reason such items are connected **in parallel,** as shown schematically in Fig. 17.8 for the case of three resistors. We note that the current in the main circuit (the battery and its connectors) divides at A among the various branches and that the separate currents reunite at C. Hence, because of conservation of charge,

$$I = I_1 + I_2 + I_3 + \cdots \qquad \qquad 17.12$$

Since all the individual resistors are connected across the common end points A and C, the potential drop is the same for each of them, and this is the same as the drop across the battery terminals. Using Ohm's law, substitute for each current appearing in Eq. 17.12:

$$\frac{V}{R} = \frac{V}{R_1} + \frac{V}{R_2} + \frac{V}{R_3} + \cdots$$

or, dividing out V,

$$\frac{1}{R} = \frac{1}{R_1} + \frac{1}{R_2} + \frac{1}{R_3} + \cdots \qquad \qquad 17.13$$

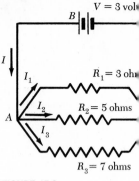

FIGURE 17.8

This relation states that *the reciprocal of the equivalent resistance of a number of conductors in parallel is equal to the sum of the reciprocals of the separate values.*

In evaluating R from this equation, it is essential to avoid the common mistake of inverting both sides of the equation *term by term.* All the fractions on the right must first be brought over a common denominator, because this is the best way to reduce a numerical problem and avoid inaccuracies caused by rounding off decimal values. After converting to a single fraction, its terms may be inverted and then divided to yield a numerical result in decimal form. The following example illustrates the procedure.

Worked example 17.3   Three resistors, 3, 5, and 7 ohms, are connected in parallel and a 3-volt battery is joined to their common terminals, as in Fig. 17.8.
    a. Find the current in the battery circuit.
    b. Find the current in each branch.

Solution   *a.* The equivalent resistance of the set is given by

$$\frac{1}{R} = \frac{1}{3} + \frac{1}{5} + \frac{1}{7} = \frac{71}{105} \qquad R = \frac{105}{71} = 1.48 \text{ ohms}$$

Notice that the value of R is *less* than that of any of the individual resistors. This is plausible, since every time an additional branch is connected in a parallel circuit another path is provided for the passage of charge. The current in the main circuit is $I = V/R = 3/1.48 = 2.03$ amperes.

    b. The current in the 3-ohm resistor is $I_1 = 3/3 = 1.00$ ampere. In the same way, $I_2 = 3/5 = 0.60$ ampere, and $I_3 = 3/7 = 0.43$ ampere. The sum of these three branch currents is 2.03 amperes, in agreement with the result of part *a.* The main current divides among the several branches in such a way that greater currents exist in the branches of lower resistance.

17.4 JOULE'S LAW

It is a matter of common observation that heat develops in a conductor carrying a current of electricity, and many useful devices such as the filament lamp, electric iron, radiant heater, etc., utilize this effect. This evolution of heat can be understood readily on the free-electron model: Between collisions with the atoms of the lattice, an electron acquires KE as it drifts under the influence of the applied field, as described on page 410. At the next impact this KE is shared with an atom, and the general increase in the random vibrational motion of the atoms results in a rise in temperature of the metal. The heat comes from the conversion of electric energy supplied by the battery or its equivalent. The situation corresponds exactly to the irreversible mechanical process in which a block of wood is dragged along a rough floor at constant speed, the work done against friction being dissipated as heat.

The amount of energy converted into heat or other forms when a charge Q passes through a conductor whose ends are at a difference of potential V is

$$W = QV \qquad\qquad 17.14$$

This follows directly from the definition of V (Eq. 16.6, page 388). The *rate* of conversion of energy is then given by dividing both terms of the equation by the time interval t during which the charge Q passes:

$$\frac{W}{t} = \frac{Q}{t} V$$

Now W/t is equal to P, the power expended, and Q/t is the current I, and so the above equation becomes

$$P = IV \qquad\qquad 17.15$$

If I is measured in amperes (C/sec) and V in volts (J/C), the product will have the units of joules per second, or watts. Thus Eq. 17.15 gives the power in watts. Even if I and V do not stay constant, this equation gives the power at any instant.

Assume that the circuit element under discussion is purely resistive—that is, it does not contain motors, chemical cells, or any other devices capable of converting electric energy into any form other than heat. Then we can introduce the resistance R of the element and get two other equivalent expressions by eliminating V and I, in turn, between Eq. 17.15 and Ohm's law, $I = V/R$. The results are

$$P = I^2 R = \frac{V^2}{R} \qquad\qquad 17.16$$

Equations 17.16 are statements of what is known as **Joule's law.** Frequently it is required to know the rate of dissipation of energy directly in heat units. Since 4,190 joules are equivalent to 1 kcal (page 258), it follows that 1 watt, which is 1 J/sec, is equivalent to 1/4190, or 2.39 × 10^{-4} kcal/sec. Then the Joule equations can be rewritten

$$H_{\text{kcal/sec}} = 2.39 \times 10^{-4}\,IV = 2.39 \times 10^{-4}\,I^2R$$
$$= 2.39 \times 10^{-4}\,\frac{V^2}{R}$$

17.17

Because the resistance of a conductor varies with its temperature, the heating effect of the current alters the value of R, and care must be taken to use the value appropriate to the temperature attained in any given experiment.

Practice set* Of two heating coils r and R, made of the same metal, the latter has the higher resistance.

1. If both are made of wire of the same diameter, which coil uses the greater length of wire?
2. With both coils joined in series in a steady-current circuit, which develops heat at a greater rate?
3. The same question for the coils connected in parallel.

17.5 ENERGY RELATIONS IN A CIRCUIT; EMF

It has been shown that energy is associated with electrical phenomena. For example, the basic concept of electric potential is directly connected with the PE of a charged body in an electric field, and charged bodies experience changes in KE in moving through such fields. Further, as described above, currents in conductors are accompanied by evolution of heat. In this section, the application of the energy concept to circuits is examined.

The preceding discussion of electric currents mentioned the use of certain devices such as batteries and electric generators for maintaining the difference of potential responsible for the current in a circuit. These are called, collectively, sources of **electromotive force.** This name is an unfortunate one, stemming from the time when several other physical entities—including energy itself—were indiscriminately called "forces." A better name, electromotance, has been proposed but is not widely used. For present purposes, the term emf (to be pronounced *ee-em-eff*) will be used in place of "electromotive force." As an algebraic symbol for emf, the script capital letter $\mathcal{E}$ will be used.

A source of emf such as a battery performs work on charge carriers. In the *external* parts of a simple circuit (see Fig. 17.7, page 412), hypothetical positive charges fall continuously in going from the positive terminal of the battery to the negative, but *inside the battery,* chemical processes must *lift* the charges to the "high" terminal again. The battery converts some other form of energy (in this case, chemical) into electric energy. This process is a general characteristic of any source of emf, and it is—in

*Ans.: R; R; r.

part, at least—a reversible one (page 262). In a battery, for instance, electric energy is changed to chemical energy in the *charging* process, whereas chemical energy is changed to electric energy in the *discharging* process.

The value of the emf, $\mathcal{E}$, of a source is defined by

$$\mathcal{E} = \frac{W}{Q} \qquad\qquad \textbf{17.18}$$

where W is the amount of energy converted when a charge Q passes through the source. Here W is measured in joules and Q in coulombs, so that $\mathcal{E}$ has the units joules per coulomb, or volts—the same as potential difference.

Although measured in identical units, emf and potential difference are not the same. To qualify as a source of emf, a circuit element must involve the *reversible* transformation of energy to and from the electrical form. Even though a potential difference exists across a pure resistor in a circuit, it is not a source of emf because there is an irreversible transformation of electric into thermal energy taking place in it. Passage of a current through a resistor heats it up, but building a fire under a resistor does not produce current.

Ohm's law will be applied to an entire circuit as an example of the use of energy methods and the concept of emf. Assume a simple series circuit of the kind shown in Fig. 17.9, containing a number of resistors as well as a number of batteries or other sources of emf. Passing through a source $\mathcal{E}$, a test charge $+Q$ gains an amount of energy $\mathcal{E}Q$ if it moves from the negative to the positive terminal and loses this amount if it goes in the opposite direction. In passing through a resistor R, such a charge always *loses* energy of amount QV, where V is the difference of potential of the two ends of R. In a complete trip around the circuit, the test charge loses exactly the same amount of energy as it gains from some of the sources of emf. Going around clockwise in Fig. 17.9, this may be expressed:

$$\mathcal{E}_a Q + \mathcal{E}_b Q - \mathcal{E}_c Q = QV_1 + QV_2 + QV_3 \qquad\qquad \textbf{17.19}$$

The third term on the left is given a minus sign because a positive test charge carried around the circuit in, say, a clockwise direction traverses $\mathcal{E}_a$ and $\mathcal{E}_b$ from the negative to the positive terminal, but in passing through $\mathcal{E}_c$ it goes from positive to negative.

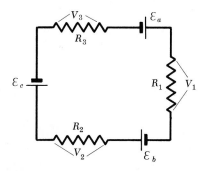

FIGURE 17.9 Applying Ohm's law to a simple circuit consisting of resistors and emf's.

On dividing through by Q and substituting IR_1 for V_1, IR_2 for V_2, and IR_3 for V_3, we get

$$\mathcal{E}_a + \mathcal{E}_b - \mathcal{E}_c = IR_1 + IR_2 + IR_3$$

$$I = \frac{\mathcal{E}_a + \mathcal{E}_b - \mathcal{E}_c}{R_1 + R_2 + R_3} \qquad\qquad \textbf{17.20}$$

If the test charge had been carried around the circuit in the opposite direction (counterclockwise), nothing would be changed, for in that case $\mathcal{E}_a$ and $\mathcal{E}_b$ would have to be given minus signs and $\mathcal{E}_c$ would then have a plus sign. Then all the V's on the right side of Eq. 17.19 would be negative since the charge would be carried to a point of higher potential in passing through each resistor. As a result, the equation would remain unchanged. In general, then, Eq. 17.20 may be written

$$I = \frac{\text{algebraic sum of } \mathcal{E}\text{'s}}{\text{arithmetic sum of } R\text{'s}} \qquad\qquad \textbf{17.21}$$

which represents Ohm's law as applied to an entire circuit.

Worked example 17.4   Two batteries and a pair of resistors in parallel are joined in a circuit as shown in Fig. 17.10.

a. Find the current in each resistor.

b. Find the potential difference across the terminals of each battery.

Solution   a. The equivalent resistance of the pair in parallel is given by Eq. 17.13 (page 414) as $1/R_\parallel = 1/4 + 1/6$, or $R = 2.40$ ohms. The zigzag line adjoining each battery symbol represents the **internal resistance** of each battery offered by the plates, chemical solutions, etc., of the cells themselves. These quantities are labeled R_a and R_b on the diagram. The internal resistances act as if they were resistors in series with the batteries, in conformity with the fact that batteries heat up when carrying current.

Apply Ohm's law to the whole circuit, using Eq. 17.21. Moving clockwise around the circuit,

$$I = \frac{-12 + 5}{2.40 + 0.30 + 0.10} = \frac{-7}{2.80} = -0.25 \text{ ampere}$$

The minus sign indicates merely that the conventional current is counter-clockwise—opposite to the direction in which the test charge was carried around the circuit.

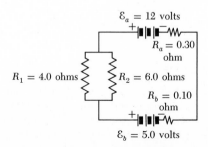

FIGURE 17.10

This is the current in the main circuit as well as in each battery. To find the current in R_1, note that the potential drop across the pair of resistors is $V_\| = IR = (0.25)(2.40) = 0.60$ volt. Then the current in R_1 will be $I_1 = V_\|/R_1 = 0.60/4 = 0.15$ ampere, and the current in R_2 will be $I_2 = V_\|/R_2 = 0.60/6 = 0.10$ ampere. The sum is $I = 0.25$ ampere, as already computed.

b. The potential drop across the terminals of $\mathcal{E}_a$ is $\mathcal{E}_a - IR_a$, or $12 - (0.25)(0.30) = 11.9$ volts, and for $\mathcal{E}_b$ it is $\mathcal{E}_b + IR_b$, or $4 + (0.25)(0.10) = 4.03$ volts. The current is backward in $\mathcal{E}_b$, which means that this battery is being *charged,* and its terminal potential difference exceeds its emf. Normally, when a battery is supplying electric energy to its circuit, as $\mathcal{E}_a$ is doing in this example, its terminal potential difference is lower than its emf by an amount that depends on the magnitude of the internal resistance as well as the current it is supplying to the circuit.

17.6 CHEMICAL SOURCES OF EMF

The Italian physician and biologist Luigi Galvani (1737–1798) found, in dissecting a frog, that its leg muscle twitched vigorously whenever it came into contact with a pair of different metals, such as brass and iron. He thought the contractions were due to some kind of ''animal electricity,'' and a great series of discussions and heated arguments arose among prominent scientists of that day.

The true explanation was finally given by a countryman of Galvani, the physicist Alessandro Volta (1745–1827), who ascribed the phenomenon to the moist contact between dissimilar metals. He tested this idea by eliminating the animal tissue and building what came to be known as a voltaic pile—a stack of alternate zinc and copper disks separated by pieces of leather or paper soaked in salt water or lye. From this device he was able to obtain effects similar to those from electrostatic generators or from charged capacitors, with the added feature that the operations could be repeated many times without replenishing any part of the system. After a long process of development, reliable and easily controlled chemical sources of electricity came into use. They dominated the field of electricity until the development of the electric generator about a century after Volta's experiments.

In addition to the electric pile, Volta had constructed a ''crown of cups'' made up of a series of vessels containing brine or lye and joined by alternate strips of two kinds of metal (Fig. 17.11). Each unit of the chain, consisting of a plate of each metal immersed in the solution, is called a **voltaic cell.** A number of interconnected cells forms a **battery.** Most of the dry cells widely used in flashlights, portable radios, etc., use zinc and carbon as the two elements and an ionized paste in place of the liquid.

17.7 ELECTROLYSIS

The key to an explanation of the voltaic cell was found during the latter part of the nineteenth century, principally through the work of the physical chemists Svante Arrhenius in Sweden and Walter Nernst in Germany. Most liquids, such as oils, organic compounds, and even water when extremely

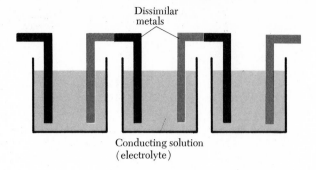

Dissimilar metals

Conducting solution (electrolyte)

FIGURE 17.11 Schematic representation of Volta's "crown of cups."

pure, do not conduct electrically to any great extent. However, the substances classed chemically as acids, bases, or salts are good conductors when in solution or melted. In all such cases, a certain proportion of the molecules of the active substance split apart (**dissociate**) into ions when the material is dissolved or melted.

These terms, which refer to electrolysis and other processes, were introduced by Faraday.

In the process called **electrolysis** a difference of potential is applied to two plates (**electrodes**) immersed in an **electrolyte** (liquid containing ions). The ions drift slowly through the liquid under the influence of the field, positive ions moving toward the negative plate, or **cathode,** and negative ions moving toward the positive plate, or **anode.** The net result is that electric charge as well as matter is transported in this process. The current in the electrolyte is the combined effect of the motion of positive and negative charges in opposite directions.

The ions drift with speeds of the order of 0.001 cm/sec, the value depending on the nature of the ion. As in the case of free electrons in metals, the drift speed is superposed on a much greater random motion which, for ions in an electrolyte, is of the order of 100 m/sec. When an ion reaches the electrode, it gives up its charge, becoming a neutral atom or atom group of the corresponding element or elements. If this is a metal, the atoms may adhere to the electrode (the cathode in this instance) in the form of a coating, in which case the process is the familiar **electroplating.** In some instances the neutralized ions may react chemically to form other products, or they may form bubbles of gas which escape from the liquid. Quantitative aspects of the process of electrolysis will be discussed later.

17.8   THE VOLTAIC CELL

In electrolysis, the application of a potential difference produces chemical changes within a cell. In the voltaic cell, the opposite occurs: chemical effects produce a usable potential difference. There is an emf associated with each of these devices. The former changes electric into chemical energy whereas the latter does the opposite.

Figure 17.12a shows, schematically, a solution of hydrochloric acid (HCl) in water. Most of the molecules of the acid have dissociated into hydrogen ions (H^+) and chloride ions (Cl^-). This dissociation in solution is explained by the high dielectric coefficient of water (see Table 16.1, page 398), which

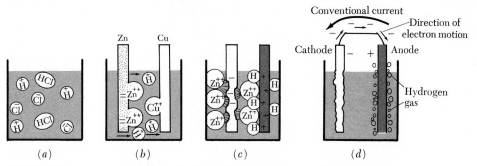

FIGURE 17.12 The action of a voltaic cell (schematic), showing the role of the ions in transporting charge within the electrolyte. In the external circuit, the charge is carried by electrons only.

is responsible for a great reduction in the electrostatic force that normally holds the ion pair together.

In Fig. 17.12b, zinc (Zn) and copper (Cu) plates have been inserted. The Zn atoms have a strong tendency to detach themselves from the plate, coming off as Zn^{2+} ions and leaving behind electrons on the plate. The accumulating zinc ions drive some of the hydrogen ions toward the copper plate. Some atoms of the copper plate have also gone into solution as Cu^{2+} ions, but these are few compared with the number of zinc ions. In Fig. 17.12c, hydrogen ions have given up their charge to the copper plate, and the neutral hydrogen atoms thus produced form bubbles of hydrogen gas which escapes from the electrolyte. The zinc ions, however, are held close to the zinc plate by electrostatic attraction, and an equilibrium is attained in which no further net action occurs. The result is a difference of potential between each plate and the electrolyte, with the zinc plate slightly negative with respect to the copper. There now exists a difference of potential between the two plates, and this is the emf of the cell. Its magnitude depends on the nature of the two plate materials.

Now the external circuit is completed by connecting the plates by means of a wire as shown in Fig. 17.12c. The electrons that accumulated on the Zn plate flow over to the Cu plate, where they neutralize the positive charges left there by the hydrogen ions. The action continues until the Zn plate is completely eaten away or until all the hydrogen is removed. While the external circuit is connected, the potential difference between the cell terminals is *less* than the emf because there is an IR drop within the cell itself, R being the internal resistance. The emf of a cell or battery is sometimes referred to as its ''no-load'' or ''open-circuit'' potential difference.

A **lead storage cell** consists of a dilute solution of sulfuric acid in which there are two lead plates, one containing lead peroxide. The action is in general similar to that of the voltaic cell, except that in the final state both plates have been converted to lead sulfate. The cell may now be recharged by passing a current through it in the opposite direction. This restores the plates to their original state, and the cell may be used repeatedly. A battery consisting of six such cells (total ''voltage'' about 12) is the usual source of electric energy in an automobile. The storage battery does not ''store electricity," but the energy of the charging current is localized in it in the form of chemical energy which is converted into electric energy when the battery is in use.

17.9 THE CHARGE ON AN ION

In Chap. 3 it was shown that the chemical discoveries made by Dalton at the beginning of the nineteenth century conclusively established the atomic nature of matter. Further evidence was later provided by Faraday's quantitative experiments on electrolysis.

In chemistry the gram is retained for measurement of atomic weights.

Chemists have determined what they call the atomic weights (page 59) of the elements. These are numbers that are proportional to the weights of individual atoms of the elements. An amount of any element equal to its atomic weight in grams actually contains $N_0/1000 = 6.023 \times 10^{23}$ atoms, where N_0 is Avogadro's number (page 234).

Dalton and others established the fact that elements combine to form a specified compound in a definite proportion by weight. Table 17.1 gives the fundamental combining weights of a number of common elements, including two clusters of atoms that behave as single entities in their usual chemical reactions but do not lead an independent existence and so cannot be isolated. Metals or units that behave chemically like metals are listed in the first group. Nonmetals or combinations like them are in the second group.

The simplest and most common chemical compounds are formed when an element in group 1 combines with one from group 2, and when such unions take place the weights involved are proportional to the combining weights listed. The number given as the **valence** is the chemist's measure

TABLE 17.1 CHEMICAL EQUIVALENTS

SUBSTANCE	SYMBOL	ATOMIC WEIGHT, g	VALENCE (NUMERICAL VALUE)	CHEMICAL GRAM-EQUIVALENT = ATOMIC WEIGHT (g) divided by VALENCE
GROUP 1: METALS				
Hydrogen	H	1.008	1	1.008
Silver	Ag	107.9	1	107.9
Copper	Cu	63.57	2	31.79
			1	63.57
Lead	Pb	207.2	4	51.80
			2	103.6
Aluminum	Al	26.97	3	8.990
GROUP 2: NONMETALS				
Oxygen	O	16.00	2	8.000
Chlorine	Cl	35.46	1	35.46
Sulfate group	SO_4	96.06*	2	48.03
Nitrate group	NO_3	62.01*	1	62.01

*Sum of atomic weights of constituents.

of the combining power of an atom or atom group of given kind as measured by the number of other atoms it can hold in combination.

Faraday's measurements showed that the mass of a given substance transported in an electrolytic cell is proportional to the total quantity of electricity that has passed through the cell. For example, if a total of 1,000 coulombs of charge is sent through a silver-plating cell, it is found that 1.118 g of the metal plates out on the cathode (and the same weight dissolves from the anode). If 10,000 coulombs passes, 11.180 g deposits, etc. This fact strongly suggests that each ion of a given element carries a definite amount of electricity through the solution. Faraday himself referred to the "atomicity of electric charge," anticipating the long sequence of corroborative discoveries and measurements that followed in related fields.

Since the quantity of electricity delivered in a given time by a steady current is $Q = It$, the result given above may be stated in the form

$$M = kIt \qquad\qquad\qquad \textbf{17.22}$$

where M is the mass of the element that deposits and k is a constant of proportionality whose value differs from one element to another.

The fundamental relation between the values of k for different elements is disclosed by passing the same quantity of electricity through a number of cells, as illustrated in Fig. 17.13. Suppose that the apparatus is allowed to run until exactly 107.9 g (one equivalent weight) of silver has been deposited in the silver cell. Then it is found that chemically equivalent weights of the various elements are liberated in the other cells as well. The interpretation of this fact leads to important conclusions. A silver ion (Ag^+) is formed by detaching a single electron from the normal silver atom. This leaves the ion with a positive charge equal in amount to the electron charge. When 107.9 g of silver is transported electrolytically, a total of 6.02×10^{23} electron charges are thus delivered, since this is the number of atoms in one gram atomic weight of any element.

For copper, on the other hand, only 31.79 g, or *half* the atomic weight, is found to be transported when the same total charge passes, which means that each copper ion carries twice as much charge as each silver ion. The copper ion is written Cu^{2+}, an ion of the element aluminum is Al^{3+}, and one of oxygen is O^{2-}, etc. The modern interpretation of valence is the number of electrons that an atom gains, loses, or shares with other atoms.

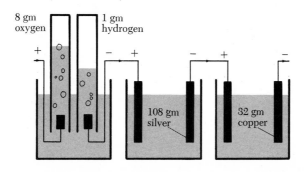

FIGURE 17.13 The passage of a given quantity of electricity releases equivalent weights of various substances.

This interpretation clearly makes chemical combination an electrical phenomenon.

Electrolysis experiments make it possible to compute the amount of electricity carried by each ion. The quantity of charge needed to transport one gram-equivalent of any substance is called one **faraday.** This is found by experiment to be about 96,500 coulombs. Since one gram-equivalent weight of an element of unit valence consists of 6.02×10^{23} atoms, the charge carried by each ion of such an element is

$$\frac{96{,}500}{6.02 \times 10^{23}} = 1.60 \times 10^{-19} \text{ coulomb}$$

This value agrees with the one given for the electron charge determined by Millikan's oil-drop experiment three-quarters of a century later (page 383). Faraday's value was not exactly the one stated above, since the value of N_0 was not known with sufficient accuracy at that time. In practice, the procedure is reversed, and the value of N_0 is computed from the more accurately known value of the electron charge.

The process of electrolysis provides an understanding of the mechanism of ionization and electric conduction in solutions, and it is important in the chemical industry as well. The plating and refining of metals has already

BIOELECTRICITY

Life processes generate electric potentials in probably all organisms, both plant and animal. **Bioelectricity** serves a variety of purposes, such as the transmission of nerve impulses, regulation of growth, etc.

Living tissue as a source of current   Nerve, muscle, or glandular tissue can be the source of electric potentials and the currents associated with them. The human heart produces potentials of the order of 0.001 to 0.002 volt (1 to 2 millivolts) at the body surface. The variations of these potentials can be picked up, amplified, and recorded as electrocardiograms for studying the action of the heart. Similarly, the brain can reveal corresponding information about its functioning through the potentials (about 2×10^{-5} to 10^{-4} volt) that it produces. There are several species of fishes capable of producing electrical pulses whose power output may be as great as 3 kW. The electrical organs are stacks of wafer-like cells, all facing the same way and suggesting a structure similar to that of a battery (see Ref. 17.5).

Nerve impulses and their propagation   The transmission of impulses along a nerve is an electrochemical process. A nerve cell is a tubular structure containing a conducting fluid and surrounded by another conducting fluid (Fig. 17.14a). The outer fluid is rich in Na^+ ions, and the inner one contains K^+ ions in great number. A still unexplained mechanism continually ejects Na^+ ions through the cell wall to the outside, maintaining a potential difference of 60 to 90 millivolts between interior and exterior. When a nerve impulse is generated, there is a momentary change in the permeability of the wall membrane. Na^+ ions enter the cell locally and, a moment later, K^+ ions leave (Fig. 17.14b). This affects the adjoining regions and a wave of depolarization moves along the fiber. Depending on a number of circumstances, the speed of travel may range from about 0.5 to over 120 m/sec. The regions where the nerve fibers are connected act as rectifiers (page 409), allowing the current to

been mentioned. Many chemical substances, such as hydrogen, chlorine, sodium hydroxide, and aluminum, are readily prepared on a commercial scale by electrolysis. In recent years engineers have perfected **fuel cells** that can produce more electric energy per unit weight of fuel than any except nuclear devices. The process in a fuel cell is the opposite of electrolysis: Oxygen and hydrogen flow into the cell on opposite sides of a membrane, where they ionize and eventually combine to form water. The ionization process causes electrons to circulate in an external circuit, making the cell a source of emf.

17.10   THERMAL ELECTRONS; CONDUCTION IN GASES

The electron gas in a metal is ordinarily unable to escape because the attraction of the positive centers prevents the electrons from leaving, even though they are in high-speed random motion. However, if the temperature is raised, the speeds increase and reach a point where electrons begin to "boil out" of the metal. This **thermal emission** of electrons becomes profuse when the metal is hot enough to glow visibly. The discovery of this phenomenon may be traced to a chance observation made in 1883 by the American inventor Thomas Edison (1847–1931) in connection with

The Edison effect.

proceed only in one direction. As in voltaic cells, the energy source for these currents is chemical.

Electrical activity of the heart   The heart maintains periodic electrical pulsations that cause the chambers to contract in sequence as they pump blood through the circulatory system. In certain heart conditions where the organ is incapable of maintaining its normal rhythm of electrical activity, a pacemaker is temporarily implanted in the chest wall. Such a unit may consist of a pulsing circuit powered by miniature mercury cells. Or, instead of chemical cells, it may use a crystal that produces potential differences by means of the contractions of the heart itself. Another recently developed pacemaker uses a small implanted coil that can be left in place permanently. There are no wires connecting this coil to the outside, and it receives power from an external coil by the process of electromagnetic induction (Chap. 19).

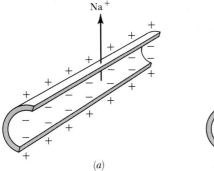 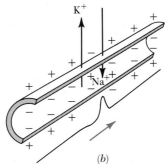

FIGURE 17.14 The propagation of a nerve impulse is an electrochemical process.

work on his early lamps. He noticed that a small current passed between the hot filament and an additional electrode sealed into the side of the bulb whenever the extra electrode was made positive, but not when it was negative (Fig. 17.15). Later, J. A. Fleming in England showed that the effect was due to the passage of electrons from the hot filament across the vacuum to the positive electrode, and the "Fleming valve" was soon used to *rectify* alternating currents (see Ref. 17.6).

Thermal electron emission is similar to the evaporation of a solid or a liquid, in which some of the more energetic atoms or molecules escape from the surface. If the vapor molecules are not removed from the neighborhood of the liquid surface, they tend to collect there, hindering the escape of other molecules. In the same way, a cloud of electrons collects around the emitter, making the escape of further electrons difficult. But if a nearby plate is given a positive charge, the tendency is to sweep the electrons across to it, and this, in effect, constitutes a current. Figure 17.4c (page 409) shows how the strength of the current varies with the applied potential difference. The current increases steeply at first but approaches a *saturation* value when the electrons are swept across the tube as fast as they are emitted from the hot surface. Since the graph is by no means a straight line through the origin, the current is not proportional to the applied potential difference and so this kind of conduction does not follow Ohm's law.

The vacuum diode.

The two-element electron vacuum tube used for rectifying alternating currents is now called a **diode** (Fig. 17.16). The cathode which emits electrons is heated by current in a filament connected to a source of emf (not shown). If an alternating emf is applied to the tube circuit, there is an electron current during the half cycle when the plate is positive with respect to the cathode but not during the half cycle that the plate is negative. Thus an alternating emf impressed on the tube produces an intermittent direct current in the circuit. By using a pair of diodes in a suitable circuit, the output current can be made quite constant.

The triode.

In 1907 the American inventor Lee De Forest (1873–1961) modified the electron tube by introducing a third electrode, called the **grid** (Fig. 17.17), and thereby opened up a great new field of application for such devices. The three-electrode tube is called a **triode.** Small changes in the potential applied to the nearby grid can produce large changes in the electron current passing from cathode to plate, so that the grid provides a sensitive control of this current. Extremely weak potentials produced in the receiving antenna by electromagnetic waves from radio or television broadcasting

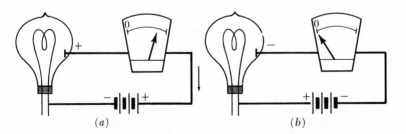

(a)   (b)

FIGURE 17.15 The Edison effec In (a) the side electrode is positiv with respect to the filament, and a electron current crosses the vac uum. In (b) the battery has bee reversed, making the side electrod negative, and there is no curren

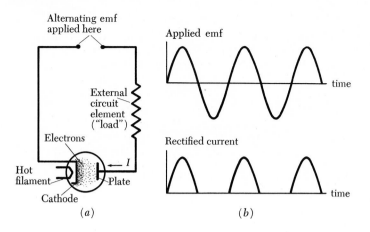

FIGURE 17.16 The diode as a rectifier. Current in the load is in one direction only.

stations are amplified by applying them in succession to the grids of a number of triodes. The resulting electrical oscillations, which may be millions of times stronger than the incoming ones, are further modified to yield audible sounds when applied to a loudspeaker or to yield pictures on a fluorescent screen.

Vacuum tubes find their widest use in the production, amplification, and control of electrical oscillations. Among their most important applications are radar, communication with and control of missiles and space vehicles, electronic computers, and other automated systems. However, since the late 1950s, a solid-state semiconductor device—the transistor—has increasingly replaced vacuum tubes in many applications. Further details will be found in Ref. 17.7.

Conduction in gases.   The molecules of gases are electrically neutral as a rule, so that these substances are good insulators. However, outside influences can ionize gases and render them conducting. Such agencies as cosmic rays and the radioactivity of the earth's crust produce a slight ionization, with the result that there are at all times a few hundred ions in each cubic centimeter of air near sea level. A charged body attracts ions whose charge is opposite to its own and, as a result, slowly discharges. It was the increased rate of leakage from charged electroscopes at high altitudes that led to the discovery of cosmic rays around 1900.

Bringing in other ionizing influences (x rays, ultraviolet light, or various

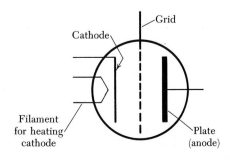

FIGURE 17.17 The triode (schematic). The grid usually consists of a wire-mesh cylinder surrounding the cathode. The grid, in turn, is within a cylindrical anode.

radiations produced in nuclear changes) increases the number of ion-electron pairs and makes a gas a conductor. If positive and negative electrodes are inserted, the ions and the free electrons move in opposite directions, constituting a current in the gas. Increasing the potential difference between the electrodes makes the current increase at first, but subsequently a saturation condition is reached (Fig. 17.4d, page 409). A further increase in potential finally leads to a new and very steep rise in the current. This happens when the charge carriers, particularly the electrons, attain enough KE so that they ionize additional gas molecules on colliding with them. The resulting ions and electrons in turn acquire enough KE to produce ionization by collision, and the process goes at an ever-increasing rate. The electrical discharge process is now able to maintain itself without the help of an external ionizing agent. Many features of the gaseous conduction process are still imperfectly understood, and further details will not be given here. It is apparent from what has been described that this process does not follow Ohm's law.

At sufficiently high gas temperatures, the molecular collisions are energetic enough to cause ionization. Flames are fairly good conductors, and at temperatures of several thousand degrees—as at the surface of the sun—most molecules disintegrate into ions and electrons. In the interior of a star where the temperature may be several tens of millions of degrees, matter is a mixture of highly ionized atoms and free electrons, called a **plasma** (page 227).

Programmed review

Instructions: See page 18.

1. What is meant by the current in a conductor?

    The net amount of charge passing through any cross-section per unit time: $I = Q/t$. [17.1]

2. State Ohm's law.

    The current in a metallic conductor is directly proportional to the applied potential difference and inversely proportional to the resistance: $I = V/R$. [17.2]

3. Describe the role of the free electrons in the existence of a current in a metal.

    The current consists of the slow drift of free electrons under the influence of the applied potential difference. [17.2]

4. Give expressions for the equivalent resistance of conductors connected in series and in parallel.

    Series: $R = R_1 + R_2 + \cdots$; parallel: $1/R = 1/R_1 + 1/R_2 + \cdots$. [17.3]

5. State Joule's law in quantitative form.

    The rate of conversion of electric to thermal energy in a resistor is given by $P = IV = I^2R = V^2/R$. [17.4]

6. What is meant by a source of emf?

    A circuit element where a reversible change between electric and some other form of energy occurs. The value of the emf is $\mathcal{E} = W/Q$.   [17.5]

7. Define anode, cathode.

    The positive and negative terminals, respectively, of a source of emf or its equivalent.   [17.7]

8. State Faraday's law of electrolysis.

    The mass of a given substance transported in an electrolytic cell is proportional to the total quantity of charge that passes.   [17.9]

9. What is meant by thermal emission of electrons?

    The release of electrons from an object at high temperature.   [17.10]

10. What is the process of ionization by collision in a gas?

    The ionization of neutral gas molecules by the impact of moving ions and electrons. The fragments so produced may in turn cause further ionization, thus continually increasing the rate of the process.   [17.10]

For further reading

17.1. *Taylor, L. W.* "Physics, the Pioneer Science." Read pp. 663–669 on the work of Ohm.
17.2. *Buchhold, T. A.* Applications of Superconductivity, *Sci. Am.,* March, 1960, pp. 74–82.
17.3. *De Santillana, G.* Alessandro Volta, *Sci. Am.,* January, 1965, pp. 82–91. An interesting, well-illustrated article on his work and on his dispute with Galvani.
17.4. *Pierce, J. R.* "Electrons and Waves." A simple introduction to electronics.
17.5. *Grundfest, H.* Electric Fishes, *Sci. Am.,* October, 1960, pp. 115–124.
17.6. *Shiers, G.* The First Electron Tube, *Sci. Am.,* March, 1969, pp. 104–112.
17.7. *Holden, A.* "Conductors and Semiconductors."

Questions and problems

17.1. Describe what would happen if the ammeter in Fig. 17.3 were removed from its present position in the circuit and inserted, instead, between the first and second cells of the battery.

17.2. In a certain proton accelerator, protons can be injected in bursts containing about 6×10^{10} protons each. If 20 bursts occur per second, what is the average current, in amperes, in the proton beam?

17.3. In the text it was stated that the resistance of all connecting wires should be negligible in comparison with the resistances of the circuit elements. Compute the total (series) resistance of four connectors, each 60 cm long and made of copper wire 1 mm in diameter. For the resistivity of copper, see Worked example 17.2 (page 411).

17.4. How many different resistance values can be obtained by using three conductors R_A, R_B, and R_C, singly and in all combinations? Sketch a diagram of each arrangement.

17.5. Two equal resistors are connected in parallel. By what factor does their combined resistance change when they are connected in series instead?

17.6. Prove that, if two resistors R_1 and R_2 connected in parallel carry a total current I, the current in R_1 is given by $I_1 = R_2 I / (R_1 + R_2)$.

17.7. Two resistors of values R_1 and R_2, respectively, have a combined resistance of R_s when connected in series and R_p when connected in parallel. Show that $R_s R_p = R_1 R_2$.

17.8. Three resistors are connected in the form of a delta, as in Fig. 17.18. Compute the resistance of the combination measured between (a) A and B; (b) A and C; (c) B and C.

17.9. An experimenter opens a "black box" marked "Resistance, 1 ohm" and finds inside a number of identical resistance coils connected as shown in Fig. 17.19. What is the resistance of each of these coils?

17.10. In Fig. 17.20, both the circle and the diameter AB are made of uniform wire of constant resistance per unit length. Prove that the ratio of the resistance measured between A and C to the resistance between A and B is $(\pi + 6)/8$.

17.11. A lamp whose nominal power rating is 60 watts is connected into a circuit where the potential difference across the filament is measured as 97.5 volts and the current as 0.610 ampere. Find the true power expended under these conditions.

17.12. The utility company sends each customer a monthly bill for the total electric energy supplied to him in that period, charging according to the number of kilowatt-hours used (power $\times$ time $=$ energy). If the rate in a certain locality is 3.2 cents per kilowatt-hour, what is the amount of the bill for a month when ten 60-watt lamps are run simultaneously for a total of 120 hr, a 550-watt washing machine for 4 hr, a 550-watt toaster for 5 hr, a 250-watt television set for 80 hr, and a 200-watt refrigerator for 320 hr?

17.13. An average office worker has a food requirement of about 2500 kcal per day (see page 248). At the rate charged in your locality, how much would the equivalent amount of electric energy cost? Compare with your estimate of the cost of the corresponding amount of food.

17.14. A 10-ohm and an 18-ohm resistor are connected in parallel in a circuit. If the current in the 10-ohm resistor is 2.0 amperes, at what rate, in watts, is heat being generated in the 18-ohm resistor?

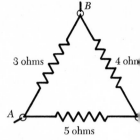

FIGURE 17.18

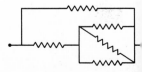

FIGURE 17.19 Resistors inside "black box."

17.15. A 25-watt lamp and a 60-watt lamp are connected *in series* to a battery. Which lamp glows more brightly? Explain. HINT: The indicated power ratings give the rate of energy transformation in each lamp when it is connected directly to the battery.

17.16. Five hundred cubic centimeters of water, initially at 20°C, is being heated by an electrical immersion heater rated 250 watts. Assuming that 85 percent of the heat supplied enters the water, how long does it take to bring the water to a boil?

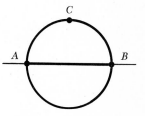

FIGURE 17.20

17.17. A certain source of emf delivers energy to a circuit at the rate of 32 watts when the current in it is 4.0 amperes. What is the value of the emf?

17.18. A 12-volt automobile battery delivers a current of 4.0 amperes to an external circuit. By how much has the total chemical PE of the battery decreased after the current is on for 8.0 min?

17.19. A dry cell whose emf is 1.48 volts has a 4.0-ohm resistor connected across its terminals, at which time a voltmeter connected between these points reads 1.40 volts. Calculate the internal resistance of the cell.

17.20. A battery of emf 4.00 volts and internal resistance 0.150 ohm is connected in series with a 6.00-ohm resistor and an 8.00-ohm resistor. Compute (a) the current, (b) the potential difference across each resistor, and (c) the difference of potential across the battery terminals.

17.21. A copper plate is the cathode in an electroplating cell in which a current of 0.040 ampere is maintained. If the metal is transported in the form of Cu^{2+} ions, how long must the cell be run in order to plate out 0.014 g of copper?

17.22. How long will it take a steady current of 0.020 ampere, applied to a copper electrolytic cell, to deposit 1.27 g of this metal?

17.23. In order to check the readings of an ammeter, it is connected in series with a silver electrolytic cell. During a run lasting 5.00 min, the ammeter reads 0.100 ampere and 0.0322 g of silver deposits. What is the error of the meter at this point of its range? Does it read too high or too low?

17.24. In a neon gas discharge tube 2.9×10^{18} Ne^+ ions move to the right through a cross-section of the tube each second, while 1.2×10^{18} electrons move to the left in this time. Find the magnitude and direction of the net current in the tube.

Chapter 18
THE MAGNETIC FIELD

The name "magnetite" may come from the ancient city of Magnesia in eastern Greece, where this mineral was found.

The first observations of magnetism can be traced back more than 25 centuries to ancient Greece. It was noticed that pieces of a mineral called magnetite, an oxide of iron, attracted bits of iron. The early Chinese discovered that a splinter of magnetite hanging from a thread would set itself in a north-south direction, and they were probably the first to use such an arrangement as a compass. It was not until the beginning of the seventeenth century that Gilbert (page 179n) gathered the known information on the behavior of magnets and published it, together with the results of his own careful experiments.

The subject of magnetism remained little more than a curiosity until the early part of the nineteenth century, when a number of observations linking magnetic phenomena with electric currents led to an understanding of the nature of magnetism and initiated the great technical developments that came afterward. The fundamental discoveries and some of their technological consequences will be discussed in this and the following chapters.

MAGNETIC FORCES

18.1 MAGNETIC EFFECT OF A CURRENT; OERSTED'S DISCOVERY

When currents are established in two straight, parallel conductors, each is found to exert a force on the other. These forces, which are equal and opposite, lie in the common plane of the two conductors in a direction perpendicular to their length (Fig. 18.1). The force is an attraction when the two currents are in the same direction, and a repulsion when the currents are in opposite directions. The magnitude of the force is proportional to the product of the two currents.

This force is not an electrostatic one, for the conductors themselves have no resultant charge even while carrying current because electrons entering each wire at one end leave in equal number at the other end. The force

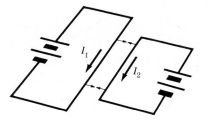

FIGURE 18.1 Currents in adjoining wires exert magnetic forces on each other.

between the pair must be ascribed to some kind of interaction resulting from the *motion* of electrons in the two conductors. Similar interactions are observed between streams of free charged particles in a vacuum, or ions in electrolytes. When such forces were first studied, no fundamental reason for their existence was known, any more than there is even now for electrostatic forces. Both had to be accepted as empirical facts describing nature. It can be said only that, in addition to electrostatic forces, there exist *magnetic* forces that arise when the charges are in *motion*. Evidently a charged particle in motion has properties that it does not have when at rest, just as a mass acquires the additional property of KE when it is set into motion. The presence of a magnetic force can be accounted for as a consequence of the special theory of relativity.

The crucial observation was made in 1819 by the Danish physicist H. C. Oersted (1777–1851). He discovered that a compass needle placed near a wire carrying a current takes a position nearly crosswise to the wire. If the current is reversed, the needle swings into a crosswise position but points to the opposite side (Fig. 18.2). The effect lasts only while the current is on.

The action on the compass needle, which is a small magnet, implies that the existence of a current of electricity produces magnetic effects. The interactions of a magnet and the current in a straight wire, as well as the interactions of a pair of current-carrying straight wires described above, are special cases of more general magnetic effects which will be described in later sections. It is now known that all magnetic effects are, fundamentally, interactions between moving charges of electricity.

As in the case of electrostatics, the entire subject of magnetism has developed from the point of view of the field concept. The presence of permanent magnets, current-carrying conductors, and streams of charged

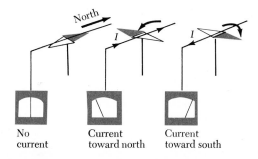

No current

Current toward north

Current toward south

FIGURE 18.2 Oersted discovered that an electric current is accompanied by magnetic effects.

particles is assumed to alter the surrounding space, giving rise to a **magnetic field** which in turn acts on magnets or moving charges placed in it. Thus, as in electrostatics, a field is invented as an intermediary agency in describing certain effects. For instance, in Oersted's experiment the current in the wire can be considered to produce a magnetic field which then exerts certain forces on a compass needle placed in it. Much of this chapter is concerned with describing these two aspects of the subject: the magnetic fields produced by various configurations of moving charges, and the effects of such fields on other systems of moving charges.

18.2 DESCRIBING THE MAGNETIC FIELD

In Chap. 16, the electrostatic field was explored by means of a test charge (Sec. 16.5). In the investigation of magnetic fields, the corresponding test object is a *moving* charge. In practice, a device originally called a **cathode-ray tube** can be used. With certain features added, this device evolved into the oscilloscope tube and the television picture tube. For present purposes, however, it consists merely of an **electron gun** mounted at one end of a highly evacuated tube. The enlarged opposite face of the tube is coated on the inside with a fluorescent material, and the fine stream of electrons from the gun produces a bright spot of light where it strikes the coating (Fig. 18.3).

A qualitative exploration of a magnetic field can be made by moving such a tube from place to place and orienting it in all possible directions. If the spot of light remains always at the same point on the screen of the tube, it can be concluded that there is no magnetic field in the region in question, at least within the limits of sensitivity of the apparatus. However, in most instances a deflection of the electron beam will be observed, but even then it is found that by turning the tube in various directions at a given place an orientation can be found where the deflection becomes zero. The direction of the longitudinal axis is then taken to be the line of action of the magnetic field at that point (Fig. 18.4a).

Next, the tube is turned so that its axis lies in any position in the plane *perpendicular* to the field (Fig. 18.4b). It is observed that the beam is bent aside *in* this plane, the deflection being always in the same directional sense and of the same amount, regardless of the orientation of the tube in this plane. Thus, when charged particles move in a plane perpendicular to the magnetic field, they experience a force that lies in this plane. In addition, it is found that the stream of particles is bent into a circular arc. This must mean that the acting force, in the nature of a centripetal force, is directed radially inward and is everywhere perpendicular to the path. To summarize

The gravitational force on electrons is negligible compared with the other forces involved. See Worked example 18.1.

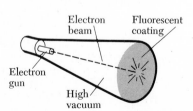

Electron beam   Fluorescent coating

Electron gun

High vacuum

FIGURE 18.3 A cathode-ray tube for exploring magnetic fields.

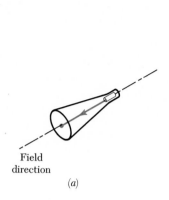

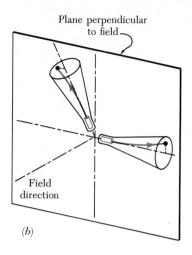

FIGURE 18.4 The direction of the electron beam for which it experiences no deflection is taken to be the line of action of the magnetic field.

the above observations: The *force acting on a moving charged particle is perpendicular to both its line of motion and the line of action of the field.*

The arrangement described in the preceding paragraph can be modified to make it more general. Instead of the particles being projected in a plane perpendicular to the field, their path may make an angle θ with the field direction (Fig. 18.5). Observation shows that only the component of the particle velocity perpendicular to the field affects the motion and that the force on each particle is directly proportional to this component. Thus if v is the particle speed, the perpendicular component is $v \sin \theta$.

A further generalization results from experimenting with particles carrying various amounts of charge. In order to carry out such experiments it is necessary to change from a cathode-ray tube to an ion tube capable of furnishing streams of particles having different charges. While making this change, only positive ions are used in these tests, making it possible to refer the observed effects to streams of positive-charged particles, corre-

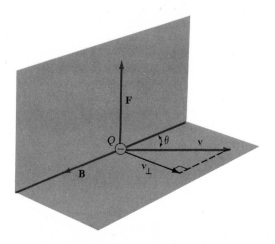

FIGURE 18.5 The force on a charged particle moving in a magnetic field depends on the component of its velocity perpendicular to the field.

sponding to the direction of the conventional current (see page 405). The results of such experiments show that *the magnitude of the force acting on a moving charged particle is directly proportional to its charge.*

In Chap. 16 an electrostatic field vector **E** was defined on the basis of the observed effects of such a field on a test charge. Similarly, we can now define a magnetic field vector **B** in terms of the observed behavior of a *stream of charge* when it is placed in such a field. For historical reasons, this vector is usually called the **magnetic induction,** although "magnetic field strength" would appear to be a more suitable name, paralleling the electrical term. However, *field strength* is traditionally used in another sense in this subject.

The magnitude of the induction at any point is taken to be

$$B = \frac{F}{Qv \sin \theta} \qquad\qquad 18.1$$

where Q is the charge on each particle, v the particle speed, θ the angle between the direction of particle motion and the field, and F the magnitude of the magnetic force on each particle. This definition merely summarizes the experimental facts described above: F is proportional to both the charge and the transverse component of the particle speed, with B the constant of proportionality.

In choosing a suitable mks unit for B, coulombs are used for Q, meters per second for v, and newtons for F. Then, from the equation, B has the units N-sec/C-m. For reasons to be explained later, this combination of units is called webers per square meter (Wb/m²). One weber per square meter is the value of the magnetic induction at a place where a particle bearing a charge of one coulomb and moving with a speed of one meter per second in a direction normal to the field experiences a magnetic force of one newton.

The units weber and gauss were named for the German physicists Wilhelm Weber (1804– 1891) and C. F. Gauss (1777–1855).

An earlier-named unit, the *gauss*, is still used in the cgs system. The relation between the two units is

1 Wb/m² $= 10^4$ gauss

Fields as high as 1,000 Wb/m² have been generated momentarily in the laboratory (see Ref. 18.4), and values of the order of 10^8 Wb/m² are predicted in collapsed neutron stars. By comparison, the earth's field, which causes a compass needle to orient itself in a north-south position, is of the order of about 5 $\times$ 10^{-5} Wb/m² at sea level.

For simplicity, Eq. 18.1 can be rewritten in the form

$$F = BQv_{\perp} \qquad\qquad 18.2$$

where $v_{\perp}$ ($=v \sin \theta$) stands for the component of v perpendicular to the field direction. The directional sense of **B** has not yet been specified. The directional sense of **B** is arbitrarily assigned so that, in turning the vector **v**$_{\perp}$ toward the position of **B,** the force **F** on a positive particle is in the direction in which a right-hand screw advances (Fig. 18.6). This is called the **right-hand-screw rule.**

From the description above, it is apparent that the magnetic-force situations are more complicated than those involving gravitational or electrostatic fields, where the force acting on a test body is in the line of action

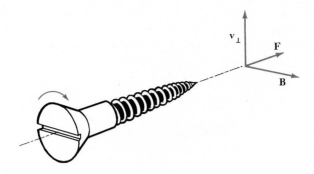

FIGURE 18.6   Right-hand-screw rule for finding the directional sense of the force on a moving positive charge. For a negative charge, the direction of $\mathbf{v}_\perp$ must be reversed.

of the field. For magnetic forces, an additional directional element (the motion of the charge) must be taken into account. This leads to relations in space for which direction rules such as the screw rule must be devised.

Practice set*

1. A proton is projected horizontally eastward in a magnetic field whose induction vector is in a horizontal southward direction. What is the direction of the magnetic deflection force?
2. The same for projection vertically downward and induction vector vertically upward.
3. Same conditions as in 1, but the particle is an electron.

Worked example 18.1

An electron is shot into a magnetic field whose induction vector has a magnitude of 1.0 Wb/m² with a speed of 5×10^7 m/sec. The angle between the velocity of the particle and the induction vector is 30°.

a. Find the magnitude of the force on the particle.

b. Find the ratio of the weight of the electron to this force. (The value of $Q = e = 1.60 \times 10^{-19}$ coulomb.)

Solution   a. According to Eq. 18.2,

$$F = (1.0)(1.6 \times 10^{-19})(5 \times 10^7)(0.5) = 4 \times 10^{-12} \text{ newton}$$

b. The mass of the electron is $m = 9.1 \times 10^{-31}$ kg; its weight is $w = mg = (9.1 \times 10^{-31})(9.8) = 8.9 \times 10^{-30}$ newton. The ratio is about $(9 \times 10^{-30})/(4 \times 10^{-12})$, or about 2×10^{-18}. This shows that the gravitational force is entirely negligible in comparison with the magnetic force, even for considerably smaller values of magnetic induction and speed than used in this example.

The concept of magnetic lines was suggested as early as 1269 by Petrus Peregrinus, a French savant.

Compare Fig. 16.5a and b, page 381.

Faraday introduced the concept of lines of force (Chap. 16) to help describe electrostatic fields. He found that similar constructions, called **lines of magnetic induction,** are useful in describing magnetic fields. In a familiar demonstration, iron filings are strewn onto a sheet of paper placed over a magnet (Fig. 18.7). The filings arrange themselves along curved paths that suggest the form of electrostatic field lines in the neighborhood of charges. The magnetic lines appear to concentrate at regions near each

Ans.: downward; no deflection; upward.

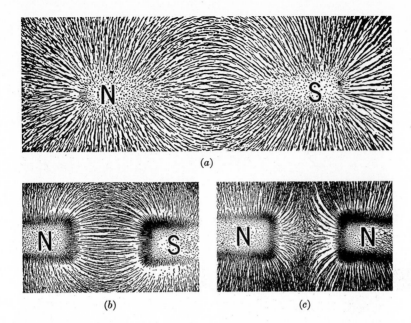

(a)

(b)   (c)

FIGURE 18.7 Iron filings line u
to portray the magnetic field lines
Patterns produced by (a) a ba
magnet, (b) opposite poles of tw
bar magnets, (c) like poles of tw
bar magnets.

end of a bar magnet, called its **poles.** If the bar is pivoted so that it is free to swing like a compass needle about a vertical axis, the pole at the end that points north is called its **N** pole. The other end is its **S** pole. As in the rule for the force between charges of electricity, like poles repel each other and unlike poles attract.

At each point in a field, the magnetic induction vector **B** is tangent to the field line passing through that point (as in electrostatics) and is in the direction of the field, as in the electrostatic situation. To agree observationally with the directional sense assigned to **B** above, this direction turns out to be that of the force on an **N** pole. Further, the magnitude of **B** at any place is represented by making the number of lines drawn through unit area of a surface perpendicular to **B** proportional to the magnitude of **B.** Thus a field of 1 Wb/m^2 is represented by drawing one line per square meter, so that "Wb" is synonymous with "line of magnetic induction."

There is further similarity with the electrostatic field. A **uniform magnetic field** is one where **B** is constant everywhere, and the lines of induction are straight and equally spaced. The field between the flat, parallel pole faces of a magnet is uniform, except at the edges.

This relation expresses B *as* Φ/A, *which is sometimes referred to as* flux density, *that is, flux per unit area.*

The total number of lines of induction piercing a given surface is called the **magnetic flux** through that surface. If **B** is uniform, the flux Φ through a surface of area A perpendicular to the field is given by

$$\Phi = BA \qquad\qquad 18.3$$

In this definition, B is measured in webers per square meter and A is in square meters, so that the flux Φ is expressed in webers.

18.3 MOTION OF A CHARGED PARTICLE IN A MAGNETIC FIELD

It was shown above that a charged particle projected into a magnetic field is acted upon by a sidewise force that deflects it from its initial direction of travel. In Fig. 18.8 a positive-charged particle is launched with a velocity **v** which lies in the plane of the diagram. A uniform magnetic field **B,** perpendicular to this plane, is indicated by the uniformly spaced dots, and its direction is *toward* the viewer. The magnetic force **F** on the particle is perpendicular to both **v** and **B** and, according to the right-hand-screw rule, its direction is the one shown. The magnitude of this side thrust is given by Eq. 18.2 and is constant as long as B, Q, and v do not change. The particle is assumed to be moving in free space, with no other forces acting on it. Since the force **F** is always perpendicular to the path, it has no component *along* the path and therefore cannot change the speed of the particle. In fact, **F** plays the part of a centripetal force, so that the path is a circle. The value of F given by Eq. 18.2 can be equated with the general expression for centripetal force (Eq. 8.5, page 172), resulting in $BQv = mv^2/R$, where m is the mass of the particle and R is the radius of the resulting circular path. This may be solved for R:

$$R = \frac{mv}{BQ} \qquad\qquad\qquad \textbf{18.4}$$

If the particle is not launched in a plane perpendicular to the field, it will have a constant velocity component $v_\parallel$ in the direction of the field, in addition to the v shown above. This $v_\parallel$ will not give rise to any magnetic force but will merely move the plane of the circular orbit uniformly along the field lines at this speed, making the actual path of the particle a helix (corkscrew) as in the photograph.

The effects of electric and magnetic fields on moving charges or currents (summarized in Fig. 18.9) are complex and should be studied carefully.

The magnetic deviation of charged particles was used by the English physicist J. J. Thomson (1856–1940) in a celebrated experiment (1897)

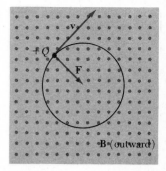

FIGURE 18.8 A charged particle projected into a magnetic field in a plane perpendicular to the induction vector describes a circular path.

Side view of the helical path of an electron in a uniform magnetic field. This cloud-chamber photograph shows the helix outlined by water drops that condensed along it. (*From Shortley-Williams, "Elements of Physics," 3d ed., Prentice-Hall, 1961.*)

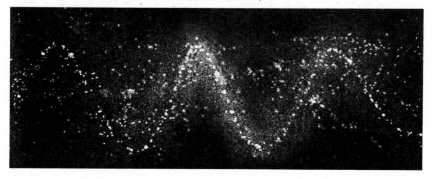

Effects of fields on moving charged particles

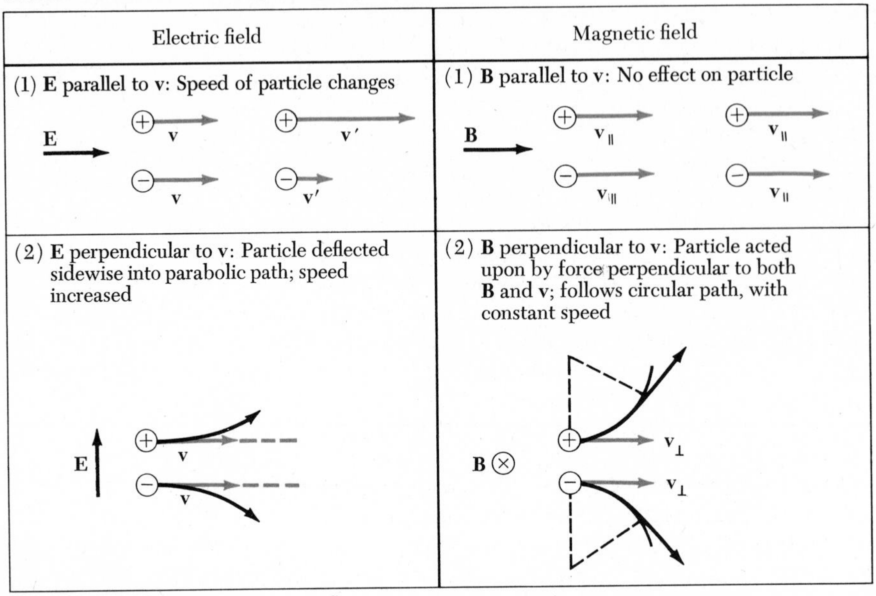

Electric field
Magnetic field
(1) E parallel to v: Speed of particle changes
(1) B parallel to v: No effect on particle
E
v
v′
B
(2) E perpendicular to v: Particle deflected sidewise into parabolic path; speed increased
(2) B perpendicular to v: Particle acted upon by force perpendicular to both B and v; follows circular path, with constant speed

FIGURE 18.9

Thomson's e/m experiment.

that first clearly indicated the existence of the electron and the smallness of its mass. The apparatus devised by Thomson is similar to the cathode-ray tube mentioned on page 434. It had a "cold cathode" instead of a heated emitter, and the nature of the "cathode rays" coming from this electrode was not known. The Thomson tube is shown schematically in Fig. 18.10. Rays originating at the cathode are accelerated by falling through a large

FIGURE 18.10 Thomson's tube for the measurement of e/m.

Cathode
Coils
Deflected beam
Anode +
Plates
Fluorescent screen

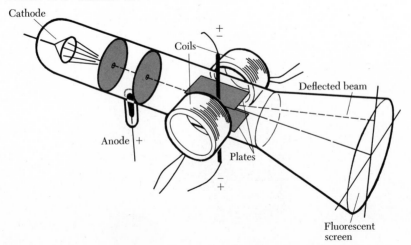

*The produc-
tion of a
field by
such coils is
described
later.*

difference of potential and are limited to a narrow beam moving along the axis of the tube. The beam passes between two parallel metal plates connected to the terminals of a high-potential battery. In the same portion of the tube the beam also passes through the uniform magnetic field produced by an external pair of coils and finally hits a fluorescent coating on the end of the tube. Depending upon the way the battery is connected, the uniform electric field between the plates is capable of deflecting the beam either upward or downward in a vertical plane. Thus the path of the rays is straight before it enters the capacitor, curved while between the plates, and again straight (but inclined to the axis of the tube) after leaving the plates (see Fig. 18.11). When the electric field is on, the bright spot produced by impact of the rays on the end of the tube no longer will be at the center but will be above or below this point.

The magnetic field is directed along the common axis of the two coils, and since the force exerted on the moving charges is normal to both the field and the direction of motion of the charges, the magnetic deflection—like the electric one—is in the up-and-down direction. By applying the electric and magnetic fields in the proper directions and by adjusting their strengths, their effects can be balanced so that the spot of light again falls exactly at the center of the screen.

If e is the charge on a cathode-ray particle, the electrostatic force acting on it when the electric field between the plates is E amounts to eE; its direction is opposite to that of the field, since the charge is negative. If the particle has a speed v, the transverse magnetic field B exerts a force Bev on it. When the two forces are made to balance, $eE = Bev$, or

$$v = \frac{E}{B} \qquad\qquad \textbf{18.5}$$

so that the speed of the particles can be found from the measured values of the two fields.

When the magnetic field alone is applied, the stream of particles now bends in a circular arc as it crosses the field and then goes on in a straight path to the screen. The radius of the circular arc is given by Eq. 18.4 (page 439) as $R = mv/Be$. Substitution of the value of v from Eq. 18.5 yields $R = mE/B^2e$. The value of R is readily found from the geometry of the apparatus and the measured deflection of the beam on the screen, so that

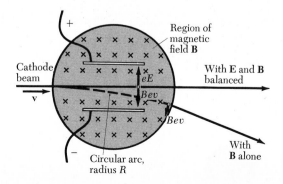

FIGURE 18.11 Paths of cathode rays in Thomson's e/m experiment.

the only unknowns in this relation are e and m. Accordingly, the ratio of these two quantities is given as

$$\frac{e}{m} = \frac{E}{B^2 R}$$ 18.6

The value obtained by Thomson is not far from the present accepted one, arrived at by a number of methods:

$$\frac{e}{m_0} = 1.7589 \times 10^{11} \, \text{C/kg}$$ 18.7

where m_0 is the rest mass of the particles (see below).

Neither e nor m separately can be found from these experiments. When the magnetic field is applied, the trace of the cathode ray on the end of the tube does not spread out but remains a small spot. This shows that the particles constituting the beam are all *alike,* all having the same value of e/m_0. Moreover, no matter what the nature of the gas originally in the tube or the material of which the electrodes are made, the same value for e/m_0 always results, showing the universal nature of the cathode-ray particles.

The large numerical value of e/m_0 impressed Thomson. By assuming that e is the same as the charge on a hydrogen ion, he found from Eq. 18.7 that the mass m of the electron must be nearly 2,000 times smaller than that of a hydrogen atom, the lightest particle then known. This was *Determining* the first clear evidence of the existence of particles smaller than an atom, *the mass of* indicating that atoms, previously considered indivisible, had a "structure." *the electron.* Thomson, working with H. A. Wilson, attempted to measure e by determining the total charge on a cloud of water droplets, but it was not until some years later that reliable values were obtained by Millikan, using the oil-drop method (page 382). Substitution of the accepted value of e into Eq. 18.7 yields the value of the rest mass of the electron:

$$m_e = 9.1072 \times 10^{-31} \, \text{kg}$$ 18.8

This number is about $1/1836$ of the mass of a hydrogen atom.

Deflection experiments with electrons revealed another result of great general importance: When e/m was measured for electrons of greater and greater speed, it was found that the value of this ratio decreased markedly, especially for the very fast electrons ejected from radioactive atoms, where the speeds approach that of light. There is no reason to assume that the charge on an electron depends on how it moves, and hence this variation in e/m must mean that *the mass of an electron increases as its speed increases.* This conclusion conflicted with previous ideas as to the nature of mass, since classical mechanics assumed that the mass of a given object was invariable. But, on the basis of electromagnetic theory, Thomson and others calculated that the inertia of a charged body should increase in a certain way with its speed of motion, and the experiments agreed with this result.

Relativistic   Later, the theory of relativity led to the same result for *any* moving body, *variation of* whether electrically charged or not. The mass m of a body moving with *mass and of* a speed v relative to the observer is given by the theory as *KE.*

$$m = \frac{m_0}{\sqrt{1 - (v/c)^2}}$$ 18.9

where m_0 is the rest mass (page 49). Figure 18.12 is a graph of this relation, showing the rapid increase of m at high speeds.

According to the special theory of relativity, the kinetic energy of a moving body is no longer correctly given by the classical expression $\frac{1}{2}mv^2$, even with the use of the altered value of m appearing in Eq. 18.9. Instead, the relation is

$$K = m_0 c^2 \left[\frac{1}{\sqrt{1 - (v/c)^2}} - 1 \right]$$

18.10

The following Worked example shows how the classical and relativity results compare.

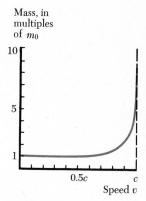

FIGURE 18.12 Variation of mass with speed according to the theory of relativity.

Worked example 18.2

An electron traversing a vacuum tube is moving with a speed $\frac{1}{2}c$.

    a. Find the ratio of its actual mass to its rest mass.

    b. Find its KE in electron volts.

Solution a. Substitution into Eq. 18.9 yields

$$\frac{m}{m_0} = \frac{1}{\sqrt{1 - (\frac{1}{2})^2}} = \frac{2}{\sqrt{3}} = 1.16$$

There is an increase in mass of about 16 percent due to the motion of the electron relative to the observer.

    b. Use Eq. 18.10:

$$K = (9.11 \times 10^{-31})(3 \times 10^8)^2 \left(\frac{2}{\sqrt{3}} - 1 \right) = 1.27 \times 10^{-14} \text{ joule}$$

or, with the conversion factor given on page 393, $K = (1.27 \times 10^{-14})/(1.60 \times 10^{-19}) = 79{,}000$ eV. Thus, an electron must fall freely through a potential difference of nearly 80,000 volts to acquire a speed of $\frac{1}{2}c$.

Mass spectrographs.

In 1907, Thomson began an investigation of positive ions, using a modification of the e/m apparatus described above. Since the magnitude of the charge on an ion is always an integral multiple of e—usually a low multiple—it is feasible to find the amount of charge for a specific ion. Then, by using the value of Q/m as measured by the apparatus, the mass is readily determined. The instrument in effect measures the masses of individual ions. Any device of this type is called a **mass spectrograph** because it sorts out ions with respect to their masses in a way that suggests the analysis of a mixture of wavelengths by an optical spectrograph. Various forms of mass spectrographs that permit measurements of atomic and molecular masses to an accuracy of a few parts in 100 million have been developed. These instruments revealed the surprising fact that all atoms of a given chemical element do not have exactly the same mass (page 567). For a discussion of modern mass spectrographs and their uses see Ref. 18.9.

When used for very precise measurements, a mass spectrograph may be called a mass spectrometer.

18.4 MAGNETOHYDRODYNAMICS

Most of the matter in the universe is in the form of a plasma (page 227), which is an electrically neutral mixture of ions and electrons. Both plasmas and magnetic fields are found in stars and nebulae and in the space be-

The solar wind is a plasma stream emitted by the sun. When this flow meets the earth's magnetic field, the Van Allen radiation belts are formed. This picture, made at NASA's Lewis Research Center, shows a simulated set of belts formed in a large vacuum chamber. The ions, generated at the right, are deflected by the magnetized sphere.

tween them. This means that magnetic forces on circulating masses of charged particles play an important part in many astronomical phenomena such as sunspots, solar flares, the internal motions of galaxies, etc. The study of the interaction of conducting fluids and magnetic fields is called **magnetohydrodynamics** (MHD). It is one of the most active branches of applied physics today.

In attempts to produce nuclear fusion (Chap. 22), laboratory experiments with plasmas make use of the principles of magnetohydrodynamics. The main concern is to confine a plasma, whose temperature may reach millions of degrees, without depending on walls made of matter. In some devices this is done by using what is called a magnetic bottle (Fig. 18.13). Electrified particles are injected into the magnetic field inside a magnetic coil consisting of a number of separately controllable coils. The high-speed particles move in tight spirals around the field lines (page 439), the field acting as a guide and confining the particles to the region near the axis. The field is strongest near the ends, so that the particles are turned back as they approach these points.

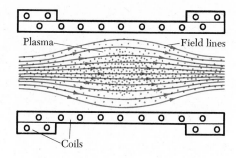

FIGURE 18.13 A "magnetic bottle" (schematic) for confining a hot plasma.

Besides the confining effect, the field can be used to compress the plasma by increasing the field strength in all the coils. This causes a rise in temperature which may lead to nuclear fusion. Decreasing the field at one end of the device while increasing it at the other moves the plasma along the axis up to speeds of 10^7 m/sec, and this phenomenon is being tested by engineers with a view to using it for jet propulsion. Another application under study is a magnetohydrodynamic generator in which a stream of conducting gas or liquid is passed through a crosswise magnetic field (Fig. 18.14). The electromagnetic forces on the moving fluid produce a difference of potential across the plates, which are connected to an external circuit and serve as a source of emf.

For a fuller description of some problems in this field, see Ref. 18.7.

18.5 CYCLOTRONS

Devices such as the Van de Graaff generator (page 394) for imparting KE to ions and other charged particles are limited in their operation by the difficulties of working with extremely high potentials. About 1931, the American experimenters E. O. Lawrence (1901–1958) and M. S. Livingston developed an ingenious scheme for accelerating particles by allowing them to fall repeatedly through moderate differences of potential. The apparatus, called a **cyclotron,** uses the principles outlined above.

Figure 18.15 shows the essential features of a cyclotron. Two metal half

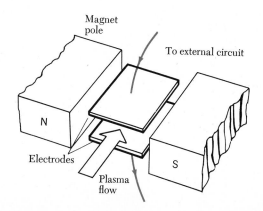

FIGURE 18.14 Principle of an MHD generator. Charge carriers of one sign are deflected to one plate; those of opposite sign go to the other plate. The result is a current in the external circuit.

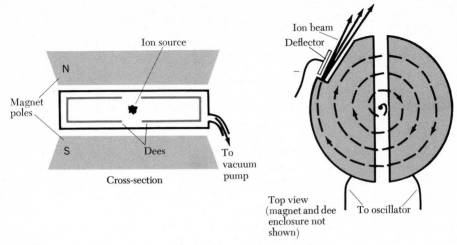

FIGURE 18.15 The cyclotron (schematic).

cylinders, called *dees* (because of their shape), are enclosed in a vacuum chamber placed between the poles of a strong magnet. At the center is a source of ions such as positive-charged nuclei of heavy hydrogen (deuterons). The dees are given alternately positive and negative charges by a powerful oscillating circuit like that of a radio transmitter. An ion moving in a horizontal plane is bent into a circular path by the magnetic field. While inside one of the dees, the ion is in a field-free region and so experiences no electric force. If a particle reaches a gap between the dees just as the charge on the dee ahead of it becomes opposite to its own charge, the particle will be accelerated. When this happens many times in succession, the particle can acquire high energy as it continues to accelerate in its orbit.

Several factors determine the path of each ion: From Eq. 18.4 (page 439), the speed of the ion is given by

$$v = \frac{BQR}{m} \hspace{4cm} \text{18.11}$$

After moving over a half circle inside one of the dees, the ion comes to the next gap. If the charges on the dees have the right signs at that time, the ion will be accelerated while crossing the gap and will move in a slightly larger circle in the second dee, since Eq. 18.11 shows that R increases with v. After traversing a half circle, the particle comes to a gap where it is once more accelerated, for in the meantime the potential difference across the dees has changed sign. The ion continues to move on increasingly larger circles as it goes from one dee to the other.

The important feature of the cyclotron is that conditions can be readily adjusted to make the revolving particles automatically maintain synchronism with the alternating potential applied to the dees, because the time required for the ion to make a half turn is independent of the size of the circle or the speed. This can be understood as follows: The time t taken by a particle to traverse a semicircle of radius R while moving at a constant speed v is given by $\pi R/v$. Substituting the value of v from Eq. 18.11 yields

$$t = \frac{\pi m}{BQ} \qquad\qquad \textbf{18.12}$$

This expression for t contains the charge and mass of the ion as well as the strength of the applied magnetic field, but it does not involve the radius of the orbit or the ion speed. Hence, t remains constant. By adjusting the value of B, t can be made the same as the half period of the oscillator. When the ions have completed a sufficient number of turns and approach the outer edge of the dees, an electric field applied to a deflecting plate swerves them away from the circular path and out through a thin window, where they are allowed to strike a target and produce the effects desired.

The potential applied to the dees may be of the order of 10,000 or 100,000 volts, so that after making a few thousand revolutions the ions emerge with energies of the order of 1 GeV or more. The oscillator frequency is of the order of 10 million cycles per second and the magnetic induction is around 1.4 Wb/m².

Modifica-tions of the cyclotron.

A limit to the particle speed attainable with the cyclotron is imposed by the relativistic mass increase. This impairs the constancy of t in Eq. 18.12, with the result that for speeds approaching c the ions get "out of step" with the oscillator and eventually receive no additional energy from it. The difficulty can be overcome by gradually decreasing the frequency of the applied potential as the ions proceed outward and acquire greater speed. This conclusion follows from Eq. 18.12, which shows that the frequency of circulation of the ion (which is inversely proportional to t) is inversely proportional to m. Hence as m increases, the frequency must be correspondingly decreased. An accelerator designed to do this is called a **synchrocyclotron.**

The attainment of very high energies with a conventional cyclotron requires a large and costly magnet. However, if both B and the oscillator frequency can be varied periodically, the radius of the particle orbit can be kept constant. This is done in a machine called a **synchrotron** which has a ring-shaped magnet instead of one with broad pole faces. Additional details are found in Refs. 18.2 and 18.4.

18.6 MECHANICAL FORCE ON A CURRENT-CARRYING CONDUCTOR

As shown above (page 435), a sidewise force is exerted on charged particles that move across a magnetic field. This is true also for the charge carriers in a conductor, in which case the forces are transmitted to the conductor itself. *A wire or other conductor carrying a current experiences a sidewise force when in a magnetic field;* this fact has many important applications. The amount and direction of the force will now be determined.

Figure 18.16 represents a portion of a straight, current-carrying conductor located in a plane perpendicular to a uniform magnetic field. Equation 18.2 (page 436) gives the magnitude of the magnetic force on each charge carrier as

$$F_1 = BQv_D$$

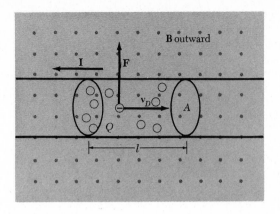

FIGURE 18.16 Sidewise force o electrons in a current-carryin conductor.

where v_D, the drift speed, is used in place of $v_\perp$. In any segment of wire of length l there are $N = nlA$ charge carriers, making the total force on the segment

$$F = F_1 N = (BQv_D)(nlA) \qquad\qquad \textbf{18.13}$$

The expression for the current (Eq. 17.2, page 407) may be written

$$I = Qnv_D A$$

When this relation is combined with Eq. 18.13 we get

$$F = IBl \qquad\qquad \textbf{18.14}$$

where the units to be used are newtons for F, amperes for I, webers per square meter for B, and meters for l. In words, the equation asserts that *each length l of a straight portion of a conductor in which there is a current I experiences a sidewise force F of amount IBl when placed crosswise to a magnetic field of induction B.* This equation can be generalized to include the case where the conductor makes an angle θ (not necessarily 90°) with the field by inserting a factor $\sin\theta$ on the right side.

The directional sense of **F** can be determined by using the right-hand-screw rule (page 436): If a right-hand screw is turned from the direction of **I** toward that of **B,** its direction of advance will indicate the direction of **F.**

Equation 18.14 can be used to compute the total force on a conductor of any shape by thinking of the conductor as a large number of very short segments, each of which can be considered straight. Then the equation (with the $\sin\theta$ factor incorporated if necessary) gives the force on each segment, and vector addition of all these forces then gives the total force. In practice, such computations are not easy to make unless the field is uniform and the conductor has a simple shape.

18.7   APPLICATION TO METERS AND MOTORS

It is an interesting fact that nearly all fundamental electrical and magnetic determinations are based on the measurement of *mechanical* quantities: Electric charge can be determined, at least in principle, through Coulomb's law by measuring a force. Electric and magnetic fields are specified by the force acting on charges at rest or in motion. Potential difference and emf are determined by work done in connection with the motion of a test charge. The most accurate method of measuring a charge is to pass it through a plating cell and weigh the deposit of metal on the cathode—a mechanical operation. Many other examples could be given.

Equation 18.14 can be applied to the determination of either current or magnetic induction by measuring a force; this is basically the principle of the ammeters and voltmeters used in steady-current circuits. In the usual form of current-measuring instruments, shown schematically in Fig. 18.17, the torque (see page 79) on a pivoted coil placed in the field of a strong permanent magnet is used as a measure of the current in the coil. The coil is usually wound on a rectangular bobbin, using a large number of turns of wire in order to increase the total mechanical effect. The net force on each side of the coil is shown in the figure. The two forces F_v have no effect on the rotation of the coil, but the pair F_h produce a torque as indicated.

Use the right-hand-screw rule to verify the directions shown for the forces.

If the coil were free to turn, it would rotate until it became perpendicular to the field and then stop, because in that position the forces F_h pass through the axis and no longer exert any torque. However, a spring opposes the turning of the coil so that, with any given current, it rotates only to the point where the opposing elastic torque of the spring becomes equal to the electromagnetic torque. A lightweight pointer attached to the coil moves over a scale to indicate the magnitude of the current in the coil. In order to read the current directly in amperes, this scale must be calibrated by comparison with a standard instrument or in some other way (see, for example, Prob. 17.23, page 431).

In most commercial meters, a stationary cylindrical piece of iron is placed inside the coil and the pole pieces of the permanent magnet are so shaped that the field is everywhere normal to the direction of motion of the vertical wires, permitting the instrument to have a scale with uniform graduations.

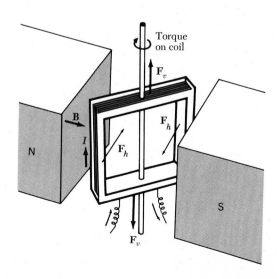

FIGURE 18.17 Principle of a moving-coil meter. The forces F_h, which are perpendicular to both **B** and the vertical sides of the coil, produce a torque around the axis as shown.

The most sensitive current-measuring instruments of the moving-coil type can respond to currents as small as 10^{-12} ampere.

An instrument of the above kind can be made into a voltmeter simply by putting a high-resistance coil in series with the moving coil. The deflection of the moving coil is proportional to the current through the meter which in turn, by Ohm's law, is proportional to the potential difference applied to the instrument. Therefore, once the scale has been properly calibrated, the instrument may be used to read potential difference directly in volts. As explained on page 407, the meter must be placed in a *side branch* connected between the two points in the circuit whose potential difference is required. The combined resistance of both fixed and movable coils must be large, so that the current diverted from the main circuit remains small.

Useful mechanical work can be obtained from the magnetic force acting on a current-bearing conductor in a magnetic field. If a pivoted coil such as the one described above is allowed to continue turning without opposition (as by the control spring), continuous conversion of electromagnetic energy into mechanical energy takes place. A practical device for doing this is an **electric motor.**

The principle of operation of a motor that uses direct currents is shown by Fig. 18.18. As in the meters described above, a loop of wire is pivoted to rotate in a magnetic field. Since continuous rotation is desired, the current can no longer be led into and out of the loop by fixed wires. Instead, this is done through a commutator—a split ring on which contacts slide. When the current is sent through as shown in Fig. 18.18, the loop turns in the indicated direction until its plane is vertical. At that point, however, the current through the loop is automatically reversed by the switching of connections as the commutator gaps pass the contacts. This reversal of the current compels the coil to make another half turn, at which point reversal occurs again, and so on. The result is continuous rotation in one direction. The design of a motor using multiple loops is shown in Fig. 18.19.

In practice, there are several coils, oriented at various angles, as shown in Fig. 18.19.

Most electric motors now used on a commercial scale operate on alternating current, and their design differs from the above description. Details can be found in books on electrical engineering.

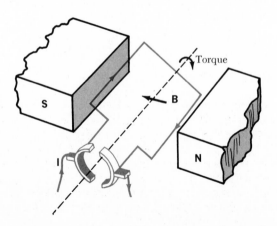

FIGURE 18.18   Principle of th
direct-current motor.

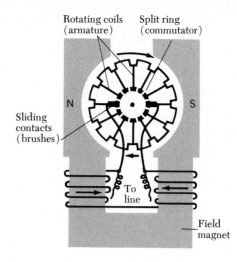

Rotating coils (armature) Split ring (commutator)

N S

Sliding contacts (brushes)

To line

Field magnet

FIGURE 18.19 Construction of a direct-current motor.

Worked example 18.3

At a certain place in Ohio the lines of the earth's magnetic field are toward the geographic north and obliquely downward at an angle of 71° with the horizontal. The induction B of the field is 6.2×10^{-5} Wb/m².

a. In what position should a straight, horizontal current-carrying wire be placed in order to experience the maximum sidewise force?

b. If the current represents the eastward drift of 2×10^{20} electrons per second through any cross-section of the wire, find the direction of the force and its magnitude per unit length of wire.

Solution *a.* The wire can be placed anywhere in a plane perpendicular to the field.

b. If the wire is to be horizontal as well as perpendicular to the field it must be oriented in the east-west direction. The current I is equal to the electron charge multiplied by 2×10^{20}, or $I = 32$ amperes. By Eq. 18.14 (page 448), the transverse force is given by $F = IBl$, or the force per unit length by $F/l = IB = (32)(6.2 \times 10^{-5}) = 0.002$ N/m. The direction is perpendicular to both B and I and, according to the right-hand-screw rule, is southward, at an angle of $90° - 71° = 19°$ downward from the horizontal.

MAGNETIC FIELDS PRODUCED BY CURRENTS

The first part of this chapter was concerned mainly with the sidewise force experienced by moving charges or by current-bearing conductors in an external magnetic field. The following sections deal with the origin of such fields and their production by currents.

One of the fundamental observations is Oersted's discovery of the magnetic field in the neighborhood of a current-carrying wire. Another is Ampère's detection of a mutual force between two adjoining current-carrying wires, which was also investigated jointly by two other French physicists,

J. B. Biot (1774–1862) and Felix Savart (1791–1841). The results are described below.

18.8   THE BIOT-SAVART LAW

Oersted's observation can be followed up in more detail by placing small compass needles at various locations near a current-carrying wire. Figure 18.20 shows a horizontal plane pierced by a vertical, straight wire that carries a current in the downward direction. Compass needles placed on this plane at any given distance from the wire all align themselves tangent to a circle drawn about the wire as a center, all pointing in the same sense around the circle. This suggests that the lines of induction of the magnetic field (page 437) are circles surrounding the wire; this is further confirmed when iron filings are scattered onto the plane (see photograph). The pattern of circular lines must be assumed to exist in every plane perpendicular to the wire, and their separation increases at greater distances from the center (Fig. 18.21). This implies a decrease of flux density in going outward from the wire.

The slight effect of the earth's magnetism can be nullified by suitable means.

If the current in the wire is reversed, the rotational direction of the lines of induction also reverses. The relation between the two may be stated

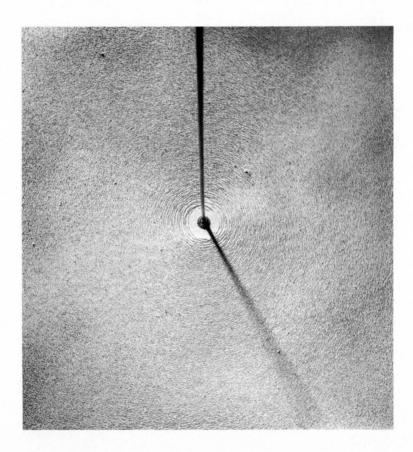

The magnetic field near a long current-carrying wire, as shown by the alignment of iron filings (*From "PSSC Physics," D. C. Heath and Company, Boston.*)

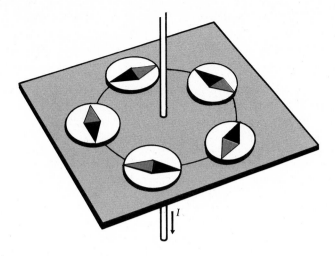

FIGURE 18.20 Direction of the magnetic field near a straight, current-carrying wire.

in the form of a "thumb rule": Imagine the wire grasped in the right hand, with the thumb extended and pointing in the direction of the conventional current. Then the fingers encircle the wire in the direction of the magnetic field vector **B.**

The right-hand-screw rule (page 436) should not be confused with the (right-hand) thumb rule described here and pictured in Fig. 18.22 on the following page.

From their observations and measurement of the force between long, straight, parallel, current-carrying wires, Biot and Savart were able to arrive at the important result that the magnitude of B at any point near a long, straight wire is directly proportional to the current I and inversely proportional to the distance r of this point from the wire:

$$B = \frac{CI}{r}$$

18.15

where C is a constant of proportionality.

At the time of these experiments the ampere had not yet been defined, and so the value of C can be arbitrarily chosen. Then the above equation can be used to define the unit of current. The choice is made by defining a new constant μ_0, called the **permeability of free space,** by setting

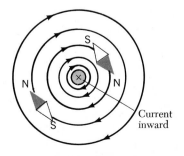

FIGURE 18.21 Circular magnetic field lines in a plane perpendicular to a current-carrying wire. Their directional sense is given by the right-hand rule.

$C = \mu_0/2\pi$. Then, if the numerical value of $\mu_0/4\pi$ is taken to be 10^{-7}, Eq. 18.15 is satisfied, with B expressed in webers per square meter, r in meters, and I in amperes. Accordingly, we set

$$\frac{\mu_0}{4\pi} = 10^{-7} \text{ Wb/A-m} \qquad \text{18.16}$$

and write Eq. 18.15 in the form

$$B = \frac{\mu_0}{2\pi}\frac{I}{r} \qquad \text{18.17}$$

FIGURE 18.22 The right-hand thumb rule gives the sense in which the magnetic field lines encircle a segment of a current-carrying conductor.

which is known as the **law of Biot-Savart.** It gives the magnitude of the induction at a distance r from an infinitely long, straight conductor carrying a current I. For strict accuracy, the surrounding medium is assumed to be a vacuum.

Earlier in this book, the ampere was defined, for convenience, in terms of the coulomb although actually the ampere is taken to be fundamental. This means that with the use of mks units for length, mass, and time, the other practical electrical units can be specified. For this reason, the international scheme of electrical units is referred to as the mksa system ("a" for ampere).

With the Biot-Savart law it is possible to compute the electromagnetic force exerted on each of a pair of straight, parallel, current-carrying conductors. This is most simply done by considering one of the conductors to be a current carrier located in the magnetic field produced by the other. Thus, in Fig. 18.23, the current I in the lower conductor produces an induction B whose magnitude at the location of the upper wire is given by Eq. 18.17 as

$$B = \frac{\mu_0}{2\pi}\frac{I}{a}$$

According to Eq. 18.14 (page 448), any length l of the upper conductor will experience a force whose magnitude is $F = I'Bl$. Inserting the value of B, and dividing through by l, we get the force on unit length:

$$\frac{F}{l} = \frac{\mu_0}{2\pi}\frac{I\,I'}{a} \qquad \text{18.18}$$

FIGURE 18.23 Interaction of two parallel, current-carrying, straight conductors.

The right-hand-screw rule shows that this force is directed upward. Similarly, the force on unit length of the lower conductor has the same magnitude and is directed downward, so that the two conductors tend to move apart. On the other hand, when the currents are in the *same* direction, the forces tend to move the conductors toward each other. It must be remembered that Eq. 18.18 is valid only for a portion of conductors of very great length. If the conductors are of finite length, the relation takes a more complex form which will not be given here.

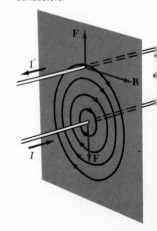

18.9 APPLICATION TO CONDUCTORS OF SPECIAL FORM

Ampère succeeded in generalizing the Biot-Savart formula, obtaining an expression for the induction produced at any point by an infinitesimal length of a current-carrying conductor oriented in any arbitrary way. The total induction produced by a finite conductor of any shape and orientation can then be calculated, at least in principle, by adding the contributions due to all the elements that make up the conductor. The process requires calculus methods, and the following are merely the results for some special cases of interest.

Circular loop The induction at the center of a current-carrying conductor in the form of a circular loop (Fig. 18.24) is given by $B = (\mu_0/2)(I/a)$. The induction vector at the center of the loop is along the axis, and its directional sense can be determined by means of the right-hand thumb rule applied to any segment of the loop. If, instead of a single loop, the circular coil consists of N turns of the same radius, placed close together, the total effect is increased by the factor N and the formula for the flat, circular coil of N turns becomes

$$B = \frac{\mu_0}{2}\frac{NI}{a} \qquad\qquad 18.19$$

It should be remembered that this formula holds only at the center of the coil. The units are as specified in the Biot-Savart law (page 454). Figure 18.24 shows the appearance of the field lines in any plane containing the axis of the coil.

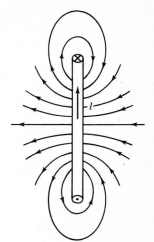

FIGURE 18.24 Field lines of a circular current-carrying turn of wire.

The term "solenoid" is from the Greek solen: a channel or pipe. The name was coined by Ampère.

Solenoid A **solenoid** is a helical coil which is usually made by winding insulated wire around a circular cylinder (Fig. 18.25). The wire is usually wound in closely spaced turns in several layers, as in winding thread on a spool. The induction near the midpoint of a long solenoid is uniform over the cross-section and is given by

$$B = \mu_0 \frac{NI}{l} \qquad\qquad 18.20$$

where I is the current, N the total number of turns of wire, and l the length of the coil. Remarkably, the radius of the turns does not enter the equation as long as it is small in comparison with the length of the coil.

The general direction of the induction vectors inside the coil is again given by the thumb rule used for the flat coil, above. Figure 18.26 shows the field lines in any plane containing the axis of the coil. Outside the solenoid the field has the same shape as that of a bar magnet, and the coil behaves as if it had two magnetic poles, an **N** pole near one end and an **S** pole near the other.

Compare Fig. 18.7a (page 438).

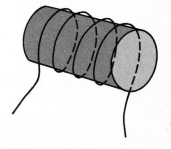

FIGURE 18.25 A solenoid.

Toroid is from the Latin word for a bulge, or a rounded molding.

Toroid If the ends of a solenoid are brought together, the result is, in effect, a coil shaped like a doughnut. It is called a **toroid.** The cross-section need not be circular, and often in practice it is rectangular (Fig. 18.27). The essential point is that by joining the ends of a solenoid, the divergence of the field lines near each end is eliminated. In the toroid, almost all the induction is confined to the interior, the field lines being circles in planes perpendicular to the axis. There is some external field because, with all its curling around the core, the entire winding is equivalent to a current

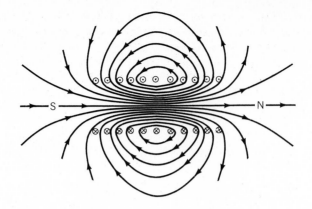

FIGURE 18.26 Section of a sole
noid made by a plane containin
the axis. The magnetic field line
are shown. Their pattern is ver
much like that of a permanent ba
magnet.

loop around the axis of symmetry. Calculation shows that the induction at points inside the toroidal surface is given by

$$B = \frac{\mu_0}{2\pi} \frac{NI}{r} \qquad\qquad 18.21$$

where I is the current, N the total number of turns, and r is the radius of the field line in question. If the radial dimension of the coil is small compared with the diameter of the entire toroid, r can be replaced by the average radius, and B may be considered essentially constant over the entire cross-section.

18.10   MAGNETIC MATERIALS

The magnetic effects discussed above exist only while current is on. A device that gives rise to magnetic flux when current exists in it is called an **electromagnet.** However, there are also *permanent* magnets—for example, pieces of magnetite, or steel bars previously subjected to strong magnetic fields—that continue to be magnets even though no current is supplied. In 1820, Ampère suggested that all magnetism is associated with electric currents, but it was not clear how the "amperian currents" could persist in magnetic materials without the continual dissipation of energy.

It is now known that *permanent magnetism is caused by the motion of electrons within atoms.* This motion may be of two kinds: the translational motion of electrons moving in closed orbits around the nucleus of the atom, and a spinning motion of each electron around its own axis. Each motion produces a magnetic field, just as the circulation of charge in a loop of wire produces such a field. In nearly every kind of atom the resultant effect of the magnetic fields of the several electrons is very small or actually zero, but in a few elements such as iron, nickel, and cobalt, the electron configuration is such that the fields do not cancel and the individual atoms act as magnets. There are also certain alloys that show strong magnetic action; all such materials are collectively termed **ferromagnetic.** They have extremely important technical applications in electric motors, generators, transformers, cyclotrons, etc.

If free to turn, a current loop like the one described on page 455 sets

*More details
are given
in Chap. 21.*

*The prefix
"ferro" is
from the
Latin word
ferrum: iron.*

FIGURE 18.27 A toroid. If the co
is closely wound, the magnetic fiel
is almost entirely within the cor

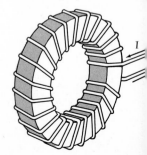

its plane perpendicular to a magnetic field in which it is placed. An atom of a ferromagnetic material behaves in a similar way and tends to line up its magnetic axis with an external field. Groups of perhaps 10^{17} to 10^{21} atoms cluster together with their magnetic axes parallel, to form what are called magnetic **domains.** Quantum-theory calculations show that in ferromagnetic substances strong forces act to preserve this alignment. In an unmagnetized body, the domains are oriented at random, but when the body is put into a magnetic field they tend to align themselves with the external field, and suitably oriented domains may grow at the expense of others. The net effect is to magnetize the body. For some materials such as hard steel alloys, the magnetization is retained after the external field is removed, and the body becomes a permanent magnet.

The magnetic properties of materials may be studied by winding a toroid on a ring made of the material in question and measuring the induction within the ring for various values of the current in the coil. For most substances, the induction is found to be very little different from the value measured when there is only a vacuum inside the coil, but for rings made of ferromagnetic materials the induction may be hundreds or thousands of times greater. This effect is expressed by replacing the permeability constant μ_0 in the ''vacuum'' induction formulas of the previous sections (Eqs. 18.17 to 18.21) by a permeability coefficient μ, characteristic of each material. This coefficient is not a *constant* because, even for a given material, its value depends on the extent of magnetization and on the previous magnetic changes to which the sample has been exposed.

In practice, ferromagnetic materials are used to increase the induction in an electromagnet. The coils are wound on cores made of certain ferromagnetic substances that do not retain appreciable magnetism after the magnetizing current is cut off. It is often important to know how much the presence of the ferromagnetic core increases the induction in comparison with an air or vacuum core. If B represents the induction when the material core is used and B_0 the induction for a vacuum core, the relative permeability is defined by the following equation:

$$\mu_r = \frac{B}{B_0} = \frac{\mu}{\mu_0} \qquad\qquad 18.22$$

The value of μ_r may be as high as 10,000 for some alloys of nickel. For air, it is for all practical purposes the same as the vacuum value of 1, the exact value at standard pressure being 1.0000004.

18.11 ELECTROSTATIC AND MAGNETIC FIELDS COMPARED

There are striking similarities between electrostatic and magnetic phenomena as described in this and the preceding chapters: the existence of forces of attraction and repulsion, the depiction of the fields by means of lines of force, the similarity of electrostatic induction to the behavior of ferromagnetic materials in magnetic fields (magnetic induction), etc. Even more important are some fundamental differences:

1. *Electric field lines always originate on positive charges and end on negative charges whereas magnetic lines are always closed loops.*
    Historically, magnetic poles (page 438) were considered to correspond

to point electrostatic charges, but the modern tendency is to recognize that poles do not have the same kind of objective existence as charges and to ascribe all magnetic effects to electric currents, as done above.

2. *Electrostatic fields are conservative* (page 205), *whereas magnetic fields are not.* For an electric field, it is possible to define a potential function V having a definite value at every point. The work done on a test charge in moving it from one point to another is independent of the particular path taken. On the other hand, path independence does not hold for magnetic fields, and it is not feasible to introduce a scalar potential function. The dependence of work on path is illustrated by the example shown in Fig. 18.28. If a "test pole" **N** moves from P_1 to P_2 in a clockwise sense along the circular field line, the magnetic field does work on this pole, but if the pole is to move from P_1 to P_2 in the opposite sense along this field line, work must be done against the field forces by some outside agency. The magnetic field here is analogous to a vortex (whirlpool) in a fluid.

Such a pole may be approximated by using one end of a very long, thin magnet.

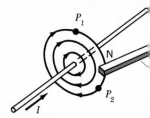

FIGURE 18.28   The work done on the test pole depends on the path taken.

Programmed review

Instructions: See page 18.

1.   Describe Oersted's discovery and state what it showed.

A compass needle placed near a current-carrying conductor takes a position crosswise to the wire. This shows that a current produces magnetic effects.
[18.1]

2.   Describe the force on a moving charged particle in a magnetic field.

The particle experiences a force that is perpendicular to both the direction of motion and the direction of the field. Its magnitude is given by $F = BQv_\perp$.
[18.2]

3.   What are magnetic lines of induction?

Curves that indicate the direction and intensity of the magnetic field at any point.
[18.2]

4.   Define magnetic flux.

The total number of lines of induction piercing a surface perpendicular to the field. If the field is uniform, the flux through a surface of area A is $\Phi = BA$.
[18.2]

5.   Describe the path of a charged particle projected into a uniform magnetic field in a plane perpendicular to the field. Name a device that uses this situation.

The path is a circle. The cyclotron.
[18.3]

6.   What did Thomson's cathode-ray experiment accomplish?

It identified the electron for the first time and gave a value for e/m.
[18.3]

7. What is the magnitude of the sidewise force on a current-carrying straight wire placed perpendicular to a magnetic field?

$F = IBl.$ [18.6]

8. Give the direction rule for the above force.

Right-hand-screw rule: Turning from the direction of the current toward that of the field gives the direction of the force as that in which a right-hand screw advances.   [18.6]

9. State a formula expressing the Biot-Savart law for the magnetic induction produced by a long, straight, current-carrying conductor.

$B = \mu_0 I / 2\pi r.$ [18.8]

10. Describe qualitatively the effect of inserting an iron core into a current-carrying coil.

The magnitude of the induction is greatly increased.   [18.10]

For further reading

18.1. *Williams, L. P.* "Michael Faraday." An excellent, comprehensive biography.
18.2. *Kernan, W. J.* "Accelerators." An informative booklet on particle accelerators, published by the U. S. Atomic Energy Commission.
18.3. *Jaffe, B.* "Men of Science in America." Read the biography of E. O. Lawrence.
18.4. *Gouiran, R.* "Particles and Accelerators." A nonmathematical presentation of high-energy physics. Particle accelerators are described in pt. 2.
18.5. *Kolm, H. H.,* and *A. J. Freeman* Intense Magnetic Fields, *Sci. Am.,* April, 1965, pp. 66–81.
18.6. *Runcorn, S. K.* The Earth's Magnetism, *Sci. Am.,* September, 1955, p. 152.
18.7. *Thompson, W. B.* Magnetohydrodynamics, *Endeavour,* May, 1964, p. 73.
18.8. *Hess, W. N.* Radiation Belts, "Encyclopedia of Physics."
18.9. *Nier, A. O.* Mass Spectroscopy—An Old Field in a New World, *Phys. Teacher,* January, 1967, p. 5.

Questions and problems

18.1. In the Oersted experiment, what happens if the wire is exactly in the magnetic east-west position, with the current directed toward the east?

18.2. Comment on the following statement: If a moving proton is deflected to one side on passing through a certain region, a magnetic field necessarily exists there.

18.3. Over a small area, the earth's magnetic field may be considered uniform. At a certain station in the tropics, the strength of the field is 3.5×10^{-5} Wb/m^2 and the lines dip downward at an angle of 30° below the horizontal. Find the magnetic flux through a closed plane loop of wire of area 150 cm^2 lying on a table.

18.4.   Prove that the radius of curvature of the path of a charged particle that moves in a plane perpendicular to a magnetic field is proportional to the momentum of the particle.

18.5.   How much work, if any, is done by a uniform magnetic field B on an electron that enters the field moving with a speed v? Explain.

18.6.   A singly ionized hydrogen atom and a doubly ionized helium atom enter the uniform magnetic field of a mass spectrometer, both moving in a plane perpendicular to the field. If the two ions have the same KE, compare the diameters of their orbits in the instrument. Take the mass of the helium ion to be 3.973 times that of the hydrogen ion and assume that the nonrelativistic expression for KE may be used.

18.7.   A doubly charged lithium ion having a speed of 5×10^6 m/sec enters the magnetic field of a mass spectrograph in a direction perpendicular to the field and is bent into a circular path of radius 22.3 cm. If $B = 0.7$ Wb/m^2, find the mass of this ion.

18.8.   Show that the speed v with which a particle must move, relative to the observer, in order to have a mass K times its rest mass is given by $v = c\sqrt{K^2 - 1}/K$.

18.9.   An electron is accelerated to a speed of $0.9998c$. What is the ratio of its mass at that speed to its rest mass? HINT: Use Eq. 18.9. In the radicand, write $(1 - 0.0002)$ in place of 0.9998; on squaring this binomial and reducing, the value of the radicand becomes very nearly 0.0004.

18.10.   In the cyclotron there is a magnetic field between the pole faces of the magnet and an electric field between the two sections of the dees. State the main function of each of these fields.

18.11.   The frequency of the oscillator of a certain cyclotron is 8.4 Mc/sec (megacycles per second). What must be the value of the magnetic induction B in order to accelerate protons? The mass of a proton is 1,840 times that of the electron.

18.12.   Find the magnitude of the force on a conductor 10 cm long carrying a current of 5 amperes if it is perpendicular to a uniform magnetic field of 1.2 Wb/m^2.

18.13.   A small body carrying a charge of 2×10^{-9} coulomb is moving perpendicular to a magnetic field of magnitude $B = 4$ Wb/m^2 with a speed of 5×10^4 m/sec. Calculate the magnitude of the sidewise force on the body.

18.14.   A wire 10 cm long, suspended horizontally from one arm of a sensitive, equal-arm balance, carries a current of 10 amperes. If the wire is in a horizontal magnetic field of induction $B = 0.020$ Wb/m^2 whose field lines are perpendicular to the wire, what mass must be added to one side of the balance to restore equilibrium?

18.15. When a positive-charged cloud passes overhead, electrons begin to flow upward along the vertical wire leading to a lightning rod and eventually stream off the pointed rod to neutralize the charge on the cloud. What is the general direction of the magnetic field caused by this electron flow at a place directly west of the wire?

18.16. Compute the magnitude of the induction B at a point 5 cm from a long, straight wire in which there is a current of 15 amperes.

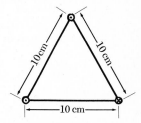

FIGURE 18.29

18.17. Figure 18.29 represents in cross section three long, straight, parallel wires, each carrying a current of 20 amperes in the indicated direction. (A dot indicates current toward the observer; a cross indicates current away.) Determine the magnitude and direction of the force, per meter of length, acting on the upper wire because of the presence of the other wires.

18.18. A flat, circular coil is suspended in the earth's magnetic field by a thread tied to one point on its rim so that it can rotate freely about its vertical diameter. A steady current is established in the coil. When the coil has come to rest, an experimenter looks through the coil toward the north. In what direction would he "see" the electrons moving around the coil?

18.19. Show that the value of B produced at the center of a hydrogen atom due to the revolution of the electron in its orbit amounts to nearly 13 Wb/m^2, comparable in magnitude with the largest steady fields so far produced by electromagnets. Assume that in a normal hydrogen atom the electron traverses a circular orbit of radius 5.3×10^{-11} m with a frequency $f = 6.6 \times 10^{15}$ cycles/sec. HINT: If e is the electron charge, the circulation of the electron is equivalent to a current of magnitude $I = ef$ in a circular loop.

18.20. A solenoid is wound uniformly along its entire length with five layers of insulated wire, each having 400 turns of wire per meter of length. All the turns may be considered to have radii that are small compared with the length of the coil. A ferromagnetic core, whose relative permeability is 600, fills the entire cross-section of the solenoid. Calculate the value of B at the midpoint of the axis of the coil when the current in the windings is 1 ampere. HINT: Use μ_r in place of μ_0 in the solenoid equation (page 455).

Chapter 19
ELECTROMAGNETISM

After Oersted's discovery that magnetism could be produced by means of electric currents, many experimenters turned their efforts to finding the reciprocal effect: the production of electric currents by means of magnets. Two men, working independently, came upon such a phenomenon within a year of each other. The first was the American physicist Joseph Henry (1797–1879), and the second was Michael Faraday. This chapter examines the consequences of their discovery which showed how to convert mechanical energy directly into electric energy and, even more significantly, led to Maxwell's electromagnetic theory of light.

ELECTROMAGNETIC INDUCTION

19.1 FUNDAMENTAL EXPERIMENTS

Many experimenters believed that a current could be established merely by placing a magnet near a wire, although it is obvious now that a magnet at rest relative to a conductor cannot give rise to a persistent current without violating the conservation of energy. It was Faraday who found that a current could be produced by the relative *motion* of a wire and a magnet.

Actually, the experiments of both Henry and Faraday led to the basic observation that *a meter connected to a coil shows a momentary deflection when a current is either started or stopped in a nearby circuit.* The arrangement is shown in principle by Fig. 19.1a. When the switch S is closed, allowing the battery to establish a current in coil C_1, a momentary "kick" of current is registered by the meter M in the circuit containing coil C_2. The circuit of coil C_2 contains no batteries or other sources of emf, and there is no metallic connection of any kind between the two circuits.

With a steady current established in circuit 1, nothing further is observed to happen in circuit 2. However, if the switch is opened, a momentary current impulse, opposite in direction to the previous one, is noted in circuit 2. Henry observed this phenomenon in August of 1830. Faraday found it

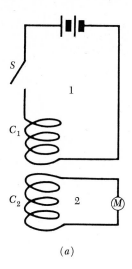

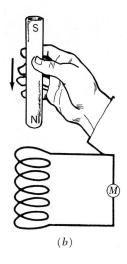

(a)   (b)

FIGURE 19.1 Two arrangements for generating induced currents.

a year later but was the first to publish the result. An entry in Faraday's notebook shows that he surmised the existence of such a connection as early as 1822.

Faraday also observed transient currents when, with circuit 1 carrying a current, the distance between the two coils was suddenly changed. He found the effects were greatly increased if the two coils were provided with iron cores. Further, in a single circuit like the one in Fig. 19.1b, he observed transient currents when one end of a bar magnet was thrust into the end of the coil and a reverse current when the magnet was withdrawn. Using the opposite pole of the magnet resulted in reversing the direction of the current in each case. The magnitude of the current was found to increase with the speed with which the magnet was moved.

In all these experiments, the current can be said to be caused by induction, but since this word has already been used for entirely different effects (pages 385, 436), the term **electromagnetic induction** is used.

Straight conductor cutting a magnetic field.

Imagine a straight conductor of length l located in a uniform magnetic field (Fig. 19.2) which is directed outward from the plane of the diagram. Suppose the conductor is moved to the right with a velocity **v** perpendicular to both the field and the wire. Because of this motion, every charged particle in the conductor will experience a force along the direction of the conductor. We have seen that the magnitude of this force is

$$F = BQv_\perp$$

Confirm these directions by applying the right-hand-screw rule (page 436).

The direction of **F** is given by the right-hand-screw rule: If there are free positive charges in the conductor, they move toward the lower end and accumulate there, whereas negative charges gather near the upper end. This can be verified directly by cutting the above conductor in two while it is passing across the field. Afterward, each piece is found to have the appropriate charge. Even if only one kind of charge is free to move, as in a metallic conductor, the separation of charge still occurs as described.

The accumulation of charge at each end of the moving wire continues until an equilibrium is reached between the electromagnetic force

FIGURE 19.2 When the conductor is moved across the magnetic field, forces arise that tend to move positive and negative charge carriers in opposite directions along the conductor.

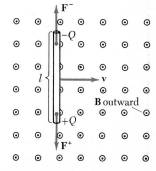

$F = BQv_\perp$ and the electrostatic force arising from other charges in the conductor. The origin of the latter force is an electrostatic field E arising from the separation of charges. This field exerts a force of magnitude EQ on a charge Q within the conductor. For equilibrium, the two forces can be equated, yielding

$$EQ = BQv_\perp$$

or

$$E = Bv_\perp \tag{19.1}$$

The existence of the electrostatic field E leads to a potential difference V between the ends of the conductor. From Eq. 16.7 (page 389), its magnitude is $V = El$, and if we substitute the value of E from Eq. 19.1 we have

$$V = Blv_\perp \tag{19.2}$$

Here, as before, V is measured in volts, B in webers per square meter, l in meters, and $v_\perp$ in meters per second.

Worked example 19.1   An airplane whose wing span is 30 m flies horizontally at a speed of 200 m/sec. If, at a certain locality, the vertical component of the earth's magnetic field has a magnitude of 5×10^{-5} Wb/m², what is the induced difference of potential between the wing tips?

Solution   The use of the vertical component of the field satisfies the condition (implied in deriving the relation 19.2) that the conductor is moved in a direction perpendicular to the field. This is equivalent to using the resultant value of the field and the component $v_\perp$ of the plane's motion perpendicular to the field. Direct substitution of the information into the equation yields

$$V = (5 \times 10^{-5})(30)(200) = 0.30 \text{ volt}$$

19.2 FARADAY'S LAW OF INDUCTION

Suppose that the moving conductor described above is sliding along the straight, parallel sides of a stationary C-shaped conductor lying in the plane of the diagram as shown in Fig. 19.3. The potential difference existing between the ends of the moving segment then acts as an emf that maintains

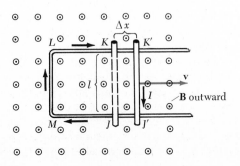

FIGURE 19.3 Deduction of Faraday's law of electromagnetic induction.

a current in the circuit $JKLM$. The direction of this current (in the conventional sense, page 405) is shown by the arrows, as may be seen by noting the direction in which the positive charges in Fig. 19.2 (page 463) have moved.

Faraday thought of the induced emf and the current which it causes as arising from the "cutting of magnetic lines of force" by the moving conductor. As Gillispie (Ref. 19.2) says of Faraday: ". . . there was given to him as to few scientists a sense of the spatial. He would almost see the moving wire slice through the lines of force and the current stir within."

Although this vivid description is adequate for a conductor in motion, there are other kinds of electromagnetic induction experiments (such as those described on page 462 in which no mechanical motion is involved) that call for a more general formulation. Figure 19.3 shows how to arrive at such a description. The conductor JK, moving transversely at the speed $v_\perp$, sweeps over an area

$$\Delta A = lv \, \Delta t \qquad\qquad \textbf{19.3}$$

in a time interval Δt. Because of this motion, the flux (the number of magnetic field lines threading through the circuit) increases by an amount $\Delta \Phi$ given by Eq. 18.3 (page 438) as

$$\Delta \Phi = B \, \Delta A = Blv_\perp \, \Delta t \qquad \text{so that} \qquad \frac{\Delta \Phi}{\Delta t} = Blv_\perp$$

But, according to Eq. 19.2, the emf induced in the moving conductor is $\mathcal{E} = Blv_\perp$, resulting in

$$\mathcal{E} = -\frac{\Delta \Phi}{\Delta t} \qquad\qquad \textbf{19.4}$$

where $\mathcal{E}$ is in volts, $\Delta \Phi$ in webers, and Δt in seconds. The origin of the negative sign will be explained below.

In order to check the units in Eq. 19.4, we note that the right side has the units Wb/sec. According to the relations on page 436,

$$1 \, \frac{\text{Wb}}{\text{m}^2} = 1 \, \frac{\text{N-sec}}{\text{C-m}}$$

and so Wb/sec becomes N-m/C which, by definition (page 388), is equivalent to volts. Hence the units stated above for the various quantities in Eq. 19.4 are consistent.

Equation 19.4 expresses **Faraday's law of electromagnetic induction:** The magnitude of the emf induced in a circuit is equal to the time rate of change of the magnetic flux through the circuit. This law was deduced above for a moving conductor, but it is valid for any situation relating to induced emf. A review of the experiments described in Sec. 19.1 reveals that whenever a current is induced in the coil there is a *change* in the magnetic flux through it. For example, when the switch is closed in the circuit of Fig. 19.1a, the coil C_1 becomes an electromagnet and lines of force spring into existence within and around it. Some of these lines thread through the turns of C_2, where no flux existed before, inducing a current in this coil. Keeping the current in C_1 constant produces no further change in flux, and there is no induced current after the first instant. However, when the switch in circuit 1 is opened, the field of C_1 vanishes, the flux through C_2 decreases to zero, and a transient current again exists in C_2. Moving the magnet toward or away from the coil in Fig. 19.1b changes the number of lines of induction passing through the coil, and this, too, constitutes a change of flux. Whenever there is a change of flux there is an induced emf, regardless of the way this change is brought about.

If the circuit consists of a coil made up of N adjoining turns, all the same size, the total induced emf is increased by a factor N, and Eq. 19.4 may be written

$$\mathcal{E} = -N \frac{\Delta \Phi}{\Delta t} \qquad\qquad 19.5$$

where Φ is the flux through the coil as a whole.

Worked example 19.2

A flat circular coil of radius 5 cm having 1,000 turns has its plane perpendicular to a uniform magnetic field of $B = 3 \times 10^{-4}$ Wb/m². If the induction is increased to 7×10^{-4} Wb/m² in an interval of 0.01 sec, find the magnitude of the induced emf.

Solution For each turn, the change in flux $\Delta \Phi = A\ \Delta B = \pi(0.05)^2(4 \times 10^{-4})$. Then, according to Eq. 19.5, the magnitude of the induced emf is given by

$$N \frac{\Delta \Phi}{\Delta t} = \frac{(1,000)(3.14)(0.05)^2(4 \times 10^{-4})}{0.01} = 0.31 \text{ volt}$$

Since there is no detailed information on the way the flux changes during the 0.01-sec interval, the above result must be interpreted as the *average* emf during that interval.

Worked example 19.3

The strength of a magnetic field may be measured by using a small, flat coil consisting of a large number of turns of fine wire connected to a suitable meter. After this search coil has been placed in the field, it is quickly removed. The reading of the meter during this process gives information on the value of B.

A search coil having 100 turns, each of area 3 cm², lies in a magnetic field with its plane perpendicular to the lines of induction. The coil is then jerked out of the field in 0.06 sec, producing an average emf of 0.20 volt during this time. Find the value of B.

Solution The flux through each turn of the coil when in the field is $\Phi = BA$, dropping to zero when the coil is removed to a field-free region. Hence the *change* in flux is also represented by BA. From Eq. 19.5 (except for the minus sign), $\Delta \Phi = \mathcal{E}_\Delta t/N$, and this is equal to BA. Then

$$B = \frac{\mathcal{E}\ \Delta t}{NA} = \frac{(0.20)(0.06)}{(100)(3 \times 10^{-4})} = 0.4 \text{ Wb/m}^2$$

19.3 DIRECTION OF INDUCED EMF; LENZ'S LAW

In the description of Faraday's experiments in Sec. 19.1 (page 462) it was pointed out that there is always a definite relation between the direction of the causative agency and that of the induced current. A generalization formulated in 1833 by the German physicist H. F. E. Lenz (1804–1865) gives the direction of the effect in any instance: *The direction of an induced current is such that its magnetic effect opposes the action that produces it.*

Lenz's law is a special statement of the conservation of energy as applied to electromagnetic induction phenomena: If the magnetic effect of an induced current were to act in a direction to reinforce rather than oppose its cause, the action would continue to increase without any outside intervention and infinite currents would be generated. This is not observed to happen, and so it must be true that the magnetic effect hinders the action that causes it. This "oppositeness" is the key feature of Lenz's law and is represented in the analytical statement by the minus sign in Eq. 19.4.

The following discussion shows how Lenz's rule applies to some of the arrangements already described for producing induced currents.

1. To maintain the motion of the crossbar in Fig. 19.3 (page 464), a force must be exerted on it toward the right. According to Lenz's law, the magnetic effect of the resulting induced current must be such as to oppose the applied force. If the direction of the conventional current is taken to be downward, as shown in Fig. 19.3, the right-hand-screw rule (page 436) indicates a magnetic force toward the left, thus opposing the motion of the bar. If the bar were not part of a complete circuit there would be no induced current, but there would still be an emf in the indicated direction.

   Figure 19.4 is a view of the crossbar apparatus of Fig. 19.3 as seen from below, with the magnetic field lines inserted. Notice how the composite field suggests opposition to the motion of the bar to the right.

   Let **F** be the force needed to move the bar uniformly. In displacing the bar a distance Δx, **F** does an amount of work $\Delta W = F\,\Delta x$. If the bar does not accelerate, this applied force must be opposed by an equal and opposite magnetic force, the magnitude of which is given by Eq. 18.14 (page 448) as $F = IBl$. Then the expression for the work becomes

$$\Delta W = F\,\Delta x = IBl\,\Delta x \qquad\qquad \textbf{19.6}$$

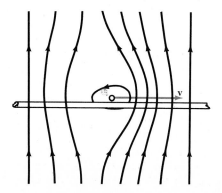

FIGURE 19.4 The external uniform field and the field produced by the current induced in the bar combine to give the configuration shown here. Faraday pictured the field lines as stretched filaments, and their crowding in front of the bar suggests opposition to its motion.

The emf induced in the bar by its motion is, by Eq. 19.2 (page 464),

$$\mathcal{E} = Blv_\perp = Bl\frac{\Delta x}{\Delta t} \qquad\qquad 19.7$$

If, now, the combination of factors $Bl\,\Delta x$ is eliminated between Eqs. 19.6 and 19.7, the result is $\Delta W = I\mathcal{E}\,\Delta t$, or the power P expended is

$$\frac{\Delta W}{\Delta t} = P = I\mathcal{E} \qquad\qquad 19.8$$

This is merely the familiar Joule's-law relation given on page 415 for the power expended in a circuit, and the derivation emphasizes the way in which electrical quantities are closely interconnected. In the crossbar apparatus, the mechanical work done by the applied force is converted into electric energy and finally, since this circuit is purely resistive, into heat.

Bar magnet moving relative to a coil.

2. In Fig. 19.1*b* (page 463), the **N** pole of the bar magnet is being moved toward the coil. According to Lenz's law, the induced current must be in the direction that makes the upper end of the coil an **N** pole so that the repulsion for the **N** pole of the magnet opposes the motion. Use of the right-hand thumb rule then shows that the current is in the direction indicated. Further, if the bar magnet is now pulled upward, the upper end of the coil must become an **S** pole so that it attracts the **N** pole of the magnet, opposing its removal. Now the induced current will be in the reverse direction from the above.

    Notice that, when the magnet approaches the coil, the magnetic field of the induced current happens to be in the opposite direction to that of the bar. This is not relevant. The only fact of importance is that the field of the coil opposes the *change* (that is, opposes the approach of the bar). When the bar is withdrawn, the two fields happen to be

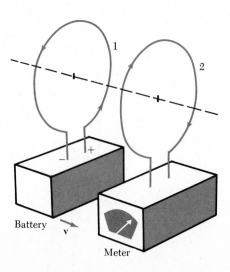

Battery

Meter

FIGURE 19.5 Lenz's law predicts the direction of the induced current in loop 2. The figure shows loop moving to the right. If, instead, is moving to the left, the direction of the induced current reverses

in the same direction, yet the field of the coil opposes the *withdrawal* of the bar.

3. The permanent bar magnet used in the preceding experiments may be replaced by a current-carrying solenoid and the results will be the same as before, since the solenoid is entirely equivalent to the bar in all its external effects. Figure 19.5 is a schematic representation of what happens. For simplicity, each solenoid has been replaced in the figure by a single loop of wire.

This experiment raises an interesting question: What happens if the current-carrying loop (circuit 1) is stationary and the other loop (circuit 2) is moved? If the theory of relativity is valid for electromagnetic phenomena as well as for mechanics, the same effects would be expected, since only *relative* motion counts. Experimentally, the anticipated result occurs, but for a subtly different reason.

In Fig. 19.6, circuit 1 is stationary while circuit 2 is moving to the left. An observer at rest relative to circuit 1 notes only the steady magnetic field due to the current in this loop. An observer moving along with circuit 2 notices a changing magnetic field in the space around him, and this changing field induces electric forces that move electrons around his loop. Thus one observer verifies the presence of a magnetic field only, and the other observes both a magnetic field and an electric field. Nevertheless, the effect is the same as if circuit 2 were at rest in the laboratory and circuit 1 were moving toward it.

This experiment, among others, indicates the impossibility of determining absolute motion through space. In this respect it is comparable with the mechanical examples given earlier and the optical test represented by the Michelson-Morley experiment. The implication that it is impossible to determine absolute translational motion by any kind of physical observation is precisely the basis of the special theory of relativity.

Only relative motion is important in induction experiments.

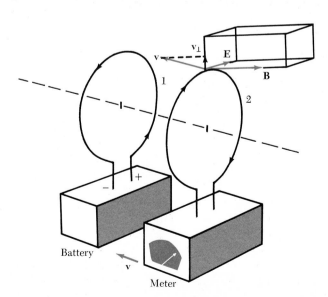

FIGURE 19.6   In contrast with Fig. 19.5, loop 2 is now in motion and experiences a changing magnetic field. This produces an electric field around the loop, and the effect is the same as in Fig. 19.5.

Practice set*

1. In Fig. 19.3 (page 464), determine the directional sense of the induced current if the direction of B is reversed and the bar is moved toward the left.

2. Two flat, circular coils lie side by side on a table, and a clockwise current is started in the coil on the left. What is the directional sense of the current induced in the other coil?

3. A flat, circular coil lies on a table in a strong magnetic field that is directed vertically upward. If the coil is suddenly distorted into an oval shape having a smaller cross-section area, find the directional sense of the current induced in it, as seen from above.

19.4   THE ELECTRIC GENERATOR

The induced currents obtained in the experiments of Henry and Faraday were feeble, short-lived, and seemed of little importance outside the laboratory, yet continuously operating generators based on these very principles are the only adequate source of the tremendous quantities of electric energy used today. As in the electric motor, the essential parts of a generator are a set of coils, a magnetic field in which the coils may rotate, and some means for connecting to an external circuit. In fact, with slight adjustments, the same machine may be used for either purpose. When a current is passed into the coils, they rotate and act as a motor; when the coils are mechanically turned, an induced current results and the machine becomes a generator.

Figure 19.7 is a schematic representation of a generator consisting of a closed rectangular loop of wire that can rotate around a horizontal axis.

Ans.: Clockwise; clockwise; counterclockwise.

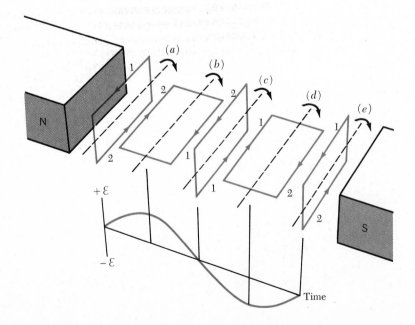

FIGURE 19.7 Operation of a simple generator. The single loop shown in several successive positions. The horizontal sides of the loop are marked 1 and 2 to distinguish them.

The loop is in a uniform horizontal magnetic field whose lines are perpendicular to this axis, and the field may be assumed to fill the space between the pole pieces **N** and **S.** The parts of Fig. 19.7 labeled *a, b, c,* etc., show, in perspective, orientations of the rotating loop at successive *times.* Opposite sides of the loop are numbered 1 and 2.

Start with the plane of the loop vertical, as in Fig. 19.7*a,* and let the loop begin to rotate in the sense shown. As soon as the loop begins to turn away from the broadside position, the flux through it decreases, giving rise to an induced current. Lenz's law can be used to find the direction of the current in the loop at any time. The current must always be in such direction as to oppose the momentary change of flux through the loop. For example, in position (*c*) the projected area of the loop is about to decrease. To counteract this, additional field must be created by the induced current. The right-hand rule shows that this current must have the direction shown in the figure.

By the time the loop has reached the horizontal position (as in Fig. 19.7*b*) the induced emf has increased to its maximum value because the *rate of change* of the flux through the loop has increased during this operation. As the loop continues to turn and approaches the position (*c*) the emf drops, passing through zero and becoming negative beyond (*c*). Notice that the arrow showing the direction of the current in wire 1 is now directed inward, where previously it was outward. Similarly, the direction of the current in wire 2 has reversed.

In the next 90° turning, the emf takes on greater and greater *negative* values, and in the final quarter turn it lapses back to zero. Then the whole sequence of events repeats. The graph at the bottom of Fig. 19.7 shows the way in which the emf or the current varies with time. The area of the wire loop projected onto a plane broadside to the field is proportional to the sine of the angle between the normal to the loop and the direction of the field, so that the emf graph is a sine curve.

If the loop is connected to an outside circuit by means of slip rings and brushes (Fig. 19.8), the current in this circuit will be alternating. An **al-**

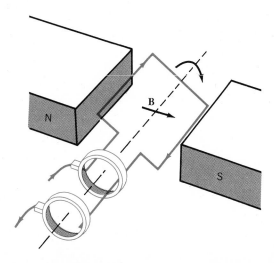

FIGURE 19.8 Brushes bearing on slip rings connect the simple generator to the external circuit. The brushes are fixed in position.

ternating current is one that reverses its direction periodically, and the frequency of alternation is the same as the number of rotations per second of the loop. An alternating current is the kind that is induced in a coil by continuous rotation in a magnetic field.

For some purposes, such as the charging of storage batteries, a direct current is required. A **direct current** is always in the same direction. To attain this kind of output current, the slip rings of the generator must be replaced by a commutator (page 450), which reverses the connections to the circuit each time the loop reaches a position perpendicular to the field. This has the effect of reversing alternate arches of the output curve (Fig. 19.9a).

The current in the outside circuit, although now in one direction, is still not a steady current and for some uses may be quite unacceptable. For a more constant current, several coil loops are set at various angles with each other. The output curve of each then reaches its maximum when the others are at intermediate positions, and the combined output shows less variation than the curve for a single loop. Figure 19.9b shows what happens when three equally spaced loops are used. With an even larger number of loops hardly any ''ripple'' remains and the output is practically constant.

19.5 THE TRANSFORMER

The wide use of alternating currents for commercial purposes is made practical by a device called a **transformer,** which is basically the combination of two coils illustrated in Fig. 19.10. Any variation of current in one coil induces a corresponding emf in the other. With an alternating current flowing in the primary coil C_1, there is a continuous variation of magnetic flux through the secondary coil C_2. As a result, an alternating current is induced in C_2, and this current has the same frequency as that in C_1.

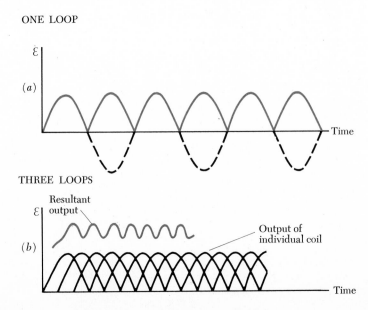

FIGURE 19.9 Commutated output of a simple generator.

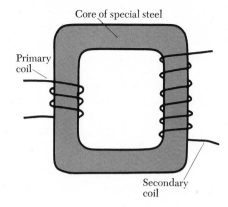

FIGURE 19.10 A transformer (schematic).

Simple air-core transformers are used in radio and television apparatus, but in power circuits the two coils are wound on a closed ring made of a high-permeability alloy which increases and concentrates the magnetic flux. At any given time, the flux is the same for all turns because practically all of it is confined to the interior of the core.

Since the flux is the same for all turns of both the primary and secondary coils, *the average emf's in the two coils are in the same ratio as the numbers of turns.* For example, if the secondary has 100 times as many turns as the primary, an emf of 10 volts in the primary will produce an emf of 1,000 volts in the secondary.

There are no mechanical moving parts in a transformer, and the energy losses in a well-designed installation are usually only a few percent. Hence the *power* developed by the secondary current is essentially equal to that of the primary. As in the case of direct currents, the power in any alternating-current circuit is proportional to the product of current and emf, so that the currents in the two coils are inversely proportional to the numbers of turns. In the above example, the effective secondary current would be 1/100 of that in the primary. Transformers may be used either to ''step up'' the emf, with consequent lowering of the current, or to ''step down'' the emf, with a corresponding increase of the current.

When electric power is to be used at a great distance from the generator, it is transmitted in the form of a high-potential alternating current, since this makes possible the reduction of the current which in turn means smaller power loss (I^2R) in the transmission line. The generator typically operates at not more than 10,000 volts, and the current is then led to a step-up transformer which puts it into the cross-country transmission line at perhaps 230,000 volts. At the edge of a city where the energy is to be used, the potential is reduced by step-down transformers to about 2,300 volts, and additional transformers along the line then reduce it further to a safe value of around 115 volts for use in individual houses.

19.6 THE BETATRON

We have seen that a changing magnetic flux through a circuit produces an emf, according to the law of electromagnetic induction. It is equally proper to say that the flux change *produces an electric field* in the circuit (page 464). Such a field is not quite the same as an *electrostatic* field arising from the presence of charged bodies: Electrostatic field lines start and end on charges, whereas field lines arising from changing magnetic flux are closed loops. Like lines of magnetic induction, they have no beginning or end. Thus an induced emf must be looked upon as existing in the entire circuit rather than belonging to any specific location such as a battery or generator.

In a conductor, closed lines of electric force cause the circulation of the charge carriers, which settle down to a steady drifting motion that constitutes the current. If, however, such electric field lines are applied to free charge carriers in a vacuum, the charges will accelerate as long as the field acts, there being no surrounding matter to hinder their motion. In effect, this is the operating principle of a type of particle accelerator devised by the American physicist D. W. Kerst in 1941. It is used mainly for accelerating electrons and is called a magnetic induction accelerator or, more commonly, a **betatron.**

Figure 19.11 shows the construction of such an instrument. A doughnut-shaped glass or ceramic tube is placed between the pole pieces of a large electromagnet. The tube is pumped to a high vacuum, and electrons are injected along the circular axis. As in the cyclotron (page 445), the strong magnetic field holds the moving electrons in a circular path. At the same time, the field is increased by raising the current in the windings of the magnet. This changes the flux through the orbit of the electrons and produces an electric field that *accelerates* them to high energies. Then the current in the windings of the electromagnet is allowed to drop back, another burst of electrons is injected, and the whole process is repeated at regular intervals. The betatron acts like a transformer whose secondary coil is replaced by a swarm of free electrons moving in a circular path.

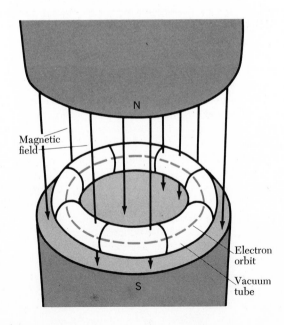

Magnetic field

N

Electron orbit

Vacuum tube

S

FIGURE 19.11 Schematic diagram of the betatron.

When the electrons have acquired the desired energy, a sudden pulse of current applied to the magnet coils makes the electrons spiral out of their circular orbit and strike a target. The impacting high-energy electrons can be scattered from the nuclei of atoms, furnishing information on the structure of the nucleus, or they may be allowed to produce x rays capable of penetrating several feet of solid metal.

The betatron is particularly suited to the acceleration of electrons, whose large relativistic mass increase would quickly drive them out of phase in a device such as the cyclotron.

Worked example 19.4   In a large betatron the orbit radius is 1 m and the average rate of change of the magnetic induction is $\Delta B/\Delta t = 90$ Wb/m^2-sec. Compute the gain in energy of the electrons after they have made 250,000 revolutions.

Solution $\mathcal{E} = -\Delta\Phi/\Delta t = -\pi r^2\,\Delta B/\Delta t$. Substituting the numerical data and neglecting the minus sign, $\mathcal{E} = (3.14)(1^2)(90) = 283$ volts. Thus, each electron acquires 283 eV of additional KE in each revolution. After 250,000 revolutions, the gain is $(283)(250,000) = 7.1 \times 10^7$ eV, or 71 MeV.

The University of Illinois betatron, built in 1950, can attain electron energies up to 340 MeV.

ELECTROMAGNETIC WAVES

19.7 SPEED OF ELECTROMAGNETIC WAVES

A charged capacitor may be "short-circuited" by connecting the ends of a conductor to the plates. The ensuing current may be either a lapse back to the uncharged state or it may be oscillatory, in which case the circuit is the counterpart of an elastic vibrating system in mechanics—for example, a tuning fork. Air resistance and internal friction in the fork ultimately bring the system to rest. In addition, the energy of the sound waves that are sent out must come from the energy originally given to the fork, and this drain is a further factor in quenching the vibrations. In a similar way, an electric circuit loses energy not only through its own resistance (I^2R heat losses) but through radiation of energy in the form of *electromagnetic waves.*

This process of radiation was predicted in 1865 by the theory developed by Maxwell which showed light and other forms of electromagnetic waves to be essentially of the same nature. Maxwell's theory is abstruse and highly mathematical, and the description that follows will simply outline some of the main physical results and their consequences.

In 1887, several years after Maxwell's death, the German experimenter H. R. Hertz (1857–1894) was able to produce electromagnetic waves by electrical means and show that they have the properties of reflection, refraction, diffraction, interference, and polarization possessed by light waves. Later, the speed of such waves in free space was found to be identical with that of light. Through the developments in technique attributable to the Italian inventor G. Marconi (1874–1937) as well as to other pioneers, transmission of electromagnetic waves over long distances formed the basis of the broad fields of modern communication.

Faraday's law of induction showed that a changing magnetic field causes an induced electric field. One of the basic accomplishments of Maxwell's theory was to demonstrate that the converse is also true: A changing electric field gives rise to an induced magnetic field. This demonstrates the remarkable symmetry of electromagnetic phenomena.

The above statements can be put into more concrete form:

1. *The field lines of a magnetic field that varies in time are encircled by electric field lines.*
2. *The field lines of an electric field that varies in time are encircled by magnetic field lines.*

The rate of variation of a field is not necessarily constant, so that the magnitude of the induced field will not be constant in time. This time-varying field in turn induces a field of the other kind, and so on. A region in which time-varying electric and magnetic fields are linked in this way is called an **electromagnetic field.** It should be emphasized that the mere super-position of unrelated electric and magnetic fields does *not* constitute an electromagnetic field.

Maxwell showed that the coupling of electric and magnetic fields in the way described above leads to a *wave motion* in which an electric field and a magnetic field move through space together, and this is the model on which the theory is based. The electromagnetic theory describes the model in more detail. For example, consider the speed of propagation of electromagnetic waves in free space. From Eq. 18.16 (page 454), the magnetic constant of free space has the value

$$\frac{\mu_0}{4\pi} = 10^{-7} \text{ Wb/A-m} \qquad\qquad \textbf{19.9}$$

and by Eq. 16.2 (page 378) the Coulomb-law constant for free space is

$$k_0 = 8.987 \times 10^9 \text{ N-m}^2/\text{C}^2 \qquad\qquad \textbf{19.10}$$

One way of writing the four fundamental equations of the theory for free space is this:

$$\text{div } \mathbf{E} = 0 \qquad\qquad \text{div } \mathbf{B} = 0$$

$$\text{curl } \mathbf{E} = -\frac{\partial \mathbf{B}}{\partial t} \qquad \text{curl } \mathbf{B} = \frac{1}{c^2}\frac{\partial \mathbf{E}}{\partial t}$$

"Div" and "curl" are symbols for certain operations to be performed on the field vectors **E** and **B** and are connected with the kind of directional relations in space that we have described above by means of thumb rules. The symbol ∂ is called a partial differential operator. Notice the presence of c, the speed of electromagnetic waves in a vacuum, in the last equation.

All the vast consequences of the theory follow from mathematical manipulation of these simple-looking statements. It may justly be said that seldom has so much been obtained from so little. This economy of expression is a feature shared by electromagnetic theory with newtonian mechanics, with thermodynamics, and with relativity; it is one of the characteristics that distinguishes a great scientific theory.

FIGURE 19.12 Maxwell's equations displayed.

From this, compute the numerical value of

$$\sqrt{\frac{k_0}{\mu_0/4\pi}} = \sqrt{\frac{8.987 \times 10^9}{10^{-7}}} = 2.998 \times 10^8$$

According to Eqs. 19.9 and 19.10, the quantity under the radical sign has the units

This reduc-
tion should
be checked
in detail at
this point.

$$\frac{\text{N-m}^2}{\text{C}^2} \frac{\text{A-m}}{\text{Wb}}$$

If Wb/m^2 is replaced by C-m/N-sec (see page 465) and A is replaced by C/sec, the above combination of units reduces to m^2/sec^2. The square root is m/sec, so that the quantity computed above is

2.998 × 10^8 m/sec

which is precisely the value of c, the speed of light in a vacuum. Thus the purely electrical quantity whose value was just calculated is found to be identical with the measured speed of light in free space:

$$\sqrt{\frac{4\pi k_0}{\mu_0}} = c \qquad\qquad \textbf{19.11}$$

This striking agreement, noted by Maxwell in 1864, is compelling evidence for the idea that light waves are electromagnetic.

19.8 PROPAGATION OF AN ELECTROMAGNETIC FIELD

An oscillating electric circuit such as the one in Fig. 19.13*b* does not ordinarily give off strong electromagnetic radiation unless it is coupled with an **antenna.** The antenna may be a length of straight wire, grounded at its lower end and pointing upward, the coupling being provided by induction effects between a coil at the midpoint of the antenna and a coil in the oscillating circuit (Fig. 19.14). The antenna follows the variations of current that take place in the adjoining circuit, setting up a time-varying electric

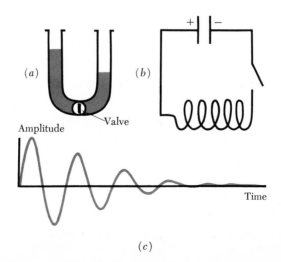

FIGURE 19.13 (a) When the valve is opened, the liquid oscillates in the U-shaped tube. (b) When the switch is closed, an oscillatory current results. (c) In both (a) and (b), dissipative effects cause a progressive decrease in amplitude.

field that leads to the propagation of an electromagnetic field, described more fully below. In this way the antenna becomes a source of electromagnetic waves.

In principle, such an antenna consists of two lengths of wire connected to some kind of generator of alternating emf that causes electrons to surge back and forth in the wires. Figure 19.15 shows the condition of the antenna several times during a cycle of current oscillation. The surrounding electric field at any instant is like that of a pair of equal and opposite charges, but the polarity is continually changing as the electrons move back and forth in the antenna. The periodic reversal of the charge makes the electric field lines pinch off and form loops that spread outward from the antenna with the speed c, as indicated. At the same time, the continually changing current in the antenna produces a time-varying magnetic field whose lines are circles drawn around the conductor (Fig. 19.16). These, too, spread outward with speed c. The combination of the interpenetrating sets of moving field lines constitutes the electromagnetic wave radiated from the antenna. Notice that the electric lines lie in planes that contain the antenna, whereas the magnetic lines are in planes perpendicular to it.

The production of radiation by the antenna is a consequence of the surging back and forth of free electrons in it. In this process, the electrons are subjected to acceleration. The electromagnetic theory shows that *radiation is emitted whenever a charged body of any kind is accelerated.* In summary: (1) the electric and magnetic field vectors vary periodically, (2) they are perpendicular to each other and to the direction of propagation of the waves, and (3) the two fields are in *phase* with each other. This can be shown to be true everywhere except very close to the source.

Figure 19.17 exhibits these relationships. The electric field vector is shown in the vertical plane, and the magnetic vector is in the horizontal plane. The vectors become zero together and attain their maximum values together. Unlike what happens in waves that travel in material media, there is nothing in an electromagnetic wave that oscillates in any mechanical sense; the wave consists merely of a pair of fields whose magnitudes change periodically. It becomes understandable, then, why this is the only kind of wave that can be propagated in empty space.

The amplitude of the waves sent out from an antenna is different in various directions. It is a maximum in any direction perpendicular to the antenna and through its midpoint, but it decreases as the waves deviate from this direction, becoming zero along the line of the antenna itself.

It should be noted that the wave represented in Fig. 19.17 is *plane-polarized* (page 361), since each of the field vectors is confined to a given orientation. Ordinary, unpolarized radiation would be a random mixture of such representations, each at a different angle around the ray. The plane in which the electric field vector lies is usually taken to be the plane of polarization, since it is found experimentally that most of the effects produced by electromagnetic radiation are ascribable to the electric rather than the magnetic vector.

The existence of an electric or magnetic field in any region of space implies the presence of potential energy, just as there is gravitational PE in the space around massive bodies. If the fields belong to an advancing electromagnetic wave, energy is being transported through space with the

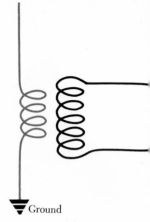

FIGURE 19.14 The oscillating circuit is joined inductively to an antenna in order to increase the intensity of the radiated energy.

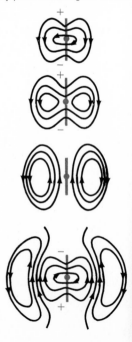

FIGURE 19.15 Electric field lines in the neighborhood of an antenna carrying an oscillating current. The lines have the same appearance in any plane containing the antenna

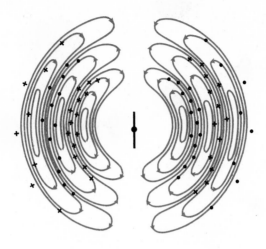

FIGURE 19.16 The outward-spreading electric field lines, shown as colored curves, are accompanied by a magnetic field. The traces of the magnetic lines as they cross through the plane of the drawing are shown by crosses (**B** inward) and dots (**B** outward).

speed c. This energy transport is a matter of familiar experience. We know, for example, that objects exposed to light become warmed by absorbing some of the radiant energy and converting this energy into heat. It is appropriate, then, to determine the *intensity* of radiation, defined on page 303 as the amount of energy passing in unit time through unit area of a surface held perpendicular to the rays. Maxwell's theory shows that this is

$$I_R = \frac{c}{8\pi k_0} E_m{}^2 \qquad\qquad \text{19.12}$$

FIGURE 19.17 Representation of a plane-polarized electromagnetic wave.

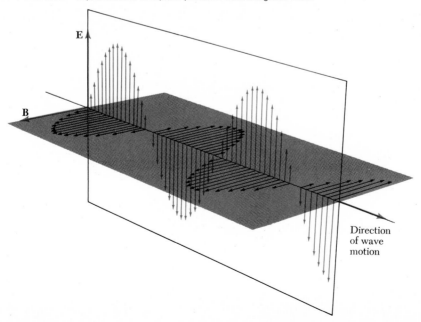

Direction of wave motion

where I_R is the wave intensity and E_m is the maximum value (amplitude) of the electric field vector in the wave. The units for I_R are watts per square meter (see Practice set below).

Another result of Maxwell's theory is that the magnitudes of the electric and magnetic vectors at any instant are proportional to each other:

$$E = cB \qquad\qquad 19.13$$

where the constant of proportionality is the speed of light c. In Eq. 19.12 the value of k_0 can be substituted from Eq. 19.11 (page 477) and the value of E from Eq. 19.13 to get an alternative expression for the intensity in terms of the magnetic vector:

$$I_R = \frac{c}{2\mu_0} B_m{}^2 \qquad\qquad 19.14$$

where B_m is the maximum value (amplitude) of the magnetic vector.

Practice set   Check the consistency of the units in Eqs. 19.12 to 19.14. The following equations will be found useful in doing this: 16.2 (page 378), 16.3 (page 379), 18.1 (page 436), and 18.15 (page 453).

19.9 ELECTROMAGNETIC MOMENTUM; RADIATION PRESSURE

The study of mechanics shows that a moving particle possesses both kinetic energy and momentum. Similarly, electromagnetic waves carry momentum as well as energy. It follows that it should be possible to discuss the momentum transported by radiation in terms of Newton's third law, and this presents a remarkable situation.

The third law describes correctly what happens in collisions, in gravitational attraction, in electrostatic interactions, etc. To discuss the interaction of two charged bodies that are in relative motion, the presence of magnetic effects must be recognized in addition. Consider the situation represented in Fig. 19.18, where two charged objects are moving past each other. The magnitude of the electrostatic force $\mathbf{F}_e$ may be calculated by using Coulomb's law. Equal forces of this amount act in opposite directions along the line joining the charges. Also, each moving charge produces a magnetic field at the position of the other charge. These two fields are indicated by $\mathbf{B}$ and $\mathbf{B'}$ in the figure and, according to the right-hand thumb rule, they have the directions shown. Then, according to the right-hand-screw

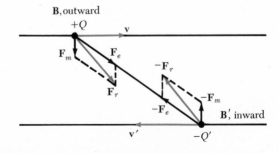

FIGURE 19.18 Two charged bodies moving relative to each other at a speed small compared with c. Newton's third law is satisfied.

rule, the magnetic force $\mathbf{F}_m$ acting on each object has the direction indicated in the figure. Finally, the resultant force on each body, $\mathbf{F}_r$, is found by combining vectorially the electrostatic and magnetic forces. The interesting result is that these two forces $\mathbf{F}_r$ are no longer in the line joining the two bodies. However, they are still equal in magnitude and oppositely directed, and so the third law is fulfilled.

Now, if the two particles are considered to be in very rapid relative motion, a new factor appears: The field produced by one of the charged bodies takes a finite time to reach the other, since electromagnetic fields travel through space with the speed c. In order to compute the field at the location of one of the bodies, the position and velocity of the other at a correspondingly earlier time must be used, which means that the action and reaction forces are not simultaneous. Consequently the rates of change of momentum are not equal and opposite, and the total momentum of the system is no longer constant.

The difficulty can be overcome by recognizing that *there is momentum associated with the magnetic field.* Inasmuch as the two bodies exert forces on each other, they accelerate. According to page 478, an accelerated charge radiates electromagnetic waves, and these waves carry off energy and momentum. Hence some of the momentum of the bodies is in transit at all times. Analysis shows that the balance is restored if, at any point in the field, the momentum crossing unit area per unit time is given by

$$P = \frac{I_R}{c} \qquad\qquad\qquad \textbf{19.15}$$

where I_R is the radiation intensity as defined on page 303. Then the total momentum of the system, electromagnetic plus mechanical, is conserved. The relation 19.15 is also justified by the special theory of relativity.

Practice set Sketch a diagram corresponding to Fig. 19.18 but with both bodies carrying positive charges. Show the directions of all vectors.

Momentum carried by radiation can be transferred to a body exposed to the radiation; as a consequence, such a body will experience a force in the direction of travel of the waves. The existence of radiation pressure, predicted by the Maxwell theory, was verified experimentally at the beginning of this century by the Russian experimenter P. N. Lebedev and by the American physicists E. F. Nichols and G. F. Hull. The pressure (force per unit area) is numerically equal to the rate of momentum transfer per unit area, according to the momentum form of Newton's second law (page 146). Hence, by Eq. 19.15, the pressure exerted on a surface that *absorbs* all the incident radiation is given, in newtons per square meter, by

$$P_a = \frac{I_R}{c} \qquad \text{(total absorption)} \qquad\qquad \textbf{19.16}$$

However, if the surface *reflects* all the radiation, the momentum change on the part of the radiation is doubled, and the radiation pressure in this instance is

$$P_r = \frac{2I_R}{c} \qquad \text{(total reflection)} \qquad\qquad \textbf{19.17}$$

Since the characteristics of real surfaces are somewhere between these two extremes, the actual pressures will fall between the values given by Eqs. 19.16 and 19.17.

Worked example 19.5

The **solar constant** is the average intensity of radiation received from the sun on a surface exposed perpendicular to the rays outside the atmosphere and has the value 1,400 watts/m².

a. Compute the pressure of sunlight on a surface at the earth's position that absorbs all the radiation.

b. Compute the magnitude of the electric and of the magnetic field amplitudes of this radiation.

Solution   *a.* From Eq. 19.16, $P_a = 1{,}400/(3 \times 10^8) = 4.7 \times 10^{-6}$ N/m², or less than 5×10^{-11} of normal atmospheric pressure.

b. By Eq. 19.12 (page 479),

$$E_m = \sqrt{\frac{8\pi k_0 S}{c}} = \sqrt{\frac{(8)(3.14)(3 \times 10^9)(1{,}400)}{3 \times 10^8}} = 1{,}030 \text{ volts/m}$$

The value of B_m can be obtained from Eq. 19.14, but it is simpler to use Eq. 19.13 (page 480):

$$B_m = \frac{E_m}{c} = \frac{1{,}030}{3 \times 10^8} = 3.4 \times 10^{-6} \text{ Wb/m}^2$$

The electric field is of considerable magnitude, but the accompanying magnetic field is very small by ordinary standards.

Particle radiation from the sun—the "solar wind"—also plays a part.

The pressure computed above is exceedingly small, yet solar radiation pressure has interesting astrophysical effects, among them the fact that comet tails point away from the sun. Radiation pressure within a star is thought to be an important factor in determining its stability (see Prob. 19.16).

19.10   THE COMPLETE ELECTROMAGNETIC SPECTRUM

In spite of the compelling nature of the results of Maxwell's calculations, the electromagnetic theory did not win wide acceptance until Hertz succeeded, more than 20 years later, in actually producing the predicted waves by means of electric circuits. In Hertz's own words, published in 1889:

We have applied the term rays of electric force to the phenomena which we have investigated. We may perhaps further designate them as rays of light of very great wavelength. The experiments described appear to me, at any rate, eminently adapted to remove any doubt as to the identity of light, radiant heat, and electromagnetic wave motion.

The work of Maxwell and the confirming observations of Hertz widened enormously the spectrum of electromagnetic waves. Ultimately, the range was increased over that of the visible region by a factor exceeding 10^{20} to give the electromagnetic spectrum now known. Regardless of differences in wavelength—and hence in frequency—all electromagnetic waves have the speed $c = 3 \times 10^8$ m/sec in free space. Characteristic names have been given to the various parts of the electromagnetic spectrum because of their special uses or means of generation or detection, but the demarcations are not sharp and a given radiation in a boundary region can often be produced from either side. The whole range, from the very shortest to the very longest waves, may be said to be known experimentally without any breaks or interruptions. The main characteristics of several of these regions are surveyed below, and Fig. 19.19 should be examined in detail in connection with this description.

Electric waves   This is a broad region of the electromagnetic spectrum that includes various kinds of waves produced and detected by means of electric circuits. The longest of these are the waves existing on alternating-current power lines. In the United States most power transmission is at a frequency of 60 cycles/sec. Since the waves on such lines have essentially the speed c, the wavelength as given by the wave equation $\lambda = c/f$ is $(3 \times 10^8)/60 = 5 \times 10^6$ m, or about 3,000 mi. Most of this energy is confined to the power line, very little being radiated into space under these circumstances. Since the frequency of a generator may be made arbitrarily small, there is in principle no upper limit to the wavelength of a wave of this kind.

Several bands of electric waves are used for communications purposes. They extend over the range 10 kc to 890 Mc, corresponding to wavelengths from 30 km down to about 30 cm. Adjoining this range and overlapping it somewhat are the **microwaves,** which have wavelengths from about 30 cm down to 0.3 mm. A wavelength of about 8 cm is used in the microwave relay networks that carry telephone and television communication across the continent. Radio waves received from astronomical bodies are the basis of the relatively new science of **radio astronomy,** at present one of the most active branches of astrophysics. Observations are confined to wavelengths from about 1 cm to 50 m, which are not too strongly absorbed by the earth's atmosphere. Earth-based radio telescopes and radio interferometers are now in use in many places.

The optical region   Infrared radiation is associated mainly with hot bodies and is sometimes inaccurately called "heat radiation" or "radiant heat." Objects at all temperatures emit infrared radiation. The earth's atmosphere absorbs from sunlight virtually all infrared waves longer than about 5.3×10^{-6} m (53,000 Å). The infrared range was first detected in 1800

The prefix "infra-" is from a Latin word meaning below, or beneath.

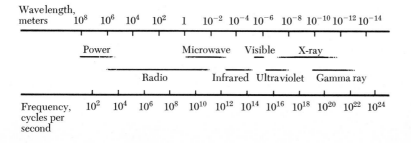

FIGURE 19.19 The electromagnetic spectrum. Note: Wavelength and frequency scales are logarithmic.

This radio telescope, located near Pretoria, South Africa, is used primarily for tracking satellites. The "dish" is 85 ft in diameter. (*South African Information Service.*)

by the German-English astronomer William Herschel (1738–1822), discoverer of the planet Uranus. He placed a thermometer in various parts of a solar spectrum formed by a prism and detected a temperature rise even when the bulb was in the dark region beyond the red. He correctly concluded that radiation was present there as well as in the visible portion of the spectrum. He went on to demonstrate the reflection and refraction of this "invisible light" and showed that it behaved in these respects like visible radiations.

Later workers had available much more sensitive means of detecting and measuring infrared waves, particularly instruments called thermocouples and bolometers which detect, by means of electrical effects, the heat resulting from absorbed radiation. Special photographic plates can record waves as long as 14,000 Å, making photography possible even in complete darkness.

Infrared radiations are strongly emitted by sources other than high-temperature bodies. From optical sources, infrared waves have been produced that are longer than the shortest electric waves. They have exactly the same properties, whether emanating from optical or from electrical sources.

Visible light is the part of the optical region detectable by the human eye. Categorically, the obvious sources of this kind of radiation may be called lamps, the active part of which is usually an incandescent body. However, there are many other kinds of nonthermal light sources using electrical, chemical, and other means. Any form of radiation that is not ascribable to heat is called **luminescence.** Much more will be said about the mechanism of light emission and absorption in Chaps. 20 and 21.

At about the same time that Herschel noted the extension of the spectrum beyond the red, an equally important discovery was made by the German physicist J. W. Ritter (1776–1810) and the English scientist-physician-inventor W. H. Wollaston (1766–1828). They found that photographic materials placed beyond the violet end of the visible spectrum were affected by a radiation falling there. This radiation, consisting of waves shorter than those of visible light, constitutes the **ultraviolet** region, whose outstanding property is the production of chemical changes, including action on photographic substances.

The prefix "ultra-" is from the Latin word meaning beyond.

In addition to its chemical effects, ultraviolet radiation can be detected and measured because it produces **fluorescence** in certain materials. Numerous substances (petroleum oil, uranium glass, chlorophyll, etc.) emit light of their own when exposed to radiation. In this process the atoms of the material absorb certain wavelengths of the incident radiation and reemit the energy, usually at longer wavelengths. The fluorescent lamps in common use are a practical application of this phenomenon. If the emission of light persists for some time after the stimulating rays have been cut off, the material is said to be **phosphorescent.** The coating materials used in fluorescent-lamp tubes show this effect to some extent (notice that the glow is still faintly visible for some time after the light is switched off). Ultraviolet radiation has important biological effects, such as the production of vitamin D in the skin. A certain range of these wavelengths is the cause of sunburn. Ultraviolet rays from a mercury-vapor lamp kill bacteria and are used for sterilizing foods and utensils.

Fluorescence and phosphorescence are forms of luminescence.

At sea level, almost all solar radiations shorter than about 2800 Å have been absorbed by the air, most of the ultraviolet absorption having taken place at altitudes of the order of 100 km, where the radiation succeeds in ionizing the air, resulting in the formation of the **ionosphere.**

The conducting layers of the ionosphere reflect electric waves of moderate wavelength, so that communications waves can travel around the curvature of the earth by repeated reflection between the ionosphere and the ground. Very long electric waves in particular tend to follow a conducting surface such as the ground or the surface of the sea, and this is the main reason for their great range. On the other hand, very short waves and microwaves penetrate the ionosphere and pass into space. As a result, they can be received only within the line of sight of their source, and for this reason television broadcasting antennas must be placed on high towers. Beginning with the launching of the Telstar satellite in 1962, television signals are now regularly relayed from one part of the world to another.

X rays   The chance discovery of x rays in 1895 by the German physicist W. K. Röntgen (1845–1923) proved to be one of the most important events in modern science, not only because of the practical applications of x rays but also because of their role in the development of physical and biological science. At least six Nobel Prizes have been awarded in this field—an indication of the significance of this branch of physics. Like most "accidental" discoveries, Röntgen's success was the culmination of a series of careful experiments, and his cautious and thorough exploration of the results of his first findings is a model of scientific enterprise (see Ref. 19.7).

Röntgen had been experimenting with cathode rays (which are now described as beams of fast electrons), using an apparatus like the one shown in Fig. 19.20 in which the glass emits a greenish fluorescent glow under the impact of fast electrons. In one experiment, Röntgen covered the tube completely with opaque paper. In the darkened room, he noticed that a fluorescent substance several feet away glowed feebly when the tube was operating. He reasoned that some type of invisible but penetrating radiation was being given off. He found that it came from the walls of the tube and was strongest near the place where the cathode rays struck the glass. It was Röntgen himself who first noted that these **x rays,** as he called them, could penetrate parts of the body, revealing internal structure. Within only a few weeks of his first production of the rays, they were being used in medicine.

X rays travel in straight lines and cast sharp shadows but, unlike cathode rays, are not deflected by a magnet. They are absorbed more by dense than by tenuous substances and produce fluorescence in suitable materials.

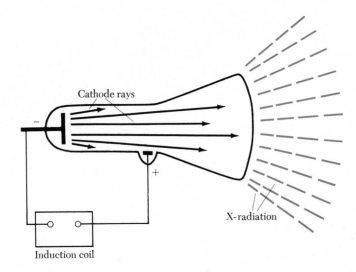

Cathode rays

X-radiation

Induction coil

FIGURE 19.20 In Röntgen's original apparatus the x-radiation was produced by the impact of cathode rays on the end of the glass tube.

Among other effects, x rays can discharge electrified bodies, initiate chemical changes, and affect photographic plates. All these properties were revealed in the original series of experiments, but neither Röntgen nor the many other experimenters who then began to explore this promising new field were able to find evidence that the rays could be reflected, refracted, or diffracted. Proof of the wave nature of the radiation was not discovered for nearly two decades. X rays are now an established region of the electromagnetic spectrum, their wavelengths extending from about 10^{-8} m down to about 10^{-12} m. The details, as well as further properties of x rays, will be presented in Chap. 20.

Refer to page 478.

X rays are produced by the sudden stopping (high acceleration) of high-energy electrons or other charged particles that are allowed to strike targets. In general, the smaller the wavelength, the more penetrating ("harder") the rays are. Most methods of detecting and measuring the intensity of x rays depend on the fluorescence, ionization, chemical, or photographic effects that the rays produce. The best-known characteristic of x rays is their ability to penetrate matter.

X-ray astronomy has become an established branch of science. In 1960, on a rocket flight, the sun was photographed for the first time in x-ray wavelengths, and other intense cosmic sources of these radiations have since been discovered. In order to work in any part of the electromagnetic spectrum unhindered by the earth's atmosphere, unmanned orbiting observatory satellites have been launched, the first one early in 1966.

Gamma (γ) is the third letter of the Greek alphabet. The reason for this designation of the radiation will appear in Chap. 22.

Gamma rays Electromagnetic waves shorter than about 10^{-10} m have most of the properties of x rays but are more penetrating. They are called **gamma rays** and arise from transformation processes occurring in the nuclei of atoms and from the conversion of matter into radiation. Two prominent sources of gamma radiation are radioactivity and cosmic rays. A detailed discussion of gamma rays appears in later chapters.

19.11 A LIMITATION OF THE MAXWELL THEORY

The electromagnetic theory provided physics with a specific model for describing optical phenomena quantitatively and in detail. Just as the ray model of light was refined and extended by the introduction of the wave concept, so Maxwell's theory made the wave model more concrete by supplying an explicit representation of the nature of these waves. In a very real sense, this theory not only effected a union between light and a vast range of other radiations but made the entire subject a branch of electricity and magnetism. Einstein was moved to remark that Maxwell's statement of the fundamental laws of electromagnetism was the most important event in physics since Newton. However, the apparent completeness and universality of the electromagnetic theory were to be challenged within half a century by the discovery of certain phenomena which the theory could not explain. As a result, it was found necessary to supplement the electromagnetic wave model with a companion hypothesis, the quantum theory, to be introduced in the next chapter.

Programmed review

Instructions: See page 18.

1. Express Faraday's law of electromagnetic induction in words and symbols.

   The emf induced in a circuit is equal to the rate of change of the magnetic flux through the circuit: $\varepsilon = -\Delta\Phi/\Delta t$.   [19.2]

2. State Lenz's law.

   The direction of an induced current is such that its magnetic effect opposes the action that produces it.   [19.3]

3. What is an electric-current generator? Of what main parts does it consist?

   A device for converting mechanical energy into electric energy. It consists of one or more loops or coils that can be rotated in a magnetic field. Provision is also made for leading the current to an outside circuit.   [19.4]

4. What is a transformer?

   A device which, by electromagnetic induction, transfers electric energy from one circuit to another. The process is usually accompanied by a change in the emf and current.   [19.5]

5. What is a betatron?

   A device for accelerating electrons, using the principle of electromagnetic induction.   [19.6]

6. Under what conditions does a charged body radiate an electromagnetic field?

   When it is accelerated.   [19.8]

7. Describe the model of a plane-polarized electromagnetic wave.

   The electric and magnetic vectors lie in a plane perpendicular to the direction of travel and are perpendicular to each other. Their magnitudes vary periodically and, except very near the source, the two vectors are in the same phase.   [19.8]

8. Give a physical reason for the existence of radiation pressure.

   Electromagnetic waves carry momentum. The pressure is caused by the transfer of some or all of this momentum to a body in the path of the radiation.   [19.9]

9. Name the principal divisions of the complete spectrum of electromagnetic waves, in order of decreasing wavelength.

   Electric waves, infrared, visible, ultraviolet, x rays, gamma rays. There is some arbitrariness in the naming of these regions or their subdivisions.   [19.10]

10. What is the commonest source of infrared radiation? Give the approximate limits in wavelength of this region.

Hot bodies. From about 5×10^{-2} to 7×10^{-7} m.   [19.10]

11. Name the principal effect produced by ultraviolet radiation. What are the approximate limits of this region in wavelength?

Chemical effects. From about 4×10^{-7} to 5×10^{-9} m.   [19.10]

12. What chief characteristic do x rays and gamma rays have in common?

Their ability to penetrate considerable thicknesses of otherwise opaque substances.   [19.10]

For further reading

19.1. *Taylor, L. W.* "Physics, the Pioneer Science." Read chap. 48 on the history of the telephone.
19.2. *Gillispie, C. C.* "The Edge of Objectivity." The work of Faraday and Maxwell is described on pp. 435–476.
19.3. *Newman, J. R.* James Clerk Maxwell, *Sci. Am.,* June, 1955, p. 58. An interesting biography.
19.4. *Roberts, M. S.* Recent Discoveries in Radioastronomy, *Phys. Today,* February, 1965, p. 28. X-ray astronomy is also discussed.
19.5. *Maran, S. P.* Radio Astronomy, "Encyclopedia of Physics." See p. 591.
19.6. *Henry, G. E.* Radiation Pressure, *Sci. Am.,* June, 1957, p. 99.
19.7. *Bleich, A. R.* "The Story of X Rays." A lively account of their discovery and applications.

Questions and problems

19.1. Calculate the emf induced in a straight wire 60 cm long which is at right angles to a magnetic field $B = 2.2$ Wb/m^2 and is moving at a speed of 1.4 m/sec in a direction perpendicular to the field and to itself.

19.2. A circus trapeze consists of a metal bar 70 cm long suspended by two ropes, each of length 7 m. The bar is drawn back until it is at the same level as the supporting hooks and is released. Compute the maximum emf induced in the bar if the vertical component of the earth's magnetic field is 4×10^{-5} Wb/m^2 at the locality in question. Neglect any effect of air resistance on the motion of the bar.

19.3. A flat, circular coil of 20 turns has a total resistance of 0.83 ohm. What current exists in the coil when the magnetic flux through it is made to change at the rate of 0.0015 Wb/sec?

19.4. When no current is being drawn from a generator, the only work needed to keep it turning is that done against friction, but as soon as the outside circuit is closed, the generator becomes very hard to turn and acts as if it were immersed in molasses. Explain.

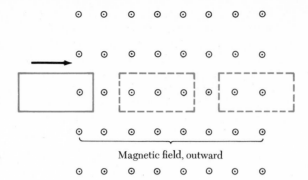

FIGURE 19.21 Wire loop moved through a field.

Magnetic field, outward

19.5.   A barrel has top and bottom hoops made of metal. A current is started in the upper hoop in the clockwise sense as seen from above. (*a*) What is the directional sense of the induced current in the lower hoop? (*b*) What is the direction of the electromagnetic force that acts on the lower hoop?

19.6.   Figure 19.21 shows three positions of a rectangular, metallic loop which is being moved continuously through a uniform magnetic field of limited extent, as shown. In which positions will there be an induced current, and what directional sense will these currents have?

19.7.   A circular coil of 100 turns, of effective radius 4 cm, is in a magnetic field of 0.010 Wb/m², the plane of the coil being perpendicular to the field. (*a*) If the coil is jerked out of the field in 0.01 sec, what emf is induced in it? (*b*) If the coil is connected into a circuit whose total resistance, including that of the coil itself, is 48 ohms, what average current exists during the process?

19.8.   A flat, circular coil of area 0.2 m² and 50 turns of wire stands with its plane perpendicular to a magnetic field $B = 0.048$ Wb/m². The coil is rotated through 180° around an axis lying in its own plane, the operation taking 0.2 sec. Compute the emf induced in the coil.

19.9.   A loop of flexible wire in the form of an overhand knot is located in a magnetic field, as shown in Fig. 19.22. The flux through the loop amounts to 0.1 Wb. (*a*) What emf is produced in the loop when the knot is drawn up tight if the process takes 0.2 sec? (*b*) When one looks along the lines of the field, what directional sense does the emf have?

19.10. The magnetic flux through a circuit of resistance R changes by an amount $\Delta\Phi$ in a time Δt. Show that the total quantity of charge Q that passes any point of the circuit is given by $Q = \Delta\Phi/R$.

19.11. *Eddy currents*. Figure 19.23 shows a copper strip being pulled to the right through the gap between the poles of a strong magnet. Consider any crosswise element of the metal, such as the one outlined by broken lines. It is a conductor moving through a mag-

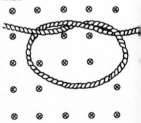

FIGURE 19.22 Tightening the wire knot.

B inward

netic field, and so an emf is induced in it. The effect is that loops of current like those shown are established in the body of the metal. These *eddy currents* have the effect of putting a drag on the motion of the strip (Lenz's law), and this force can be used to damp out vibrations in machines and even to serve as a brake for electric railway cars. (*a*) Can such a brake be used to hold a car at rest on a slight incline? Explain. (*b*) Sketch the directional sense of the eddy currents in each of the two sets of loops in Fig. 19.23.

19.12. What will be the effect on the output emf of a generator of (*a*) doubling the value of the induction B of the field; (*b*) doubling the rate of rotation of the machine?

19.13. A "step-up" transformer used on a power line where the average emf is 110 volts delivers a current of 2 amperes at 770 volts from its secondary. (*a*) If the primary winding has 50 turns, how many turns are there in the secondary? (*b*) What average current is drawn from the line?

19.14. A radio wave has a maximum electric-vector amplitude of $E_m = 10^{-4}$ volt/m. Calculate (*a*) the value of the maximum magnetic induction B_m; (*b*) the wave intensity I_R; (*c*) the radiation pressure P on a surface that absorbs all the wave energy.

19.15. Beam intensities up to 3×10^{18} watts/m² have been attained with gas-dynamic lasers. Calculate (*a*) the order of magnitude of the average electric field E of these waves, expressed in volts per meter; (*b*) the radiation pressure on a totally absorbing surface, in newtons per square meter. How much is this in terms of normal atmospheric pressure?

19.16. Show that the order of magnitude of the radiation pressure at the base of the sun's chromosphere (its outermost layer) is 0.1 N/m². Start with the information on solar radiation received by the earth, given in Worked example 19.5 (page 482), and assume the inverse-square law holds. Useful data: distance, earth to sun, 1.5×10^{11} m; distance, base of chromosphere from center of sun, 7×10^8 m.

19.17. *Micropulsations.* Electromagnetic waves with periods as long as 100 sec have been detected on earth by their magnetic effects. These *micropulsations* are of unknown origin. (*a*) What is the wavelength of these radiations, in terms of the diameter of the earth? (*b*) How does their frequency compare with that of the 60-cycle/sec waves on power lines?

19.18. If the longest usable electric waves have a wavelength of 10^7 m and the shortest gamma rays have a wavelength of 10^{-15} m, what is the wavelength of a radiation that is *logarithmically* halfway between these two? In what region of the spectrum does it lie?

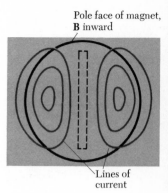

Pole face of magnet, **B** inward

Lines of current

FIGURE 19.23 Eddy currents.

Part Five
THE FUNDAMENTAL STRUCTURE OF MATTER

Chapter 20
QUANTA, PHOTONS, AND MATTER WAVES

By the end of the nineteenth century the great theories underlying the main divisions of what is called classical physics had been brought to a high state of perfection and the subject had attained a remarkable unity and consistency. Mechanics, as developed by Galileo and Newton, gave a rational explanation of the motions of bodies ranging in size from ordinary objects to planetary systems. The kinetic theory of matter placed the atomic concept on a firm basis and brought the entire subject of heat within the scope of mechanics. The principle of conservation of energy became the fundamental unifying idea of all of physics. Additional phenomena that were apparently distinct and unrelated were brought together under the head of electromagnetism, leading to momentous practical developments. Perhaps most outstanding of all was Maxwell's electromagnetic theory of radiation, which made all of wave optics a branch of electromagnetism.

It is not surprising that several prominent scientists of the day expressed the opinion that all the fundamental laws of physics had already been discovered and were not likely to be supplanted in the future, but this prediction turned out to be completely unwarranted. Just before the end of the century, a series of remarkable discoveries were made in rapid succession concerning phenomena entirely unknown to earlier science. Röntgen (page 486) produced x rays in November 1895 and published his findings early the following year. Only a few weeks later, the French physicist Henri Becquerel (1852–1908) discovered natural radioactivity, followed within a few months by the identification of the electron by Thomson. These discoveries, and the theoretical structures erected to account for them, touched off a chain reaction of activity that has made the present century a time of unprecedented scientific and technological growth.

Although it was in 1905 that Einstein formulated the special theory of relativity, it may well be considered part of classical physics because it left many concepts unchanged, and the innovations it brought about were only in the space-time structure of the science. By contrast, another revolution in physics that began in 1900 went far deeper because it struck at the very core of the subject and demanded a radically new formulation of the structure of natural laws. This doctrine is the quantum theory. It deals with the phenomena of the subatomic world and with certain elementary events occurring there—events that defy analysis in classical terms and hence appear as discontinuities.

The remainder of the book is concerned mainly with the development of the quantum idea. This chapter traces the historical origin of this concept and describes its application to the interpretation of a number of important phenomena.

20.1 ORIGIN OF THE QUANTUM HYPOTHESIS

Ideal thermal radiation.

The idea of quanta arose in an attempt to explain certain aspects of **ideal thermal radiation,** a phenomenon that was familiar experimentally and fairly well understood theoretically long before the beginning of this century.

When a solid or liquid sample of matter is heated to incandescence and the light from it is dispersed by a prism or grating, a **continuous spectrum** is observed. In a continuous spectrum, all wavelengths are present (page 347), although the intensity falls off rapidly at very short and at very long wavelengths. Besides depending strongly on the temperature of the body, the distribution of intensity in the spectrum also varies with the nature of the surface. Dull, dark-colored surfaces emit more strongly than shiny, light-colored ones at the same temperature.

About a century ago, the German physicist G. Kirchhoff (1824–1887) discovered that the emitting power of a surface is directly proportional to its absorbing power. An object that could absorb all the radiation falling on it would appear perfectly black. Hence a body that would radiate more strongly than any other at the same temperature is called an **ideal black body.**

Actual surfaces absorb radiation at widely different rates. Dull black velvet fabric and lampblack absorb 97 to 99 percent of the radiation falling on them, whereas polished metals absorb only a few percent. The behavior of an ideal black surface, on the other hand, is independent of any material, a factor that makes it suitable for theoretical description.

The nearest approach to an ideal black body is a small hole in the side of a cavity that has rough interior walls (Fig. 20.1). Here, any radiation that finds its way into the cavity from the outside is soon absorbed by repeated reflections within and has negligible chance of being reflected out through the opening. As the walls absorb the incoming radiation, their temperature rises, they begin to radiate in turn, and the cavity becomes filled with radiation that is in equilibrium with the interior walls. Any of this radiation that diffuses out through the hole is **ideal cavity radiation** corresponding to the temperature of the interior. *The hole itself* represents an ideal black body.

The radiative power of a cavity as a function of temperature was examined experimentally by the Austrian physicist J. Stefan in 1879, and shortly thereafter his findings were confirmed by a theoretical calculation made by Boltzmann (page 236n). The **Stefan-Boltzmann radiation law**

FIGURE 20.1 Radiation entering the cavity is absorbed by the rough walls. Only radiation reemitted by the walls can find its way out again, and the intensity distribution of this radiation depends only on the cavity temperature.

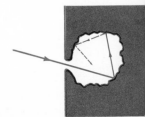

*The Stefan-
Boltzmann
radiation
law.*

states that the rate of emission of energy from an ideal black body is proportional to the fourth power of its absolute temperature:

$$R = \sigma T^4 \qquad\qquad \textbf{20.1}$$

where R represents the total energy of all wavelengths emitted per unit time and per unit area of surface of the radiator. It is called the **total emittance.** T is the absolute temperature, and σ (sigma) is a constant. When R is measured in watts per square meter and T in kelvins, the value of σ is 5.672×10^{-8} W/m²-deg⁴.

The equation shows that a cavity at any finite temperature emits some radiation and that the occurrence of T to the fourth power indicates that the rate of emission increases very rapidly as the temperature is increased. The same is true of any radiating body, except that the value of R is some fraction of the ideal value given by Eq. 20.1, depending on the nature of the surface. At high temperatures many actual surfaces approximate to ideal radiators, and Eq. 20.1 can be used to determine the temperature from a measurement of the total emittance.

Instead of determining the total emittance over all wavelengths, we may measure the energy radiated per unit time and per unit area over a small range of wavelengths $\Delta\lambda$ at various places in the spectrum. Each such measurement, after division by $\Delta\lambda$, becomes the **monochromatic emittance** R_λ. The curves in Fig. 20.2 show monochromatic emittance plotted against wavelength for a cavity radiator at various temperatures. For each temperature, the energy radiated changes markedly with wavelength, and

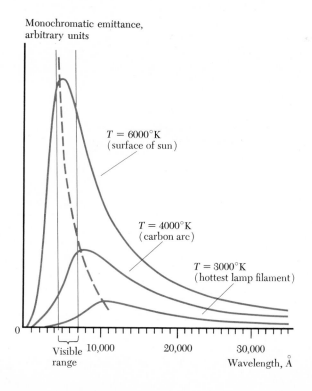

FIGURE 20.2 Radiation curves of an ideal cavity radiator at various temperatures. The total radiation, for any given temperature, is represented by the area under the curve. The dashed line is drawn through the peaks of the curves.

each curve has a maximum at a definite wavelength. As the temperature increases, not only does the total emittance increase (Stefan-Boltzmann law), but the wavelength at which the maximum emittance occurs shifts progressively to smaller wavelengths. In 1893, the German physicist Wilhelm Wien (1864–1928) derived a simple connection between the wavelength of the peak of the curve and the absolute temperature. His relation, which was confirmed experimentally by other workers, is

$$\lambda_M T = \text{constant} \qquad\qquad\qquad \textbf{20.2}$$

where λ_M is the wavelength of the maximum of a given radiation curve and T is the corresponding absolute temperature. When λ_M is measured in angstroms and T in kelvins, the constant has the numerical value 2.898×10^7 Å-deg. The relation 20.2 is known as **Wien's displacement law.**

Photographs of the sun's spectrum, taken from rockets above the earth's atmosphere, show that the maximum radiation occurs at a wavelength of about 4650 Å. Using Wien's law, the surface temperature of the sun can be estimated (Eq. 20.2) to be about $2.9 \times 10^7/4650$, or 6200 K. The sun is not an ideal black body, and the surface temperature is actually some 300 to 400° lower than this figure. The application of the Stefan-Boltzmann or the Wien law to radiation measurements on the stars shows that the surface temperatures of blue-white stars are as high as 20,000 K.

The above two laws were deduced from well-known classical principles such as the laws of thermodynamics and electromagnetic theory. Encouraged by this accomplishment, physicists attempted to construct a theory that would yield the form of the emittance curves shown in Fig. 20.2, but their efforts were not successful. One of the more promising theories, devised by Wien, led to a formula which fits the experimental curves sufficiently well at short wavelengths but departs markedly from them elsewhere. Another theoretical attempt by the English physicists Rayleigh and Jeans was successful at very long wavelengths but not at shorter ones (Fig. 20.3). It seemed that nothing further could be accomplished using only classical laws.

Among the scientists engaged in these attempts was the German theoret-

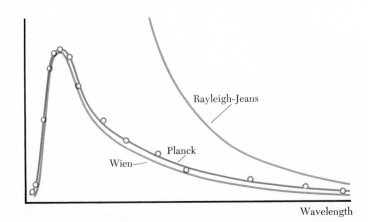

FIGURE 20.3 The experimental points (open circles) can be fitted very well by a Planck curve over the entire spectral range. Wien's curve fits only at short wavelengths whereas that of Rayleigh and Jeans agrees with the observations only at very long wavelengths.

Planck. (*German Information Center.*)

Planck became a distinguished professor of physics at Berlin and won the Nobel Prize in 1918 for the work described in this section.

ical physicist Max Planck (1858–1947). In 1900, using the earlier theories as a guide, he succeeded in finding, by trial, a mathematical formula that fitted the experimental radiation curves with satisfactory accuracy. This relation has the form

$$R_\lambda = \frac{C_1}{\lambda^5(e^{C_2/\lambda T} - 1)}$$ **20.3**

Here C_1 and C_2 are constants that can be calculated from the experimental data; each has the same value for the whole set of curves in Fig. 20.2. The symbol e represents the *base of natural logarithms,* a number that arises in higher mathematics. It has the value 2.718. . . .

Meanwhile, more exact measurements extending over a greater range of wavelengths became available, and these too could be represented accurately by the empirical formula 20.3.

Planck set about constructing a physical model that would yield this relation. It seemed natural to make the assumption that an ideal radiating body contains in effect a large number of classical electrical *oscillators,* each having a definite frequency. According to classical physics, such an oscillator sends out radiation of the same frequency with which it is vibrating. The individual oscillators receive their energy from the source that is maintaining the temperature of the radiating body, and they dispose of this energy as electromagnetic radiation. In present terminology these oscillators represent the various possible modes of vibration of the electrons, atoms, and molecules of the material making up the radiating body.

The next task was to compute the various ways in which the energy could be apportioned among groups of such oscillators. According to classical principles, each oscillator behaves like a swinging pendulum or a weight suspended by a spring and can take on and give off energy in any amount. Further, an oscillating electric charge has accelerated motion and so it must be expected to radiate energy continuously (page 478).

Planck's quantum hypothesis. Planck soon saw that it was impossible to retain these classical principles and yet arrive at the empirical formula he had discovered. His detailed calculation showed the necessity for making a radical new assumption: *The energy of each oscillator must vary discontinuously, or in "jumps," and consequently the absorption and reemission of the energy must also be a discontinuous process.*

The details of Planck's computation are complicated and are not given here. The incorporation of Wien's law into the theory led to the conclusion that the energy of an oscillator can change only by an amount proportional to the frequency. The minimum possible energy change E may be written

$$E = hf \qquad\qquad 20.4$$

where f is the frequency of oscillation and h is a universal constant that has come to be called **Planck's constant.** The accepted value, in mks units, is

$$h = 6.6256 \times 10^{-34} \text{ J-sec} \qquad\qquad 20.5$$

This expression indicates that h has the dimensions of energy multiplied by time, and this follows from the defining equation 20.4:

$$\frac{\text{Energy}}{\text{Frequency}} = \frac{\text{energy}}{\text{cycles/time}} = \frac{\text{energy} \times \text{time}}{\text{cycles}} = \text{energy} \times \text{time}$$

since *cycles* is merely a pure number.

Planck's constant appears in connection with many phenomena besides thermal radiation, and as a result there are a number of independent ways of computing the value of this quantity, which now ranks in importance with other universal physical constants such as G, c, and e. By comparing his formula with radiation data, Planck obtained a first approximate value for h of about 6.5×10^{-34} J-sec.

To the amount of energy $E = hf$ Planck gave the name one **quantum** (plural, *quanta*). Thus his hypothesis may be expressed by saying that the amount of energy possessed by an oscillator can only be an integral multiple of one quantum. The concept of quantization of the energy of an oscillator can be represented graphically by an **energy-level diagram** such as Fig. 20.4. The possible amounts of energy that the oscillator can possess are depicted by equally spaced horizontal lines, like the rungs of a ladder. The energy of the oscillator must be located on one of these levels, except when in transition. The energy-level representation will be used later where it permits the discussion of the behavior of more complex systems without requiring a detailed picture of their structures.

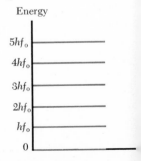

FIGURE 20.4 Energy-level dia gram for a Planck oscillator of fre quency f_o. Only those energy values that are integral multiples of hf_o occur.

Practice set*   Consider two solid bodies emitting thermal radiation: A is a block of carbon at a temperature where it just begins to glow dull red (1000 K), and B is an incandescent lamp filament (3000 K). Assuming both to be ideal radiators, find the ratio of:

1. The wavelength of maximum monochromatic emittance of A to that of B
2. Their total emittances
3. The energy of one quantum of radiation from each at 5000 Å

20.2 QUANTIZATION IN PHYSICS

At the time it was proposed, Planck's quantum hypothesis seemed bizarre and at variance with "common sense." Physicists were reluctant to embrace the idea, and for some years Planck himself had misgivings that drove him to look industriously for a possible alternative. These efforts were unsuccessful.

Although the quantum idea as applied to energy represents a frank break with the principles of classical physics, it is not without precedent, as shown by some examples of "quantum ladders" that are familiar and accepted:

1. The first and perhaps best known is the atomic structure of matter: The rest masses of the atoms do not form a continuous range but have only certain discrete values, as indicated in Fig. 20.5a. In Dalton's time it was thought that the atomic masses were all *integral multiples* of a certain smallest mass, but it is now known that this is not true. In any

*Ans.: 3; 1/81; 1.

FIGURE 20.5 Instances of quantization in classical physics.

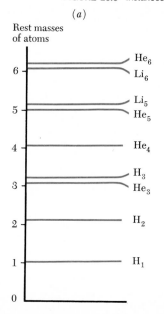

(a)

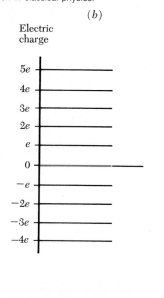

(b)

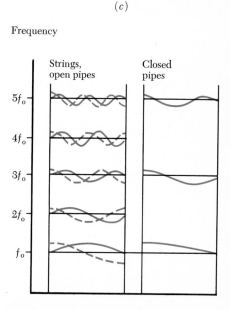

(c)

event, the rest masses of the various species of atoms in their normal states form a discrete (although not equally spaced) set.

2. Electric charge is another quantity that is "atomic." The amount of any charge, positive or negative, is always an integral multiple of e, as represented in Fig. 20.5b. No exception to this quantization of charge has ever been found.

3. A surprising instance of quantization manifests itself in stationary waves in vibrating systems. It was shown in Chap. 13 that the possible frequencies of steady vibration of a stretched string or of an air column in a pipe are always integral multiples of the fundamental frequency (Fig. 20.5c). It will be seen later that precisely this quantization associated with stationary waves became an important feature of modern quantum theory.

Recall that for an open pipe only the odd integral multiples are present.

In addition to its application to Planck's oscillators, does energy quantization apply to macroscopic systems as well? There are no grounds for believing otherwise, and the only reason that the discrete nature of energy is not evident in dealing with objects of ordinary size is that the energy of an individual quantum is exceedingly small compared with the total energy of such systems. The following Worked example brings out this point.

Worked example 20.1

A pendulum bob has a mass of 1 kg, and its period is 1 sec. At the extreme ends of its initial swings the bob is 2 cm above its lowest position.

a. What is the total energy of the pendulum, referred to its lowest position, in joules?

b. As the pendulum continues to swing, it loses energy through air resistance. By how much does its energy decrease when it loses 1 quantum?

c. How many quanta has it lost when the maximum height has been reduced by 0.001 cm?

Solution *a.* At the end of a swing, the energy is entirely gravitational, and it amounts to GPE $= mgh = (1)(9.8)(0.02) = 0.2$ joule.

b. The energy of 1 quantum for this system is $E = hf = (6.6 \times 10^{-34})(1) = 6.6 \times 10^{-34}$ joule, which is extremely small in comparison with the total energy.

c. From the result of part *a*, the energy decrease is $(0.2)(0.001)/2 = 1 \times 10^{-4}$ joule. Dividing this by the energy per quantum yields (1×10^{-4}) (6.6×10^{-34}), or about 1.5×10^{29} quanta. From this it is apparent that the discontinuities in the height of the bob, as its swings die down, are far too small to be noticed. Similar conclusions can be drawn for any macroscopic system.

20.3 THE CORRESPONDENCE PRINCIPLE

The computation in the above Worked example involved a swinging pendulum, but the result emphasizes a fact of far broader significance. In order to discuss this, it is helpful, first, to consider the experimental basis of scientific theories in general.

Every physical theory is deduced from observational data that extend over a limited range of values, leaving the possibility that the theory may fail to predict correctly in circumstances outside this range, particularly if conditions are extreme. For instance, it may well be questioned whether the law of gravitation holds in the far reaches of the universe. It is known that calculations based on this law are in agreement with observations for the planets and their satellites, and even up to the much larger distances between the components of double-star systems, but there is no direct evidence of the validity of the law for any greater distances. Another example: Coulomb's law was deduced from experiments on charged bodies in the laboratory, but is it applicable to the prediction of the forces between charged particles in an atomic nucleus? In this case it is known that the answer is ''No'' because additional forces, to be described later, become effective.

In the history of science, a previously successful theory sometimes fails to make accurate predictions for a new experimental situation. The phenomenon of thermal radiation, discussed above, is a prime example. In such an instance, how can the old theory be modified in order to restore harmony? A guide that is often used is the **correspondence principle,** which was conceived in a more special setting by the Danish theoretical physicist Bohr in 1923. The correspondence principle asserts that the predictions of a more general theory must reduce to those of a valid special theory in all cases where the restricted conditions of the latter apply.

Several examples of this principle have already been presented. In optics, the casting of shadows, reflection, refraction, etc., were adequately described by drawing rays. However, in order to describe properly diffraction, interference, and polarization, a more general theory based on a wave model was needed. The correspondence principle requires that the predictions of the wave theory reduce to those of the ray theory in all circumstances where the wavelength is small compared with the other dimensions involved, as, for instance, when radiation of very short wavelength passes through a wide aperture. Then diffraction effects are negligible, and the wave nature of light can be ignored in favor of the more simple ray description.

Relativity provides a more general formulation of the laws of mechanics than the newtonian form and gives applicable results even with speeds approaching that of light. If the correspondence principle is valid, the relativistic relations must convert to the classical ones for speeds that are small compared with c. Formulas for distance, mass, momentum, energy, etc., all reduce to the newtonian ones in the limiting condition of low speeds.

All forms of quantization become undistinguishable from continuity when applied to macroscopic situations. For example, the atomic nature of matter is not perceived in dealing with bodies that are large enough to be seen and handled. An analogy is furnished by the pouring of sand from a jar. Viewed from some distance, the material appears to be continuous rather than granular. Similarly, the discrete nature of electric charge is not evident when the amounts of charge involved are large multiples of e. In the pendulum experiment of Worked example 20.1 (opposite), the motion of a macroscopic system appears to be perfectly continuous because its quan-

Niels Bohr (1885–1962) was one of the most distinguished and imaginative scientists of the modern age. He was awarded the Nobel Prize in physics in 1922. Other aspects of his work will be described in chapters to follow.

tum structure is so fine-grained. The discrete nature of energy becomes evident only in microphysical systems.

20.4   THE PHOTOELECTRIC EFFECT

The Latin phrase ad hoc, applied to a scientific hypothesis, signifies that it was devised to account for a specific case.

In spite of the success of Planck's quantum hypothesis in explaining ideal thermal radiation, it had to be regarded as a purely *ad hoc* assumption. This circumstance, plus the fact that the whole idea was of such a radical nature, accounted for the hostility it met from the start. However, within 5 years from the time it was announced, the quantum concept received one of its strongest confirmations when Einstein showed that it gave a direct and convincing explanation of another phenomenon, the **photoelectric effect.**

In 1887, in the course of Hertz's experiments with electromagnetic waves (page 475), he noticed that a spark would jump a gap more readily when the electrodes were exposed to the light from another spark. Subsequent experiments, viewed in terms of Thomson's discovery of the electron a decade later, led to the correct description of what happened: Electrons can be released from metals and other substances when radiation falls on them. This is the **photoelectric effect.**

The ejected electrons are referred to as photo-electrons, although they are no different from any other electrons, free or bound.

Figure 20.6 is a schematic diagram of a photocell suitable for detecting and measuring the photoelectric effect. Incoming radiation releases electrons from the coating on the inside of the tube. A battery maintains the central electrode (anode) positive with respect to the coating (cathode), and the electric field between them causes the photoelectrons to move to the anode, giving rise to a current through the detecting meter.

In investigating the effect experimentally, provision is made for illuminating the cell with monochromatic light of various wavelengths and for varying the intensity of the light. Also, in order to measure the speeds with which the photoelectrons are emitted, *retarding* fields may be applied by reversing the battery (see Fig. 20.7). Emitted electrons are then slowed down as they approach the collecting electrode. If the battery potential is gradually increased until the current drops to zero, even the fastest photoelectrons are turned back before they reach the collector. The corresponding value of the potential is then a measure of the initial KE of the fastest electrons. Details will be given below.

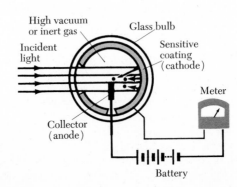

FIGURE 20.6 A photocell (schematic).

The main observational facts are:

1. For each substance there is a certain minimum frequency of radiation that must be exceeded before any electrons are emitted. This is called the **threshold frequency** f_0. No electrons are released for any frequency less than f_0, no matter how strong the light. For most common metals, f_0 lies in the ultraviolet region. For the alkali metals (sodium, potassium, etc.) and some of their compounds, the threshold frequencies are in the visible and even in the infrared.

Existence of a threshold frequency.

2. Once the threshold frequency is reached, electron emission begins immediately, no matter how weak the light. The time between the arrival of the radiation and the ejection of electrons is of the order of 10^{-9} sec.

Immediate ejection of electrons.

3. For radiation having a given frequency, the measured current of photoelectrons is directly proportional to the light intensity, as long as the frequency is above the threshold value. This means that the number of photoelectrons released per second is proportional to the light intensity, since all electrons carry the same charge e. This is the basis for using a photocell as a light meter (see page 508).

Photocurrent is proportional to light intensity.

4. As long as the radiation frequency exceeds the threshold value, the maximum KE of the photoelectrons is independent of the intensity of the radiation. This observation is obtained from experiments using a retarding potential on the cell, as described above.

Maximum electron KE is independent of intensity.

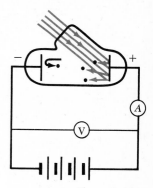

FIGURE 20.7 Photocell (schematic) for experimental investigation of the photoelectric effect. Provision is also made for varying the potential across the cell, and for reversing it.

An examination of the above experimental facts shows that only item 3 can be explained on a classical basis, whereas the other three cannot. For example, in item 1 (existence of a threshold frequency), the release of electrons should depend only on the rate at which energy reaches the surface, that is, on the light intensity but not on its frequency. Experimentally, the production of photoelectrons is found to depend only on whether or not f exceeds a certain value.

Regarding item 2, according to classical principles electrons should not be emitted until enough energy can be accumulated from the incident light waves, yet such emission is actually observed to occur immediately. The magnitude of the discrepancy, which is especially great when weak radiation is used, is illustrated by the following Worked example:

Worked example 20.2 Assuming that radiant energy is carried by light waves, how long would a typical photoelectric material have to be exposed to the light of the full moon before electrons begin to emerge? Use the following data: The intensity of full moonlight is 3×10^{-4} W/m², about the same as the illumination from a 60-watt lamp on a surface about 60 ft away. The electrons are bound to the surface with an energy of about 5 eV.

Solution Suppose that each electron collects its energy from a square area of about one atom diameter, or 10^{-10} m on a side. Then the effective area for each electron is 10^{-20} m². This means that it receives energy at the rate of $(3 \times 10^{-4})(10^{-20}) = 3 \times 10^{-24}$ watt (or J/sec). Further, 5 eV is equivalent (page 393) to $(5)(1.6 \times 10^{-19}) = 8 \times 10^{-19}$ joule. If it is assumed that all this energy is absorbed, the time required is $t = (8 \times 10^{-19})/(3 \times 10^{-24}) = 2.7 \times 10^5$ sec, or about 75 hr, compared

TABLE 20.1   THE PHOTOELECTRIC EFFECT—SAMPLE DATA

FREQUENCY OF LIGHT, cycles/sec	INTENSITY, ft-c*	PHOTOCURRENT, amperes	MAXIMUM SPEED OF PHOTOELECTRONS, m/sec
7.3×10^{14} (blue)	1	1.0×10^{-7}	8×10^5
	2	2.0×10^{-7}	8×10^5
5.7×10^{14} (yellow)	1	0.8×10^{-7}	6×10^5
5.1×10^{14} (red)	Any value	None	

*The unit is the *foot-candle,* the commonly used unit of illumination in the English system.

with an actual time of about a billionth of a second even when the incident radiation is a million times weaker than moonlight.

Table 20.1 may be helpful for keeping in mind the observed relations between the intensity and frequency of the radiation, the rate of ejection of photoelectrons, and their maximum energy.

Now consider the experimental basis of item 4, above. Suppose that the sensitive surface is illuminated by light of a given frequency and that retarding potentials of increasing magnitude are applied, as described on page 504. The current in the tube decreases and becomes zero when a certain *stopping potential* V_0 is reached. At that point, the work done by the retarding electric field on the fastest electrons released is just equal to their initial KE (see Eq. 16.11, page 393). In symbols, if v_M represents the maximum speed of photoelectrons under the given conditions,

$$\tfrac{1}{2}mv_M{}^2 = V_0 e \qquad\qquad 20.6$$

The use of the classical, nonrelativistic formula for KE is allowable, since photoelectrons released by light in the optical region have low speeds compared with c.

Next, a series of experiments is performed, using a number of different frequencies for the incoming light. The value of V_0 is measured for each, and when the results are plotted (circled points, Fig. 20.8), the points fall along a straight line. Similar data for other photosensitive materials fall on lines that are parallel to this one. A point on any such line represents the smallest applied potential that will stop completely the passage of photoelectrons across the tube when light of the corresponding frequency is used. In no case do the values depend on the *intensity* of the light used.

Examine the form of the algebraic equation of any one of the lines in Fig. 20.8. Drop a perpendicular PP' from any point (not necessarily a data point) on one of the lines to the frequency axis. Since the slope of a straight line is constant, the ratio of PP' to the segment cut off on the frequency axis is constant, or $V_0/(f - f_0) = C$, where C is the value of the slope. This relation can be written

$$V_0 = Cf - D \qquad\qquad 20.7$$

where $D = Cf_0$ is a new constant.

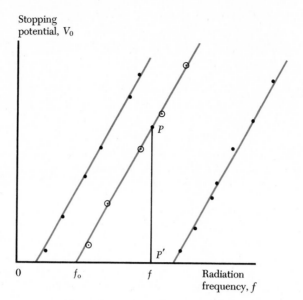

FIGURE 20.8 The stopping potential is a linear function of the frequency of the incident radiation and is independent of the intensity of the radiation.

Einstein's explanation of photo-electricity.

A brilliant yet simple explanation of the complicated experimental facts of photoelectricity was given by Einstein in 1905. Realizing that the wave properties of light could not explain the observations, he made the bold assumption that *a beam of light must for some purposes be thought of as a stream of discrete packets of energy.* This description immediately suggests quanta and was unquestionably inspired by Planck's idea. However, Planck did not go quite this far in his theory of cavity radiation. He believed that, although the oscillator energy was quantized, the emitted radiation traveled through space in the form of electromagnetic waves. Einstein took the additional step of assuming that light energy in transit is in the form of discrete entities (which he referred to as "light quanta") rather than spreading waves. Later, the name **photon** was given to a quantum of radiant energy. The amount of energy of a photon is $E = hf$, where f is the frequency of the associated radiation.

Einstein found the key to an explanation of the photoelectric effect by assuming that *each photon,* on striking an atom, *can release one electron:* a photon, if it is absorbed at all, never shares its energy among a number of electrons but delivers all of it to a single electron. This simple idea leads at once to an explanation of Eq. 20.7 and the experimental curves in Fig. 20.8 in the following way:

Part of the energy of an incoming photon is used in removing an electron from the material, and the rest goes into the KE of the freed electron. If the electron happens to be at the surface, an amount of energy called the **work function** is needed to detach it from the metal. For most materials the work function, whose value depends only on the kind of material, is a few electron volts. However, the incoming radiation can penetrate several layers of atoms of the substance, and some photoelectrons emerge from below the surface. These electrons lose small amounts of energy in collisions before reaching the surface, so that it is the electrons released at the surface that have the *maximum* KE afterward.

The conclusion becomes apparent upon writing the energy equation for these electrons. The energy of the incident photon, hf, can be equated to the KE of the fastest emerging electron plus the work function w:

$$hf = \tfrac{1}{2}mv_M^2 + w \qquad \text{or} \qquad \tfrac{1}{2}mv_M^2 = hf - w \qquad\qquad \textbf{20.8}$$

If we multiply Eq. 20.7 through by e and substitute in it the value of V_0e from Eq. 20.6, the result is

$$\tfrac{1}{2}mv_M^2 = eCf - eD \qquad\qquad \textbf{20.9}$$

This equation has exactly the same form as Einstein's theoretical relation, Eq. 20.8. Comparison of Eqs. 20.8 and 20.9 identifies the values of the constants:

$$C = \frac{h}{e} \qquad \text{and} \qquad D = \frac{w}{e} \qquad\qquad \textbf{20.10}$$

It appears, then, that the value of h and the value of the work function w can be obtained from the experimental constants C and D (see Worked example 20.3 following).

The first accurate experimental data of the kind shown in Fig. 20.8 were obtained by Millikan in 1916. His measurements verified the Einstein equation and also led to a confirmation of the value of h as first computed from data on cavity radiation. Two Nobel physics prizes were based at least in part on investigations connected with the photoelectric effect: Einstein's in 1921 and Millikan's in 1923.

Worked example 20.3   The threshold frequency for sodium, as deduced from Millikan's data, is 4.4×10^{14} cycles/sec, and the data fall on a line whose slope is 4.1×10^{-15} volt-sec/cycle (refer to Fig. 20.8 to justify these units).

a. Find the work function for this metal.

b. Find the value of Planck's constant.

Solution   *a.* From Eqs. 20.7 and 20.10, $D = Cf_0 = w/e$, or $w = eCf_0$, C being the slope of the data line. To get the value of w directly in *electron volts* rather than in regular energy units (joules), this must be divided by e, so that $w_{eV} = (4.9 \times 10^{-15})(4.4 \times 10^{14}) = 1.8$ eV.

b. From Eq. 20.10, $C = h/e$, or $h = eC$. Substitution of the numbers yields $h = (1.6 \times 10^{-19})(4.1 \times 10^{-15}) = 6.2 \times 10^{-34}$ J-sec. This agrees, to two digits, with the value given on page 500.

20.5 PHOTOELECTRICITY IN PRACTICE

Applications of the photoelectric effect have become too common to list in detail. Photocells are indispensable parts of the equipment used for sound motion pictures (Fig. 20.9), television, and picture transmission (Ref. 20.7). The current from a photocell may be used to actuate a burglar alarm, automatic door opener, inspecting and sorting mechanisms, etc.

The ordinary photographic exposure meter is operated by a **photovoltaic cell,** not by a vacuum photocell of the kind discussed above. The sensitive

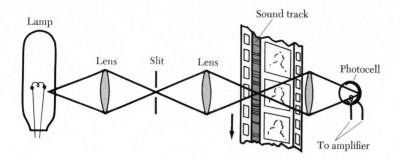

FIGURE 20.9 Use of a photocell in projecting sound motion pictures.

element consists of a junction of two different materials which may be a layer of copper oxide deposited on a copper plate. The junction is connected directly to a meter calibrated to read illumination values directly.

The components of some photovoltaic cells are semiconductors. An example is the **solar battery,** which uses sunlight to generate moderate currents for supplying energy for spacecraft, telephone lines, etc. Solar cells can convert about one-seventh of the incident radiation into electric energy, and twice this efficiency is in prospect.

Another type of photocell makes use of the **photoconductive** effect shown by such semiconductors as cadmium sulfide and germanium, whose electrical resistance decreases markedly when they are illuminated. Incoming photons are absorbed, producing free electrons within the material. This makes the resistance drop to a fraction of its previous value. Photoconductivity is basic in the photocopying process called **xerography.**

A device called a **photomultiplier** is important in physics research. It is used to register and amplify very weak light. For example, in nuclear physics, ionizing particles or radiation can be detected by the flashes of light they produce in certain crystals. This exceedingly weak light detaches electrons from the sensitive cathode of the tube. These electrons are then accelerated toward a plate where they release additional electrons by impact. The process is repeated on a succession of plates (Fig. 20.10) until the electron current is strong enough to operate a recording device. Amplifications of up to 10^8 are feasible.

20.6 PRODUCTION OF X RAYS; BREMSSTRAHLUNG

The production of x rays by the impact of fast electrons on a solid target (page 486) is the inverse of the photoelectric effect. In an x-ray tube, part of the KE of the moving electron is changed to electromagnetic energy (now believed to be in the form of photons) whereas in a photocell part of the energy of incident radiation is converted into the KE of emitted electrons.

Examination of the spectrum of x rays confirms the photon concept. When a fast electron strikes the target in an x-ray tube it undergoes collisions with several atoms of the target material and is rapidly brought to rest. According to classical electromagnetic theory, the large (negative) acceleration of the electron is accompanied by the emission of radiation (page 478). More generally, this happens in any instance where an electron

FIGURE 20.10 A photomultiplier tube having five stages (schematic).

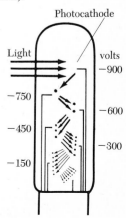

"Brems-
strahlung"
is a German
word:
Bremse,
a brake;
Strahlung,
radiation.

passing close to the nucleus of an atom is swerved aside, experiencing a change in its (vector) velocity. The accompanying radiation is called *Bremsstrahlung.* Classically, such radiation would be expected to consist of a continuous range of frequencies of all possible values. In quantum terms, it consists of one or more photons, each having a definite frequency. Having emitted radiant energy, the electron leaves with its KE diminished (Fig. 20.11). The nucleus acquires KE in the process, but the amount is negligible because a nucleus is thousands of times more massive than an electron. Thus, for the emission of a single photon, the conservation-of-energy equation may be written

$$\tfrac{1}{2}mv^2 - \tfrac{1}{2}mv'^2 = hf \qquad\qquad \textbf{20.11}$$

where v is the initial speed and v' the final speed of the electron and hf represents the energy of the created photon. Photons of various frequencies are produced, depending on how close the electrons come to the nuclei, and hence how much their speeds are changed as a result. The maximum photon frequency results when the electron is brought completely to rest in a single event. In this case, the above equation becomes

$$\tfrac{1}{2}mv^2 = hf_{\text{max}} \qquad\qquad \textbf{20.12}$$

The expected continuous distribution of photon frequencies should have an abrupt cutoff at a frequency given by Eq. 20.12. Such a *short-wave limit* of the continuous x-ray spectrum is actually observed (Fig. 20.12), and its value is found to depend only on the potential applied to the tube. The energy that binds the electron to a target atom is negligible compared with the initial electron energies used in an x-ray tube. Thus, if w is dropped from Eq. 20.8 (page 508), this equation becomes identical with 20.12 above. Any Bremsstrahlung process, such as the production of x rays, may be considered the inverse of the photoelectric effect.

20.7   THE COMPTON EFFECT

It has been shown that the quantum theory provides a satisfactory account of cavity radiation for which it was originally proposed, as well as for the photoelectric effect and the production of x rays. The application to the

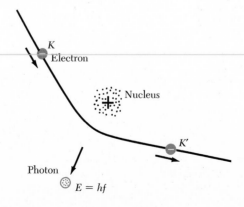

FIGURE 20.11 Schematic diagram showing Bremsstrahlung occurring when an electron passes close to an atomic nucleus. In the case shown, a single photon is created.

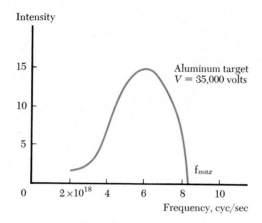

FIGURE 20.12 Distribution of intensity in the continuous x-ray spectrum from an aluminum target at 35,000 volts. The cutoff frequency here is 8.46 × 10¹⁸ cycles/sec, corresponding to a wavelength of 3.55 Å.

latter two phenomena confirmed an additional feature, Einstein's photon model, that Planck himself refused to accept even a decade after it was proposed. This concept, however, received additional support in connection with a new phenomenon discovered in 1922 by the American physicist A. H. Compton. He came upon the effect that bears his name while studying what happens when a beam of x rays strikes a piece of solid matter. He observed that, when x rays are scattered from carbon, their wavelength is slightly increased. This sounds extremely simple and matter-of-fact, yet it proved to be of the greatest importance because of the support it gave to the photon concept.

The experimental arrangement is represented in Fig. 20.13. A narrow beam of hard (high-frequency) x rays of a single wavelength is allowed to strike a small block of carbon. Some of the radiation is scattered in all

Arthur H. Compton (1892–1962) is noted for his investigation of x-ray scattering and also for his extensive work on cosmic rays. He was corecipient of the Nobel Prize in physics in 1927.

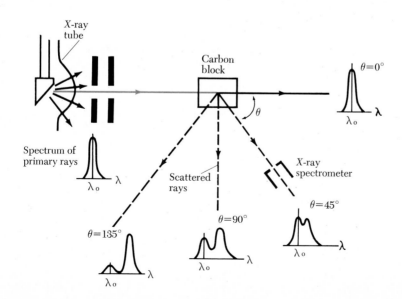

FIGURE 20.13 The Compton effect. Except at zero scattering angle, the unmodified line is accompanied by a line of slightly greater wavelength.

directions in much the same way as light from a searchlight is scattered by dust in the air. The radiation scattered in any given direction is received in an x-ray spectrometer, where its wavelength is accurately measured.

According to classical theory, when an electromagnetic wave strikes a free charged particle such as an electron, the oscillating electric field makes the particle oscillate with the same frequency as that of the wave. The oscillating particle, in turn, radiates electromagnetic energy in all directions, much like an antenna (page 478), and this scattered radiation has the same frequency as that of the particle and the original radiation. Although this classical view of the scattering process is in full agreement with observations made with optical and electric waves, Compton found that it fails for high-frequency radiation such as x rays or gamma rays. His measurements showed that the scattered radiation contained, in addition to the original wavelength, another radiation of slightly greater wavelength. The difference between these two components of the scattered radiation increases with the angle of scattering, as indicated by the small graphs in Fig. 20.13.

No explanation could be devised on the basis of classical wave theory, but Compton conjectured that the photon concept might provide a solution, and his calculations confirmed this. It is known that a photon has an amount of energy $E = hf$, where f is the frequency of the radiation. Moreover, this must be *kinetic* energy, since photons exist only when in motion with speed c and have no rest mass. It is then reasonable to assume that such a packet of energy also possesses linear momentum. According to electromagnetic theory, the momentum carried by radiation in free space is equal to the energy transported divided by the speed of light. Thus, a photon of energy hf has a momentum

$$p = \frac{hf}{c} \qquad\qquad \textbf{20.13}$$

See Prob. 20.15.   If the incident photon in the Compton effect is considered to be an entity having kinetic energy and momentum, it may also be regarded as able to collide with and rebound from matter. Compton considered what happens if a photon strikes a free electron and rebounds elastically. The photon communicates some of its energy to the electron and so must have less energy afterward. But if its energy is less, its frequency is also less after collision. This means that the scattered photons have lower frequency—and so a greater wavelength—than the incident ones, as observed.

The mechanics of this "billiard-ball" problem was worked out quantitatively by applying the laws of conservation of momentum and energy. Relativistic expressions for these quantities had to be used, since the mechanical speeds involved are high. The calculation yielded an equation for the change in wavelength, $\Delta\lambda$, of the incident photons that are scattered at any given angle θ:

$$\Delta\lambda = \frac{h}{m_e c}(1 - \cos\theta) \qquad\qquad \textbf{20.14}$$

where m_e is the rest mass of the electron and θ is the angle of scattering (see Fig. 20.13). In computing with this equation, θ is always measured from the direction of the transmitted beam, so that when θ exceeds 90° the cosine term must be given a negative sign. Notice that the change in wavelength is independent of the initial wavelength and the nature of the target material and that this change is always positive, which means an increase in wavelength.

In addition to the shifted wave, some radiation having the original wavelength is always present. It consists of photons scattered from electrons that are firmly attached to atoms, whereas the displaced component consists of photons scattered from free, or very lightly bound, electrons. When a strongly bound electron is struck by a photon, the entire atom recoils. It takes up essentially no KE because of the large mass involved. The photon rebounds with the same energy and hence the same frequency it had originally. Thus radiation of the initial frequency is always present along with the shifted frequency (see Fig. 20.13).

This can be seen also from Eq. 20.14, where replacement of m_e by m_{atom} makes $\Delta\lambda$ very small.

Vapor trails of the recoil electrons can be observed in a cloud chamber, and the measured speed of recoil agrees with the value computed from the Compton equations. The Compton shift is observed only with hard x rays and gamma rays because the increase in wavelength is independent of the original wavelength, and so the change in wavelength $\Delta\lambda$ is an appreciable fraction of the wavelength λ itself only when the waves are short. For optical and longer wavelengths, the change is far too small to detect.

Inverse Compton effect.

In 1965, the Russian physicist O. F. Kulikov and his associates discovered the inverse of the Compton effect. Photons from a laser source, allowed to collide head on with fast electrons, rebounded with their frequency increased. These experiments gave further support to the applicability of the photon concept.

The Compton effect differs from the photoelectric effect in an important way: In the photoelectric effect the *entire* energy of a photon is given to a bound electron and the photon disappears in the process, whereas in the Compton effect, *part* of the energy of the initial photon is given to a free electron, leaving another photon of lower energy.

Compton's purely mechanical theory accounts quantitatively for the observations, and so it can be considered an experimental check on the validity of conservation of momentum in the subatomic realm. The mechanical theory was later supplanted by a deduction based on the more general quantum theory (to be discussed later), but Eq. 20.14 follows from the new theory just as it did from the old.

Practice set*

1. For what scattering angle does the wavelength shift $\Delta\lambda$ have its maximum value?
2. What is the ratio of the $\Delta\lambda$ for radiation scattered at $135°$ to that scattered at $45°$?
3. For a given angle of scattering, how does $\Delta\lambda$ for incident radiation of wavelength 0.5 Å compare with that for 1.0 Å?

20.8 THE POSITRON; PAIR EVENTS

The photoelectric effect and the Compton effect, and their inverses, show that exchanges can take place between the electromagnetic energy of photons and the mechanical energy of existing material particles such as electrons. Around 1928, certain developments in quantum theory suggested

*Ans.: 180°; 5.8; the same.

that photon energy could be "frozen" in the form of newly created matter and, conversely, matter could disappear in the process of its conversion into photon energy. In particular, it was predicted that if a high-energy photon (hard gamma ray) were to pass into the strong electric field near an atomic nucleus, the photon would vanish and in its place would appear a pair of oppositely charged particles having equal masses. Experimenters began to look for such particle pairs, and the American physicist C. D. Anderson and P. M. S. Blackett in England first found them, in 1932, in cloud-chamber photographs.

Pair production.

For his discovery of particle pairs, Anderson shared the 1936 Nobel Prize in physics.

Figure 20.14 represents what is typically observed. A gamma ray, entering a cloud chamber in the direction shown, may give rise to two vapor tracks that start at some point inside the vessel. The chamber is in a strong magnetic field applied in a direction perpendicular to the page. The two trails bend in opposite directions, which shows that they were made by particles having opposite kinds of charge. The equality of the curvatures indicates that they had about the same mass (Eq. 18.4, page 439). Further, the density of the droplets along each trail suggests that it was made by a particle having roughly the mass of an electron. The negative particle is indeed an electron. Its positive counterpart is a new particle called the **positron.**

Other experimenters confirmed the existence of the new particles and found additional sources in certain nuclear reactions. Positrons can now be produced in the laboratory in the form of intense beams. It has been established that these particles have the same mass as the electron and an electric charge of the same magnitude but of opposite sign. The positron is the **antiparticle** of the electron. Many other examples of particle-antiparticle pairs are now known; some of them will be discussed in Chap. 23.

The term "electron," unless otherwise qualified, continues to mean the negative kind.

The creation of an electron-positron pair from a photon would be expected to satisfy the conservation laws for (1) electric charge, (2) momentum, and (3) mass-energy (page 517). The observed equality of the charges on the two particles satisfies charge conservation, since the original photon had no charge and the positron charge $+e$ and the electron charge $-e$ together equal zero.

Conservation of linear momentum requires that the two created particles have a net momentum along that of the initial photon since the photon possesses momentum. Mass-energy conservation requires that the energy of the original photon, hf, be equal to the total rest energy of the two created particles plus any KE that they may have. Further, a strict analysis shows that the presence of a nucleus is essential for the conservation of momentum and mass-energy.

A photon having the minimum frequency for pair production furnishes just enough energy to create the equivalent of two electron rest masses, with no energy left over in the form of KE of these particles. Thus its frequency f_{min} is given by

$$hf_{min} = 2m_0c^2 \qquad\qquad \textbf{20.15}$$

Substitution of the numerical values of h, m_0, and c leads to a frequency corresponding to a wavelength $\lambda_{max} = 0.012$ Å, which is in the gamma-ray region. A photon of longer wavelength cannot give rise to pair production.

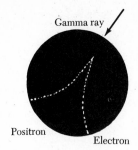

FIGURE 20.14 Pair production. Vapor trails of electron and positron as seen in a cloud chamber placed in a strong magnetic field.

Practice
set
Check the numerical value of λ_{max} given above.

After a positron is created in a nuclear reaction or by pair production it loses its KE by collisions with other particles. When its speed has become low enough it often forms a bound system with an electron, the two particles revolving around each other like a double-star. The system is called **positronium.** In a time of the order of a ten-billionth of a second, the particles disappear in a burst of radiation. This process, the opposite of pair production, is called **pair annihilation.** Thus the positron—unlike the electron—is not a permanent constituent of matter; it has only a transient existence because of the presence of electrons with which it can unite.

The two converse processes differ in one respect: In pair annihilation, at least *two* photons must be produced. In a positron-electron pair at rest, the total linear momentum of the system is zero. Therefore, if photons are to be emitted, there must be at least two, moving in opposite directions, to ensure that the total vector momentum is zero after the event as well as before. Three or more photons having different frequencies can also be created in such a process, but this is found to happen only once in several thousand events.

20.9   THE WAVE-PHOTON DUALISM

The quantum interpretation of the phenomena described in this chapter, as well as others to be discussed later, brought about a profound change in ideas concerning the nature of radiation. The pre-quantum view of light as a wave motion was one of the best-supported ideas in science. In 1889, Hertz remarked:

Ever since the time of Young and Fresnel we know that light is a wave motion. We know the speed of the waves, we know their length, we know that the waves are transverse; in a word, we know completely the geometric relationships of this motion. These things no longer permit of any doubt, and a refutation of this view is unthinkable to the physicist. Insofar as human beings can know truth, the wave theory is certainty.

Ironically, a minor observation made by Hertz in his own experiments on electric waves (page 504) was probably the first evidence for a quantum phenomenon.

Compare the quota-tion from Russell on page 375.
What is the fundamental nature of the dilemma concerning the nature of light? It is not a matter of the wave model being wrong. Rather, it is too *limited in scope* to account for all aspects of radiation that are observed. Light phenomena seem to permit of two mutually contradictory interpretations. Considered logically, the contradiction arises because the opposing assumptions, wave and photon, presume to be able to tell what light *really is*. However, physics is in no way competent to make such a statement. It can only describe the laws according to which optical phenomena unfold. These laws must be able to predict what will happen in any conceivable experiment involving light, and nothing more.

Light behaves in two quite different ways, depending on the type of optical experiment under consideration. Where propagation alone is concerned, light obeys the laws of wave motion. This is not the same as saying that light *is* a wave but only that the space-time relations of any changes caused by light in the material world can be understood through the mathematical laws of wave motion. On the other hand, in all instances where light interacts with individual atoms, molecules, or subatomic systems, it appears to be of a discrete nature, behaving as though it consisted of indivisible photons. Here also it must not be asserted that light consists of such photons, but only that certain processes are representable by a mathematical formalism that coincides with the particle description.

This concept of the dual nature of light is accepted reluctantly by some theorists, and attempts have been made to reconcile the two descriptions. Einstein suggested that light waves act in a certain sense as guides for the travel of photons. Further, he speculated that the electric and magnetic fields in the wave determine only the *probability* of finding photons at given places. There are inherent difficulties with such interpretations and with others that have been proposed.

At one stage in the development of these ideas it was thought that interference experiments using very weak light might show some unusual effects—for example, that each of the quanta used in a diffraction experiment with a double slit would have to choose which of the two slits to pass through. However, no unusual effects are found in such tests, and it must be concluded that in some way each photon is affected by both slits, in a sense interfering with itself.

The search for a unified description of all optical phenomena is difficult because the concepts taken over from classical physics are not adequate to the task. Classical physics derived its concepts from observations of macroscopic phenomena. Physical concepts are meaningful only when their content can be described in terms of measurement, so that an idea pertinent to the macro-world may lose all meaning when applied to a microscopic system, where the measurements on which its definition rests are impossible to carry out. The general forms of quantum theory evolved by adopting concepts from classical physics and redefining them in accordance with new methods of measurement appropriate to the micro-world.

Terms using the prefixes "macro" and "micro" (from the Greek words for large and small, respectively) are useful and are self-explanatory.

The processes occurring in the micro-world do not lend themselves to a description based on the principles of classical physics. Every attempt at such a description seems to be thwarted by experience. If, in spite of this, the concepts are stubbornly retained simply to make it possible to visualize phenomena in familiar terms, there is apparently no recourse but to accept two mutually contradictory pictures, wave and corpuscle, employing one or the other according to the nature of the phenomena. The only other course would be to avoid the use of misleading labels such as "wave" and "corpuscle" altogether.

20.10 MATTER WAVES

Mass-energy equivalence.

The existence of the wave-photon dilemma is remarkable enough in its own right, but the curious fact is that further developments revealed an exactly parallel dualism for matter. In order to see how this came about, Einstein's celebrated mass-energy relation $E = mc^2$ is brought into the discussion.

On page 512 it was shown that a photon of energy E has linear momentum of amount $p = E/c$. If a photon could be regarded as having a dynamic mass m, it would be necessary to assign to it a momentum mc, since this entity travels with speed c. Then, equating these two expressions for photon momentum, $E/c = mc$, or

$$E = mc^2 \qquad\qquad\qquad \textbf{20.16}$$

The special theory of relativity can prove that this relation has a significance much broader than its application to photons. It applies as well to any material body. The **principle of mass-energy equivalence** states: To every quantity of energy E there corresponds an equivalent mass m, and with every mass m there is associated a certain intrinsic energy E. Supplying energy in any form (KE, PE, chemical energy, etc.) to a body does not necessarily change the total number of atoms or other particles in it. Rather, it can increase the inertia of the body in harmony with the relativistic mass increase described by Eq. 18.9 (page 442). However, in certain cases the actual number of material particles may increase (as in pair production) or may decrease (as in pair annihilation).

In any method of supplying energy to a macroscopic body, the mass increase is far too small to detect by ordinary means, such as weighing. The large numerical value of the factor c^2 in Eq. 20.16 shows why: Communicating 1 joule of energy to an object will increase its mass by only $1/(3 \times 10^8)^2 = 1.1 \times 10^{-17}$ kg. It is only in subatomic processes, where large amounts of energy can be communicated to bodies of microscopic mass, that the mass changes become appreciable. In the opposite process, the conversion of even a minute mass can furnish an enormous amount of energy (''atomic'' energy, Chap. 22). The mass-energy interchanges in the phenomena of pair production and annihilation are excellent examples, and further instances will be described later in connection with nuclear transformations.

If the idea of mass-energy equivalence is applied to a closed system of any kind,

$$\text{Total rest mass} + \frac{\text{total energy}}{c^2} = \text{constant} \qquad\qquad \textbf{20.17}$$

This **principle of conservation of mass-energy** replaces the separate pre-relativity principles of conservation of mass and conservation of energy. Unquestionably, it is one of the most general propositions in all of science.

De Broglie's hypothesis of the dualism of matter.

In 1924, the French theoretical physicist Louis de Broglie proposed the idea that matter, as well as radiation, may possess both wave and particle aspects. He reasoned that if Einstein's idea of photons guided by or accompanied by light waves is tenable, then perhaps material particles are also guided by some kind of wave. Temporarily sidestepping the unexplained wave-corpuscle dualism for radiation, he boldly tried to explore the possibility that a similar dualism might exist for matter.

In order to determine what characteristics the hypothetical ''matter waves'' might have, de Broglie recalled that the linear momentum of a *photon* is given by $p = hf/c$ (Eq. 20.13, page 512). Using the wave equation $c = f\lambda$, we can eliminate f between the two relations and get

$$\lambda = \frac{h}{p} \qquad\qquad\qquad \textbf{20.18}$$

De Broglie. (*French Embassy Press and Information Division.*)

Suppose, now, that electrons and other particles of matter behave in this respect like photons and that, associated with each particle, there are matter waves whose wavelength is given by the above relation. In this case, since a material particle—unlike a photon—has a speed v that is less than c, the symbol p in the equation can be taken to be equal to mv, and the relation becomes

$$\lambda = \frac{h}{mv}$$ 20.19

Before seeking a description of the physical nature of these matter waves (sometimes called de Broglie waves), we examine the order of size of the typical wavelength expected. For a body of macroscopic size moving with moderate speed (say a 1-g object traveling 10 m/sec), the de Broglie wavelength is, by Eq. 20.19, $\lambda = (6.6 \times 10^{-34})/(10^{-3})(10)$, or 6.6×10^{-32} m. This is very many orders of magnitude smaller than any waves known by experience, and there is no known way of measuring waves that are so short. However, for a body of microscopic size, the denominator in the preceding expression becomes much smaller, and so the value of λ can be made large enough to come within the physical region.

Suppose momentum is imparted to the smallest permanent charged particle, the electron, by letting it fall through a difference of electric potential in an electron gun. If the speed given to the particle is not too high, the classical mechanical formulas can be used: The momentum $p = mv$, and the kinetic energy $K = \frac{1}{2}mv^2$.

Let the electron fall through 100 volts. Then its KE will be 100 eV, and its speed is given by $v^2 = 2K/m$. (On substituting numbers, remember to change K from electron volts to joules by multiplying by the factor 1.6×10^{-19}.) Then

$$v = \sqrt{\frac{(2)(100)(1.6 \times 10^{-19})}{9.1 \times 10^{-31}}} = 5.9 \times 10^6 \text{ m/sec}$$

This is less than 2 percent of c, showing that the use of the classical formulas is warranted.

The linear momentum of this electron is

$$p = mv = (9.1 \times 10^{-31})(5.9 \times 10^6) = 5.4 \times 10^{-24} \text{ kg-m/sec}$$

Finally, the length of the accompanying matter waves is, according to Eq. 20.19,

$$\lambda = \frac{h}{p} = \frac{6.6 \times 10^{-34}}{5.4 \times 10^{-24}} = 1.2 \times 10^{-10} \text{ m, or } 1.2 \text{ Å}$$

Experimental detection of matter waves.

This number is of the order of size of typical x-ray wavelengths, suggesting the possibility of detecting matter waves by diffraction by a crystal.

Such experimental verification was not long in coming. In 1927, during the course of another investigation, American physicists C. J. Davisson and L. H. Germer found that electrons striking a crystal were reflected only in certain directions. The Davisson-Germer apparatus, shown schematically in Fig. 20.15, is in a high vacuum. Electrons from an electron gun strike a crystal of nickel, and the reflected electrons are detected by a collecting chamber which can be swung in various directions and is connected to a sensitive meter. It is found that at certain angles the number of reflected electrons increases greatly. Calculation of the length of waves selectively diffracted at these angles yields a value agreeing with the de Broglie expression.

In 1929 de Broglie was awarded the Nobel Prize in physics for his concept. Davisson and G. P. Thomson shared the 1937 Prize for their experimental work.

Additional evidence for matter waves was soon found by G. P. Thomson, the son of J. J. Thomson, when he succeeded in getting diffraction patterns by shooting electrons through thin metal foils. It is a curious circumstance of history that, 30 years earlier, the elder Thomson's experiments with cathode rays established the particlelike nature of electrons. Other experimenters were able to produce diffraction using neutrons and even entire

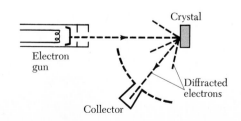

FIGURE 20.15 Detecting the wavelike properties of electrons: the Davisson-Germer experiment.

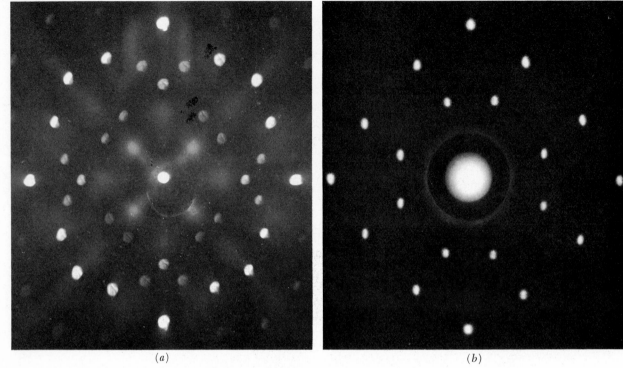

(a)   (b)

Diffraction by a crystal of rock salt of (a) x rays and (b) neutrons.

atoms and molecules. This is good evidence that all bodies, charged or uncharged, exhibit wave properties.

The de Broglie hypothesis and its experimental verification led to the development of a new field of technology called **electron optics.** It had been known for a long time that the paths of moving electrons can be changed in direction by electrostatic or magnetic means in much the same way that light rays—the paths of photons—can be deflected by lenses. Considered as charged particles, electrons can be focused by means of electrostatic or magnetic lenses (Fig. 20.16), and the characteristics of the images formed can be determined just as in ordinary optics.

Electron microscope.

A combination of electron lenses can be arranged to form an **electron microscope.** An instrument using magnetic lenses is shown diagrammati-

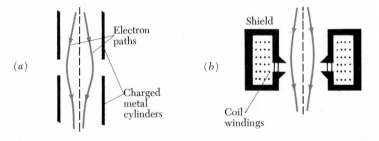

FIGURE 20.16 Examples of (a) electrostatic and (b) magnetic electron lenses. The electrons are strongly deflected when they pass through a region where the electric or magnetic field is highly inhomogeneous.

See lower photograph on page 57.

cally in Fig. 20.17. The form and structure of objects placed in the path of the electron beam can be seen in an enlarged final image formed on a fluorescent screen or photographic plate. Magnifying powers of the order of 100,000 can be attained; this is 50 to 100 times larger than the capability of light microscopes.

In any kind of image formation, an even more important consideration than magnification is the **resolving power** of the instrument. This determines the size of the smallest details that can be observed. The resolving power is governed by the wave nature of the radiation used, since it is fundamentally connected with diffraction effects (see page 356).

The ability of an electron microscope to reveal fine details can be attributed to the fact that it is possible to work with short wavelengths (and so attain high resolving power) by employing high potentials to accelerate the electrons. Potentials of from 40,000 to 100,000 volts are now used in practice, and details as small as 1 to 2 Å can be seen in the final image. In this way, features of certain crystals, viruses, and even large molecules that are far beyond the resolving power of optical microscopes have been revealed. High-resolution electron microscopes are assuming importance in molecular biology, where there is need to find out the arrangement of chemical groups in giant molecules.

For a more complete discussion of electron optics, see Ref. 20.8.

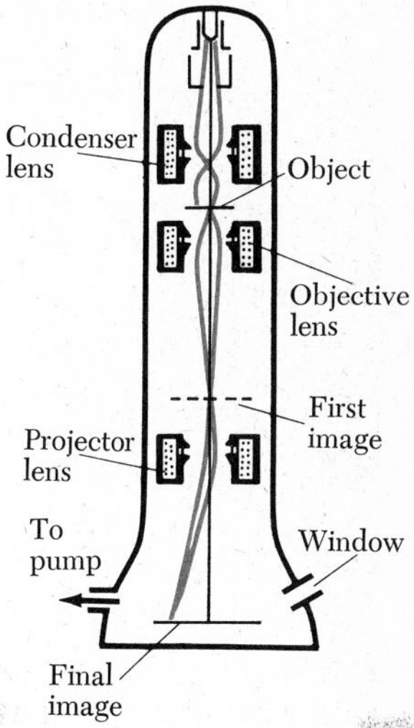

FIGURE 20.17 Schematic diagram of a magnetic electron microscope.

20.11 PHOTONS AND PARTICLES

After the discovery of electrons and x rays, physicists took great pains to distinguish carefully between entities that were identified as particles of matter and others that were observed to have the properties of waves. But de Broglie's suggestion, revolutionary as it was, has been verified experimentally beyond all question, and now they are forced to admit that a clear distinction between these two qualities cannot be made. Both matter and radiation have wavelike and particlelike aspects.

In this view, the descriptions of radiation and matter acquire a curious similarity: Both have *energy* and *momentum,* which are quantities originally associated exclusively with matter, and both have a *wavelength* (or frequency), a property formerly ascribed only to radiation. The connection in each instance is through either the *quantum constant h* or the "relativity constant" *c,* as shown, for example, by $\lambda = h/p$ and $E = mc^2$.

Within a short time of de Broglie's proposal concerning the wave nature of matter, his idea was taken up and enlarged by others, to become the dominant theory of the micro-world. These developments will be traced in the next chapter.

Programmed review

Instructions: See page 18.

1. What characterizes an ideal (black-body) thermal radiator?

   It radiates more strongly, at any given temperature, than any other thermal radiator. [20.1]

2. How does the total radiant energy emitted by an ideal black body in a given time depend on its temperature?

It is proportional to the fourth power of the absolute temperature.   [20.1]

3. Describe Planck's hypothesis regarding the emission or absorption of energy by an ideal thermal radiator.

The energy is exchanged discontinuously in amounts that are integral multiples of $E = hf$, the quantum of energy corresponding to the given frequency.   [20.1]

4. What is the correspondence principle?

The general proposition that the predictions of a broad theory reduce to those of a more special, experimentally established theory whenever the circumstances of the special theory apply.   [20.3]

5. Define what is meant by the photoelectric effect.

The release of electrons from substances when illuminated by radiation.   [20.4]

6. State Einstein's photoelectric relation in words.

The KE of each of the fastest photoelectrons is equal to the energy of one quantum of the effective radiation minus the work function of the material.   [20.4]

7. What determines the wavelength of the shortest waves produced by an x-ray tube?

The potential difference applied to the tube.   [20.6]

8. What is meant by the Compton effect?

The slight increase in wavelength of x rays scattered from matter.   [20.7]

9. Describe a positron.

A fundamental particle having the same mass as the electron and an equal but opposite charge. It is not a permanent constituent of matter.   [20.8]

10. What is pair production?

The formation of an electron-positron pair by the materialization of a photon when it passes close to an atomic nucleus.   [20.8]

11. State the principle of the conservation of mass-energy.

In a closed system, the sum of the total rest mass and the total energy (expressed in mass units) is constant.   [20.10]

12. What is meant by saying that a moving object is accompanied by matter waves?

Any object in motion possesses a wavelike property; the wavelength is given by $\lambda = h/p$, where p is the linear momentum of the body.   [20.10]

For further reading

20.1. *Taylor, L. W.* "Physics, the Pioneer Science." Black-body radiation and the photoelectric effect are treated on pp. 808–814. Read also pp. 767 and 768 on the mathematical basis of physical concepts.

20.2. *Compton, A. H.* The Scattering of X-rays as Particles, *Am. J. Phys.,* December, 1961, p. 817.

20.3. *Hoffman, B.* "The Strange Story of the Quantum." An imaginative, nonmathematical treatment. Read chaps. 2, 3, and 4 at this point.

20.4. *Frisch, O. R.* "Atomic Physics Today." Summarizes most of the content of this chapter in an informal, stimulating way. Read chap. 6 on the nature of quantum theory and the wave-particle dualism.

20.5. *Semat, H.* "Introduction to Atomic and Nuclear Physics." Details of particle diffraction experiments are found in chap. 6.

20.6. *Shamos, M. H.* (ed.) "Great Experiments in Physics." Read Einstein's paper on the photoelectric effect on pp. 235–237.

20.7. *Zworykin, V.,* and *E. Ramberg* "Photoelectricity and Its Applications." For details on practical devices using the effect.

20.8. *George, B. A.* Electron Optics, "Encyclopedia of Physics." See p. 216.

Questions and problems

20.1. A heated solid begins to glow visibly when its temperature approaches 800 K. Examine the portions of the curves of Fig. 20.2 (page 497) within the visible region and explain why a heated body first appears dull red, then orange, yellow, and finally "white hot" as its temperature is raised.

20.2. An *optical pyrometer* is a device that determines the temperature of an incandescent body by measuring its emittance. A small opening in the side of a furnace emits radiant energy at the rate of 28.7 J/(sec)(cm²). Find the temperature of the interior, in kelvins.

20.3. The intensity of solar radiation at the earth's distance from the sun is about 1,400 W/m². Allowing for the radiation reflected by clouds and atmosphere, as well as for the effect of oblique incidence of the rays, it is estimated that each square meter of the earth's surface reradiates about 400 watts into space. Assuming ideal radiation, use the Stefan-Boltzmann law to estimate the average temperature of the earth, in kelvins.

20.4. A violin string vibrates with a frequency of 440 cycles/sec. What is the smallest amount, in joules, by which its vibrational energy can change?

20.5. Discuss some civil, political, or geographical examples of the quantization of space or time. The division of land areas into countries, states, etc., is an instance of the former, and the establishment of international time zones is an example of the latter. Can you think of any others?

20.6. A brick is placed on a table. (a) When the brick rests in any of its stable positions, is its GPE quantized? Explain. (b) How many "quantum states" does it have? Remember that the brick may not be homogeneous.

20.7. The intensity of the solar radiation received by the earth is 1.4 kW/m² (see Prob. 20.3). Assume that the average wavelength of the radiation is 7000 Å, and estimate how many incoming photons cross a square meter each second.

20.8. Under special conditions the human eye can respond to light intensities as low as 10^{-16} watt. If all the incident light is at the wavelength to which the eye is most sensitive, 5500 Å, compute the minimum number of photons per second that the eye can detect.

20.9. Compute the wavelength, in angstroms, of a photon having an energy of 1 MeV.

20.10. The photoelectric work function for sodium is 2.3 eV. Will green light of wavelength 5600 Å release electrons from a sodium surface? Explain.

20.11. Radiant energy of wavelength 4000 Å is falling on a photosensitive surface at the rate of 1.6×10^{-8} watt. (a) How many photons arrive per second? (b) If the work function of the material is 2.0 eV, what is the speed of the fastest electrons released by this radiation?

20.12. The photoelectric threshold for a certain metal is 3400 Å. What is the maximum energy, in electron volts, of the electrons ejected by radiation of wavelength 1100 Å?

20.13. Electrons in an x-ray tube operating at 50,000 volts are stopped in a distance of 5×10^{-4} cm when they strike the target. Estimate their average acceleration while being brought to rest.

20.14. The short-wave limit of the continuous x-ray spectrum produced by an x-ray tube operated at 15,000 volts is 0.828 Å. From this, compute the numerical value of the combination of fundamental constants hc/e, and show that its units may be written as meter-volts.

20.15. Show that for a photon the relation $p = hf/c$ follows from Eq. 19.15 (page 481). HINT: The equation is for the total momentum per unit area per unit time, and for energy per unit area per unit time. Apply it, instead, to a single photon.

20.16. When x rays of wavelength 2.174 Å are scattered in a certain direction from a carbon block they are found to be accompanied by waves of length 2.210 Å. Find the angle of scattering.

20.17. X rays of wavelength 1.2 Å are scattered from a carbon target at an angle of 34° from the transmitted beam. Find the Compton shift, in angstroms.

20.18. Photons of two distinct wavelengths, 0.011 and 0.013 Å, pass close to a heavy nucleus. (a) Is pair production possible for one or both of these? (b) Calculate the speed of each particle produced, making the special assumption that the members of a pair have equal KE's.

20.19. Show that, when a positronium system at rest decays into two photons, each has a wavelength of 0.024 Å. HINT: Write an equation similar to 20.15 (page 514).

20.20. On page 517 it was shown that the mass change accompanying a change in the energy of a macroscopic body is not detectable. Discuss this fact from the point of view of the correspondence principle.

20.21. Give reasons to show that, without the qualifying adjective *rest,* the naïve definition of mass as "quantity of matter," mentioned on page 49, cannot be strictly maintained in view of the principle of mass-energy equivalence.

20.22. The dot over the letter "i" contains about 3×10^{18} carbon atoms, each of mass 2×10^{-27} kg. Show that, if this mass could be converted completely into electric energy, it would keep a 100-watt lamp lighted for about two months.

20.23. By how many kilograms does the mass of a 1-kg iron ball increase when its temperature is raised from $20°C$ to its melting point, $1535°C$? Take the average specific heat to be 0.13. Is there any possibility of being able to detect this increase?

20.24. Baseballs are thrown at a picket fence. Would you expect them to be observably diffracted? Assume that the ball has a mass of 0.15 kg and is thrown with a speed of 80 ft/sec. Compute the length of its matter waves.

20.25. The mass of an oxygen molecule is 5.3×10^{-26} kg. (a) Calculate the de Broglie wavelength of this molecule when it is moving with a thermal speed of 500 m/sec. (b) How could these waves be detected?

20.26. In the original Davisson-Germer experiments, the matter waves of the electrons had a wavelength of 1.65 Å. Find (a) the potential through which the electrons fell and (b) their speed, expressed as a fraction of c.

20.27. On page 517, the relation $\lambda = h/p$ for photons was derived from the principle of mass-energy equivalence. Show that this relation also follows from Eq. 20.13 and the wave equation.

20.28. In calculating the length of matter waves for fast-moving particles, the relativity expression $m = m_0/\sqrt{1 - (v/c)^2}$ must be used in place of m_0. Draw a graph of λ versus v based on this relation.

20.29. An electron and a proton are each moving so that their matter waves have a length of 1 Å. Compare their (a) momenta; (b) kinetic energies.

Chapter 21
THE QUANTUM STRUCTURE OF THE ATOM

During most of the nineteenth century the atom continued to be regarded as an indivisible entity, although such phenomena as ionization, thermal emission, and the photoelectric effect strongly suggested its composite nature. Then, toward the end of the century, Thomson's identification of the electron showed convincingly that atoms have internal structure.

Following this discovery, the detailed structure of the atom and the laws governing its internal affairs became matters of concern. The spectra obtained from various substances (Sec. 21.2) were believed to have some relation to atomic structure, but no real connection was apparent. Attempts that were made to explain the observations were a mixture of classical and quantum principles until eventually it became clear that a workable general model could be based only on quantum theory. This chapter traces the development of ideas about the structure of the atom and discusses the theories advanced to explain atomic phenomena.

ATOMIC STRUCTURE AND THE EARLY QUANTUM THEORY

21.1 RUTHERFORD'S NUCLEAR ATOM

In the course of his electron experiments, around 1900, Thomson considered the possibility that ". . . the atom consists of a number of corpuscles [electrons] moving about in a sphere of uniform positive electrification. . . ." He imagined the electrons occupying certain equilibrium positions while capable of vibrating around these positions when slightly disturbed. It was generally believed that the emission of light by atoms could be accounted for in this way.

A few years later, the British physicist Ernest Rutherford (1871–1937) proposed a different model as a result of some experiments conducted by two of his associates, Hans Geiger and E. G. Marsden. Their observations suggested to Rutherford an entirely different picture of the atom, one in which the positive charge is confined to a very small region—the atomic **nucleus**—with the electrons distributed at relatively great distances from it.

Rutherford, perhaps the greatest experimental physicist of modern times, was a pupil of J. J. Thomson. He is noted for his work in radioactivity and atomic structure and received the Nobel Prize (in chemistry) in 1908. For a biography, see Ref. 21.1.

The Geiger-Marsden experiment used a fine stream of fast alpha particles from a bit of radioactive material. An alpha particle, described in Chap. 22, is the same as a helium atom from which the two electrons have been removed. Rutherford believed that if the alphas were allowed to hit a thin metal foil, the subsequent paths of the particles might reveal something about the nature of the atoms of the foil. From earlier experiments, he knew

Rutherford in the Cavendish Laboratory, Cambridge University, in 1935. (Photo courtesy of C. E. Wynn-Williams.)

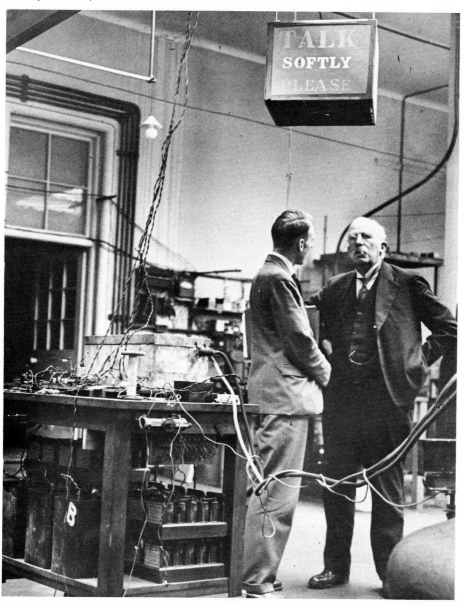

that most of the particles pass straight through the foil, but some are deviated (scattered) by a small amount. Geiger and Marsden found that, in addition, a few particles were deflected through very large angles and that an occasional one almost doubled back in the direction from which it came. For example, with platinum foil, about 1 particle in 8,000 was deviated through 90° or more. In their laboratory arrangement, shown schematically in Fig. 21.1, each alpha particle produces a tiny scintillation (pinpoint flash of light) when it strikes a small fluorescent screen. When viewed through a low-powered microscope, the flashes can be seen individually. The device can be rotated so that particles scattered at various angles can be counted.

With a modern scintillation counter, as many as a million events per second can be counted and recorded electronically.

With characteristic insight Rutherford realized that the large-angle scattering was important, for if a sheet of metal only a few ten-thousandths of a millimeter thick can turn back a projectile that is relatively as massive as an alpha particle, moving at a speed of more than 10,000 km/sec, the nature of the impact must differ fundamentally from that of a baseball striking a brick wall. As Rutherford expressed it some years later, the large-angle scattering observations were "about as credible as if you had fired a 15-inch shell at a piece of tissue paper and it came back and hit you."

Rutherford saw the futility of trying to ascribe large-angle scattering of the alphas to the electrons, or even to a distributed positive charge as used in Thomson's model (page 526). Calculations that he made convinced him that "it was impossible to get anything of that order of magnitude unless you took a system in which the greater part of the mass of the atom was concentrated in a minute nucleus. It was then that I had the idea of an atom with a minute massive center carrying a charge."

An alpha particle is about 7,300 times as massive as an electron.

Accordingly, he made a calculation of the change in direction of an alpha particle on passing close to a compact, electrically charged massive particle (nucleus). The computation was based on the assumption that Coulomb's law holds for the electrostatic force between the two objects. At first, Rutherford was undecided as to the sign of the central charge. However, this does not affect the general character or the numerical value of the result. In the light of other evidence, he later concluded that the nuclear charge was positive and of an amount equal to the total charge on the electrons in the normally neutral atom.

The problem is essentially that of determining the orbits of particles under an inverse-square central repelling force. Because of the high speed of the incoming particles, the orbits are not closed curves (ellipses), as in

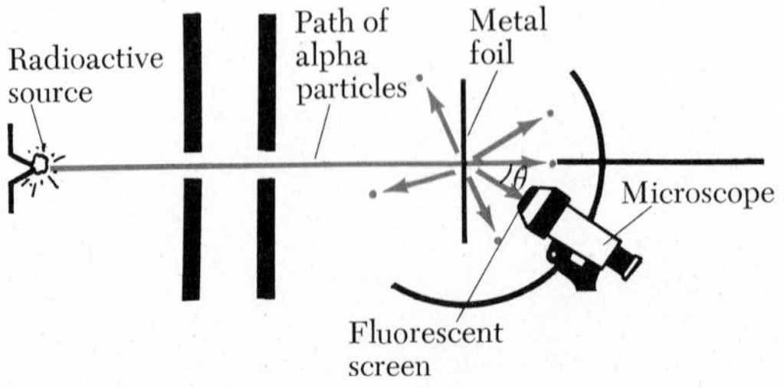

FIGURE 21.1 Geiger and Marsden found that alpha particles impinging on a thin metal foil are scattered in all directions.

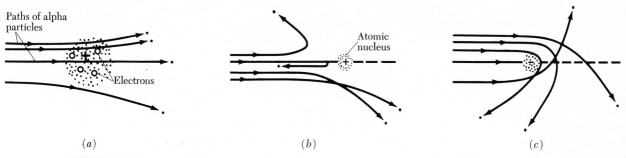

Paths of alpha particles

Electrons

Atomic nucleus

(a)   (b)   (c)

FIGURE 21.2 Scattering of alpha particles by (a) a Thomson atom, (b) the nucleus of a Rutherford atom, and (c) a concentrated negative charge. The smallness of the deviations in (a) shows that this cannot be the correct model.

the case of the planets of the solar system. The alpha particles are merely swerved aside and move off again to an indefinite distance, the extent of the deflection depending on how close the particle passes to the nucleus. The paths are curves called **hyperbolas** (Fig. 21.2).

One of the relations that evolves from the theory gives the number of particles scattered at any given angle. Figure 21.3 shows the excellent agreement with the observations. On the other hand, the theory of the Thomson atom yields results that differ hopelessly from the measured values, especially at large scattering angles, where for the Thomson atom there would be no scattering at all.

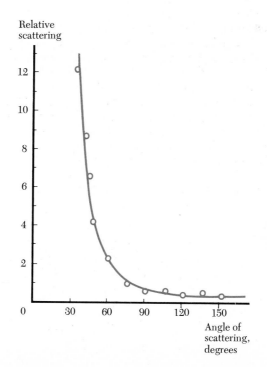

Relative scattering

Angle of scattering, degrees

FIGURE 21.3 Observed values of the relative numbers of particles scattered at various angles lie very near the theoretical curve.

Further comparison of theory and experiment led to the conclusion that *all the positive charge of an atom, as well as practically all its mass, is confined to a region less than about 10^{-14} m in diameter* (see Worked example 21.1, below). Evidence from chemistry and from the kinetic theory shows that the diameter of an atom is of the order of 10^{-10} m, so that the nucleus has a diameter only about one ten-thousandth that of the whole atom.

Rutherford's picture of a "nuclear atom" represents the concept of the structure of the atom today. The electrons must be assumed to be distributed within a globular region of the order of 1 Å across, and, particularly if the electrons are in motion, this would be a part of space from which any similar configuration (another atom) would be excluded. However, the atom is a very "open" structure, and an energetic projectile such as an alpha particle can pass through many thousands of atoms without being stopped or even appreciably deflected unless it happens to pass very close to a nucleus. Our solar system is open in the same degree to the passage of comets and meteoroids, which are not deviated unless they pass close to the sun or to a major planet.

The scattering experiments also showed that the magnitude of the positive charge on the nucleus increases with atomic weight. An improved version of the experiment, performed later by James Chadwick (page 571n) in Rutherford's laboratory, showed that the nuclear charge is proportional to the atomic number of the chemical element in question. The **atomic number** is the number, Z, that gives the position of the element in the periodic table. Thus $Z = 1$ for hydrogen, 2 for helium, 3 for lithium, etc.

In the meantime, the same conclusion concerning nuclear charge had been reached by comparing the x-ray spectra of a number of the elements. It will be shown, in the next chapter, that each positive charge, of magnitude equal to the electron charge e, is carried by a **proton.** For this reason, the term "atomic number" can be supplanted by **proton number.**

Worked example 21.1

Estimate the order of size of a nucleus by considering the head-on collision of an alpha particle with it.

Solution   The alpha particle approaches the nucleus until all its initial KE is converted into electrostatic PE, after which the particle retraces its path and goes off again to a great distance. The distance of closest approach may be identified with the "radius of the nucleus," although it is evidently not a fixed boundary even for a given nucleus.

Equation 16.9 (page 390) gave the electrostatic potential at a distance r from a point charge $+Q$ as $V_p = kQ/r$. If a charge $+Q'$ is placed at this distance, the system acquires an amount of electrostatic PE equal to $Q'V = kQQ'/r$. Equating the initial KE of the approaching alpha particle to the electrostatic PE at the turnaround point, we have

$$\tfrac{1}{2}m_\alpha v_0{}^2 = \frac{kQQ'}{r} = \frac{k(Ze)(2e)}{r}$$

where Ze represents the charge on a nucleus of proton number Z and

$2e$ the charge on an alpha particle. When solved for r, the above becomes

$$r = \frac{4kZe^2}{m_\alpha v_0^2}$$

Numerical values may now be inserted: Take $Z = 79$ (for gold atoms) and $m_\alpha = 6.6 \times 10^{-27}$ kg, $v_0 = 2 \times 10^7$ m/sec (a reasonable speed for alphas), and the standard values for k and e. The result is $r = 2.7 \times 10^{-14}$ m, which is of the order of size quoted above.

The agreement between this figure and other estimates of nuclear size may be considered a check on the validity of Coulomb's law down to distances of this order of magnitude, although even in the Geiger-Marsden experiments the numbers of particles scattered at very large angles began to show slight departures from the predicted values. The next chapter shows that incident particles of greater energy actually can enter and amalgamate with the target nucleus and that within the nucleus itself the force law changes to one of *attraction*.

The indications that emerged from the alpha-particle scattering experiments, together with supporting evidence from other fields, gradually gave shape to the present picture of atomic structure. The periodic table now emerges as something more than an empirical arrangement of the elements according to their qualitative chemical properties. It becomes, rather, an orderly array of structures of increasing complexity, all of them combinations of certain numbers of a few kinds of basic particles. The details were filled in by later developments connected with the application of quantum theory to the atom.

The following is a summary of facts about the composition of the atom:

1. Every atom consists of a compact, massive, positive-charged nucleus, around which electrons are distributed.
2. All the atoms of a given element are identical as to their complement of charged particles (protons and electrons).
3. In its normal state, an atom is electrically neutral, and the number of protons in the nucleus is equal to the number of electrons surrounding the nucleus. Each of these numbers is equal to the proton number Z of the element in question:

Number of nuclear protons = number of external electrons = Z

At the time of the scattering experiments, it was believed that protons and electrons were the sole constituents of atoms. With the discovery of the neutron in 1932, the new particle had to be incorporated into the scheme. It will be shown in the next chapter how this modified and extended the model of the atom. However, the statements in the above summary remain strictly true.

Practice set*

The proton number Z of lithium is 3. For a normal atom of this kind, what is (*a*) the maximum number of electrons that can be removed in the process of ionization; (*b*) the total number of charged particles; (*c*) the total electric charge?

* See footnote on the next page.

21.2   ATOMIC SPECTRA

The continuous spectra observed for radiation from incandescent solid and liquid sources were described in Chap. 20. Although such spectra can furnish information on the temperature of the source, they give no indication of the composition of the emitting material. However, if the source is in the form of a gas or vapor, the emitted light, after dispersion by a prism or grating, yields a spectrum consisting of a number of distinct bright lines against a dark background. This means that such sources radiate a limited number of different wavelengths rather than an infinite set, since each position in the spectrum corresponds to a given wavelength of light.

The dark gaps that correspond to the spaces between spectrum lines were observed by a Scottish experimenter, Thomas Melvill (1726–1753), the first significant observation on spectra since Newton's pioneering work nearly a century earlier. Spectrum lines were observed by Wollaston (page 485) and more systematically by Fraunhofer (page 358), beginning about 1814. Both men also noticed, for the first time, that the spectrum of sunlight was not continuous but was crossed by numerous *dark* lines located in positions corresponding exactly to bright lines from laboratory sources (Fig. 21.4). Fraunhofer measured the wavelengths of the more prominent dark lines and noted the same pattern in the spectrum of moonlight as well as in the light from the planets.

The explanation of the dark lines was given by Kirchhoff (page 496) in 1859, in the course of an extensive study of spectra which he conducted with the German chemist R. W. Bunsen (1811–1899). Fraunhofer had noticed that the spectrum of a candle flame contained a bright yellow line (actually two lines close together) which coincided in position exactly with the dark line labeled D in his solar spectrum. Kirchhoff identified the yellow line as belonging to the element sodium. But going further, he sent a beam of white light through a flame heavily charged with salt (sodium chloride) and found that a dark line appeared precisely where the sodium flame alone had previously given a bright line (Fig. 21.5).

After experimenting in a similar way with the spectra of other chemical elements, Kirchhoff drew his interpretation of what was happening: When white light enters the flame containing sodium vapor, all the vibrations pass through practically undiminished except those whose frequency happens to coincide with natural frequencies of the sodium atom, that is, with fre-

Ans.: 3; 6; 0.

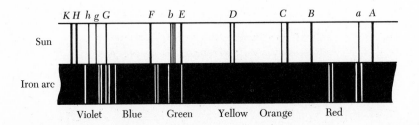

FIGURE 21.4 Fraunhofer lines (above), and spectrum of iron (below). Only the brightest lines are shown. Coincidence of lines G and E with the iron lines shows that iron is present in the outer part of the sun. See also color insert inside the book's covers.

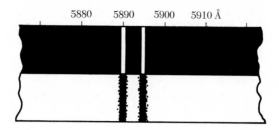

FIGURE 21.5 The yellow-line pair of sodium (Fraunhofer D lines). Above: Bright lines from a sodium-vapor lamp. Below: The same lines in absorption.

quencies that this atom can emit. This process is one of *resonance* (page 287). The sodium atoms, having absorbed energy from the white light at certain definite frequencies, reradiate it. However, since this energy will be thrown out again in *any* direction, the light going in the direction of the original beam is weakened at the particular places in the spectrum corresponding to these frequencies. The result is the observed pattern of dark lines on a bright background, called an **absorption-line** spectrum. The lines are not absolutely black but appear dark in contrast with the surrounding continuous spectrum. Figure 21.6 gives a schematic representation of the process.

The scattered light is found to consist of just those wavelengths that are missing in the transmitted light.

After studying the spectra produced by vaporizing various substances in a gas flame, Kirchhoff and Bunsen concluded, about 1860, that *each chemical element, when stimulated to give off light, produces a spectrum consisting of a definite pattern of bright lines. This pattern is different for each element and may be used to identify its presence in the source of light,* even when the element in question is part of a chemical compound. As little as a millionth of a gram of the material is often enough for this purpose.

In practice, some qualifications are necessary: The spectrum of the most abundant element in a mixture is usually the strongest, and elements present in very small amounts may show only a few lines of their spectra. Also, the nature of the source often determines the extent to which the spectrum develops. In fairly low-temperature sources such as flames, only a few elements show any spectrum at all and even then only one or two of the strongest lines may appear. When the spectra of the same elements are

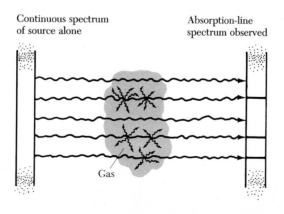

Continuous spectrum
of source alone

Absorption-line
spectrum observed

Gas

FIGURE 21.6 The dark lines originate when the cooler gas cloud absorbs light of certain characteristic frequencies.

produced in an electric arc or in a high-voltage spark, they may contain hundreds of lines.

In spite of such practical difficulties the identification of chemical elements by means of their spectra has become one of the most valuable procedures in experimental science, with applications to industry, medicine, criminology, astronomy, etc. (see next section).

Techniques have been developed for making **quantitative spectro-chemical analyses,** so that a chemist can not only tell the nature of an "unknown" substance from the pattern of lines in the spectrum but determine within satisfactory limits the percentage composition from the intensities of the lines. Moreover, analysis by means of the spectrum usually takes much less time than with chemical methods, and the results can be recorded automatically. This is especially important in industry. In a steel mill, for example, repeated determinations of the composition of a batch of molten metal may be made as its processing continues.

21.3 ASTROPHYSICAL SPECTRA

Not only the spectrum of the sun but the spectra of most types of stars are absorption spectra. It is known that the stars, including the sun, consist of incandescent gases—hotter and denser in the interior, cooler and more tenuous near the surface. Light from the interior originates in a dense gas at high temperature and so, if directly observable, would show a continuous spectrum. However, when this radiation passes through the cooler, thinner outer regions, the frequencies corresponding to atoms present there are absorbed, and the light reaching an observer on earth gives an absorption-line spectrum by which the gases present in the outer region can be identified. More than two-thirds of the elements known on earth have been identified in the sun and in the stars by means of their spectra, so that on the whole the universe must be made up of familiar materials. The existence of a previously unknown element, helium, was shown by an otherwise unidentified set of lines in the solar spectrum. Later, helium was detected and isolated on earth.

One of the most important uses of spectra in astrophysics is in the determination of motion by means of the Doppler effect (Sec. 12.10, page 297). This manifests itself as a shift in position of the spectrum lines from a source that is moving with respect to the observer. Figure 21.7 shows part of the visible spectrum of a star as recorded on earth. Adjoining is the spectrum of a laboratory source producing some of the same lines. The displacement of the lines toward the violet—the higher-frequency end of the spectrum—shows that this star and the earth are approaching each other. A shift toward the red means recession. Equations 12.15 and 12.16

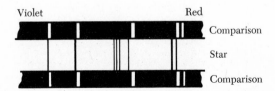

FIGURE 21.7   Doppler shift c stellar spectrum lines.

are applicable here, and from the measured shift in wavelength of any line the relative speed of approach or recession can be computed.

The application of Doppler's principle to astronomy was first suggested by Fizeau, and the earliest successful measurements were made by the English astronomer Huggins around 1870. For most stars of our own galaxy the relative speeds are less than about 50 mi/sec. The method is used in determining the rate of rotation of the sun, planets, and double stars. The spectra of the spiral nebulae and of other sources far outside our own galaxy show, almost without exception, very large line displacements toward the red that increase in proportion to the distances of these objects. The Doppler effect seems to be the most straightforward interpretation at this time, and if relative motion is responsible for the observed shift, it must mean that these bodies are rushing apart in conformity with the idea of an *expanding universe.*

In recent years, astronomers have discovered **quasars** (quasi-stellar radio sources) whose Doppler speeds range as high as $0.8c$, corresponding to a distance of about 15 billion light-years.

21.4 THE SPECTRUM OF HYDROGEN; BALMER'S FORMULA

In the years following the pioneering work of Kirchhoff and Bunsen, data on spectrum-line wavelengths and intensities continued to accumulate. The general feeling that spectra might hold the key to an understanding of the microstructure of matter stimulated many attempts to find some order among the myriad measurements. Not much was accomplished, which is not surprising since typical spectra give little impression of regularity of arrangement of the lines.

A notable exception is the spectrum of hydrogen, the lightest element and presumably the simplest in structure. Under certain conditions, this spectrum consists of a moderate number of lines whose spacing decreases systematically toward shorter wavelengths, the set approaching a limit in the near ultraviolet (Fig. 21.8). In 1884 a Swiss science teacher, J. J. Balmer, devised an empirical formula that matched the wavelengths of the four hydrogen lines then known. It had the unusual form

$$\lambda_A = 3645.6 \frac{n^2}{n^2 - 4} \qquad n = 3, 4, 5, \ldots \qquad \textbf{21.1}$$

FIGURE 21.8 The Balmer series in the spectrum of hydrogen. (See also color insert inside the book's covers.)

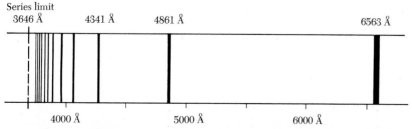

Here $\lambda_{\text{Å}}$ is the wavelength in angstroms, and the sequential lines of the series are obtained by setting n equal to successive integers, beginning with 3.

Balmer found that his formula represented not only the four original lines but a large number of additional ones discovered in star spectra. Observed and computed wavelengths agreed to within about one-tenth of a percent. Several other series of hydrogen lines were discovered later, some of which lie in the infrared and one in the ultraviolet (Fig. 21.9). These series can be represented by an expression like that of Eq. 21.1 but of a more general form which became the prototype of equations representing the line spectra of all elements.

It is customary to write the general relation for the hydrogen spectra in terms of the *reciprocal* of the wavelength, in which case it becomes

$$\frac{1}{\lambda} = R\left(\frac{1}{p^2} - \frac{1}{q^2}\right) \qquad \textbf{21.2}$$

Rydberg's constant was named for the Swedish spectroscopist J. R. Rydberg (1854–1919).

Here R is **Rydberg's constant,** whose numerical value is 1.097×10^7 when λ is measured in meters. The number p has a given integral value for each series. Within the series, q takes successive integral values beginning with $q = p + 1$, to give the wavelengths of the individual lines. Thus, $p = 2$ for the **Balmer series,** and q is given the value 3 for computing the wavelength of the first line, 4 for the second, and so on.

If Eq. 21.2 is multiplied throughout by the speed of light, c, it gives directly the frequency f of each line:

$$\frac{c}{\lambda} = f = \frac{cR}{p^2} - \frac{cR}{q^2} \qquad \textbf{21.3}$$

This form shows that the frequency of each line can be expressed by taking the difference between two numbers, called spectrum **terms.** Even in a very complicated spectrum, the frequency of any observed line can be expressed as the difference between two of a relatively limited set of terms, although in such spectra the denominators in Eq. 21.3 are no longer integers. The important point is that theory later showed that the array of terms represents a fundamental physical characteristic of the light-emitting substance.

Worked example 21.2

Compute the wavelength (a) of the third line of Balmer's series and (b) of the series limit.

Solution   a. Here $p = 2$ and $q = 5$. Use the above value for the Rydberg constant, $R = 1.097 \times 10^7$, and substitute into Eq. 21.2:

$$\frac{1}{\lambda} = 1.097 \times 10^7 \left(\frac{1}{4} - \frac{1}{25}\right) \quad \text{or} \quad \lambda = 4.3408 \times 10^{-7} \text{ m} = 4340.8 \text{ Å}$$

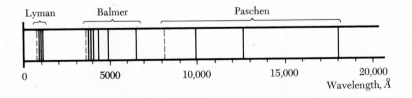

Lyman   Balmer   Paschen

0   5000   10,000   15,000   20,000
                                          Wavelength, Å

FIGURE 21.9 Some of the stronge[st] lines of the Lyman, Balmer, an[d] Paschen series of hydrogen. Line[s] belonging to two additional pre[-]dicted series are found farther [in] the infrared. The series limits a[re] shown here in color.

This is in very good agreement with the measured value of 4340.5 Å.

b. The series limit corresponds to $q = \infty$, so that

$$\lambda_{\text{lim}} = \frac{4}{1.097 \times 10^7} = 3.646 \times 10^{-7} \text{ m} = 3646 \text{ Å}$$

Compare this value with that of the constant in the Balmer equation 21.1.

21.5 BOHR'S THEORY

The realization that all frequencies in the hydrogen spectrum can be expressed in the general form represented by Eq. 21.3 prompted experimenters to devise a model of the hydrogen atom that would account for such a relation, but such efforts did not succeed. The frequencies emitted by a given atom are evidently related in a more complex way than are the partial tones of a simple acoustic system, and no model that was proposed seemed to satisfy the observations.

The Rutherford nuclear atom appeared to offer promise, but there was a grave difficulty. If the electrons were assumed to be at rest, electrostatic attraction would quickly make them fall into the nucleus. On the other hand, if they are assumed to move in orbits around the nucleus they would be subject to acceleration, and since classical theory predicts that an accelerated charged body must radiate electromagnetic energy, these orbiting electrons would continuously lose energy. As a consequence, such an electron would spiral into the nucleus in a time of the order of 10^{-8} sec, and an atom of this kind could not have a permanent existence.

In 1913, a bold and entirely unprecedented theory of atomic structure and spectra was proposed by the Danish physicist Niels Bohr (page 503n), who at that time was one of Rutherford's group of "bright young men." He was able to account for the spectrum formula by applying quantum principles to the Rutherford model of the atom, but in order to do this he

Bohr. (*Courtesy of Niels Bohr Library, American Institute of Physics.*)

had to make assumptions that were in direct contradiction to certain classical ideas, at the same time making free use of others. The Bohr theory, as will be seen later, ultimately proved too restricted in outlook and had to give way to a more general approach. Although it may be considered mainly of historical interest today, this theory represents the first successful application of quantum ideas to the radiation from individual atoms and molecules and is worth examination for its contribution to the broader formulation of later quantum theory. The steps in the development of the Bohr theory will now be traced.

Bohr started with hydrogen, the atom having the simplest structure, and assumed that the single electron moves in a circular orbit around the nucleus, which in this case ($Z = 1$) consists of a proton (Fig. 21.10). The general similarity with the motion of a planet around the sun suggests why this structure is sometimes called a *planetary* atom.

Perrin (page 228n) suggested a similar model in 1901.

The centripetal force holding a planet in its orbit is the gravitational attraction between planet and sun. In the atomic counterpart it is the electrostatic attraction between the charge ^-e of the electron and the charge ^+e of the proton. The gravitational attraction inside an atom, if indeed it exists there at all, must be weaker by a factor 10^{39} (see page 378) and so can be neglected. Since the alpha-scattering experiments showed the inverse-square Coulomb force to be valid even at small distances from the nucleus, Bohr equated the general expression for centripetal force (Eq. 8.5, page 172) to the Coulomb force: $mv^2/r = k_0 e^2/r^2$, which reduces to

$$v^2 = \frac{k_0 e^2}{mr} \qquad\qquad 21.4$$

Here e is the magnitude of the charge on either particle, m is the mass of the electron, v is the speed in its orbit, and r is the radius of the orbit. If the electron is to continue in stable motion in its circle, this relation must hold between the electron speed v and the orbit radius r.

Although Eq. 21.4 is perfectly valid from the point of view of classical *mechanics,* it contradicts the classical *electromagnetic* principle that a charged body cannot move in a curved path without radiating energy. At this point Bohr found himself compelled to break with classical physics and assume that *there exist certain orbits in which the electron can continue to move without radiating energy.* The existence of a discrete set of stable, nonradiating orbits suggests (or, possibly, was suggested by) the quantum idea.

Stable orbits.

Bohr chose to assume that the stable orbits are characterized by a condition analogous to the one that Planck found for the individual oscillators in his theory of cavity radiation. To see what this is, it should be noted that in advanced mechanics it is convenient to define a quantity called rotational momentum, or **angular momentum,** which is merely the rotational counterpart of linear momentum. It is represented by a vector drawn along the axis of rotation in a direction given by a right-hand rule (Fig. 21.11).

For a particle of mass m moving in a circle of radius r with speed v the magnitude of the angular momentum p_θ is

$$p_\theta = mvr \qquad\qquad 21.5$$

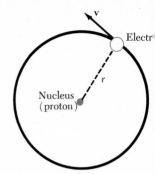

FIGURE 21.10 Bohr's conception of the structure of the hydrogen atom.

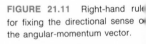

FIGURE 21.11 Right-hand rule for fixing the directional sense of the angular-momentum vector.

For an extended physical body in rotation, such as a flywheel, the total angular momentum must be calculated by adding expressions of the form of Eq. 21.5, one for each particle of the wheel. If the body is of relatively simple shape, this summation can often be carried out by calculus processes.

There is a fundamental mechanical law, not mentioned so far, concerning angular momentum: The principle of the **conservation of angular momentum** parallels the conservation of linear momentum discussed in Sec. 7.5 *Review the* (page 154) and states that if the resultant externally applied torque on a *definition of* system is zero, the total angular momentum remains constant.

torque on
page 79.   At this point in the development of his theory, Bohr introduced the idea of quantization of orbital angular momentum in a way analogous to Planck's quantization of energy in the thermal-radiation problem.

Angular   Bohr made the assumption that the stable orbits are those and only those *momentum* for which the angular momentum is given by
is quantized.

$$p_\theta = mvr = \frac{nh}{2\pi} \qquad\qquad \textbf{21.6}$$

where n is an integer. It should be noticed that, according to Eq. 21.5, angular momentum, in mks units, is expressed in kg(m/sec)m. This may be written (kg-m/sec²)(m)(sec), which is then equivalent to (N)(m)(sec), or J-sec—the same units used for h (page 500). Thus the quantum constant h provides a natural unit of angular momentum.

Check this   In order to find the size of each of these allowed orbits, solve Eq. 21.6 *algebraic* for v and substitute the value into Eq. 21.4. The result, solved for r, is *reduction.*

$$r = \frac{n^2 h^2}{4\pi^2 k_0 m e^2} \qquad\qquad \textbf{21.7}$$

This expression shows that the radii of the stable orbits in which the electron can circulate without losing energy are proportional to 1^2, 2^2, 3^2, . . . , or 1, 4, 9, ... Some of these orbits are shown to scale in Fig. 21.12. If the numerical values of the fundamental constants are substituted into the expression 21.7, the result, with $n = 1$, is $r = 0.53 \times 10^{-10}$ m, which agrees surprisingly well with the radius of a hydrogen atom as determined from kinetic theory. It is accordingly assumed that when the hydrogen atom is in its normal state the electron occupies the smallest of its allowed orbits, which will be shown below to be the one having least energy and therefore the greatest stability.

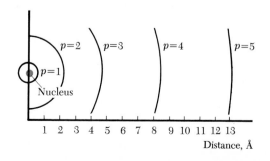

FIGURE 21.12 Portions of the first five stable circular orbits for hydrogen, shown to scale.

Worked example 21.3

Find the speed of the electron in the lowest orbit of the hydrogen atom.

Solution   Equation 21.4 (page 538) may be written $mv^2r = k_0e^2$, and Eq. 21.6 is $mvr = nh/2\pi$. If we divide the former by the latter and set $n = 1$ for the lowest orbit, the result is

$$v_1 = \frac{2\pi k_0 e^2}{h}$$

The ratio of this speed to the speed of light is $\alpha = v_1/c = 2\pi k_0 e^2/hc$, and substitution of the values of the constants yields $\alpha = 1/137.0$. Thus the electron speed is less than 1 percent of c, and the use of the classical mechanical expressions in deriving the equations is justified. There are, however, certain observable relativistic effects of a more subtle kind that must be taken into account in a more general theory.

The quantity α is an important dimensionless natural constant. It plays a prominent part in **quantum electrodynamics,** which is the theory of the interaction of radiation with matter, subject to quantum laws.

When Bohr went on to compute the energy relations in the atom, an even more impressive agreement with known facts was revealed. The crucial test that had to be met by the theory was the prediction of the positions of the observed spectrum lines.

The next step is the computation of the energy of the electron in any of its stable orbits. The total energy is made up of KE and electrostatic PE:

$$E = K + P$$

where K is given in classical form by $\frac{1}{2}mv^2$ and P is, by the reasoning of Worked example 21.1 (page 530),

$$P = -\frac{k_0e^2}{r} \qquad\qquad 21.8$$

The minus sign arises from the fact that the charge on the electron is negative. The total energy is then

$$E = \tfrac{1}{2}mv^2 - \frac{k_0e^2}{r}$$

Substitution of the value of v^2 from Eq. 21.4 (page 538) then yields

$$E = \frac{1}{2}\frac{k_0e^2}{r} - \frac{k_0e^2}{r} = -\frac{k_0e^2}{2r} \qquad\qquad 21.9$$

Notice that the KE is equal to minus one-half the PE and that the sum is a *negative* quantity. The negative sign (which is merely a reflection of the sign of the electron charge) implies that the total energy is *algebraically greater* for the larger orbits, approaching a maximum value of *zero* for very large orbital radius. Thus the larger the orbit, the greater is the energy.

Bohr went on to assume that *the electron can jump spontaneously from a larger to a smaller orbit* and that in this process it radiates energy. The difference in energy ΔE in the initial and final orbits is sent out in the form of a photon whose frequency f is given by $\Delta E = hf$.

Where does this radiated energy originate? In most of the atoms of an undisturbed sample of hydrogen gas at ordinary temperature the electron is in its lowest orbit. In order for such an atom to radiate energy, the electron must first be boosted to a larger orbit having higher energy. This can be done in various ways. In an electric discharge tube this so-called **excitation** is accomplished by the impact of fast electrons moving through the gas. Or if light is sent through the gas, a photon will be absorbed if it has the right amount of energy needed to raise the electron to a higher orbit. The transmitted beam of light will show an absorption line (page 533) at this frequency, whereas photons having any other frequency will be strictly ignored by the atom. In any event, once the electron in the atom has been elevated to one of the larger permitted orbits (it cannot stop anywhere in between), it is in a condition to fall back into a smaller orbit.

The frequency formula.

Later work showed that an electron ordinarily spends only about 10^{-8} sec in any one of its excited states before spontaneously dropping to a lower state. When it descends, it may not go to the smallest circle ($n = 1$) in a single jump but may do this in several successive steps. Each transition, however, is assumed to produce radiation of a single frequency f determined by $\Delta E = hf$, where ΔE is the difference in energy of the initial and final orbits involved. The frequency emitted by the atom when its electron goes from an initial orbit q to a final orbit p is then given by

$$f = \frac{\Delta E}{h} = \frac{E_p - E_q}{h} \qquad\qquad \textbf{21.10}$$

where E_p and E_q are the energies in the final and initial states, respectively.

The concluding step in Bohr's calculation was to substitute the values of E_p and E_q that resulted from his assumptions. The value of E in any orbit is given by Eq. 21.9. When the expression for the radius that is given by Eq. 21.7 is inserted, the result is

$$E = -\frac{2\pi^2 k_0{}^2 m e^4}{n^2 h^2} \qquad\qquad \textbf{21.11}$$

and if this value is used in Eq. 21.10, the expression for the frequency of the radiation becomes

$$f = \frac{2\pi^2 k_0{}^2 m e^4}{h^3}\left(\frac{1}{p^2} - \frac{1}{q^2}\right) \qquad\qquad \textbf{21.12}$$

This relation has exactly the same algebraic form as the spectrum formula 21.2 (page 536). Further, if Eq. 21.2 is written to express the frequency f instead of the reciprocal of the wavelength, it becomes $f = c/\lambda = Rc(1/p^2 - 1/q^2)$. Comparison with Eq. 21.12 then shows that the expression for the Rydberg constant is

$$R = \frac{2\pi^2 k_0{}^2 m e^4}{h^3 c} \qquad\qquad \textbf{21.13}$$

and substitution of the numerical constants yields a value for R that is identical, to four significant digits, with the one obtained earlier by fitting the observed spectrum lines into the Balmer formula. This agreement must be regarded as one of the impressive achievements of the Bohr theory.

The specific form of Eq. 21.12 that holds for the Balmer series of hydrogen is obtained by setting $p = 2$. On the Bohr model, this means that all lines of this series are produced by electron jumps that end in the second orbit. For example, a hydrogen atom radiates light of a frequency corresponding to the first line of this series when, after excitation, its electron falls from the third to the second orbit; the second line of the series results from a jump from $q = 4$ to $p = 2$, and so on. Such events, taking place in myriads of atoms throughout the gas, produce the observed set of spectrum lines.

The existence of other possible series of hydrogen, corresponding to final orbits for which $p = 1, 3, 4, 5, \ldots$, was surmised by Bohr, and members of four such series were actually found (Fig. 21.9, page 536). Thus, his theory, by its surprisingly accurate numerical agreement with observation and its prediction of new facts, claimed serious attention throughout the scientific world and stimulated an enormous amount of investigation.

Refinements of the Bohr model of the atom.

The simple situation on which Bohr's theory was based soon was generalized and refined in attempts to explain additional details of spectra. It was recognized, for example, that the preceding computation took no account of the fact that the nucleus as well as the electron must have a motion of revolution. Although a proton is 1,840 times as massive as an electron, the former is not fixed in position; instead, both objects move around the center of mass of the pair. When this circumstance is allowed for, the value of the Rydberg constant is slightly reduced (by less than 0.1 percent), but this change improves the agreement with experimental results.

This refinement of the theory made it applicable to *hydrogenlike* atoms as well as to hydrogen itself. Although the helium atom normally has two orbital electrons, one of them may be temporarily removed by ionization. The resulting structure consists of a single electron revolving about a nucleus—the same *type* of structure as an atom of hydrogen. In the same way, a doubly ionized lithium atom, a trebly ionized beryllium atom, etc., are hydrogenlike, and their spectrum frequencies should be calculable by the Bohr theory. This proves to be the case; each such atom has its own value of the Rydberg constant. The differences are very small, but they are readily detectable by spectroscopic means.

The existence of "heavy hydrogen" **(deuterium)** was verified by means of its spectrum. Deuterium is an **isotope** (Chap. 22) of hydrogen whose nucleus is now known to consist of a proton bound to a neutron and so is about twice as massive as the nucleus of ordinary hydrogen. Deuterium is present in natural hydrogen gas only to the extent of about 1 part in 6,000. In 1932 the American chemist and physicist H. C. Urey and his collaborators found it possible to enrich the natural mixture. They obtained a spectrum that showed a faint line adjoining one of the regular hydrogen lines at the position given by theory for deuterium.

Another extension of the original circular-orbit theory, developed mainly by the German theorist Arnold Sommerfeld (1868–1951), took into account the possibility of *elliptical orbits* for the electron. In the solar system the planets move in orbits of this shape, and the same possibility exists in the Bohr atom. When the calculation is carried through, it is found that each hydrogen line is composed of a close set of lines ("fine structure"), but the agreement with the measured wavelengths is not satisfactory. However, a further extension of the theory that attributes to the electron an intrinsic *spinning motion* (page 557) about its own axis accounts for the observed line patterns more accurately.

21.6   ACCOMPLISHMENTS AND LIMITATIONS OF THE BOHR THEORY

One of the achievements of the Bohr theory is its emphasis on the basic importance of the energy levels of the atom. The theory brought out the fact that the empirical spectrum terms referred to above are merely *energy states* of the corresponding atom. They are purely observational quantities, independent of any particular model of the atom, a point of view that was to be fully supported by the more general form of quantum theory developed later, where one of the main goals is still the computation of energy levels.

Excitation and ionization.   Figure 21.13 shows the energy-level diagram for the hydrogen atom, with the fine structure of the levels omitted for simplicity. All the levels have *negative* energy values, as brought out by Eq. 21.9 (page 540). The various observed series of spectra are shown as vertical lines indicating transitions between energy levels. All lines belonging to a given series correspond to transitions ending in the same level. The lowest state of the atom, designated by $n = 1$, is called the **ground state;** the levels above this are

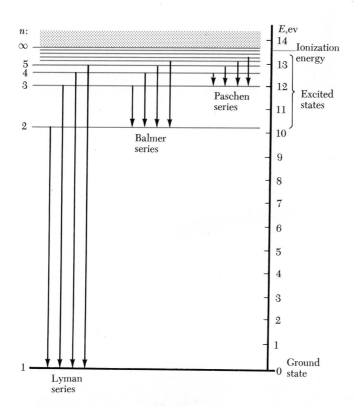

FIGURE 21.13 Energy levels and transitions for some of the series of the hydrogen spectrum.

excited states. The energy that must be supplied to the electron in the ground state in order to raise it to one of the excited states is called the **excitation energy** corresponding to the higher state. In particular, the energy needed to remove the electron completely from the atom is the **ionization energy.** Its value may be calculated from Eq. 21.11 (page 541) by putting $n = 1$. Thus

$$E_I = \frac{2\pi^2 k_0{}^2 m e^4}{h^2}$$

21.14

Substituting the values of the constants yields $E_I = 2.17 \times 10^{-18}$ joule, or $(2.17 \times 10^{-18})/(1.6 \times 10^{-19}) = 13.6$ eV. Experiment shows that electrons that have fallen through a potential of 13.6 volts have just enough energy to ionize hydrogen atoms on colliding inelastically with them, in complete agreement with the above computation. The ionization energy is a specific example of what is called the **binding energy.** The **binding energy** of a system is the energy that must be supplied to separate the constituents and leave them at rest an infinite distance apart.

Another prediction made by any quantized atom model is confirmed by observation: Suppose that a free electron having a large amount of KE drops into the lowest level of a previously ionized hydrogen atom. In this transition the atom should emit radiant energy whose frequency is greater than that of any member of the Lyman series (see Fig. 21.13) whose lines have the common final state $n = 1$. Since the initial energy of the free electron is not quantized, the radiation can have *any* frequency greater than that of the series limit, giving a *continuous* spectrum that extends from the series limit toward shorter wavelengths. Similarly, a continuous spectrum may be expected beyond the Balmer series limit. These predicted continuous spectra were looked for and actually observed.

Series-limit continuum.

The existence of continuous spectra at the series limits is to be expected on the basis of Bohr's correspondence principle (Sec. 20.3, page 502). The very high energy levels of an atom are closely spaced, approaching a continuous distribution of energy values, so that for large values of n, the quantum results and the classical ones merge, as predicted by the correspondence principle. In fact, calculation shows that the frequency of the radiation from an electron transition between two adjoining very large orbits approaches the mechanical frequency of the electron in its orbit. This identity of the two frequencies is the result to be expected according to classical electromagnetic theory. In this connection, Bohr was able to prove that the quantization of angular momentum follows from the correspondence principle.

Despite its initial achievements, the Bohr model of the atom soon revealed limitations which suggested that it might be too literal and too narrow in scope to become the basis of a general theory of atomic structure. In attempting to account for the details of the hydrogen spectrum, physicists had already found it necessary to resort to successive extensions of the original theory by introducing elliptic orbits, motion of the nucleus, electron spin, etc. Even with these modifications, such features as fine structure and spectrum-line intensities remained unexplained. Moreover, the theory failed completely for atomic systems with more than a single electron. Another questionable aspect of the theory is the mysterious nature of the

Some difficulties of the Bohr model of the atom.

assumed electron jumps; yet many physicists in their qualitative thinking about subatomic phenomena continue to visualize electrons springing from one place to another.

Other criticisms have been leveled against the theory on logical and philosophical grounds. In constructing his theory, Bohr used a peculiar mixture of classical and quantum concepts, retaining some of the classical laws of dynamics and electrostatics but discarding the electromagnetic principle concerning radiation by an accelerated charged body. Later Sommerfeld, in developing the elliptic-orbit modification, introduced the relativistic variability of mass (page 442) as well. Thus it may be rightly contended that the basic postulates of the theory are a somewhat arbitrary mixture of classical, quantum, and relativity concepts.

Can any theory be depended on to represent final truth? See Chap. 1.

Bohr's theory is not "wrong." It merely is too specialized and so can explain only part of what is observed. Its greatest contribution was that it struck out in a significant new direction, providing a basis for the more general quantum theory, which is described below. A detailed discussion of the newer theory requires much more advanced mathematical methods than did the Bohr theory, and so the presentation here will be on a more descriptive level.

THE NEWER QUANTUM THEORY AND ITS CONSEQUENCES

21.7 MATRIX MECHANICS

Heisenberg's profound and penetrating work in quantum theory was begun when he was 23 years old. It won him the Nobel Prize in physics in 1932.

The first successful formulation of a more comprehensive and powerful quantum theory was attained by the German physicist Werner Heisenberg in 1924. Proceeding on the assumption that the purpose of a physical theory is to provide the simplest possible connection between experimental data, he tried to develop a scheme in which only *directly observable* quantities are used. In this category he placed the frequencies and intensities of spectrum lines, while he refrained from introducing such concepts as electron orbits, positions, and speeds that he regarded as mere constructs and not as observable data.

Heisenberg's theory furnished a set of equations for finding the frequencies and intensities of spectrum lines and the energy levels in an atom. It accounted, for example, for the details of the helium spectrum—something that the Bohr theory was unable to do. It also described correctly other more subtle phenomena such as the fine structure of the hydrogen lines and the splitting of spectrum lines that occurs when the source is put into strong electric or magnetic fields. Above all, Heisenberg freed physical theory from the restriction of treating atomic processes in terms of a mechanistic, nonoperational model.

In his formulation, Heisenberg made use of a kind of mathematical quantity called a matrix, and the theory is usually referred to as **matrix mechanics.** The use of the word *mechanics* in this connection is unfortunate, since the theory is not bound to a mechanical model in any sense. Quite the contrary: The new theory is incomparably more abstract than the Bohr theory, and it makes no concessions at all to intuition. Nevertheless, it is possible to get at least a general idea of the matrix method without following the mathematical details. Chapters 9 and 10 of Ref. 21.4 are excellent for this purpose.

21.8   WAVE MECHANICS

Within a few months of Heisenberg's announcement of his matrix mechanics, the Austrian theoretical physicist Erwin Schrödinger (1887–1961) formulated another approach to quantum theory that has become known as **wave mechanics.** Schrödinger started from the idea of matter waves proposed by de Broglie (page 517). This was still some time before the existence of such waves had been verified experimentally.

Schrödinger shared the 1933 Nobel Prize in physics with the English theorist P. A. M. Dirac, who also contributed outstandingly to the development of wave mechanics.

Schrödinger's thinking, like that of de Broglie, was guided by a comparison of the behavior of light with that of material particles. It is true that, for many purposes, light can be regarded as something that proceeds in straight lines (rays), but more subtle aspects such as diffraction and interference make it necessary to replace this concept by the travel of waves. Schrödinger likened the classical mechanics of Newton to ray optics and asked himself what would correspond to the general *wave optics of matter.*

Certain features of classical wave theory suggest a connection between particles and waves. A single wave train of frequency f can have infinite extension in space and thus carries no implication of any properties other than the ones usually associated with waves. On the other hand, as noted in Chap. 12, if *two* wave trains of different frequencies but the same speed travel in the same direction through a medium, the result is a "bunching" that produces local regions of large amplitude separated by other regions of small amplitude. This is the phenomenon of *beats,* described on pages 296–298 and depicted in Fig. 12.19.

Now consider a number of wave trains of various frequencies, traveling as above. If all the wave trains are in the same phase at a given place, the resultant disturbance will be large at that point. However, in departing from that place in either the forward or backward direction, the phase relations become quite random and the resulting displacement becomes small, as a mathematical analysis confirms. The combined disturbance is called a **wave packet.** It is almost entirely confined to a region of finite width, in much the same way as we picture a particle occupying a rather definite portion of space. Thus *there is a similarity between a moving particle and a wave packet.* The larger the spread in frequency of the waves making up the set, the narrower the packet becomes, and vice versa.

In developing his electromagnetic theory of radiation, Maxwell arrived at a set of mathematical equations describing the behavior of electromagnetic waves (see page 476). In a similar way, Schrödinger's pursuit of the ideas mentioned above led him to a form of wave equation governing the motion of matter. Like Maxwell's equations, this relation is a *differential equation.* We exhibit it here in a special *one-dimensional time-independent* form for a particle of mass m, total energy E, and potential energy V traveling in the x direction; d represents an operator called a *differential*:

$$\frac{h^2}{8\pi^2 m}\frac{d^2\psi}{dx^2} + (E - V)\psi = 0 \qquad\qquad \textbf{21.15}$$

The heart of this relation is the variable ψ (psi), called the **wave function,** whose possible values in any given problem are found by solving the equation by the prescribed methods of the theory of differential equations. Using the term "possible" values of ψ implies that if the equation is to represent *quantum* states of a system, it must yield a discrete set of solutions corresponding to such states, and no others. This can be understood by examining the example of waves on a string (Chap. 12). A string is a one-dimensional system, but similar conclusions apply to three-dimensional systems such as atoms.

Transverse waves of any frequency or wavelength can be propagated along a string of infinite length. The speed of such waves is determined jointly by the tension in the string and its mass per unit length (Eq. 12.14, page 295), but there is no limitation on the frequency. This is not true for a string of finite length. If two points are fixed in position, as in a violin string, the possible stable states of vibration are at once limited to a discrete set, corresponding to the *stationary waves* on such a string. No other frequencies of vibration are present. Thus, by the imposition of **boundary conditions** on the medium, its possible modes of vibration are reduced from an infinite number to a finite set.

The same thing happens with matter waves as described by the Schrödinger equation. Insertion of boundary conditions appropriate to a given system, such as a hydrogen atom, is accomplished by putting the appropriate values of E and V into the equation. The result is that a discrete set of solutions is obtained, and these correspond to the energy states allowed by quantum theory. The Bohr theory had to inject quantization of angular momentum arbitrarily, whereas wave mechanics attains the same result in a natural way that parallels classical mechanics.

The vibrating-string analogy, applied to the hydrogen atom, leads directly to the quantization of angular momentum, starting from the wave property of the orbiting electron. According to Eq. 20.19 (page 518), the matter waves associated with such a body have a length

$$\lambda = \frac{h}{mv} \qquad\qquad \textbf{21.16}$$

Assuming that, in agreement with the case of the vibrating string, the stable orbits are those onto which an integral number of waves can fit (see Fig. 21.14), it becomes possible to write

$$2\pi r = n\lambda \qquad\qquad \textbf{21.17}$$

where r is the radius of any permitted orbit and n is an integer. Then, eliminating λ from the two preceding equations,

$$mvr = \frac{nh}{2\pi}$$

which is the quantization of angular momentum as given by Eq. 21.6.

Within a few months of the discovery of wave mechanics, its mathematical equivalence with matrix mechanics was proved. In 1965 Dirac found that this equivalence is not complete for certain problems of quantum electrodynamics, although most physicists believe that the two treatments are capable of solving much the same problems and yielding the same

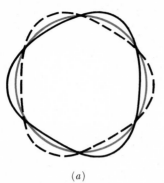
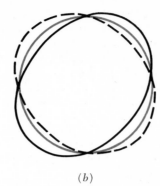

(a) (b) (c)

FIGURE 21.14 A circular loop of elastic wire vibrating in its own plane. In (a) and (b) an integral number of wavelengths fits around the ring, resulting in stationary waves; in (c) this is not true and the waves interfere destructively.

results. Matrix and wave mechanics are referred to collectively as **quantum mechanics.**

With the continued development of quantum mechanics, its superiority to the mechanistic Bohr theory and its much wider applicability became apparent. It succeeds in giving quantitatively the frequencies, intensities, states of polarization, etc., of spectrum lines in a systematic and unified way without the introduction of semiempirical or entirely arbitrary assumptions, as in the Bohr theory. In addition, the Bohr theory was powerless to solve many problems which have yielded to the new quantum procedures. An example is the question of the forces holding together chemical molecules such as hydrogen (H_2) and the problem of radioactive disintegration (Chap. 22). Although numerous mathematical difficulties exist in attempting to apply quantum mechanics to certain situations, it is generally felt that there are in principle no obstacles and that quantum mechanics represents a universal method of describing the microphysical world.

21.9 INDETERMINACY AND COMPLEMENTARITY

The wave-mechanics picture of the hydrogen atom outlined in the preceding section replaces the particlelike electron by a ring of waves. Precisely where is the electron? This is a question that has proved to be of profound philosophical interest and, moreover, one that has no logical answer.

It has been shown that elementary processes of all kinds exhibit a discontinuity characterized by Planck's constant h and that only for macroscopic systems consisting of large numbers of elementary particles do these quantum characteristics give a semblance of continuity. In 1927, Heisenberg arrived at a highly significant generalization concerning the observation and measurement of elementary processes. It is called the **principle of indeterminacy.**

Any measurement made on a macroscopic system is unavoidably accompanied by errors. The extent of the inexactness depends on the nature of the situation, but it would seem that the error could be made vanishingly

A somewhat anthropomorphic and less desirable name, uncertainty, is sometimes used instead of indeterminacy.

small by refining the procedure. Heisenberg contends that such an assumption is wrong in principle. To show this, he pointed out that in every process of measurement or observation there must be an actual physical interaction between the observer and what is observed. This interaction is far too small to take into consideration for macroscopic objects, but it is an entirely different matter in seeking to observe, say, an electron in an atom. In order to "see" an electron, ordinary visible light cannot be used because an object can produce an appreciable effect (such as reflection) on a set of waves only if the dimensions of the object are not too small compared with the wavelength. Hence, in order to "observe" an object of subatomic dimensions, radiation of very short wavelength—gamma radiation—would have to be used.

Refer to the discussion in Sec. 13.10, page 322.

What does this imply on the basis of the photon concept? In order to observe the electron, at least one photon must be reflected from it. In view of the extremely high frequency of gamma rays these photons have high energy, since $E = hf$. Consequently, when a photon strikes the electron there is a Compton effect in which the electron recoils with considerable speed, so that the very process of observation alters what one is trying to observe. Moreover, there is no way of telling in advance the direction in which the recoil will take place. If long waves are used, the *position* of an elementary particle cannot be determined accurately; if short waves are used, the exact *velocity* (or the momentum) of the particle prior to the act of observation remains unknown.

Considerations of this kind suggest that *there is a fundamental inexactness, an uncontrollable or undefined element, inherent in all measurements connected with elementary processes.* Heisenberg and Bohr analyzed many other situations by means of "thought experiments" such as the preceding one. In every case they found the above generalization to hold. The conclusion seems to be that in every experiment that is designed to determine the value of a given quantity it is impossible to avoid a loss of information on the value of some other quantity.

Heisenberg's indeterminacy relation.

Heisenberg became convinced that this indeterminacy is inherent in the nature of things and that it is responsible for the inability to visualize quantum processes. His considerations led him to an algebraic relation expressing the degree of indeterminacy involved in specific instances. This result may be obtained somewhat intuitively by formulating more precisely the process described above. If radiation of wavelength λ is used in a measurement of length or position, the smallest distance Δx that can be observed will be of the order of size of λ. According to the simple wave equation, $\lambda = c/f$, where f is the frequency; hence

$$\Delta x \geq \frac{c}{f} \qquad\qquad \textbf{21.18}$$

The symbol $\geq$ is read "is equal to or greater than" or "cannot be less than." Thus the condition of observability sets a minimum value for Δx. Further, the indeterminacy in the momentum of an elementary particle is of the same order of size as the momentum that can be delivered to it by a photon. From Eq. 20.13 (page 512), the photon momentum is given by hf/c, so that the indeterminacy in the momentum of the observed particle is of the order of

$$\Delta p = \frac{hf}{c} \qquad\qquad 21.19$$

Multiplying the respective members of the last two equations together gives

$$\Delta x\, \Delta p \geq h \qquad\qquad 21.20$$

This is one form of Heisenberg's **indeterminacy relation.** It expresses the fact that *the more closely the position of a particle is determined, the less accurately can its momentum be specified,* and vice versa. Because of this inherent indeterminacy, the concept of a particle as an extensionless point had to be abandoned, and any physical particle must be considered to be "smeared out" in space.

The similarity between a particle and a wave packet was pointed out on page 546, and it was stated that a large spread in the frequencies of the waves making up the packet is associated with a small extension of the packet, and vice versa. This circumstance is closely related to the principle of indeterminacy in the following way: There is a value of momentum corresponding to each wave belonging to the packet, the connection being given by the de Broglie expression $\lambda = h/p$. If the component waves extend over a large spread of frequencies, the momentum values will also be widely distributed. Then, according to what has been said above, the extension of the wave packet in space will be small. Conversely, if the wave frequencies—and hence the momentum values—fall within narrow limits, the wave packet will be spread out. The results are completely in accord with the indeterminacy relation in the form of Eq. 21.20.

The indeterminacy relation may be written in various forms, using other combinations of suitably paired quantities. The criterion is that the dimensions of the product must be those of h. Since h can be measured in joule-seconds, energy and time furnish one example:

$$\Delta E\, \Delta t \geq h \qquad\qquad 21.21$$

This says that, if a system is known to be in an energy state E for a period of time Δt, the value of the energy is undetermined by an amount that is at least $h/\Delta t$. Conversely, if the energy of a system is known within limits ΔE, it is impossible to tell *how long* it was in this state any more accurately than $h/\Delta E$.

The indeterminacy relation brings out the broad significance of Planck's constant: h is something more than the factor that relates the energy and frequency of radiation; more generally, it is a universal constant that limits quantitative knowledge of nature. If the value of h were exactly zero rather than finite, indeterminacy would disappear. The correspondence principle indicates how a situation equivalent to this but restricted to macroscopic bodies is approached. In making measurements on such bodies, the instrumental errors cover up the inherent quantum indeterminacies to the point where their existence is not apparent. The following Worked example illustrates this point.

Worked example 21.4

Compute the indeterminacy of the linear momentum in each of the following cases and compare its value with the momentum itself:

 a. An electron whose KE is 2,000 eV, confined to a space of atomic dimensions, 10^{-10} m.

 b. A 1-kg body having a speed of 1 m/sec whose position can be determined to 10^{-4} m.

Solution   *a.* In ordinary energy units, the KE of the electron is $(2 \times 10^3)(1.6 \times 10^{-19}) = 3.2 \times 10^{-16}$ joule. The linear momentum is given by $p = mv$ and, if the energy is not too large, $K = \frac{1}{2}mv^2$. Eliminating v, $p^2 = 2mK$, or

$$p = \sqrt{2mK} = \sqrt{2(9.1 \times 10^{-31})(3.2 \times 10^{-16})} = 2.4 \times 10^{-23} \text{ kg-m/sec}$$

Now, the order of size of the indeterminacy in the momentum is given by

$$\Delta p = \frac{h}{\Delta x} = \frac{6.6 \times 10^{-34}}{10^{-10}} = 6.6 \times 10^{-24} \text{ kg-m/sec}$$

and this is comparable with p itself.

 b. Here the momentum $p = 1$ kg-m/sec. Its indeterminacy is

$$\Delta p = \frac{h}{\Delta x} = \frac{6.6 \times 10^{-34}}{10^{-4}} = 6.6 \times 10^{-30} \text{ kg-m/sec}$$

which is entirely negligible in comparison with the value of p.

The above statement concerning particles is equally applicable to light. Attempts to fix the path of photons can be made only by restricting the beam of light by means of narrow slits. However, this introduces diffraction effects in which the wavelike aspects of light become noticeable.

In 1928, Bohr formulated such conclusions in what has been called one of the most significant philosophical statements of modern physics. It is *Bohr's principle of complementarity.* known as the **principle of complementarity:** *Every experiment that allows us to observe one aspect of a phenomenon deprives us of the possibility of observing a complementary aspect.* The two complementary features may be the position and momentum of a particle, the wave and particle characteristics of radiation or of matter, etc. In selecting an experiment that reveals one aspect of a phenomenon, the possibility of observing the complementary aspect must be ruled out. Notice that indeterminacy is included in the more general complementary principle.

Considerable caution is needed in wording and interpreting a statement of such generality as the principle of complementarity. According to Frank (Ref. 21.7), care must be used to avoid saying that "it is impossible to measure the position and velocity of a particle simultaneously." This statement implies that there exist *definite* positions and velocities but that science is powerless to find them—a statement that has no operational meaning. A better approach is provided by Bohr's later formulation of the complementarity idea in which he discusses experimental arrangements where the expressions "position of a particle" and "velocity of a particle" can never be used at the same time. One or the other may be employed but never simultaneously; the two are *complementary* descriptions. This formulation avoids the error of thinking of a "real" world whose details are somehow elusive.

21.10 CAUSALITY AND STATISTICS

The constancy of natural laws, which was referred to in Chap. 1 as the uniformity of nature, is closely bound up with the concepts of **causality** and **determinism.** These terms have many connotations extremely difficult to define, and instead of attempting this, it may be more useful at this point to examine the implications for physics.

When modern physicists began to investigate the micro-world, they tried to proceed in much the same way as their predecessors did in classical physics in the belief that the same possibilities for observation existed in the field of elementary phenomena that had long been used with success in exploring large-scale processes. Classical physics evolved largely from mechanics, and it was tacitly assumed that all the events taking place in space and time had the objective character associated with ordinary mechanical phenomena. The aim of physical science was taken to be the setting up of laws that would predict exactly the future configurations of the universe. As already pointed out, this idea came to require drastic modification.

Early in the nineteenth century Laplace (page 295) made the much-quoted statement that an omniscient mind, knowing the position and velocity of every atom at any one instant, could deduce the entire course of events in the universe, both past and future. Acceptance of the Heisenberg indeterminacy concept acknowledges this view to be no longer valid even in principle. The Laplace statement assumes that the universe, even down to its smallest units, is a machine in which all the parts are deterministically interlocked—that *strict causality* holds in the sense that every event is rigorously determined by the preceding one. The indeterminacy principle frees modern physics from such a restriction, although it cannot actually *disprove* the existence of strict causality.

If strict causality is not valid in microphysics, what takes its place? It is known that large-scale physical phenomena appear to follow rigorous laws that make possible reliable, quantitative predictions, and so it must be that nature possesses some kind of regularity in spite of the indeterminacy of elementary processes. But what is the character of this regularity? The answer given by the majority of physicists is that *the orderliness observed in macroscopic experiments is of a statistical nature,* and it may be that all physical laws, without exception, are based on probability.

The application of probability considerations to physical phenomena is not a new idea. It goes back to the work of Bernoulli and Joule on the kinetic theory of gases and, more particularly, to the introduction of statistical mechanics (page 232) by Maxwell and Boltzmann. This led to the conclusion that, although it is hopeless to try to follow the career of each molecule in a gas, it is nevertheless possible to describe accurately certain *average* properties of large ensembles of molecules. Since that time, statistical methods have been applied successfully to a wide variety of physical phenomena such as the brownian motion, radioactivity, turbulence in fluids, etc. Metaphorically it may be said that in many complex physical situations the *climate* can be described even if the state of the *weather* is not known.

But there is a fundamental difference between the meaning of the probability concept as applied to macroscopic phenomena and its meaning for elementary processes. In the kinetic theory, for example, the need for using statistical methods is a consequence of the very large numbers of molecules involved and the practical impossibility of following the course of each one. However, no doubt was felt as to the possibility of computing the motion of every molecule at least in *principle;* it was held merely to be too complicated. However, the statistical nature of quantum physics is a different matter because *the processes themselves are statistical in character.* Elementary events are governed by pure chance, and it is only in large-scale phenomena that myriads of separate occurrences combine to give results of such high probability that they can be regarded as practically certain.

Although most scientists and philosophers would agree in principle with the English physicist K. Mendelssohn that "the fundamental ideas of modern physics are based on the fact that truth is statistical," some have reservations regarding the extent to which this is valid in microphysics. More will be said about this in Chap. 23. There follows, in the next section, a brief discussion of the Schrödinger wave function ψ (page 547), showing that it has a statistical interpretation.

21.11   INTERPRETATION OF THE WAVE FUNCTION

From the beginning of wave mechanics, there was speculation about the possibility of attaching some physical meaning to the wave function ψ in the Schrödinger equation. Although certain formal correspondences exist between photons and light waves on one hand and matter waves on the other, there are also some decided differences. Light waves may be described in terms of electric and magnetic fields, whereas the behavior of matter waves is governed by the wave function, which cannot be observed directly. It is merely a quantity connected with the wave equation. Nevertheless, physicists continued to look for a model.

Schrödinger originally conjectured that the wave function represents the density of electric charge at various places. In this view, the charge of an electron, for example, is not confined to a definite region but is "smeared out" through all space; the charge density is large in certain locations but thins out rapidly in different directions in a way prescribed by the wave function. According to this formulation, the electron in any of its energy states may be pictured as a fuzzy cloud, most of whose bulk is found within a rather limited but not sharply bounded region of space. Each permitted distribution is based on a particular solution of the wave equation.

An alternative interpretation was suggested in 1926 by the German theoretical physicist Max Born (1882–1970). He suggested that the electron is a localized particle rather than an attenuated cloud and assumed that *the wave function determines the probability of finding the particle at various locations* in space. It can be shown that Born's idea is quite analogous to Einstein's interpretation of electromagnetic waves as guides for the distribution of photons. Consider, for example, the illumination produced on a surface by a uniform beam of light of given wavelength and of constant intensity. On the wave model, this represents a steady stream of wave energy, whereas on the photon model it must be pictured as a hail of corpuscles on the surface. It is not possible to tell exactly where the photons are located within the light beam, but as long as the light is not too weak, the overall effect of the random impacts of the myriad photons will be perceived as a constant, uniform illumination.

On the wave model, the intensity of the beam is measured by the square of the amplitude of the electric field vector, as given by Eq. 19.12 (page 479). On the photon model, the intensity is proportional to the number of photons per unit volume of the beam. Comparison of the two views shows that the photon density must be proportional to $E_m{}^2$. Since "photon density" in a statistical sense means merely the probability of finding a photon in any chosen small volume of space, it can be concluded that this probability is proportional to $E_m{}^2$.

Born's interpretation of the wave function ψ makes exactly the same assumption for material particles: It says that *the probability of finding a particle,* for example, the electron of a hydrogen atom, *in any given small volume of space is proportional to ψ^2.* On this basis, the distributions referred to above should be taken to represent clouds of probability. They do not claim to show where the electron is but merely where it is *likely to be found.* They correspond to hypothetical time exposures of the electron.

With this interpretation, the dematerialization of the electron is complete. The indeterminacy principle suggests the impossibility of fixing the position of an electron in its Bohr orbit, and the above considerations make the orbits themselves dissolve in a cloud of probability. If this view seems strange, it must be remembered that the achievements of quantum mechanics, which go far beyond the more literal early theory, fully justify the confidence that physicists have in it.

21.12   QUANTUM NUMBERS AND THEIR SIGNIFICANCE

Quantum mechanics not only supplies a comprehensive and detailed description of the spectra of complex atoms but also accounts for the whole scheme of the system of elements that had been revealed earlier by chemical means. It is no exaggeration to say that, in principle, quantum mechanics embraces all chemistry. Some aspects of these applications will now be presented in a descriptive way. Without the detailed mathematical development to support it, the procedure suggests a sort of numbers game. It is, however, a game that works remarkably well.

In the original Bohr description of the hydrogen atom, one variable, the number n, was sufficient to describe the possible energy states. This number appears in the energy equation 21.11 (page 541), as well as in the relation for the orbit radius, Eq. 21.7, and for the angular momentum, Eq. 21.6 (both on page 539). Quantum mechanics requires four numbers to describe accurately each state of each electron of the atom. Instead of representing an electron orbit, each quantum state corresponds to a cloud (probability distribution) of given size and shape, associated with a definite wave function ψ. The four numbers needed are called **quantum numbers** and are designated in this way:

n is the **principal quantum number.**
l is the **orbital angular momentum quantum number.**
m_l is the **orbital magnetic quantum number.**
m_s is the **spin magnetic quantum number.**

Each of these numbers can take on certain discrete values, subject to restrictions described below.

In the Bohr model, the total energy in any state is determined entirely by the value of n, but in quantum mechanics there are additional small contributions that depend on the other quantum numbers. The result is that a single Bohr energy state is replaced by a set of states, quite close together, and this multiplicity of values gives rise to the fine structure (page 542) of the spectrum lines. In addition to being the main factor in determining the energy, the principal quantum number n determines the value of r in Eq. 21.7 (page 539), which is no longer to be interpreted as an orbit radius but as the most probable distance of the electron from the nucleus. In the wave-mechanics view, each spectrum line is produced by photons emitted when the probability cloud changes from one configuration to another. As before, the principal quantum number n can take all integer values from 1 upward:

$$n = 1, 2, 3, \ldots \qquad \text{21.22}$$

Another result of quantum mechanics that differs from that of the Bohr theory relates to the *angular momentum* (page 538). Although quantum mechanics dispensed with the literal picture of electrons revolving about the nucleus, the definition of angular momentum has been generalized so that it applies to the new description. Here again, angular momentum is always quantized. In the new theory, the orbital quantum number determines the permitted values of the orbital angular momentum L, whose magnitude is given by

$$L = \sqrt{l(l + 1)} \, \frac{h}{2\pi} \qquad \text{21.23}$$

where l can take any integer value from 0 to $n - 1$:

$$l = 0, 1, 2, \ldots, n - 1 \qquad \text{21.24}$$

A further refinement is necessary. Since **L** is a vector, there is a limitation on its possible directions in space as well as on its magnitude. This **space quantization** is related to the fact that an atomic electron always finds itself in a magnetic field. Even if a field is not intentionally applied, one always exists because of the apparent motion of the charged nucleus around the electron. In the laboratory frame of reference, the circulation of the electron is equivalent to an electric current loop, and this gives the orbit the property of a small magnet. It follows that the total energy of the atom must depend on the orientation of this magnet with respect to the external field. Because the circulating charge is negative, its magnetic effect must be represented by a vector **m** whose direction is *opposite* to that of **L.** This vector **m** designates the orbital **magnetic moment.**

The orientation of **L** can be specified by its component along the line of the magnetic field **B.** In its normal position, **L** sets itself as close as possible to the direction of **B,** and this corresponds to the lowest state of the magnetic PE. To turn the orbit into any other orientation, work must be done. This increases the magnetic PE of the atom, which becomes a maximum when **L** has been turned around so that it points as nearly as possible in a direction *opposite* to **B.**

The possible values of L_B, the component of **L** along the direction of the magnetic field, are given by

$$L_B = m_l \frac{h}{2\pi} \qquad\qquad \textbf{21.25}$$

where the magnetic quantum number m_l can take any integer value from $-l$ through 0 to $+l$:

$$m_l = -l,\ -(l-1),\ \ldots,\ 0,\ \ldots,\ (l-1),\ l \qquad\qquad \textbf{21.26}$$

As an example, the allowed orientations of L corresponding to $l = 2$ are shown in Fig. 21.15.

Although the quantum numbers l and m_1 can have any of the integer values specified by Eqs. 21.24 and 21.26, atomic transitions between states are limited by what are called selection rules. Such rules, based on observation, were already part of the old quantum theory. From the many conceivable transitions that might occur, they pick out the ones that are actually observed. For the allowable changes Δl and Δm_l, the rules are:

$$\Delta l = \pm 1 \qquad\qquad \textbf{21.27}$$
$$\Delta m_l = 0 \text{ or } \pm 1$$

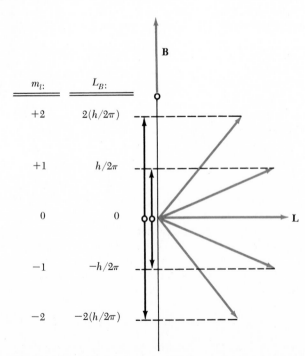

FIGURE 21.15 Possible orientations of the orbital angular-momentum vector **L** with respect to a reference direction in space when $l = 2$.

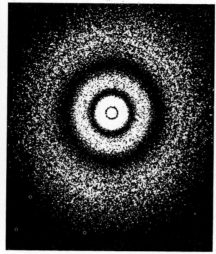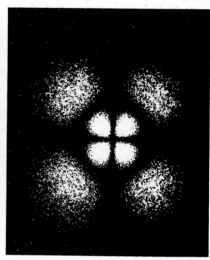

Cross-sections of computer-generated representations of the wave function of the hydrogen atom. The sharp Bohr orbits have been replaced by blurred shells of probability. Each picture corresponds to a specified set of quantum numbers. (*Courtesy of Dr. Robert Ehrlich.*)

Zeeman and his countryman H. A. Lorentz shared the 1902 Nobel Prize in physics for their work on the phenomenon of energy-level splitting.

According to modern quantum theory, these rules do not forbid transitions other than those described by Eq. 21.27; they merely imply that such transitions are of exceedingly low probability. Weak "forbidden" spectral lines are observed under certain conditions.

Since m_l can have a number of possible values for any n other than $n = 1$ (which corresponds to $l = 0$), an external magnetic field would be expected to divide the energy levels of an atom into several **sublevels.** It is actually observed that a strong field can split the spectrum lines into a number of close components whose spacing depends on the value of B. The phenomenon was discovered in 1896 by the Dutch spectroscopist P. Zeeman (1865–1943).

Practice set*

For an atom in the state $n = 3$, how many possible values are there (*a*) for l; (*b*) for m_l? (*c*) Find an algebraic expression for the value of L corresponding to $l = 2$.

The scheme outlined above still left unexplained certain subtle details of the fine structure and of Zeeman splitting, and it was soon recognized that a further variable property of the state of the electron must be involved. In 1925, the American physicists S. Goudsmit and G. E. Uhlenbeck suggested that the electron makes an angular-momentum contribution of its own, much as if it were a macroscopic body such as a planet, spinning on its axis. The magnitude of this **electron spin angular momentum,** L_s, is given by an expression like that of Eq. 21.25:

$$L_s = m_s \frac{h}{2\pi}$$

21.28

*Ans.: 3; 9; $\sqrt{6}\, h/2\pi$.

where m_s is the spin magnetic quantum number, which can take only the two values

$$m_s = -\tfrac{1}{2}, +\tfrac{1}{2} \qquad\qquad\qquad 21.29$$

This restriction on the values of m_s implies that an atomic electron may have one of two possible orientations in space: The spin angular-momentum vector may be either *parallel* or *antiparallel* to the magnetic field (Fig. 21.16). The slightly different energies of the atom corresponding to these two positions explain certain aspects of the observed structure of spectrum lines and of the Zeeman effect.

Even before the prediction of electron spin, direct experimental evidence for space quantization was found by O. Stern and W. Gerlach in Germany. They observed that, when a beam of silver atoms passed through a non-uniform magnetic field, the atoms were deflected in two sharply defined directions corresponding to opposite orientations of the angular-momentum vector of the atom. It was remarkable that, in 1928 when Dirac worked out a generalized wave equation that incorporates relativity, he found that the property described in the naïve picture as electron spin is a direct mathematical outcome of the theory.

Other kinds of elementary particles have spin (Chap. 23). Unlike the angular momentum of a system of finite size, spin is independent of any mechanical torque. It is a characteristic, unalterable property of each particle, like charge or rest mass.

21.13   ATOMIC STRUCTURE AND THE PERIODIC TABLE

When the chemical elements are arranged in the order of increasing atomic number, elements having similar properties are found to occur at regular intervals. A scheme of this kind, worked out in detail for the first time by the Russian chemist D. I. Mendeléef around 1869, is known as the **periodic system** of the elements. Mendeléef's work was based on atomic weight rather than atomic (proton) number. Only about 60 elements were known a century ago, but those discovered since that time have been successfully incorporated into the general arrangement.

The empirical features of the periodic system are discussed in some detail in Ref. 21.9.

One version of the periodic table of the elements in its present form is given in Table A.7 (page 635). The atomic (proton) number precedes the symbol of each element, with its gram atomic weight shown below. Elements in any vertical column constitute a **group** having similar physical and chemical properties. Across any horizontal row, the properties change progressively, and this progression begins again in the row following. For this reason, each row is known as a **period.** Interrupting the sequence of the eight principal groups are the groups of the **transition elements,** which share certain properties. The members of a given group are usually very similar. Those in the "lanthanide" and "actinide" groups are especially so and are placed outside the main body of the table.

The periodic system succeeded in bringing order into modern chemistry. It exhibited the existence of natural families of elements—the alkali metals, the halogen elements, the rare-earth metals, etc. It predicted the existence of undiscovered elements and even indicated what their properties would be. And yet for almost half a century the periodic classification remained

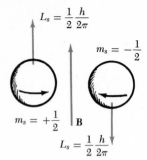

FIGURE 21.16 Schematic representation of the two possible orientations of the spinning electron relative to an indicated direction, such as that of a magnetic field.

an empirical scheme, lacking any fundamental explanation. Then, with the development of the Rutherford-Bohr model of the atom, and particularly with the interpretation of atomic number, scientists could begin to work out a correlation between chemical behavior and atomic structure.

The observed spectra of the elements provide the key to an understanding of the periodic arrangement. The experimentally obtained energy levels indicate that no more than one electron occupies any given state in any one atom. This generalization was stated by Pauli in 1925, in the form of his celebrated **exclusion principle:** *No two electrons in an atom can have the same set of values of their quantum numbers.* For example, if two electrons in a given atom have the same values of n, l, and m_l, they cannot have the same value for the remaining quantum number m_s. One of the electrons must have $m_s = +\frac{1}{2}$, the other $m_s = -\frac{1}{2}$, so that the two must be spinning in opposite directions. It should be mentioned that the exclusion principle can be applied to electrons in any small volume of space, even if this region is not the interior of an atom.

The periodic system of elements can be accounted for in detail with the guidance of the Pauli principle and energy considerations. In an atom, those

Wolfgang Pauli (1900–1958) was born in Austria but did most of his scientific work in the U.S. His brilliant theoretical contributions won him the Nobel Prize in physics in 1945.

Pauli (right), with Enrico Fermi (left) and Werner Heisenberg at a conference at Como, Italy, in 1927. (*Courtesy of Niels Bohr Library, American Institute of Physics.*)

electrons having the same value of the principal quantum number n are said to occupy the same **electron shell.** This suggests that they have about the same energy and are likely to be found at about the same distance from the nucleus. Of these electrons, some may have the same value of the orbital quantum number l and are described as belonging to the same

Note that the word orbital used in this sense is a noun.

subshell, or **orbital.** The orbitals are three-dimensional regions where the probability of finding the electron is high. Their size and shape are found by solving the wave equation.

It is now possible to determine the electron occupancy of the various orbitals by noting that in an atom in its normal (ground) state (page 543), the electrons tend to occupy the lowest energy levels available to them. For convenience, Eqs. 21.22 and 21.24 (page 555), and 21.29 (page 558) are collected in Table 21.1 (below).

Assignment of electron quantum numbers.

Begin with the hydrogen atom (proton number $Z = 1$). In the ground (unexcited) state, the single electron has $n = 1$, $l = 0$, $m_l = 0$. What about m_s? According to the given condition, the value could be either $-\frac{1}{2}$ or $+\frac{1}{2}$. However, the state with $-\frac{1}{2}$ is more likely, since it has lower energy.

In the construction of the next atom, helium ($Z = 2$), the exclusion principle prevents the second electron from being in the same low energy state as the first one. It can occupy the next higher state by having $n = 1$, $l = 0$, $m_l = 0$, as does the former, but with $m_s = +\frac{1}{2}$ instead of $-\frac{1}{2}$, and so the two electrons in a helium atom have oppositely directed spins. All the possibilities for $n = 1$ have now been used. The lowest shell of helium is an example of a **closed** (filled) **shell,** characterized by a spherical probability cloud. Such a configuration is especially stable. Helium and the other elements in the same column of the periodic table all have this property, which accounts for the almost complete chemical inertness of these atoms.

The next atom, lithium ($Z = 3$), has 3 electrons. The first two completely occupy the first shell, and so the third one must go into the $n = 2$ shell. Of the two possible subshells $l = 0$ and $l = 1$, it goes into the former because this one has lower energy. The table shows that $m_l = 0$; also, the added electron has $m_s = -\frac{1}{2}$. The chemical behavior of an atom is known to depend on the number of electrons in its outermost shell. The atoms of lithium and the other alkali elements have but one electron in the largest shell. This accounts for the high chemical activity of these elements, since the single electron is loosely bound and so is readily shared with atoms of other elements.

In the next atom in the periodic table, beryllium ($Z = 4$), a fourth electron having $m_s = +\frac{1}{2}$ is added; this exhausts all the possibilities for the $l = 1$ subshell of the shell characterized by $n = 2$. In the boron atom ($Z = 5$), the fifth electron must have $n = 2$ and $l = 1$. Here there are $2l + 1$, or 3, possible values for m_l and, corresponding to each of these, 2 values

---

TABLE 21.1   ALLOWED VALUES OF THE FOUR ATOMIC QUANTUM NUMBERS

1. Principal quantum number: $n = 1, 2, 3, \ldots$ (no upper limit)
2. For each n: $l = 0, 1, 2, \ldots, n - 1$ (n possible values)
3. For each l: $m_l = -l, -(l - 1), \ldots, 0, \ldots, (l - 1), l$ ($2l + 1$ possible values)
4. For each m_l: $m_s = -\frac{1}{2}, +\frac{1}{2}$ (two possible values)

for m_s. This makes a total of six configurations, which are incorporated one after another from boron through neon. In all, 8 electrons are allowed in the second shell, accounting for the fact that there are 8 elements in the second period of the chemical table. With neon there is again a closed outer shell. Like helium, this element is chemically almost completely inactive.

Fluorine, the element preceding neon, is one electron short of having a closed outer shell and so would be expected to have a great tendency to gain an electron to complete this shell. This explains the very energetic reactions observed when fluorine is allowed to combine with hydrogen or the alkali metals. The element following neon is sodium. The last of its 9 electrons is alone in the third shell. Thus sodium is like lithium in being an alkali element, except that the lone, outermost electron of sodium has two closed shells beneath it, whereas that of lithium has one.

Table 21.2 shows, with some omissions, the electron arrangement for the atoms of the first four periods. The third period is complete with argon (Ar), which has 8 electrons in its outermost shell. With the next element, potassium (K), there occurs a break in the sequence of adding electrons: This atom has an electron in its fourth shell, although its third shell is still not completely filled according to what is allowed by the exclusion principle. This occurs in the groups of transition elements four times, and it is found that these violations in the order of filling the shells are explainable in terms of energy considerations. Further, the unique magnetic behavior of iron, cobalt, and nickel is accounted for by an unfilled subshell in which several electrons have their spins aligned in the same direction.

TABLE 21.2 PART OF THE PERIODIC TABLE, SHOWING ELECTRON CONFIGURATIONS

PERIOD	SHELL	GROUP I	II	III	IV	V	VI	VII	VIII
1	1st	1 H 1							2 He 2
2	1st 2d	3 Li 2 1	4 Be 2 2	5 B 2 3	6 C 2 4	7 N 2 5	8 O 2 6	9 F 2 7	10 Ne 2 8
3	1st 2d 3d	11 Na 2 8 1	12 Mg 2 8 2	13 Al 2 8 3	14 Si 2 8 4	15 P 2 8 5	16 S 2 8 6	17 Cl 2 8 7	18 Ar 2 8 8
4	1st 2d 3d 4th	19 K 2 8 8 1	20 Ca 2 8 8 2					35 Br 2 8 18 7	36 Kr 2 8 18 8

Notes: Each box gives the atomic number and symbol of the element and the number of electrons in the 1st, 2d, etc., shell. The number of electrons in the outermost shell is the same as the group number, and the total number of electrons in each box is equal to the atomic number. For a list of the names corresponding to the above symbols see Table A.6 (page 634).

Even without a more detailed discussion it becomes apparent that the quantum theory is a powerful tool for the study of the atom and for understanding the material world.

Programmed review

Instructions: See page 18.

1. Describe Rutherford's concept of the structure of an atom.

   Most of the mass is concentrated in a relatively small nucleus, around which the electrons are distributed. [21.1]

2. What is meant by the proton (atomic) number of an element?

   The number giving the position of the element in the periodic table. It is also equal to the number of protons in the nucleus and to the number of electrons present in the atom in its normal, undisturbed state. [21.1]

3. Describe briefly how Bohr made use of the quantum concept in his theory of the hydrogen atom.

   He assumed (*a*) that the angular momentum of the electron is quantized and (*b*) that when the electron drops from one orbit to a lower one the difference in energy is radiated in the form of a photon. [21.5]

4. Define the ionization energy of an atom.

   The energy required to remove an electron from an atom in the ground state and leave the electron at rest a great distance away. This quantity is usually stated in electron volts. [21.6]

5. What was Heisenberg's aim in setting up his matrix theory?

   To represent the states of an atom exclusively in terms of observable quantities such as spectrum-line frequencies and intensities rather than nonoperational concepts such as the orbits and electron "jumps" used in the Bohr theory. [21.7]

6. Describe the basic nature of Schrödinger's wave mechanics.

   Starting with the de Broglie waves associated with the electron, he developed a wave equation. Solution of this equation for a particular system then yields the energy states of the system. [21.8]

7. State, in words, the principle of indeterminacy as applied to the position and momentum of a particle.

   In any given experiment, the more precisely the position is determined, the less precisely can the momentum be specified. In particular, the value of the product of the indeterminacies of the two quantities cannot be less than h. [21.9]

8. What is meant by statistical causality?

   The concept that, although microphysical events are not governed by strict causality, the average effects observed for macroscopic events can be described by precise laws.   [21.10]

9. State the interpretation of the Schrödinger wave function that is most widely accepted at present.

   The value, at any location, of the wave function of a particle gives the probability of finding the particle there.   [21.11]

10. Name the four quantum numbers that are needed to specify the energy state of an atom.

    Principal (n), orbital (l), magnetic (m_l), and spin (m_s).   [21.12]

11. What is meant by electron spin?

    The electron itself makes a small contribution to the total angular momentum of a given state, much as if it were spinning on its axis. The spin angular momentum is proportional to m_s.   [21.12]

12. State Pauli's exclusion principle.

    No two electrons in an atom can have the same set of values of their quantum numbers.   [21.13]

For further reading

21.1. *Andrade, E. N.* ''Rutherford and the Nature of the Atom.'' A charming account of the man and of the work with which he was associated, written by one of his students.

21.2. *Cline, B. L.* ''The Questioners: Physicists and the Quantum Theory.'' Bohr is the key figure in this well-told story of the contributors to the theory.

21.3. *Gamow, G.* ''Mr. Tompkins in Paperback.'' An updated edition of two of Gamow's witty, highly original fantasies. Chapters 7 and 8 deal with the quantum concept.

21.4. *Hoffman, B.* ''The Strange Story of the Quantum.'' Chapters 8, 9, and 13 give the flavor of quantum mechanics, but the rest of the book can be read profitably at this point. (Earlier sections were recommended in the reading reference on page 523.)

21.5. *Einstein, A.* ''The World as I See It.'' On pp. 66 and 67 the author expresses his hope for a return to strict determinism.

21.6. *Bohm, D.* ''Causality and Chance in Modern Physics.'' A criticism of the generally accepted view of determinism.

21.7. *Frank, P.* ''The Philosophy of Science.'' The question of free will is discussed on pp. 249–259.

21.8. *Gamow, G.* The Exclusion Principle, *Sci. Am.*, July, 1959, p. 74.

21.9. *Holton, G.*, and *D. H. Roller* ''Foundations of Modern Physical Science.'' Chapter 24 deals with the periodic table of the elements.

Questions and problems

21.1. An atom of gold has a diameter of about 2.6×10^{-10} m. If it is assumed that the atoms are practically in contact, (a) how many atom layers are there in the gold foil of thickness 6×10^{-7} m used in the Geiger-Marsden experiments? (b) Making use of the figure for the size of the gold nucleus computed in Worked example 21.1 (page 530), calculate the ratio of the cross-section area of the nucleus to that of the entire atom. Show from these results that the scattering of an alpha from more than one nucleus of a foil is extremely improbable.

21.2. Outline a qualitative reason for expecting a substance to show a line spectrum when in the gaseous state and a continuous spectrum when in the more condensed phases (liquid, solid).

21.3. Discuss the validity of the analogy between a Bohr atom and the solar system. Are there "permitted" orbits in the latter; spontaneous "jumps" between orbits; "excited states"? To what extent does the spinning electron have a counterpart in the solar system?

21.4. Show that the circulation of the electron in the first Bohr orbit of hydrogen is equivalent to a current of 1.06×10^{-3} ampere. What is the magnitude of the magnetic induction B produced at the nucleus by this current?

21.5. Compute the number of revolutions per second of the electron in the lowest Bohr orbit of hydrogen. How does the rate in the second orbit compare with this value?

21.6. The interval between the excitation of a typical atom and the emission of a photon is called the lifetime of the excited state and is of the order of 10^{-8} sec. According to the Bohr theory, about how many revolutions are made by the electron in the first excited state of a hydrogen atom before it returns to the ground state?

21.7. Calculate the radius of the fourth circular Bohr orbit of a hydrogen atom. Compare the result with Fig. 21.12.

21.8. The hydrogen spectrum series for which $p = 1$ and q takes the values 2, 3, 4, . . . is called the Lyman series, after the American spectroscopist who discovered it. Use Eq. 21.2 to compute the wavelengths of the first two lines of this series.

21.9. Compute the diameter of a hydrogen atom whose electron is in the orbit $n = 50$. Would a "particle" the size of such an atom be visible in an ordinary optical microscope? Would quantum theory have to be used to describe it? Show that this situation is an example of the correspondence principle.

21.10. On page 544 it was shown that the ionization energy of a normal hydrogen atom amounts to 13.6 eV, which is equivalent to saying that the absolute value of the energy of the $n = 1$ level of such an atom is -13.6 eV. Using Eq. 21.11 (page 541), determine the absolute energy values for the second and third levels.

21.11.   *The Franck-Hertz experiment.* From the result of Prob. 21.10, how much KE, in electron volts, must an electron have in order to raise a hydrogen atom from its lowest state to the first excited state on colliding inelastically with it? What is the speed of such an electron? Experiments of this kind performed by the German physicists J. Franck and G. Hertz in 1914 gave the first direct evidence for the quantization of the energy levels of atoms.

21.12.   The electrons in the second shell of an atom are shielded from the positive charge Ze of the nucleus by the presence of the two electrons that occupy the first shell. In effect, this means that the $n = 2$ electrons are held by a central charge $+(Z - 2)e$. Using this idea, modify Eq. 21.7 so that it gives the radius of the second shell of any atom of atomic number Z. Use the result to calculate the value for an atom of beryllium ($Z = 4$). Note that the calculation is oversimplified because it fails to take account of the mutual repulsion of any other electrons that may be present in the second shell. This effect, however, is relatively small.

21.13.   A free hydrogen atom, initially at rest, emits a photon corresponding to the second line of the Balmer series. (*a*) What is the energy of the photon, in electron volts? (*b*) What is its momentum? (See page 512.) (*c*) Using conservation of momentum, compute the recoil speed of the atom.

21.14.   Work out the Bohr theory for positronium (page 515). Figure 21.17 shows the electron-positron pair bound together in the $n = 1$ orbit as they revolve about their common center of mass. Equate the total angular momentum of the pair to $nh/2\pi$, and compute the radius of the positronium atom in its lowest energy state. Compare with that of the hydrogen atom.

21.15.   If the position of an electron can be determined only to the nearest angstrom (10^{-10} m) in a certain experiment, find the order of magnitude of the uncertainty in the speed of this particle, expressed as a fraction of c.

21.16.   In a given experimental arrangement, an elementary particle can be observed at intervals as short as 10^{-7} sec. What is the order of the indeterminacy of the energy of the particle, expressed in electron volts?

21.17.   Using the information given in Prob. 21.6, and the fact that the numerical magnitude of the energy of an atomic state is a few electron volts, estimate the ratio of the indeterminacy of the energy of the electron to its total energy. Is this consistent with the fact that spectrum lines are normally quite sharp?

21.18.   Describe the position and directional sense of the earth's spin angular momentum vector. How many atomic units of spin ($h/2\pi$) does the earth have? Take the earth's mass to be 6×10^{24} kg, its radius 6.4×10^6 m, and the time for one rotation 8.6×10^4 sec. For purposes of computing the spin angular momentum, the

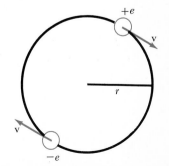

FIGURE 21.17 The positronium atom.

entire mass may be considered concentrated at a distance from the axis equal to 0.63 of the radius.

21.19. An atomic electron is in the quantum state $n = 3$. (a) How many subshells does this state contain? (b) How many possible values of m_l are there altogether? (c) How many distinct energy values are possible for this electron?

21.20. From the statement of conservation of angular momentum on page 539, show that Kepler's second law (page 177) is equivalent to this principle.

21.21. By comparing Eqs. 21.23 and 21.15 show that the orbital angular-momentum vector **L** can never be exactly parallel or antiparallel to the magnetic field **B,** even when $m_l = l$. Physically, this is a consequence of the inherent indeterminacy in direction of **L.**

21.22. Using Table 21.2, make rough sketches, similar to Fig. 21.10 (page 538), showing the electron arrangement in a normal atom of (a) sodium, (b) magnesium, (c) neon.

Chapter 22
THE NUCLEUS

In the preceding chapter it was shown that the optical and chemical properties of the elements can be accounted for on the basis of the electron configurations of their atoms as described by quantum mechanics. The atomic nucleus, through its positive charge, determines the electron energies, which are of the order of only a few electron volts, so that electron configurations can be changed or disrupted by moderate amounts of energy such as those available in electrical discharge tubes. To alter the nucleus, however, usually requires energy expenditures of much higher orders of magnitude, as furnished typically by high-energy particle accelerators.

This chapter outlines the main features of nuclear structure and behavior, based on the quantum theory. However, in contrast with the completeness of the description of the outer portions of the atom, many fundamental questions relating to the nucleus remain without satisfactory answers.

NUCLEAR STRUCTURE

22.1 ISOTOPES

Beginning with the work of Dalton and of Avogadro it became possible for chemists to estimate the relative masses of the atoms of the elements (their atomic weights) by ordinary chemical procedures. Such methods necessarily make use of finite amounts of material, which consist of myriads of atoms; hence, in making atomic-mass determinations in this way it was assumed that the masses of all atoms of a given element are identical. In 1907, J. J. Thomson modified his e/m apparatus (page 440) and found a way to measure the masses of individual atoms and molecules. The unexpected result, described below, was that *the atoms of a given element do not all have the same mass.*

Thomson's apparatus is shown schematically in Fig. 22.1. One of the more recent instruments, designed and used by the American physicist K. T. Bainbridge, is represented in Fig. 22.2. In the Bainbridge device, positive

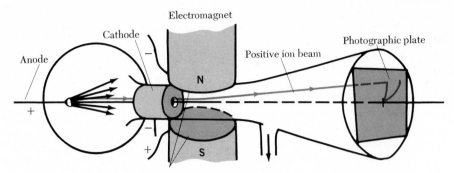

FIGURE 22.1 In Thomson's mass spectrograph, electric and magnetic fields detect the positive ions in two perpendicular directions. All ions having a given charge-to-mass ratio hit the plate on a given parabolic curve. The mass of the ion can be computed from the measured spread of the curve.

ions of the desired kind are produced in an electric discharge (not shown in the diagram) and accelerated in an electric field. Then they enter the "velocity selector" which allows ions having only a particular speed to continue along the axis and enter the main part of the instrument where the magnetic field sends them into circular orbits (page 439). If particles of various masses are present in the ion beam, each kind is directed into a circle whose radius is proportional to the mass (see Prob. 22.4). After moving through a half circle the particles hit a photographic plate, and the position of each trace gives the value of the radius, from which the mass of the ion can be calculated. The relative amounts of each ion present in the sample can be found by measuring the darkness of each image or, more accurately, by replacing the plate with collecting electrodes that measure the ion currents. All these devices have the function of sorting out groups of ions according to mass, analogous to the separation of a mixture of wavelengths by spectroscopic instruments. This is how the ion devices came to be called **mass spectrographs.** Sometimes the term **mass spectrometer** is used for the kind that records the ions electrically rather than photographically.

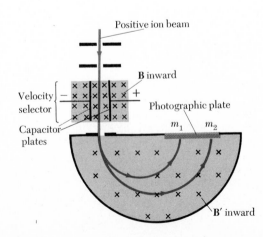

FIGURE 22.2 The Bainbridge mass spectrometer (schematic). All ions of a given charge-to-mass ratio strike the plate at a given place.

F. W. Aston (1877–1945), a student of Thomson, received the 1922 Nobel Prize in chemistry for his atomic-mass measurements.

In his pioneering experiments, Thomson detected such particles as singly ionized hydrogen atoms (H^+) and molecules (H_2^+), oxygen atoms (O^+) and molecules (O_2^+), singly ionized carbon monoxide molecules (CO^+), and many others. Aston used his modification of the Thomson apparatus in an attempt to determine the mass of the atoms of neon (chemical atomic weight 20.183 g/mol), but the plate showed no trace corresponding to this mass. Instead, there was a well-defined image for particles of mass 20 and a fainter one for mass 22. Thomson and Aston were forced to conclude that neon is composed of *two kinds* of atoms that are not distinguishable by their spectra but occur in nature in a fixed ratio, and that the chemically determined atomic weight of 20.183 obtained from bulk measurements is merely the average value for the natural mixture of the two.

Subsequent experiments fully confirmed the idea that *most elements are mixtures of atoms of different masses,* which are a result of differences in the constitution of the nuclei. In order to discuss conveniently the various distinct kinds of atoms, whether they belong to the same element or not, a definition of the atom species is needed. The term **nuclide** is used for any atom species characterized by the composition of its nucleus. The various nuclides belonging to a given element are called **isotopes** of that element.

"Isotope" comes from the Greek words isos, the same, and topos, place.

About 20 elements have more than 6 isotopes each. Tin (Sn) has 10 isotopes and, counting unstable (radioactive) nuclides, some elements have as many as 30. Of all known nuclides, 280 are stable, and about 1,400 others are radioactive. Before the discovery of isotopes, scientists believed that there were fewer than a hundred different atom varieties, corresponding to one for each of the known elements. Now nearly 2,000 are known. This complicates matters from the fundamental structural point of view but leaves chemistry virtually undisturbed, because the several isotopes of any given element have the same electron configuration, which determines their chemical properties. However, the chemical reaction *rates* of the various isotopes are observed to be slightly different because of the difference in nuclear mass.

With regard to the physical properties, the slight differences in mass can be used to bring about a partial separation of the isotopes. For instance, the lighter isotopes evaporate somewhat more profusely from a liquid, or diffuse more rapidly through a porous wall, or are transported in greater number in an electrolytic cell. By passing the material repeatedly through such processes, greater and greater separation is attained, but the techniques are time-consuming and expensive. Another possibility is to use an electromagnetic means of separation, as in the mass spectrometer. Although it, too, has great difficulties, this method and a diffusion process were the two schemes used to separate the isotopes of uranium for the first atomic bombs.

The precise determination of nuclide masses is important in modern physics. Knowledge of high-precision values led to the first observational proof of mass-energy conservation (Sec. 22.9) and to the use of nuclear energy. In 1960, an international commission established a basis for an atomic (nuclidic) mass scale by taking the mass of the neutral carbon-12 atom to be exactly 12. Earlier, two separate scales based on oxygen were in use (Ref. 22.1). They differed from the new scale by only about 0.03 percent.

TABLE 22.1   STABLE ISOTOPES OF SOME LIGHT ELEMENTS

NUCLIDE	MASS OF NEUTRAL ATOM, IN u	RELATIVE ABUNDANCE, %
$_1H^1$	1.0078252	99.98
$_1H^2$	2.0141022	0.02
$_2He^3$	3.0160299	0.00015
$_2He^4$	4.0026036	≈100.
$_3Li^6$	6.015126	7.5
$_3Li^7$	7.016005.	92.5
$_4Be^9$	9.012186	≈100.
$_5B^{10}$	10.012939	18.4
$_5B^{11}$	11.0093051	81.6
$_6C^{12}$	12 exactly	98.9
$_6C^{13}$	13.0033543	1.1
$_7N^{14}$	14.0030744	99.62
$_7N^{15}$	15.000108	0.38
$_8O^{16}$	15.9949149	99.76
$_8O^{17}$	16.999133	0.04
$_8O^{18}$	17.9991598	0.20

One-twelfth of the mass of the carbon-12 atom is defined as one **unified atomic mass unit,** abbreviated u. Its relation to the kilogram is given by

$$1 \text{ u} = 1.6598 \times 10^{-27} \text{ kg} \qquad\qquad 22.1$$

Table 22.1 gives the relative abundances and masses of the nuclides belonging to the lighter elements, up to and including oxygen.

In order to specify a given nuclide, two integers are attached to the chemical symbol of the atom. For instance, in place of the symbol Li for a lithium atom, either $_3Li^6$ or $_3Li^7$ is used, depending on which isotope is intended. The *subscript,* customarily written to the left of the symbol, is the proton number Z (page 530) of the element; the *superscript,* on the right, is the mass number A. The **mass number** A is the nearest integer to the actual mass value expressed in atomic mass units. A complete symbol, written as described here, stands either for the specified nuclide as a substance or, more particularly, for a single atom or ion of this substance. The exact meaning in any given case must be clearly indicated. Sometimes, in written or spoken descriptions, a given nuclide is indicated by stating the name of the element followed by the mass number of the isotope under discussion, as, for example, boron 10 or radium 226. Even though it does not appear explicitly, the atomic (proton) number can be found by consulting the periodic table (see Table A.7, page 635).

An alternative meaning of the mass number is given in Table 22.3, page 572.

The atomic masses are for the most part quite close to whole numbers, particularly for the lighter elements—a generalization called Aston's **whole-number rule.** The existing departures, known as **mass defects,** cannot be attributed to the presence of still other isotopes, since the figures are obtained from mass spectra and thus refer to individual atoms. The reason for the whole-number rule and mass defects will be brought out below.

22.2 COMPOSITION OF THE NUCLEUS; THE NEUTRON

The whole-number rule strongly suggests that the nucleus consists of a number of protons sufficient to account for the nuclear mass plus enough electrons to make the total charge of the nucleus $+Ze$. If this conclusion were true, the nucleus of $_3\text{Li}^7$ should contain 7 protons to make the nuclear mass of about 7 u, and 4 electrons to reduce the effective nuclear charge to $+3e$. There are, however, several objections to this view, one of them coming from the indeterminacy principle: If an electron is to be confined to a nucleus, whose linear dimension is of the order of 10^{-15} m, the indeterminacy of its momentum would have to be (Eq. 21.20, page 550)

$$\Delta p \approx \frac{h}{\Delta q} \approx \frac{6.6 \times 10^{-34}}{10^{-15}}$$

which is of the order of 10^{-18} kg-m/sec, corresponding to an energy of around 10^3 MeV. If this is the *indeterminacy* in the energy of a nuclear electron, the energy values themselves must be of this order of size. However, electrons in atoms are never observed to have energies of this magnitude, and so the presence of electrons in the nucleus must be ruled out.

Chadwick won the Nobel Prize in physics in 1935 for his discovery of the neutron.

The problem of nuclear constitution was solved with the discovery of the neutron by the British experimenter James Chadwick in 1932. Previously, it was known that when a light metal such as beryllium is struck by alpha particles, a highly penetrating radiation is produced. It was assumed that this was of the nature of hard gamma radiation. However, when paraffin or water was placed in the path of these rays, fast protons were thrown off, and the measured energy of these particles was much greater than the computed gamma-ray energy.

As early as 1920 a U.S. chemist, W. D. Harkins (1873–1951), predicted the existence of a neutral subatomic particle.

After performing additional experiments, Chadwick found agreement with the observations by assuming that the penetrating radiation consisted of particles rather than electromagnetic waves. It was necessary to infer that these particles (1) were electrically neutral, since charged particles would be slowed down drastically as they produced ionization in passing through matter, and (2) were of about the same mass as protons, since they passed on most of their KE to them in a collision (see Worked example 22.1 on the next page).

The neutron accounts not only for the experimental results described above but also for the observed properties of nuclei. It leads to a model which depicts any nucleus as an aggregate of two kinds of particles, protons and neutrons. Neutrons and protons collectively are called **nucleons.** The rest masses of these particles are not very different:

$$m_n = 1.008665 \text{ u} \quad \text{(neutron)}$$
$$m_p = 1.007277 \text{ u} \quad \text{(proton)}$$

22.2

The number of protons, Z, determines the charge on the nucleus and hence the number of external electrons in the neutral atom. The neutrons that may be present in a nucleus add to its mass but do not affect its total charge (see Table 22.2). Although the neutron by itself cannot be an atomic nucleus, it may be considered the "zeroth" element, with the symbol $_0n^1$.

The **neutron number** N of a given nuclide, together with the proton number Z, determines the mass number A defined on page 570. Because

TABLE 22.2 COMPOSITION OF SELECTED NUCLIDES

NUCLIDE	COMPOSITION OF NUCLEUS		ELECTRON COMPLEMENT
	Protons	Neutrons	Number of electrons
$_1H^1$	1	0	1
$_1H^2$	1	1	1
$_2He^3$	2	1	2
$_{26}Fe^{56}$	26	30	26
$_{92}U^{238}$	92	146	92

the mass of a neutron and the mass of a proton are both very close to unity, the mass number A also represents the total number of nucleons in the nucleus:

$$A = N + Z \qquad\qquad 22.3$$

The meanings and functions of these numbers are summarized in Table 22.3.
Electrons are only about 1/1840 as massive as nucleons, the electron mass being

$$m_e = 0.0005486 \text{ u} \qquad\qquad 22.4$$

Since the neutron cannot be given the form of an ion, its mass cannot be measured by mass spectroscopy. The most accurate determinations come from nuclear changes in which these particles are involved. See Worked example 22.2, page 574.

Practice set*   State the number of (a) neutrons, (b) protons, and (c) external electrons in a neutral atom of $_7N^{15}$.

Worked example 22.1   Give a qualitative argument to support the statement on page 571 that, in an elastic collision with a body at rest, the greatest energy transfer occurs when the two bodies are of comparable mass.

Solution   This statement can readily be given an exact proof by applying the laws of conservation of momentum and energy, but it can be made plausible by intuitive reasoning in this way: Maximum recoil speed of the struck body can result only if the collision takes place centrally. A light

Ans.: 8; 7; 7.

TABLE 22.3 CHARACTERISTIC NUCLEAR NUMBERS

Proton number, Z	Number of protons in the nucleus; or
	Number of external electrons in a neutral atom of the element in question
Neutron number, N	Number of neutrons in the nucleus
Mass number, A	Nearest integer to the nuclear mass, in u; or total number of nucleons in the nucleus: $A = Z + N$

object, such as a marble, makes a direct, elastic collision with a heavy object, such as a bowling ball, that is at rest. The marble rebounds in the opposite direction with virtually no change in speed and the bowling ball acquires practically no KE. In the opposite extreme, suppose a bowling ball makes a direct hit on a stationary marble. The marble is knocked forward, but since its mass is small the KE acquired is small. Consequently, it would appear that the most efficient transfer of energy occurs when the two bodies have comparable masses. The exact analysis shows that, for example, when the striking body has one-tenth the mass of the other, the KE transferred is about one-third of the original amount. When the striking body has 10 times the mass of the other, only about one-fortieth is transferred. But when the masses are equal, *all* the initial KE goes to the struck body, and the striking body stops dead.

22.3 NUCLEAR BINDING ENERGY; MASS DEFECTS

A $_2$He^4 nucleus is the same as an alpha particle (page 527).

The existence of mass defects among the known stable nuclides (page 570) is revealed through the fact that the nuclear masses are not exactly equal to the sums of the masses of the constituent protons and neutrons. For example, when the values given by Eq. 22.2 are used, the total mass of the 2 protons and 2 neutrons that make up a $_2$He^4 nucleus is $2(1.008665 + 1.007277) = 4.031884$ u. However, the measured mass of this nucleus is equal to the value given for the complete atom in Table 22.1, with two electron masses (Eq. 22.4) subtracted from it; this amounts to $4.0026036 - 2(0.0005486) = 4.001506$ u. The difference is the mass defect of about 0.03 u. The mass of the aggregate is *less than* the sum of the masses of its constituents.

The binding energy of a system was defined on page 544.

Discrepancies of the kind described are connected with the *binding energy* of the nucleus. Like the atom as a whole, the nucleus has a binding energy of its own, representing the PE that is lost when the constituents come together. It will be shown later that this field of force is many orders of magnitude stronger than the Coulomb field that binds electrons to an atom. The binding energy comes at the expense of the total mass of the system, as described by Einstein's principle of mass-energy equivalence (page 517): The existence of a binding energy ΔE is accompanied by a decrease Δm in the total rest mass of the particles, such that

$$\Delta E = (\Delta m)c^2 \hspace{4cm} \textbf{22.5}$$

Binding energies are usually expressed on an electron-volt scale rather than in joules, and so it is useful to compute the equivalent of one atomic mass unit on such a scale. According to Eq. 22.1, $1 \text{ u} = 1.66 \times 10^{-27}$ kg, and so the equivalent energy amounts to $\Delta E = (1.66 \times 10^{-27})(3 \times 10^8)^2 = 1.49 \times 10^{-10}$ joule. Then, since $1 \text{ MeV} = 1.60 \times 10^{-13}$ joule,

$$1 \text{ u} = \frac{1.49 \times 10^{-10}}{1.60 \times 10^{-13}} \text{ MeV, or } 1 \text{ u} = 931 \text{ MeV} \hspace{2cm} \textbf{22.6}$$

Thus the binding energy of the alpha particle is the equivalent of the 0.03 u found above, or $(0.03)(931) = 28$ MeV. This is of the order of a million times the binding energy of an electron in an atom. For instance (see page 544) the ionization energy of the hydrogen atom is only 13.6 eV.

Worked
example
22.2

One of the most accurate ways of measuring the neutron mass is by *photodisintegration* of the **deuteron,** or deuterium nucleus (page 542). It is found that the deuteron can be split into its constituents, a proton and a neutron, by a gamma-ray photon that can supply an amount of energy of at least $E_B = 2.22$ MeV. What value does this give for the neutron mass? (The mass equivalent of the recoil energy of the particles is negligible.)

Solution   If the particle masses are expressed in u and the binding energy in MeV,

$$m_p + m_n - m_D = \frac{2.22}{931}$$

Next, substitute the numbers for m_p and m_D: Take m_p from Eqs. 22.2, and get m_D by subtracting the electron mass from that of $_1H^2$ in Table 22.1. The result is $m_n = 1.00867$ u, which agrees with the value quoted in Eqs. 22.2.

22.4   THE NUCLEAR FORCE

Perhaps the most remarkable fact about the atomic nucleus is that it stays intact, because, in the absence of other influences, a collection of protons and neutrons would not be expected to remain closely crowded together in a small region of space. In order to account for the existence of stable nuclei, physicists had to postulate a nuclear force whose nature is still not adequately understood. In 1704, Newton made the remarkably prophetic statement: "The attractions of gravity, magnetism and electricity, reach to very sensible distances, and so have been observed by vulgar eyes; and there may be others which reach to so small distances as hitherto escape detection." From observation, it is known that nuclides of various degrees of stability exist. This information, together with the scheme of nuclear constitution described in the preceding sections, leads to certain models of the nucleus that account for many observed features in a satisfactory way.

There is clear evidence that the nuclear force is essentially different from gravitational or electrostatic forces. Experiments on the scattering of protons and of neutrons from nuclei show that exceedingly strong forces of attraction arise between nucleons when they come within distances of the order of the nuclear radius, and that the force is of about the same strength whether the nucleons concerned are charged or neutral. On the other hand, the Coulomb force between protons operates to disperse these particles rather than hold them together, and the gravitational attraction between nucleons is much too feeble to account for the observed binding energies. This point about the nuclear force can be summarized by saying that as long as the particles are within a nucleus, the (attractive) forces between two protons, between two neutrons, and between a proton and a neutron are very nearly equal. For this reason the nuclear force is said to be *charge-independent*.

Forces between nucleons.

Another characteristic of the nuclear force that distinguishes it from either gravitational or electrostatic force is its *short range*. The range is, in effect, not much greater than the closest distance to which nucleons can be made to approach.

If the distance between two ordinary, macroscopic particles is doubled, their gravitational attraction decreases to one-fourth its former value, and so does their mutual electrostatic force if they are charged. However, if the distance between two bound nucleons is doubled, the nuclear force between them becomes negligibly small. Experimental evidence shows that it requires about the same expenditure of energy to remove a neutron from a light nucleus containing only a dozen or so nucleons, as from a heavy one containing a hundred or more. Such observations lead to the conclusion that a given nucleon is attracted only by its nearest neighbors and not by the other nucleons in the nucleus.

22.5 NUCLEAR MODELS

Although the discussion above was concerned mainly with forces, it is preferable for most purposes to consider the energies associated with these forces. For example, it is useful to compute *the average binding energy per nucleon* for the various stable nuclides. In each case, this is the binding energy (as computed from the mass defect) divided by the total number of nucleons present. A graph of this quantity against mass number A is shown in Fig. 22.3. With a few exceptions, the plotted points fall on a smooth curve that rises steeply over about the first 20 nuclei, after which it levels off and declines gradually at higher mass numbers. The nuclei near the crown of the curve are those requiring the greatest amount of energy for their disruption and hence are the most stable ones. On the other hand, elements near the two extremes of the mass range are the least stable, and it will be shown later that certain of these elements are connected with processes for the large-scale release of nuclear energy.

The graph indicates that for all except the first 20 or so nuclides the binding energy per nucleon amounts to roughly 8 MeV. This, then, is the

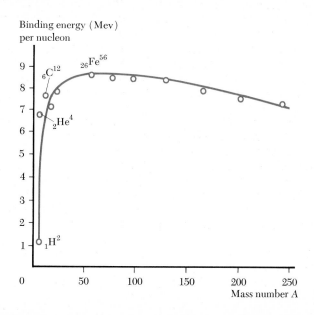

FIGURE 22.3 Binding energy per nucleon of the stable nuclides, plotted against mass number. Small circles represent data for a few nuclides.

average amount of work needed to remove a nucleon from an intermediate or heavy nucleus. The gradual decrease of the binding energy per nucleon toward large values of A is due to the net Coulomb repulsion between a proton in the outer part of the nucleus and all the others. For larger nuclei, the magnitude of this force increases. A calculation shows how large such a force can be:

Worked example 22.3

Compute the magnitude of the Coulomb force on a proton at the "surface" of the nucleus of an atom of $_{92}U^{238}$. Assume that the remaining protons are uniformly distributed throughout the nucleus, which may be taken to be a sphere of radius 9×10^{-15} m.

Solution   The remaining 91 protons have a total charge of (91) (1.6×10^{-19}) coulomb, and this charge may be considered to be concentrated at the center of the nucleus. According to Coulomb's law (page 377), the force of repulsion is

$$F = \frac{(9 \times 10^9)(91)(1.6 \times 10^{-19})^2}{(9 \times 10^{-15})^2} = 260 \text{ newtons}$$

or more than 58 lb force. This is equivalent to a pressure over the cross-section of the proton of $260/(10^{-15})^2 = 2.6 \times 10^{32}$ N/m², or about 10^{25} lb/in².

Additional regularities among the stable nuclei are shown by plotting the neutron number N against the proton number Z, as in Fig. 22.4. Here, the first 20 or so elements stand apart from the rest: They lie close to the sloping line $N = Z$, which means that these nuclei, which are the most stable, contain nearly equal numbers of neutrons and protons. For example, the highly stable nuclide $_8O^{16}$ has $N = Z = 8$. For the intermediate and heavier nuclides, the path of stability veers progressively farther from the sloping line, with N increasing faster than Z. For instance, for $_{82}Pb^{207}$, $N = 125$, whereas $Z = 82$; this nucleus contains more than 1.5 times as many neutrons as protons.

The near equality of N and Z for the light nuclides and the excess of neutrons for the heavier ones can be accounted for by the **shell model** of the nucleus, which assumes that the nucleons are arranged in energy groups, analogous to the shell structure of the electron complement of the atom. The Pauli exclusion principle (page 559) is assumed to apply to nucleons, with the result that only certain energy states are possible. The principle governs the motions of the protons and, independently, those of the neutrons. One important feature of the shell model is that it explains why nuclei having either N or Z equal to 2, 8, 20, 50, or 82 are especially stable. These "magic numbers" apparently represent closed shells.

Certain nuclear states are favored because of their low energy, and if each of these states can accommodate one of each kind of nucleon, there is a tendency for neutrons and protons to be present in equal numbers, as in the lighter nuclei. In the heavier nuclei, each proton is strongly repelled by the aggregate of the others, while the nuclear force, which would tend to hold things together, is only of short range. Hence the heavier nuclei

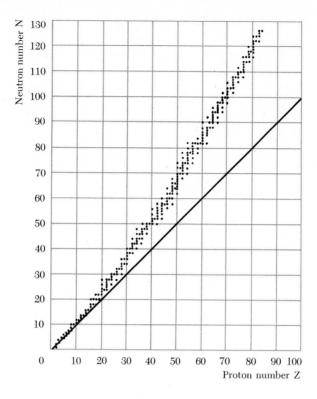

FIGURE 22.4 A plot of neutron number against proton number for the stable nuclei. Points lying above the sloping line represent nuclides having more neutrons than protons.

would be disrupted if it were not for the presence of excess neutrons acting as a binder for the protons. When the atomic number approaches 90, the nuclear structure becomes quite unstable, regardless of the number of neutrons that are present. As a result, the nuclei of the heaviest elements break down spontaneously, a process called **radioactivity** (Sec. 22.7).

An alternative model of the nucleus, called the **liquid-drop model,** was proposed by Bohr. This model treats the motions of the nucleons statistically, and it explores the correspondence between the nuclear force and the cohesive forces of the molecules of a liquid. Each nucleon, surrounded by other nucleons, interacts primarily with its nearest neighbors in a way quite analogous to what happens on the molecular scale in a liquid. The model is especially suitable for describing the nucleus while it is in process of changing its composition or state (see photograph on page 596).

22.6 POTENTIAL BARRIER; EXCHANGE FORCES

About 1928 the Russian-born American physicist George Gamow (1904–1968) proposed thinking of the nucleus in terms of the potential energy of a positively charged particle approaching from the outside. As in Rutherford's scattering experiments, such a particle, moving toward a nucleus, would be opposed by a Coulomb force of increasing magnitude as long as the distance was not too small. This force law is represented graphically by the dashed curve of Fig. 22.5. However, this law, which indicates that the repulsive force becomes very large for close approach (as shown by the calculation in Worked example 22.3, page 576) cannot continue to represent the net effect of what happens down to distances as small as 10^{-15} m. Gamow adopted the idea that, at a certain distance of approach, the curve representing the net force begins to depart from the Coulomb form and dips over sharply, as represented by the curve in color in Fig. 22.5. The part of the curve below the horizontal axis represents a force of *attraction.*

Experiments on the scattering of neutrons yield the best values for the nuclear force, since the approach of these particles to a nucleus is not complicated by the presence of a Coulomb force. The range of the nuclear force may be taken to be about the distance at which the resultant force curve dips below the axis, and the nuclear radius may be taken to be of this same order of size. Experimental values of the radius differ slightly, depending on the method used, but all results may be represented approximately by the relation

$$r = (1.2 \times 10^{-15})A^{1/3} \quad \text{m} \qquad\qquad \textbf{22.7}$$

where A is the mass number. The fact that the radius increases only with the $\frac{1}{3}$ power of the mass number shows that the range of nuclear sizes is not very great. Thus, from hydrogen to uranium, the nuclear radius increases by only a factor $(238/1)^{1/3} = 6.2$.

Practice set   Identify the nuclide that has a nuclear radius twice that of carbon 12. Make use of the periodic table (Table A.7, page 635).

An interesting conclusion follows from Eq. 22.7. Assume the nucleus to be a roughly spherical cluster of nucleons. The volume of a sphere is proportional to the cube of its radius, and by cubing both sides of the equation we see that the volume must be proportional to A. Then, since

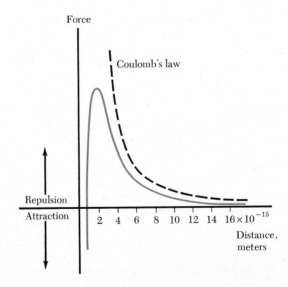

FIGURE 22.5 Force between a nucleus and a positive particle.

all nucleons have about the same mass, it can be concluded that they are packed into the nucleus uniformly. There is another way of saying this: In all nuclides, the *density* of nuclear matter is constant. The numerical value is enormous. One teaspoonful of nuclear matter would weigh over a billion tons (see Prob. 22.23).

In discussing nuclear structure, the nucleons should not be thought of as concentrated pellets but as smeared-out wave functions. In place of describing the variation of force with distance, it is more useful to represent the variation of the PE of an incoming particle in the field of the nucleus. The curves representing this energy have the same general shapes as the force curves shown in Fig. 22.5.

Certain aspects of the behavior of the nuclear field may be visualized with the help of a mechanical analog. Imagine a curve similar to Fig. 22.5 rotated around the vertical axis, so that it generates a surface resembling a volcanic cone and its central crater (Fig. 22.6a). A ball rolled toward the crater would be the counterpart of a positive particle approaching a nucleus. If the approaching ball is not moving very fast or not heading directly for the crater, it will roll part way up the side of the cone and roll off again in a different direction, in the manner of an alpha particle in the Rutherford experiments (Fig. 21.2b, page 529). On the other hand, if the ball is projected directly toward the crater with enough energy to surmount the potential barrier, it can drop into the crater and become part of the nucleus. This mechanical analog is incomplete in some respects because it is based on classical rather than on quantum principles. It will be shown later, for example, that there is some chance of a positive particle entering or leaving the nucleus even if its initial energy happens to be less than that corresponding to the height of the **potential barrier.**

By contrast with the behavior of a positive particle, a neutron approaching the nucleus is not confronted with a potential barrier and, if headed in the right direction, can drop into the central well (Fig. 22.6b) even if its KE is small. This suggests why neutrons are very effective agents for bringing about nuclear transformations.

Mechanical models of the kind described are useful in recalling and correlating the observations but they fail to provide a more fundamental description of the nature of the nuclear force (or potential). This question remains largely unanswered, but some speculations are noteworthy. Heisenberg found a significant clue when he examined mathematically the corresponding question of the nature of Coulomb forces. Theory indicates that the electrostatic force between two charged particles can be described formally by assuming that photons are continually passing back and forth between the two. This suggestion prompted the Japanese physicist H. Yukawa to investigate the possibility of accounting for nuclear forces, too, as **exchange forces** and led to his prediction that a new particle, about 200 times as massive as the electron, was involved. Such a particle, now known as a π meson (or **pion**), was later discovered in cosmic radiation and can be produced in great numbers in particle accelerators.

Since about 1955, the American physicist Robert Hofstader and others have been examining the structure of nucleons by scattering high-energy electrons from them. They find that a nucleon may be considered to have a core surrounded by a cloud of pions. These exchange

Two Nobel physics awards were connected with these problems: Yukawa's in 1949 and Hofstader's in 1961.

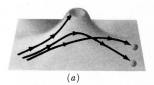

(a)

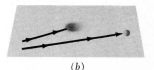

(b)

FIGURE 22.6 Mechanical analogs of the behavior of a nucleus toward incoming particles. (a) For a positive particle, the nucleus acts like a mound with a central crater. (b) For a neutral particle, it behaves like a well in a level plane.

particles are called *virtual* pions since they are not directly observable. The rapid exchange of virtual pions between two nucleons is, in this view, responsible for the strong attractive force. Thus the proton and neutron may be thought of as two states of the same particle.

NUCLEAR TRANSFORMATIONS

So far, this chapter has dealt mainly with stable atomic nuclei, which preserve their composition for indefinitely long times. It was mentioned, however, that hundreds of unstable nuclides are known. In the following sections it will be shown that certain naturally occurring nuclei are inherently of this kind, and others can be made unstable by means at our disposal. Changes that follow nuclear instabilities are responsible for the release of "atomic energy."

22.7   NATURAL RADIOACTIVITY

The discovery of radioactivity is an excellent example of how wrong assumptions sometimes lead to results of value. When Röntgen's first experiments with x rays were reported in France in January 1896, it occurred to the physicist A. H. Becquerel (1852–1908) that penetrating radiations might be connected with luminescent materials, since x rays were produced where the glass of a cathode-ray tube showed its strongest fluorescent glow. Becquerel began a series of tests of minerals that were known to show phosphorescence after exposure to light. Samples were placed in contact with a photographic plate wrapped in black paper, and the whole arrangement was placed in sunlight. Among the materials tested, the only one that gave positive results was a compound of the element uranium. Since this mineral produced the effect even when not previously exposed to light, Becquerel was convinced that it could emit some kind of penetrating radiation quite independent of its ability to show phosphorescence.

An extensive investigation of the radiation was begun by Marie Curie (1867–1934), using pitchblende, a natural ore rich in compounds of uranium. This material showed activity, measured by its ability to produce ionization, that suggested the emission of particles or electromagnetic radiation, or both. The activity was far greater than could be expected from the amount of uranium it contained, and Mme. Curie suspected that this might be due to a new and much more highly active substance present in the ore. She and her husband, Pierre Curie (1859–1906), undertook a long series of repeated crystallizations of salts derived from pitchblende in order to concentrate the active ingredient. Finally, from a ton of the ore, they obtained a small quantity of a new active element, which they named polonium, and, later, a minute amount (less than 0.02 g) of a still more active element, radium. Further experiments by the Curies and by others soon led to the discovery of many other substances that showed natural radioactivity. For the most part, they are isotopes of the elements from lead to uranium ($Z = 82$ to 92).

The radiation from radioactive materials was investigated very thoroughly by Rutherford and his collaborators, and it was found that the rays are

Becquerel and the Curies shared the 1903 Nobel Prize in physics, and Marie Curie won the 1911 chemistry prize as well.

Marie and Pierre Curie in 1896. (*Courtesy of Niels Bohr Library, American Institute of Physics.*)

of three types. In passing into a strong magnetic field, one kind is deflected slightly to one side, another kind much more strongly in the opposite direction, and a third component is not deflected at all. This behavior of what are called **alpha** (α), **beta** (β), and **gamma** (γ) **rays,** respectively, is shown schematically in Fig. 22.7. Experiments of this kind showed that alpha rays consist of positive-charged particles, beta rays consist of

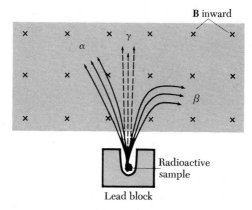

FIGURE 22.7 Effect of a magnetic field on the three types of rays from radioactive materials (schematic). Similar effects would be produced by an electric field directed toward the left in the figure.

negative-charged particles, and gamma rays must be either electro-magnetic waves or neutral particles.

Alpha radiation   These rays consist of alpha particles (helium nuclei). Rutherford put an alpha-radiating substance in a vacuum tube and found that, after some time, an electric discharge revealed the presence of helium gas. The particles are found to have initial speeds of the order of 10^7 m/sec, corresponding to energies of several MeV. The magnitude of this energy shows that the particles must come from the nucleus of the atom. Moreover, some radioactive materials eject several groups of alphas, each group having a quite definite range in air, and hence a definite initial KE (see photograph). This suggests the existence of discrete energy levels in the nucleus. Alpha particles can penetrate several centimeters of air, losing KE by ionizing the atoms of the gas in their path, but are stopped completely by thin metal foils.

The large value of the binding energy of an alpha particle (page 573) shows that it is a very stable configuration and makes plausible the fact that it is ejected as an entity by about 160 different natural radioactive nuclides, whereas individual protons or neutrons are never observed to be emitted spontaneously.

Quantum mechanics is able to explain how an alpha particle, although it may have much less energy than that corresponding to the potential barrier, has a finite chance of breaking out. The theory was worked out in 1928 by George Gamow (1904–1968) and independently by R. W. Gurney and E. U. Condon. In this treatment the position of the particle is represented by its wave function ψ. For the potential distribution of the nucleus, represented typically by the crater model, the ψ function has the form shown by the colored curve in Fig. 22.8: It is oscillatory inside the

Alpha particles from a radioactive source show two distinct ranges in this cloud-chamber photograph. (*From Rutherford, Chadwick, and Ellis, "Radiations from Radioactive Substances." By permission of Cambridge University Press, Publishers.*)

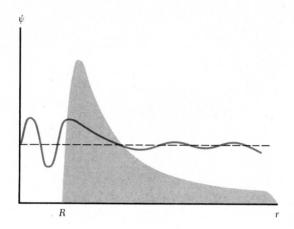

ψ

R

r

FIGURE 22.8 The tunnel effect. Since the wave function has a finite (but very small) amplitude outside the nucleus, the alpha particle has a chance of emerging. Similar considerations apply to the entry of such a particle.

crater, drops rapidly in amplitude as it passes through the wall, and is again oscillatory—but with very small amplitude—on the outside. The fact that the ψ wave is present beyond the wall shows that there is always some chance of finding the alpha particle outside the nucleus, which means that such a particle, although it may not have enough energy to surmount the barrier, can nevertheless escape. This is often called the quantum-mechanics **tunnel effect,** since the alpha particle seems to escape by finding a passage through the wall rather than over it. The probability of escape depends on the height and thickness of the barrier and on the energy of the particle when it is inside the nucleus.

Beta radiation   This radiation consists of electrons whose speeds may be within 1 percent of c. However, because the mass is so small, the initial KE of a beta particle is ordinarily less than that of an alpha. A beta electron can penetrate a lead plate about 1 mm thick.

In contrast to the discrete energy distribution of alphas, beta particles from a given nucleus are found to have a *continuous* energy distribution, and this gives rise to a puzzling situation: The measured energy of most of the beta particles is *less* than the equivalent of the mass difference of the initial and final nuclei. What happens to the missing energy and its associated linear momentum? Moreover, there is trouble with regard to the constancy of the angular momentum of the system. In 1930, Pauli attempted to resolve these difficulties by suggesting that the missing energy and linear momentum were carried off by a hypothetical uncharged particle, emitted at the same time as the beta, and that this particle also had the right amount of intrinsic spin to conserve angular momentum. This hypothetical particle was named the **neutrino** by Enrico Fermi.

In Italian, the coined word neutrino *would mean "little neutral one."*

In spite of an intensive search by many experimenters the neutrino remained undetected; its existence was a mere supposition designed to save the conservation of energy and of momentum for the case of beta emission. From its elusive nature, physicists concluded that a neutrino would have virtually no interaction with matter and that its rest mass would be essentially zero. It was estimated that a neutrino could cross the entire universe with a chance of only 1 in 10^{25} of being absorbed. Finally, after more than a quarter of a century, neutrinos were detected in the intense radiation from a nuclear reactor and, in 1965, neutrinos produced by cosmic rays were recorded in a deep mine shaft. These particles have energies of more than 10 billion electron volts. Pauli's bold conjecture and the conservation laws are thus confirmed.

For the reason given in the next paragraph, the neutrino accompanying the electron should technically be called an antineutrino.

It is now possible to see in detail what happens in the process of the beta decay of a nucleus: A neutron changes to a proton (which remains in the nucleus), and the resulting electron and neutrino are ejected (Fig. 22.9a). Nuclei of high atomic number, which contain more neutrons than protons, tend to change to a more stable state by this process.

There are some artificially produced nuclei which have an excess of protons. In such an instance, greater stability results by the changing of a proton to a neutron, accompanied by the ejection of a positron and a neutrino (Fig. 22.9b). This process, too, is usually included in the term "beta radiation." In assigning the terminology, the ghost particle accompanying the positron was called the neutrino, and the one accompanying the electron (which is the antiparticle of the positron) was called the antineutrino.

Gamma radiation   These rays have already been described (page 487) as very penetrating electromagnetic waves, shorter than about 10^{-10} m. They may be equally well represented as streams of high-energy photons. As an example, the radioactive nucleus protactinium 234 emits gamma radiation of wavelength 0.01 Å, which corresponds to a photon energy of about 0.8 MeV. Of the three types of radiations from radioactive materials, the gammas are the most penetrating by far; some can pass through several centimeters of lead. The emission of an alpha or beta particle usually leaves the nucleus in an excited state. In lapsing back to its condition of greatest stability, it emits high-energy gamma radiation. Like the alphas, the gammas from a given nuclide have discrete energies.

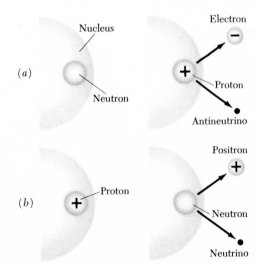

FIGURE 22.9 Beta decay (schematic). In (a) a neutron in the nucleus ejects an electron and an antineutrino and is thereby converted into a proton. In (b) a nuclear proton ejects a positron and a neutrino, leaving a neutron behind. In either case, the nucleus recoils.

22.8 RADIOACTIVE TRANSFORMATIONS; THE DECAY LAW

The scheme for designating nuclides is described on page 570.

A radioactive nuclide usually undergoes a sequence of changes before it attains a condition of stability. Each particle emission changes the effective charge of the nucleus, or changes the mass, or both. For example, when a radium nucleus $_{88}Ra^{226}$ decays by emitting an alpha particle, it becomes about 4 mass units lighter and also loses 2 positive charges. This reduces the mass number by 4 and the proton number by 2, leaving the nucleus $_{86}Rn^{222}$.

The ejection of a beta particle leaves the mass of a nucleus practically unchanged but alters the nuclear charge by 1 unit. If an electron is emitted, the proton number (the effective plus charge of the nucleus) is increased by 1 unit by the electron's departure. If a positron is emitted, it is *decreased* by 1 unit. In both instances the mass number is unchanged. For example, when the radium isotope $_{88}Ra^{228}$ emits an electron it becomes an isotope of actinium, $_{89}Ac^{228}$. Also, the synthetic nuclide of silver $_{46}Ag^{106}$ can emit a positron and become an isotope of palladium, $_{45}Pd^{106}$. Gamma radiation produces no change in either the mass number or the atomic number of the nucleus from which it comes. Table 22.4 summarizes the changes and their consequences.

Practice set*

The radioactive nuclide $_{92}U^{235}$ decays by emitting an alpha particle, and the product nucleus then emits a beta particle (electron). State the net change in (*a*) mass number, (*b*) atomic number, and (*c*) neutron number. Identify the resulting element.

The decay law.

It is found that the rate at which a given radioactive substance disintegrates is virtually independent of external conditions and is a characteristic of the substance itself. Further, the rate at which material is transformed is not constant. Starting with a specified amount of radioactive nuclide, the rate of decay is found to be greatest at first, falling off steadily as the amount of the original substance decreases. The number of nuclei decaying in any short interval of time is proportional to the number present. The

*Ans.: -4; -1; -2; $_{91}U^{231}$.

TABLE 22.4 RADIOACTIVE PROCESSES AND THEIR EFFECTS ON THE EMITTING NUCLEUS

RADIATION EMITTED	RESULTING CHANGE IN	
	MASS NUMBER A	PROTON NUMBER Z
ALPHA: (He nucleus)	-4	-2
BETA:		
a. Electron	0	$+1$
b. Positron	0	-1
GAMMA: (electromagnetic radiation)	0	0

interpretation of this fact is that *the breakdown of the nucleus is governed by pure chance.* The amount remaining at any specified time is shown in Fig. 22.10. One-half of the original amount is left after the passage of a certain time T; half of this (one-fourth of the original amount) remains after a total time $2T$, one-eighth after $3T$, and so on.

Since the decay process is a statistical one, there is no way of predicting when any particular nucleus will break down. However, it is possible to measure one characteristic property and compute the total number of nuclei that will be left after a given time has elapsed, provided that the number of remaining nuclei is not too small for statistical purposes. The time T required for any initial amount of a substance to decrease to half that amount is called its **half-life.** Half-life values may be specified for either alpha or beta decays. The numerical magnitudes vary enormously from one radioactive nuclide to another, ranging from a few ten-millionths of a second to more than 100 billion years. Radium has a half-life of about 1,620 yr. The determination of a half-life value that is either extremely short or extremely long must be undertaken by indirect methods.

Radioactive series. Rutherford and his colleagues found that the decay product of a given nuclide may itself be radioactive and that the process of breakdown can be traced through a series of elements. Three such sequences were dis-covered. They are called the thorium series, the uranium series, and the actinium series, each after its longest-lived member. After going through a chain of alpha and beta emissions, each series ends as a stable isotope of lead. Figure 22.11 is a diagram tracing the course of the thorium series. Notice that $Z = 83$ has two different decay modes. The branching that occurs there has counterparts in each of the other series. A fourth chain, the neptunium series, was discovered in 1942. It is headed by $_{94}Pu^{241}$, which is not found in nature. Unlike the other three series, this one ends with an isotope of bismuth rather than one of lead.

In addition to the members of the four series, certain isotopes of a few of the lighter elements are known to be radioactive, and these nuclides

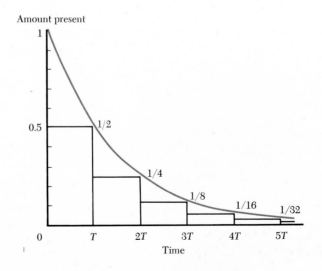

FIGURE 22.10 Decay curve of a radioactive nuclide, showing the fractions of the original amount remaining after various multiples of the half-life T.

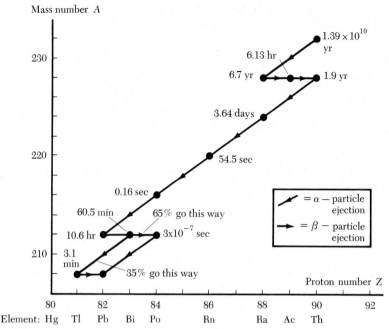

FIGURE 22.11 The decay scheme of the thorium series. The time appearing opposite each nuclide is its half-life.

decay in a single step. For instance, $_{19}K^{40}$ emits a beta electron and becomes the stable nuclide $_{20}Ca^{40}$. All these isolated active nuclides have half-lives comparable with the estimated age of the universe, about 10 billion years; this accounts for the fact that they are still present in natural materials in appreciable amounts.

22.9 NUCLEAR CHANGES PRODUCED ARTIFICIALLY

The interpretation of natural radioactivity as the spontaneous breakdown of atomic nuclei suggested to Rutherford the possibility of causing nuclear changes artificially by letting energetic particles strike atoms of matter. In a program of experiments begun in 1917 Rutherford used as projectiles alpha particles produced by the natural radioactivity of an isotope of polonium. The apparatus is shown schematically in Fig. 22.12. The alpha source was located in a chamber that could be filled with various gases, and a small opening opposite the source was covered with a metal foil thick enough to stop all the alpha particles. A fluorescent screen attached to a microscope was placed just outside this window for observing scintillations.

With oxygen or carbon dioxide in the chamber, no scintillations were seen. But when nitrogen was used, scintillations were detected even when the distance of the source from the window was as much as six times as great as the range of the alpha particles. Rutherford was compelled to conclude that the alphas in some way altered the nitrogen nuclei and that

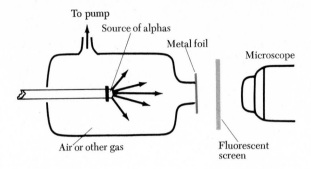

FIGURE 22.12 Simple apparatus devised by Rutherford for observing the effect of alpha particles on a gas.

To pump
Source of alphas
Metal foil
Microscope
Air or other gas
Fluorescent screen

Recall that protons are not emitted in natural radioactivity.

long-range particles were emitted in the process. Magnetic-deflection experiments showed that these particles were probably protons; this was later confirmed by other work. In a short time it was found that 10 or more other elements in the early part of the periodic table could be made to give off protons.

More detailed investigations with cloud chambers revealed something of the dynamics of the process. Eight photographs in four hundred thousand showed an alpha particle colliding with a nitrogen nucleus (see photograph). The alpha track branches sharply, splitting into a short, heavy spur caused by the rebound of the struck nucleus and a long, thin track produced by the emerging proton. The fact that the alpha track ends at the point of impact shows that this particle enters the nucleus and remains part of it, ejecting a proton in the process. This interaction was the first instance of a ''man-made'' nuclear transformation.

Cloud-chamber photographs are usually taken in stereoscopic pairs: Two pictures are taken simultaneously from different angles in order to be able to determine the true length of the paths and their relative position in space. Measurement of such photographs showed that momentum was conserved in these processes. In some cases the total KE of all the particles after impact was less than that of the incident alpha particle, but in certain others it was greater. Rutherford surmised that the extra KE in the latter case came from the binding energy of the struck nucleus.

All nuclear changes may be written in a form similar to chemical reaction equations. Thus, schematically, alpha particle + nitrogen nucleus $\longrightarrow$ product nucleus + proton. The product nucleus can be identified by writing a nuclear symbol for each particle and invoking two conservation principles. These are (1) conservation of nucleons (mass number, page 570), and (2) conservation of nuclear charge (proton number). The symbol for the alpha is written $_2He^4$, since it is a helium nucleus, and the proton is $_1H^1$ since it is a hydrogen nucleus. Then, for reasons about to be given, the reaction equation becomes

$$_2He^4 + _7N^{14} \longrightarrow _8O^{17} + _1H^1 \qquad\qquad 22.8$$

Conservation of mass number shows that the product nucleus must have $A = 17$ in order to make the sum of the superscripts the same on both sides of the equation. Similarly, conservation of charge shows that it must have $Z = 8$. Thus the product nucleus is identified as the relatively rare oxygen isotope of mass 17 and is given the chemical symbol O in the above equation.

Reaction energy.

Equation 22.8 gives only the participants in the reaction. More information can be obtained by including the *energy changes* that are involved. Consider the principle of the conservation of mass number, as used above. The mass numbers of the particles are, of course, not identical with the actual nuclidic masses, and it must be remembered that what is exactly conserved is the combination of mass and energy (page 517). Thus, to an equation such as 22.8 must be added a relation of the form

$$M + \frac{K}{c^2} = M' + \frac{K'}{c^2} \qquad\qquad \textbf{22.9}$$

where M represents the total mass of all substances present before the reaction and M' after the reaction, and K and K' are their total kinetic energies, before and after. A common form for such an equation is

$$Mc^2 = M'c^2 + Q \qquad\qquad \textbf{22.10}$$

where $Q = K' - K$. The quantity Q is called the **nuclear reaction energy** and is usually stated in MeV.

In a given process, Q may have a positive value indicating that energy is set free in the reaction, or it may be negative, which means that some of the initial KE of the particles has been converted into mass. If mass-energy equivalence is satisfied in nuclear reactions, Q should be equivalent to the difference in mass of the original particles and the final particles, which proves to be quite accurately true in all the many instances where the energies have been measured. Nuclear reactions furnish one of the best experimental confirmations of the correctness of $E = mc^2$. In fact, many of the tabulated isotopic mass values, especially of the scarcer nuclides, do not come from mass-spectrometer determinations but are computed from mass-energy conservation, using the measured Q and the masses of the other nuclides taking part in the reaction. See Prob. 22.19.

Reaction energies also occur in connection with chemical changes, but they are negligibly small compared with the masses involved, amounting

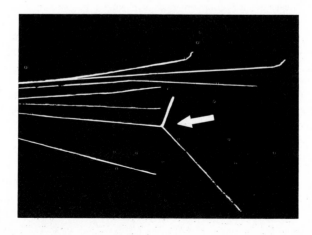

Collision of an alpha particle with a nitrogen nucleus in the cloud chamber. The alphas come in from the left.

to only a few electron volts, whereas the Q values encountered in nuclear reactions are measured in MeV. Even so, the energy changes of considerable size that occur in nuclear reactions are accompanied by only minute changes in mass because of the large value of the factor c^2 in $\Delta E = (\Delta m)c^2$. For a typical Q value of 20 MeV the net change in mass is, according to Eq. 22.6, only about 0.02 u. There is no known reaction where the magnitude of the mass change is large enough to be comparable with the mass of a nucleon, which is the reason the atomic mass is always a measure of the nucleon number.

After succeeding in producing a variety of nuclear changes by using particles ejected from natural radioactive substances, Rutherford considered the possibility of using artificially accelerated particles for this purpose. At his suggestion, T. S. Walton and J. D. Cockcroft built an electrical particle accelerator capable of allowing protons to fall through potential differences of several hundred thousand volts, and in 1932 they succeeded in transforming lithium nuclei by this means. The nuclear reaction equation may be written

Cockcroft (1897–1967) and Walton shared the 1951 Nobel Prize in physics.

$$_1H^1 + {_3}Li^7 \longrightarrow {_2}He^4 + {_2}He^4 + 17.3 \text{ MeV} \qquad \textbf{22.11}$$

The last term is the positive Q value representing the observed excess of the total KE of the two alpha particles over the KE of the incoming proton.

Now use Table 22.1 (page 570) to write the mass values involved in the reaction. These would properly be the masses of the nuclei themselves, but the result is the same if the isotopic masses are used since they are always tabulated to include the masses of the electrons. The substitution is justified by the conservation of nuclear charge, which implies that the total number of electrons is the same on both sides of the equation, so that when the change in mass in a nuclear reaction is calculated, the masses of the electrons cancel out in any case. However, reactions in which positrons are produced are exceptions: For each positron produced, an electron is dissociated from the parent atom.

Returning to the details of the present reaction equation, the net change in mass is computed:

INITIAL MASSES	FINAL MASSES
$_1H^1$: 1.00782 u	$_2He^4$: 4.00260 u
$_3Li^7$: 7.01600	× 2
8.02382 u	8.00520 u

There has been a net decrease in mass amounting to 0.01862 u. At the same time, there has been an increase in the energy of the system of $Q = 17.3$ MeV, which is equivalent (Eq. 22.6) to $17.3/931 = 0.01860$ u. The two numbers are in complete agreement to three significant digits, well within the experimental limits.

Cockcroft and Walton were able to use high-energy protons for producing nuclear reactions in a number of elements. Their success stimulated numerous investigations that used protons, deuterons, neutrons, etc., as the incident particles. Other workers concentrated on the design of higher-energy machines, culminating in the invention of the cyclotron and related devices capable of giving particle energies of hundreds of billions of electron volts. The availability of such machines has led to the finding of new particles.

22.10 INDUCED RADIOACTIVITY

One of the most significant discoveries in modern physics was made in 1933 by Frédéric Joliot-Curie (1900–1958) and his wife Irène (1897–1956), the daughter of Pierre and Marie Curie. In experiments in which some of the light elements were bombarded by alpha particles, they found that positrons appeared but the positron emission did not stop when the source of alphas was removed. The positron activity was observed to fall off as time passed exactly in the way that a natural radioactive element decays, the half-life being of the order of several minutes. The experimenters concluded that the alpha particles had converted the target nuclei into radioactive isotopes of other elements, which then decayed into stable atoms of another kind by ejecting positrons. Such processes are referred to as **artificial,** or **induced,** radioactivity.

With an aluminum target, for example, the initial reaction is

$$_{13}Al^{27} + {}_2He^4 \longrightarrow [{}_{15}P^{30}] + {}_0n^1 \qquad\qquad \textbf{22.12}$$

where the brackets indicate that the phosphorus isotope that forms is radioactive. Its decay then proceeds, with a half-life of 2.5 min, according to the equation

$$[{}_{15}P^{30}] \longrightarrow {}_{14}Si^{30} + {}_{+1}e^0 + \nu \qquad\qquad \textbf{22.13}$$

Here the symbol ${}_{+1}e^0$ denotes a positron and the Greek letter ν (nu) a neutrino. The P^{30} nucleus, being radioactive with a short half-life, is not found in nature, but the stable Si^{30} nucleus is.

Following the discovery of induced radioactivity by the Joliot-Curies, Lawrence exposed common salt (sodium chloride) to a beam of high-energy deuterons from the cyclotron and found that the sodium became radioactive. The reaction is

$$_1H^2 + {}_{11}Na^{23} \longrightarrow [{}_{11}Na^{24}] + {}_1H^1 \qquad\qquad \textbf{22.14}$$

The target material, after removal from the deuteron beam, was found to be emitting electrons and gamma radiation:

$$[{}_{11}Na^{24}] \longrightarrow {}_{12}Mg^{24} + {}_{-1}e^0 + \gamma \qquad\qquad \textbf{22.15}$$

The symbol of an accompanying neutrino or antineutrino is often omitted from the reaction equation.

with a half-life of 15 hr. The nuclide ${}_{11}Na^{24}$ is called **radiosodium.** Because of its short half-life, it is not found in nature.

Many of the isotopes formed as a result of induced radioactivity began to find significant applications in biology and medicine almost from the time of their discovery. Radioactive isotopes (often called **radio-isotopes**) of nearly all the common chemical elements can now be obtained. The active atoms enter into the same chemical reactions as the normal atoms, since they have the same electron configuration. They can be built into compounds and allowed to take part in biochemical processes, and the distribution of the element in space and time can then be followed by means

See Ref.
22.3.

of sensitive Geiger counters or other detectors of radiation. Radio-isotopes thus furnish the biologist with **tracers** for investigating fundamental processes occurring in living matter and have also found extensive applications in industry and engineering. Radio-isotopes are sent all over the world from the Oak Ridge Laboratories of the U.S. Atomic Energy Commission and from the Radio-chemical Centre at Amersham, England. Some of these materials are prepared in a cyclotron, but the most prolific source is the nuclear reactor (Sec. 22.12), where neutrons are extremely plentiful.

Radiocarbon dating.

The action of cosmic rays continuously produces a number of radio-nuclides in the atmosphere. One of the most interesting of these is radio-

RADIATION AND LIFE

High-energy radiation, whether particle or electromagnetic, can damage living cells. A biologically important molecule such as the DNA in the cell nucleus may be shattered by direct particle impact, or ions may be produced that recombine to form new compounds which threaten the delicate physical and chemical processes going on in normal cells. Cells that are not killed outright may have their genetic material changed so that mutations occur when the cells reproduce.

The damage done to an organism by radiation is found to be related to the energy absorbed per unit mass of living tissue. The commonly accepted unit for measuring this damage is called a **rad** (for *r*adiation *a*bsorbed *d*ose). One rad is equivalent to the amount of radiation that delivers 0.01 joule of energy per kilogram of living material.

The average American is exposed to radiation from various sources (Table 22.5). The total is slightly less than 200 millirads (1 millirad $= 10^{-3}$ rad), of which about one-third is from man-made sources. Authorities consider as much as 10 rads/yr to be acceptable. However, since exposures are likely to increase greatly in the future, they warn against taking too much comfort in the factor of about 50 by which the safe limit exceeds the present exposure.

Damage to the body by radiation depends not only on the total amount absorbed, but on the rate at which the radiation is applied. A large dose absorbed over a short interval of time can make growing tissues lose their capacity for cell division, resulting in death. However, the same total dose administered over a longer interval allows time for cell replacement and no serious symptoms may appear (see Table 22.6).

TABLE 22.5   ANNUAL RADIATION EXPOSURE (IN MILLIRADS) IN THE UNITED STATES*

Natural Sources		Man-made Sources	
Outside the body		*Medical uses of radiation*	
Cosmic radiation	50	(diagnosis, therapy, etc.)	60
The earth	50	*Atomic energy installations*	0.2
Inside the body		*Radiation from appliances,*	
Breathing of air	5	*etc.* (TV tubes, watch dials,	
Natural radioactive		radioactive wastes)	2
materials in tissues	20	*Radioactive fallout*	4

* Rough estimates based on a report of the Federal Radiation Council.

carbon, C^{14}, which results from the impact of high-energy neutrons on nitrogen nuclei according to the equation

$$_7N^{14} + {_0}n^1 \longrightarrow [_6C^{14}] + {_1}H^1$$

The radiocarbon immediately begins to break down, with a half-life of about 5,600 yr:

$$[_6C^{14}] \longrightarrow {_7}N^{14} + {_{-1}}e^0$$

Meanwhile, radiocarbon atoms have been incorporated into molecules of carbon dioxide in the air. Some of these active molecules are taken in by

TABLE 22.6 EFFECTS ON THE HUMAN BODY OF VARIOUS INTENSIVE DOSES OF RADIATION

DOSE, rads	EFFECT
Local exposure	
100,000	Complete breakdown of nervous system, followed by convulsions. Fatal.
10,000–3,000	Skin destruction. Fatal.
2,000–600	Itching and burning of skin; some loss of hair. Most recover.
Whole-body exposure	
700–300	Acute radiation sickness, followed by blood changes. Some recover.
150–50	Blood changes, but usually no severe acute injury.
Below 50	Permanent damage rare.
0.1	Maximum weekly permissible dose rate for radiation workers.
0.004	Weekly natural dose rate in the United States.

On the other hand, genetic effects seem to be cumulative, at least for some organisms. Impaired reproductive cells continue to produce others that carry the same mutation; every exposure, no matter how small, contributes to the total damage.

In contrast with its harmful effects, radiation has become the basis for the new field of nuclear medicine, concerned with the use of radio-isotopes for both diagnosis and therapy. The increasing use of radiations from high-energy machines succeeded in doubling the cancer cure rate between 1940 and 1960.

Public awareness and understanding of the dangers of radiation exposure must become universal if the human race is to survive. Regulations are needed to protect the individual from nuclear accidents, exposure to unshielded radiation sources, and overexposure during radiation therapy and diagnosis. Finally, it is obvious that the nations of the world must act to prevent nuclear war, with its catastrophic and perhaps irreversible consequences.

References Booklets in the series "Understanding the Atom," published by the U.S. Atomic Energy Commission. Titles of particular interest here are Genetic Effects of Radiation, Radioisotopes and Life Processes, Radioisotopes in Medicine, and Your Body and Radiation. Single copies are available without charge from USAEC, Box 62, Oak Ridge, Tenn. 37830.

plants, and thus by animals that feed on these plants. The concentration in each organism eventually reaches an equilibrium value. It is estimated that there are enough C^{14} atoms built into the structure of the human body to constitute one of the major sources of radiation to which we are normally exposed. This is far below the danger level (page 592), although still enough to produce some gene damage.

As soon as an organism dies, the exchange with the atmosphere stops and the concentration of radiocarbon decreases steadily as time passes. By measuring the ratio of C^{14} to C^{12} in a sample of ancient wood, a radiochemist can determine with fair accuracy the date when the tree died or was cut down. The American chemist W. F. Libby has made extensive studies of this kind, and the results have enabled him to date recent geological events and to fix the age of a variety of archeological artifacts. For this work, he was awarded the Nobel Prize in chemistry in 1960. Further details may be found in Ref. 22.4.

22.11   NUCLEAR FISSION

With the availability of much information on neutron-produced nuclear reactions, Fermi and his coworkers decided, in 1934, to try the effect of neutrons on heavy elements such as thorium and uranium. The very first trials using uranium gave rise to beta activity with several different half-life

Fermi. (*Courtesy of University of California, Los Alamos, New Mexico.*)

Enrico Fermi (1901–1954), one of the outstanding physicists of modern times was equally gifted as a theorist and as an experimenter. Some of his achievements, for which he was awarded the Nobel Prize in 1938, are described below.

periods, and the experimenters thought that they had succeeded in producing nuclei of elements heavier than uranium, the last element in the periodic table at that time. Recall (Table 22.4, page 585) that when a nucleus gives off a beta electron, its nuclear charge (and thus its proton number Z) effectively increases by 1 and it becomes a nucleus of the succeeding element in the table. Fermi and his group attributed the observed beta activities to successive disintegrations of a given atom, which would lead to **transuranic** ("beyond uranium") elements for which $Z = 93, 94, 95$, etc. Similar observations were soon made in France and in Germany, and the formation of transuranic elements was generally accepted as the explanation of all these results.

In 1939, careful work by the German chemists O. Hahn and F. Strassmann on the chemical separation of the products showed that the above view was correct only in part. Tests revealed that one of the product materials was either radium ($Z = 88$) or barium ($Z = 56$). Both elements are found in the same column of the periodic table and hence are similar in their chemical behavior. However, they can be separated at least partly by a long series of crystallizations. When this was done, it was found that the beta activity went along with the barium rather than with the radium. Other radioactive materials present were identified as various elements that, like barium, are not far from the middle of the periodic table. The results led to the striking conclusion that *neutron bombardment can cause a uranium nucleus to break apart, producing two or more fragments of moderate size.*

The novelty of this view lies in the fact that all the nuclear disintegrations previously encountered among the heavier elements, including those of natural radioactivity, involved only the sloughing off of relatively small portions of the original nucleus, whereas here the nucleus splits into pieces of comparable mass. The term **fission** (from the Latin *fissio*, cleaving) was suggested for such changes. The word is well chosen, for the process suggests the biological action of the same name by which certain plants and animals, such as paramecia, divide to form new individuals. The fission reaction can take place in many ways, and radio-isotopes of dozens of elements have been identified in the fission products from thorium and uranium. One possibility is shown in Fig. 22.13. The primary fission products invariably have too many neutrons for stability, and so they decay by emitting neutrons or electrons until a stable configuration is reached.

The liquid-drop model of the atomic nucleus (page 577) provides a visualization of fission. Because of the existence of the nuclear force a

FIGURE 22.13 A typical fission reaction (schematic). In (*a*) a neutron enters a U²³⁵ nucleus and amalgamates with it. The resulting U²³⁶ is unstable (*b*) and forms a drop consisting of two lobes of comparable size. In (*c*) the drop has pinched off, and two neutrons are emitted as well. The resulting nuclei of yttrium and iodine are themselves unstable and will undergo further disintegrations, but not fissions. Gray circles represent neutrons; black circles represent protons.

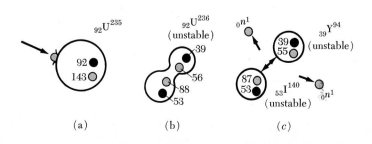

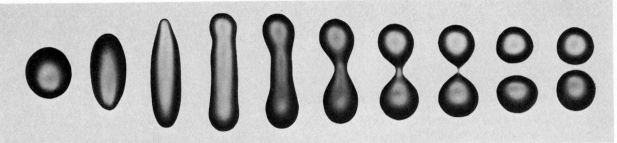

Fission of a suspended drop. An electric field applied in the vertical direction caused the drop to elongate. With sufficient deformation it became unstable and pinched off, after which the resulting fragments rounded out to a state of minimum energy. (*Courtesy of Dr. R. B. Leachman, Los Alamos Scientific Laboratory.*)

nucleus would normally be expected to be roughly spherical in shape, much as surface tension makes a small raindrop take this form. The entry of a neutron into a heavy nucleus causes an instability that deforms the "drop," and if the disturbance is violent enough, the nucleus may elongate and pinch off into two globules (see photograph).

The most impressive fact about nuclear fission, recognized almost from the start, is the large amount of energy released in the process—around 200 MeV (see Worked example 22.4 below), fully an order of magnitude greater than the energies associated with the nuclear reactions considered above. This makes fission a practical source of energy, especially since it can be made to take place in almost all the nuclei of a sample rather than merely among a few.

The chronicle of the discovery of fission and its exploitation for military and other purposes is a vital part of the history of our times. For more detailed accounts, see Refs. 22.5 and 22.6.

Worked example 22.4   One way in which the U^{235} nucleus can undergo fission is according to the reaction observed by Hahn and Strassmann. The equation is

$$_{92}U^{235} + {_0}n^1 \longrightarrow [_{56}Ba^{141}] + [_{36}Kr^{92}] + 3{_0}n^1$$

Typically, a neutron must supply an energy of 5 or 6 MeV to the target nucleus to produce fission, and the total energy of the emitted neutrons has about this same value, so that the net contribution to the Q of the reaction is from the mass change of the participating nuclei. Find the value of Q for this process.

Solution   The total masses of initial and product nuclei are computed by using values taken from tables of nuclidic masses:

INITIAL MASSES	FINAL MASSES
U^{235}: 235.0439 u	Ba^{141}: 140.9139 u
n: 1.0087	Kr^{92}: 91.8973
236.0526 u	$3n$: 3.0261
	235.8373 u

The difference, 0.2153 u, is equivalent to $(931)(0.215) = 200$ MeV (see inside back cover), which is the Q of the fission reaction of the U^{235} nucleus.

The above isotopes of Ba and Kr are unstable. They have too many neutrons and get rid of the excess by direct ejection and also by a series of beta decays.

22.12 CHAIN REACTIONS; THE NUCLEAR REACTOR

Natural uranium consists of about 99.3 percent of U^{238}, 0.7 percent of U^{235}, and an insignificant amount of U^{234}. U^{238} can be made to undergo fission by means of fast neutrons, although neutron capture is more likely (page 598). Also, it is difficult to maintain an adequate supply of fast neutrons in a sample because they rapidly lose energy by collisions with atoms. On the other hand, fission of U^{235} can be produced readily by very slow neutrons which have come to temperature equilibrium with the atoms of the sample (''thermal'' neutrons).

A single thermal neutron can cause a U^{235} nucleus to undergo fission, but the product nuclei are too rich in neutrons to be in equilibrium and tend to eject some directly. The significant fact is that, in the process, *more than one* neutron, on an average, is emitted.

Fission can also be initiated by other particles and by hard gamma rays.

This makes a **chain reaction** possible, with consequent release of energy on a large scale (Fig. 22.14). The average number of neutrons set free in the thermal fission of a U^{235} nucleus is now known to be about 2.42. If more than one of these, on the average, were itself able to cause fission in another U^{235} nucleus, the number of neutrons would increase rapidly, fission would proceed at an ever-increasing rate, and in a very short time the whole sample of fissile material would be transformed.

Fermi suggested that a controlled fission device could be made by disposing small pieces of uranium throughout a mass of some material capable of slowing down the fission-produced neutrons so that they could cause fission in additional nuclei. Such a material, called a **moderator,** must consist of lightweight atoms so that the neutrons can lose KE in repeated collisions with them according to the mechanism discussed in Worked

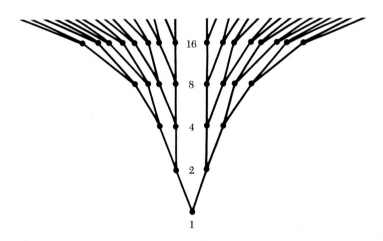

FIGURE 22.14 Schematic representation of a particular chain reaction in which each event always succeeds in producing two new events. Reading upward, the number occurring at each stage increases geometrically, that is, by the constant factor 2.

example 22.1. Neutrons originating in one of the lumps of uranium would find their way into the surrounding moderator, where they are slowed without danger of capture by U^{238} nuclei, after which there is a good chance of their entering another slug of uranium where they can produce U^{235} fissions.

The first such **nuclear reactor** was successfully operated by Fermi and his group in 1942. It contained over 6 tons of uranium and used highly purified graphite (carbon) as a moderator. The energy output could be readily controlled because of the fortunate circumstance that there is some delay in time before some of the fission-produced neutrons are released from their parent nuclei. Projecting into the reactor are several movable control rods made of cadmium metal, which is a strong absorber of neutrons. When these rods are pulled out beyond a critical point, the activity of the reaction begins to increase, as measured by the amount of heat and the number of neutrons produced. The activity can be held at a desired level by manipulating the rods. The possibility of using the heat generated in a nuclear reactor as a commercial source of power was soon recognized.

The capture of neutrons by U^{238} is detrimental to the operation of a power reactor because these neutrons can no longer contribute to the fission process. However, this circumstance proved of great value in realizing the military goal of the World War II American atomic project, which was not directed to the generation of power but to the production of a weapon. The action of low-energy neutrons on U^{238} produced the first of the transuranic elements, neptunium ($Z = 93$), by the following sequence of reactions:

$$_{92}U^{238} + {}_{0}n^1 \longrightarrow [_{92}U^{239}] + \gamma$$

The U^{239} then decays by emitting a beta particle, leaving a nucleus of neptunium:

$$[_{92}U^{239}] \longrightarrow [_{93}Np^{239}] + {}_{-1}e^0 \qquad \text{(half-life, 23 min)}$$

As indicated by the brackets, neptunium is also radioactive. On emitting a beta particle it becomes a nucleus with $Z = 94$, called **plutonium,** according to the following reaction:

$$[_{93}Np^{239}] \longrightarrow [_{94}Pu^{239}] + {}_{-1}e^0 \qquad \text{(half-life, 2.3 days)}$$

This isotope of plutonium is alpha-active, the product nucleus being U^{235}, but the period is extremely long:

$$[_{94}Pu^{239}] \longrightarrow {}_{92}U^{235} + {}_{2}He^4 \qquad \text{(half-life, 24,400 yr)}$$

Pu^{239} readily undergoes fission. Although it is practically nonexistent in nature, considerable amounts of this nuclide have been produced in uranium-fueled reactors for use in weapons. This is an example of a **breeder** reaction, in which more nuclear fuel is produced than is consumed (see Ref. 22.7). Another fissile substance, U^{233}, was produced from thorium by the American chemist G. T. Seaborg and his associates.

Usually, when a nucleus undergoes fission, it breaks into two roughly equal pieces plus a few neutrons, but there are rare instances where an alpha particle is also released. This happens about once in 400 events. There is also evidence that, in perhaps one event in 100,000, three or four fragments of comparable mass are produced.

The name "plurission" was proposed for this phenomenon by the Albanian physicist Ari N'améerf.

Stray neutrons are always present, because of cosmic rays, radioactivity in the earth's crust, and other causes. Such a neutron, entering a mass of fissile material could conceivably start a reaction that would spread explosively through the whole mass. How can we be sure that a sample of fissile material, once manufactured, will not be accidentally touched off?

The answer is that the probability of fission increases with the bulk of vulnerable material on hand and decreases with the chance of escape of neutrons from the sample, which in turn depends on the surface area. If a sample is very small, its surface will be relatively large for the amount of material present and an explosive reaction will be prevented by the large rate of escape of neutrons to the surroundings. There is a certain **critical size** below which the rate of escape and loss of nonfission capture of neutrons are less than their rate of production by fission. Such a sample is safe, but if enough of such smaller specimens are brought together to form one piece, the number of neutrons will multiply rapidly and an explosion will result. Thus there is no such thing as a scale-model fission bomb; the critical size must be attained before it will work at all. In a fission bomb, the firing of a charge of ordinary chemical explosive either compresses a single mass of fissile material or rams together two or more masses of subcritical size. The chain reaction then follows.

Nuclear power.

Although the production of useful radio-isotopes in a nuclear reactor is important, it must be remembered that the primary product is energy. The kinetic energy of fast fission and secondary fragments is merely heat energy. Delivered at a steady, controlled rate, it can be used for industrial and technological power production through the use of some conventional type of heat engine (such as a steam engine or turbine) coupled with an electric generator (Fig. 22.15).

A single pound of U^{235} can give as much energy as the combustion of 1,300 tons of coal. This makes the prospect of nuclear power an attractive one, since the world's supply of coal, oil, and natural gas may not last for more than a few hundred years at present rates of consumption.

The fuel costs and even the expense of periodic purification of the active materials are negligible fractions of the total value of power so produced.

FIGURE 22.15 A nuclear reactor used for power generation.

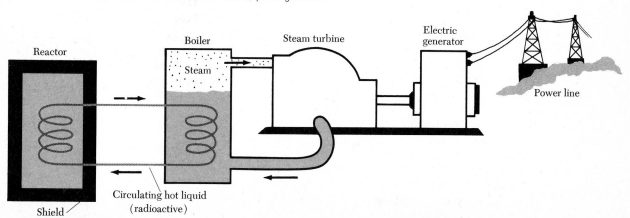

Reactor

Boiler

Steam turbine

Steam

Electric generator

Power line

Shield

Circulating hot liquid (radioactive)

The original cost of a nuclear plant is considerably more than that of a conventional steam plant using coal, but this gap is being closed. In regions where natural fuels are expensive, nuclear plants are generating power at costs below those of coal-fired units. Numerous industrial nuclear-power installations are now in operation or under construction, and the number is estimated to exceed 300 by the year 2000. The accompanying environmental problems may be formidable (see pages 219 and 592–593).

Another obstacle to the rapid development of commercial nuclear power is the radiation hazard. All active materials must be handled by remote control, and operating personnel and others must be assured adequate protection from harmful neutron and gamma radiations and from the consequences of accidents. Nuclear propulsion for ships and submarines is a reality and should be possible for trains and space installations. However, it may be some time before nuclear power becomes feasible for automobiles and aircraft.

22.13 NUCLEAR FUSION

The discussion of the binding energy per nucleon for the stable nuclides (Sec. 22.5) led to the conclusion that the elements near the two extremes of the mass range are the least stable of all. This is plausible, because (1) in light nuclei, which contain only a small number of nucleons, each interacts with only a few neighbors, and so the binding energy is not large, and (2) in heavy nuclei the electrostatic repulsion between protons causes a considerable decrease in binding energy. There is a large energy yield in fission, when some of the heaviest nuclei convert to nuclei nearer the crown of the curve in Fig. 22.3 (page 575). This figure suggests that even greater amounts of energy would be released if light nuclei along the steeper side of the curve could be made to combine to form heavier nuclei. Such a reaction is called **nuclear fusion.** Fusion was attained on a small scale in laboratory experiments as early as 1932, and the first ''H-bomb'' was detonated in 1952. A controlled fusion process is being actively sought for practical power generation.

Hydrogen fusion has probably been going on for billions of years in the interior of the sun and stars (see below).

The direct combination of four hydrogen atoms to form a helium atom (or, essentially, four protons to form an alpha particle) is a fusion reaction that would yield a large amount of energy:

$$4_1\text{H}^1 \longrightarrow {}_2\text{He}^4 + 2_{+1}e^0 + Q \qquad\qquad 22.16$$

The value of Q, computed from the net mass change, is about 26 MeV.

This is not the reaction taking place in the H-bomb. See remark on the following page.

There is an important difference between fission and fusion with regard to the methods of starting and maintaining the reactions. The presence of a single stray neutron in a mass of fissile material can start a chain reaction which spreads through the whole sample. However, in order to bring nuclei close enough together to fuse, they must have very high speeds to overcome the Coulomb repulsion; this means that the material must be brought to a very high temperature, at least locally. For this reason, fusion is called a **thermonuclear reaction.** It becomes plausible that fusion can proceed in the interior of a star, where temperatures reach many million degrees.

How can a thermonuclear reaction be started in an experiment conducted on earth? An uncontrolled fusion reaction (H-bomb) can be triggered by detonating a fission bomb placed within a mass of fusible material, since the temperature attained momentarily in fission is probably greater than at the center of the sun. There are good technical reasons (see below) for using materials other than H^1 in a thermonuclear bomb, and H^2 and H^3, as well as lithium, have presumably been used in devices already tested.

Intensive experimentation is now going on toward producing the controlled release of fusion energy, analogous to that attained with fission in the nuclear reactor. The controlled fusion reaction with the best prospects is believed to be

$$_1H^2 + {}_1H^3 \longrightarrow {}_2He^4 + {}_0n^1 + 17.6 \text{ MeV}$$

(see Ref. 22.9). A promising line of attack seems to be the conversion of light materials to a plasma (page 227), whose constituents can then be heated electrically or compressed magnetically. The requirements are formidable: production of a plasma having around 10^{14} nuclei per cm^3 at temperatures of the order of 10^8 K which must remain confined for about one second. Confinement methods such as the magnetic bottle (pages 444–445) are being worked on in laboratories in several countries.

22.14 THE SOURCE OF STELLAR ENERGY

Nuclear physics can now give a plausible answer to an interesting question of long standing: What is the origin of the tremendous quantities of energy that the sun and the stars are continually pouring into space? Each kilogram of the sun's mass is releasing energy at the rate of nearly 2×10^{-4} watt. This does not sound impressive, since radium and its decay products evolve energy, per unit mass, at nearly a million times this rate. However, because of the great mass of the sun, its total energy output is enormous, amounting to nearly 4×10^{26} watts, so that, even at the earth's distance, each square meter of area exposed perpendicularly to the sun's rays receives energy at the rate of nearly 1.4 kW.

Many attempts have been made to account for the production of this great power, but none of the suggested sources proved adequate. With the discovery of nuclear transmutations, however, it was realized that here was available an energy source of the correct order of magnitude. In 1929, R. Atkinson and F. Houtermans suggested nuclear fusion as a possible source of stellar energy, but nearly a decade elapsed before enough detailed knowledge of nuclear processes accumulated to lead to a definite theory. In 1938, H. Bethe in the United States and C. von Weizsäcker in Germany succeeded almost simultaneously in explaining stellar energy production by the process of combination of hydrogen atoms to form helium. The 1967 Nobel Prize in physics was awarded to Bethe for his contribution.

Even earlier, in 1915, Harkins (page 571n) had proposed the possibility of hydrogen fusion.

The production of helium from hydrogen cannot take place in one step because the probability of simultaneous collision of four protons is infinitesimal, and so it must proceed through a number of stages. Two different sequences are probably involved. In the cooler stars, a deuteron, an He^3, and finally an He^4 are formed. This is the process that accounts for most of the sun's energy.

The cube of granite and the lump of coal have the same weight, but nearly 100 times as much nuclear energy could be extracted from the granite as can be obtained by burning the coal. Breeder reactors suitable for using the small amounts of thorium and uranium in rocks may be perfected in the near future.

In the hotter stars, nitrogen or carbon nuclei are involved in the intermediate reactions but are left in their original state at the end of the process. They serve merely the function of **catalysts**—agents that promote a reaction without themselves undergoing permanent change. The net effect is the formation of helium nuclei from hydrogen as given by Eq. 22.16.

With an energy output rate of 2×10^{-4} W/kg, the corresponding loss of solar mass, computed from $E = mc^2$, is found to be over $4\frac{1}{2}$ million tons per second. Nevertheless, even if the present rate were maintained, there is enough hydrogen fuel in the sun to make its energy production last some tens of billions of years. For further details on stellar energy production see Ref. 22.8.

22.15 ENERGY SOURCES COMPARED

It is interesting to compare the amounts of energy that can be made available from various sources, both chemical and nuclear, through the conversion of rest mass into kinetic energy in accordance with the Einstein relation $E = mc^2$. The results, computed on an equal-mass basis, are collected in Table 22.7. The first entry represents, in order of magnitude, the energy obtainable from chemical processes such as the burning of fuel or the metabolism (combustion) of food by an animal. This quantity is insignificant compared with the other items, since it originates in changes affecting only the outer parts of the atom, the electrons. The next two items refer to nuclear energy and are of the order of a million or more times greater. Notice that, although the Q of the hydrogen fusion reaction is only 26 MeV whereas that of fission is 200 MeV, the energy yield *per unit mass* is nearly a full order of magnitude greater for fusion. In practice, the time element is also important. There are ordinary radioactivities where Q is large but the rate in time is low.

The most energetic process conceivable is the complete annihilation of a sample of matter, represented by the last entry in Table 22.7. In our part of the universe, this is observed to occur only in the mutual destruction of isolated particle-antiparticle pairs: electron and positron, proton and

TABLE 22.7   ENERGY OBTAINABLE FROM ONE KILOGRAM OF MATTER BY VARIOUS PROCESSES

FUEL	ENERGY PRODUCED, joules	RATIO TO PRECEDING NUMBER	EXAMPLE OF A COMPARABLE AMOUNT OF ENERGY
Chemical change: combustion of coal	3.2×10^7	. . . .	Solar energy falling on a house each day
Nuclear fission of U^{235}	8.2×10^{13}	2.5×10^6	Energy released in a severe earthquake
Nuclear fusion of H^1	6.3×10^{14}	7.3	Energy value of food consumed by world population in one month
Complete annihilation of matter	9.0×10^{16}	140	About one-fifth of daily energy needs of mankind

antiproton, etc. Such events are so diluted in space and time that they do not produce results of great consequence. However, there may be distant parts of the universe composed entirely of antimatter. The atoms of such matter would have nuclei composed of antiprotons and antineutrons accompanied by clouds of positrons, and such atoms would probably obey much the same laws that are observed for ordinary matter. If, however, antimatter should come into contact with ordinary matter, the annihilation would be complete except for the production of certain mesons which have no distinguishable antiparticles (see Chap. 23). In any case, these mesons decay into betas, which are then annihilated.

Programmed review

Instructions: See page 18.

1. Define (*a*) nuclide; (*b*) isotope.

    (*a*) Any atom species, characterized by the composition of its nucleus; (*b*) each variety of nuclide belonging to a given element is called an isotope of that element.   [22.1]

2. What is a mass spectrograph?

    An electromagnetic device for separating atoms or molecules according to their masses.   [22.1]

3. What is meant by the mass number of a nuclide?

    The nearest integer to the actual mass expressed in atomic mass units, or the total number of nucleons in the nucleus.   [22.2]

4. Of what does an atomic nucleus consist?

    A number of protons equal to the atomic (proton) number of the element, together with a number of neutrons that depends on the isotope under consideration.   [22.2]

5. Explain why the mass of a bound system is less than the sum of the masses of the constituents.

    Part of the total mass-energy of the system is present in the form of (potential) energy of binding and so the total mass is less than before the system was assembled.   [22.3]

6. What is meant by the nuclear force? How does it differ from the Coulomb force?

    The strong force binding the constituents of a nucleus together. Unlike the Coulomb force, it is of short range and is independent of the charges on the particles.   [22.4]

7. What is meant by the half-life of a radioactive substance?

    The time during which any amount of the substance decays to half its original value.   [22.8]

8. Define nuclear reaction energy.

    The energy equivalent of the difference between the initial and final values of the total mass of the system. The reaction energy is positive if there is a mass decrease and negative if there is a mass increase during the course of the reaction.   [22.9]

9. What is a radio-isotope?

    A radioactive isotope, usually artificially produced, of an ordinary chemical element. It may be used as a tracer, as a source of radiation, etc.   [22.10]

10. Define nuclear fission.

    The division of the nucleus of a heavy atom into two fragments of comparable mass, together with some neutrons.   [22.11]

11. Define nuclear fusion.

    A reaction in which lightweight nuclei are made to unite to form heavier nuclei.   [22.13]

For further reading

22.1. *Nier, A. O.* New Unified Scale for Atomic Masses and Weights, *Phys. Teacher*, April, 1963, p. 11.

22.2. *Shamos, M. H.* (ed.) "Great Experiments in Physics." Read Rutherford's account of his discovery of artificial transmutation on pp. 250–265.

22.3.   *Robinson, G.,* and others "The New Force of Atomic Energy."
        A pamphlet prepared by the U.S. Atomic Energy Commission.
        Pages 46–58 deal with radio-isotopes and their uses.

22.4.   *Deevey, E. S.* Radiocarbon Dating, *Sci. Am.,* February, 1952, pp.
        24–28.

22.5.   *Segré, E.* "Enrico Fermi—Physicist." An engaging account of his
        life and work, written by a colleague and friend.

22.6.   *Smyth, H. D.* "Atomic Energy for Military Purposes." The famous
        Smyth Report, giving a detailed history of the development of the
        atom bomb.

22.7.   *Seaborg, G.,* and *J. Bloom* Fast Breeder Reactors, *Sci. Am.,*
        November, 1970, pp. 13–21.

22.8.   *Menzel, D. H.* "Our Sun." An exposition of solar physics, includ-
        ing details of the thermonuclear processes.

22.9.   *Gough, W.,* and *B. Eastlund* The Prospects of Fusion Power, *Sci.
        Am.,* February, 1971, pp. 50–64.

22.10. *Clark, R. W.* "Einstein, the Life and Times." An excellent biogra-
        phy, noteworthy for the examination of Einstein's role in the devel-
        opment of atomic energy.

Questions and problems

22.1.   What is the exact mass, in u, of an O^{16} *nucleus?* Use Table 22.1
        (page 570).

22.2.   A certain atomic nucleus contains 79 protons and 118 neutrons.
        What is (*a*) its "atomic" number; (*b*) its mass number? (*c*) What
        element is it?

22.3.   Assuming that neon consists essentially of the two isotopes 20
        and 22, use the information on page 569 to calculate the ratio
        of the amount of Ne^{20} to Ne^{22} in natural neon. HINT: The chemical
        atomic weight is the "weighted average" of the isotope weights.

22.4.   Verify the statement made on page 568 about the proportionality
        between the mass of an ion and the radius of its path in a mass
        spectrograph by combining Eqs. 18.4 (page 439) and 18.5 (page
        441). Write the former for a magnetic field B_1 which bends the
        ions into circular paths and the latter for an electric field E_1 and
        a magnetic field B_2 used in the velocity selector. Assume that all
        ions have the same charge Q_i.

22.5.   Explain why it is possible for a low-energy neutron to enter a given
        nucleus whereas an alpha particle needs a large KE in order to
        do so.

22.6.   Show that, unlike an electron, a nucleon can be confined within
        a region of the order of size of a nucleus (10^{-15} m) and still have
        energy within the average binding energy per nucleon (page 575).
        HINT: Use the indeterminacy principle.

22.7.   The amount of a certain member of the thorium series that remains
        after 2.72 min is exactly one-eighth of the original amount. (*a*) What

is the half-life of this nuclide? (*b*) From Fig. 22.11 (page 587), which element is it?

22.8.   For every gram of radium existing now, how many grams were there in 6000 B.C. when the first settlements on the Nile were established?

22.9.   Suppose that an experimenter has a means of watching a single atom of a certain radioactive material continuously. He knows that the half-life is 2.7 hr. Can he predict when this atom will decay? Explain.

22.10.   *Internal conversion.* Sometimes a beta electron, instead of originating in the nucleus, is an orbital electron that gets its KE from a gamma-ray photon emanating from the decaying nucleus. The process is a photoelectric effect. Would the energies of the beta electrons be expected to have discrete energy values or a continuous distribution? Give reasons for your answer.

22.11.   A neutron, when outside a nucleus, is unstable and undergoes beta decay with a half-life of about 12 min. For this process, compute (*a*) the maximum energy of the beta electrons emitted; (*b*) the minimum energy. Account for the difference. HINT: In writing the reaction, be sure to include all the participants.

22.12.   Which of the following are *not* possible nuclear reactions? Give reasons for your answers.

a. $_6C^{12} + {}_6C^{12} \longrightarrow {}_{10}Ne^{20} + {}_2He^4$
b. $_1H^2 + {}_{20}Ca^{40} \longrightarrow {}_0n^1 + {}_{20}Ca^{42}$
c. $_1H^2 + {}_1H^2 \longrightarrow {}_2He^3 + {}_0n^1$
d. $_{15}P^{30} \longrightarrow {}_{14}Si^{30} + {}_{+1}e^0$
e. $_1H^1 + {}_5B^{11} \longrightarrow {}_4Be^9 + {}_2He^4$

22.13.   The gamma radiation from cobalt 60 is used in therapy as a substitute for radium and x rays. Write the nuclear reaction equation that represents the formation of this nuclide from stable Co^{59} by neutron irradiation.

22.14.   Is the C^{12} nucleus apt to disintegrate into three alpha particles spontaneously, or must energy be supplied to bring this about? Compute the reaction energy.

22.15.   Calculate the minimum energy required to disintegrate an alpha particle into two deuterons.

22.16.   What is the maximum wavelength a photon can have if it is to disintegrate a C^{12} nucleus into three alpha particles? See Prob. 22.14.

22.17.   How much energy is released when a nucleus of Pu^{239} emits an alpha particle? The mass of Pu^{239} is 4.0082 u greater than that of U^{235}.

22.18.   When Li^6 nuclei are struck by fast deuterons, two different reactions can occur: In one case, protons are ejected, and in the other case, alpha particles. Write the reaction equations, and compute the Q for each.

22.19. Compute the mass of the O^{17} atom, using Eq. 22.8 (page 588) and the measured value of $Q = -1.26$ MeV. Compare the result with the value given in Table 22.1 (page 570).

22.20. When molecular oxygen and hydrogen combine chemically to form water, the measured heat of combustion, per mole of water, is 60 kcal. How much energy is liberated, per molecule? Convert this to electron volts, and compare the result with the ionization energy of the hydrogen atom (page 544), which is of the same order of size as the energies involved in chemical reactions.

22.21. What is the maximum wavelength of a gamma ray that can produce photodisintegration of a deuteron? See Worked example 22.2 (page 574).

22.22. Compute the reaction energy Q for each of the following processes. Be sure to include the algebraic sign of the result.

a. $_7N^{14} + {}_1H^2 \longrightarrow {}_1H^1 + {}_7N^{15}$
b. $_3Li^7 + {}_1H^1 \longrightarrow {}_2He^4 + {}_2He^4$
c. $_6C^{12} + {}_2He^4 \longrightarrow {}_1H^1 + {}_7N^{15}$

22.23. Verify by calculation the statement on page 579 about the density of nuclear matter by applying Eq. 22.7 to a nucleus of intermediate weight, such as silver. Show that the order of magnitude of the result is consistent with the facts that the density of a solid is around 10^4 kg/m^3, the diameter of an atom is about 10^4 times the diameter of the nucleus, and virtually all the mass of the atom is that of the nucleus.

22.24. A nuclear power generator designed for use on the moon employs Cm^{242} as a fuel. This nuclide emits an alpha particle when it decays, and part of the energy of the alphas is converted directly into electric energy by means of thermoelectric or thermionic devices. Write the reaction equation for this alpha decay.

22.25. The silver nucleus Ag^{107}, when struck by a neutron of the proper energy, becomes the artificially radioactive nucleus Ag^{108}, and gamma rays are emitted. The Ag^{108} decays by emitting a beta electron. Write the reaction for each of these two processes, and identify the residual nucleus in the second reaction.

22.26. Give as many examples as you can of chain reactions other than the nuclear kind.

22.27. In a certain fission reaction a neutron is captured by a U^{235}, and two neutrons and two fission fragments are formed. If one of the latter is $_{38}Sr^{95}$, identify the other fragment.

22.28. The critical mass in air of an isolated spherical sample of Pu^{239} metal is about 16 kg. If the density of the metal is 2×10^4 kg/m^3, what is the approximate diameter of the sphere of critical size?

22.29. The volume of a sphere of radius R is proportional to R^3, and the surface area is proportional to R^2. In exploding a fission bomb, suppose that two equal spherical masses of fissile material are rammed together to form a single spherical mass. Assuming no change in the total volume, compare the volume-to-area ratio for the combined mass with that of the pair of original masses. If the mass of each of the original objects is just less than half the critical mass, is an explosion likely to result?

22.30. Justify the assertion that, when we use the energy of fossil fuels such as coal, petroleum, etc., on earth, we are using nuclear energy indirectly.

22.31. Verify the figure for the rate of decrease of solar mass as given on page 601. The mass of the sun is 2×10^{30} kg, and 1 ton = 910 kg.

22.32. Estimate the mass of hydrogen that must undergo fusion, each second, in the sun in order to account for the observed output of radiant energy. Do not confuse this with the *loss* of solar mass per second, referred to in the preceding problem.

Chapter 23
PARTICLES AND PROSPECTS

In the long search for the basic constituents of the universe, our understanding of the term elementary (or fundamental) particle has had to undergo several drastic revisions. Toward the beginning of the present century, atoms and molecules as the ultimate material entities gave way to electrons and protons. Then, in a single year (1932), the neutron and the positron were added to the list. The photon had already arrived on the scene, and the neutrino had been conjectured although not yet observed. Later, numerous additions to the group resulted from work on cosmic rays as well as from experiments using high-energy particle accelerators.

Hundreds of particles are now known, and in spite of intensive worldwide research aimed at understanding their properties and classifying them, what has been accomplished so far appears to be only a beginning. The great number of these particles suggests that some important unifying and simplifying principle may still be eluding us, presenting a problem which is at the frontier of our understanding of the physical world.

23.1 COSMIC RAYS

It has been known for a long time that an electroscope gradually loses its charge at a rate faster than can be accounted for by leakage across the insulating support. This discharge has been traced to the existence of ions in the surrounding air, which normally contains several hundred per cubic centimeter.

At the beginning of this century it was believed that the ionization was caused solely by radioactive materials in the earth's crust. However, electroscopes carried aloft in balloons showed that the ionization increases with height. The Austrian-born physicist V. F. Hess (1883–1964) found that at an altitude of $5\frac{1}{2}$ mi the effect was eight times as great as at sea level. After studying his observations, he announced: "The results . . . are best explained by the assumption that a radiation of very great penetrating power enters our atmosphere from above. . . ." The phenomenon was soon named **cosmic radiation.**

Hess was the winner, with C. D. Anderson, of the Nobel Prize for physics in 1936.

Intensive investigations were begun by many physicists. Measurements of the radiation which were made at ground stations, beneath the sea, and in the atmosphere with detectors carried by high-altitude balloons and rockets yielded information on the nature and origin of the rays. It is now established that the primary rays coming in from space consist of charged particles. About 90 percent are protons, 9 percent alpha particles, and 1 percent heavier nuclei. A few electrons and positrons are also present. The energy of the incoming particles is usually below about 15 GeV, but on rare occasions particles arrive with energies up to 10^{11} GeV, which is more than a billion times the energy attainable in the largest existing particle accelerators. Some of the primary particles originate in the sun, but most of them appear to come from other sources within our galaxy or even beyond it. It was first suggested by Fermi that they get their energy from fields connected with flares on the sun and on other stars, and from magnetic fields in interstellar space. Also, recent observations show that the explosion of certain stars or galaxies could provide particles of sufficient energy.

1 GeV (giga-electron volt) equals 10^9 eV (see page 393).

Primary cosmic-ray particles seldom reach the surface of the earth because they interact violently with nuclei of the gases in the air, producing a variety of changes (see Fig. 23.1). Through pair production, Bremsstrahlung, etc., a single primary proton may give rise to thousands of photons, electrons, positrons, and other secondary cosmic-ray particles (cosmic-ray "showers"). Certain kinds of secondary particles are unstable, decaying into other varieties of elementary particles and neutrinos after average lifetimes that range from 10^{-20} up to 10^{-6} sec. These particles, when classified primarily according to their mass, belong to families called **leptons, mesons,** and **hyperons.** For some purposes it is convenient to group

Names of particle families come from the Greek leptos, small; mesos; middle; hyper, beyond; bary, heavy.

An "exploding" galaxy, such as M 82 shown here, may be a prime source of cosmic rays. (*Photograph from the Mount Wilson & Mount Palomar Observatories.*)

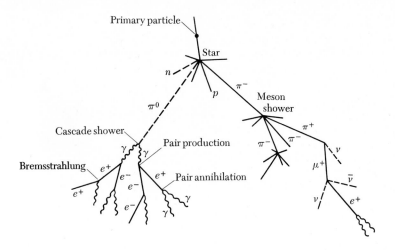

Primary particle

Star

n

p

π^-

π^0

Meson shower

Cascade shower

γ γ

π^- π^-

π^+

Pair production

ν

Bremsstrahlung e^+

e^-

e^+ Pair annihilation

μ^+

e^+

e^-

e^-

γ

$\bar{\nu}$

γ

ν e^+

FIGURE 23.1 Typical events resulting from the entry of a high-energy primary particle into the atmosphere. The primary collides with a nucleus and produces a variety of fragments. Neutral pi mesons that form decay into high-energy gamma rays which then produce cascade showers. Charged pi's decay into mu mesons which then break down into electrons of both signs. The radiation reaching sea level is made up of about 70 percent mu's, about 30 percent electrons and positrons, with about 1 percent remaining pi's. Paths shown by the dashed lines are those of neutral particles, which leave no visible tracks in detectors.

protons and neutrons together with the hyperons, making a larger aggregation called **baryons.**

Table 23.1, which should be examined closely, represents a simplified collection of the measured and assigned properties of the elementary particles and their antiparticles and shows that a certain measure of success has been attained in their classification. Electric charges on these particles are found to be only $+e$, $-e$, or 0, and spin values are restricted to the multiples 1, $\frac{1}{2}$, or 0 of the basic quantity $h/2\pi$. Other characteristic numbers, called **baryon number** and **isotopic spin,** have been omitted from the representation shown here.

In addition to the particles listed, there exist a great many **resonances.** These appear to be excited states of known elementary particles, and they persist only for times of the order of 10^{-23} sec. The rest masses of the resonances thus far observed range as high as about 6,350 times the electron mass, and two varieties of resonance particles have charges of $+2e$. Because the resonances are not directly observable and their existence can only be inferred from their decay products, it is not clear at present whether they deserve to be classified as particles. They have ultrashort lives, even in comparison with the particles in the table.

All the accepted particles listed in Table 23.1, as well as the resonance particles known at present, can be produced under controlled conditions in particle accelerators, particularly by proton beams that have been given energies of several GeV. In fact, the 6.2-GeV proton accelerator put into operation at the University of California in 1953 was especially designed for searching for the antiproton. In 1955, O. Chamberlain, E. Segré, and their collaborators succeeded in producing this particle. They shared the 1959 Nobel physics prize for their accomplishment.

The new accelerator at Weston, Illinois, is expected to produce beams of protons with energies up to 500 GeV.

23.2 QUANTUM FIELDS AND PARTICLES

In earlier chapters of this book, the concept of a field has often been used as a means of representing the transmission of forces through space. The actions of gravitational, electric, and magnetic forces were depicted in this

TABLE 23.1 ELEMENTARY PARTICLES

FAMILY	PARTICLE	MASS, MULTIPLES OF m_e	AVERAGE LIFETIME, sec	PARTICLES — SPIN 1/2 — +1	0	−1	s[4] +1	0	−1	ANTIPARTICLES — 1/2 — +1	0	−1
				CHARGE, MULTIPLES OF e								
Photon	Photon	0	Stable					γ				
Graviton	Graviton	0	Stable					(−)				
Leptons	Neutrino[5]	0	Stable		ν						$\bar{\nu}$	
	Electron, positron	1	Stable			e^-				e^+		
	Muon	206.8	2.2×10^{-6}			μ^-				μ^+		
Mesons	Pion	264.2	8.9×10^{-17}					π^0				
		273.2	2.6×10^{-8}				π^+		π^-			
	Kaon	966.6	1.2×10^{-8}				K^+		K^-			
		974.2	[6] $\begin{cases} 8.7 \times 10^{-11} \\ 5.3 \times 10^{-8} \end{cases}$					$K^0, \bar{K}^0$				
Baryons	(nucleons) Proton	1,836.12	Stable	p^+								$\bar{p}$
	Neutron	1,838.65	930		n						$\bar{n}$	
	(hyperons) Lambda	2,183	2.5×10^{-10}		Λ^0						$\bar{\Lambda}^0$	
	Sigma	2,327	8×10^{-11}	Σ^+								$\bar{\Sigma}^+$
		2,333	$<1 \times 10^{-14}$		Σ^0						$\bar{\Sigma}^0$	
		2,342	1.6×10^{-10}			Σ^-				$\bar{\Sigma}^+$		
	Xi	2,573	3×10^{-10}		Ξ^0						$\bar{\Xi}^0$	
		2,585	1.7×10^{-10}			Ξ^-				$\bar{\Xi}^-$		
	Omega	3,272	1.1×10^{-10}				$\bar{\Omega}^+$		Ω^-			

Notes:
1. The members of a particle-antiparticle pair are identical with regard to mass, spin, and average lifetime.
2. Where otherwise not distinguishable from the symbol for its particle, that of the antiparticle has a bar over it.
3. A neutral particle differs from its antiparticle with regard to the direction of spin.
(4). The spin is 0 for mesons, 1 for the photon, $\frac{3}{2}$ for omega, and 2 for the graviton. (This last particle has been postulated but not yet observed.)
(5). There are two neutrinos. The one involved in beta decay is not identical in all respects with the one taking part in mu decay.
(6). The neutral kaon is unique in consisting of two components having different decay rates.
7. Resonance particles (page 611) have not been listed here.

way. In recent years, an alternative model—**quantum field theory**— has been developed. This theory assumes (Sec. 22.6, page 579) that interactions are produced by the passing of energy packets back and forth between the interacting particles. This view was suggested by the discovery of new kinds of forces, such as the nuclear force, which had no explanation on the basis of classical field theory.

According to quantum field theory, the interaction particle (or quantum) of the electromagnetic field is the photon, and charged particles are assumed to exert forces on each other by continual exchange of photons. Since these photons do not manifest themselves as ordinary radiation, they are referred to as *virtual* photons. It should be remembered that this model is merely an attempt at visualizing a situation best described by the abstract mathematical formalism of the theory.

When the existence of the nuclear force became apparent, it was natural that quantum field theorists should attempt to account for it on an exchange basis also. In 1935, Japanese physicist H. Yukawa calculated that the exchange particle of the nuclear force should be about 200 times as massive as an electron. No such entity was known at that time, but in 1936 the American experimenters C. D. Anderson and S. H. Neddermeyer found a particle of about the computed mass in their observations on cosmic rays. Curiously enough, this did not turn out to be the Yukawa particle because it did not interact vigorously with atomic nuclei but passed unchanged through appreciable thicknesses of matter. The observed particle is called the **muon.** About 70 percent of the cosmic-ray particles observed at sea level are muons.

In 1947, English physicist C. F. Powell and his group discovered a particle of somewhat greater mass than the muon, the π meson, or **pion,** which has the properties required to furnish the nuclear force. Thus the nuclear force, which binds protons and neutrons within a nucleus, is ascribed to the exchange of (virtual) pions. Outside the nucleus, pions quickly break down into muons, and for this reason they remained undiscovered for some time.

Two kinds of forces between elementary particles have already been discussed: (1) *electromagnetic* interactions, ascribed in quantum field theory to the exchange of virtual photons between charged particles, and (2) *strong* (nuclear) interactions, arising from the interchange of virtual pions among nucleons and other baryons as well as mesons. A third type, the *weak* interaction, includes such processes as beta decay (Sec. 22.7, page 583). A fourth category is *gravitational* interaction, which presumably exists between particles of any kind. If gravitational forces are to be included in quantum field theory, a suitable exchange particle must exist. Such a particle, which is referred to as the **graviton,** has not yet been detected.

The four types of particle interactions have widely different relative strengths, characterized by the numbers appearing in Table 23.2. These numbers may be thought of as the relative magnitudes of the four forces existing between two protons placed a fixed distance apart within a nucleus. Note from the table that the stronger interactions proceed tremendously more rapidly than the weaker ones. A complication in the systematics of elementary particles is that a given kind of particle can in most cases respond to more than one of the four types of interparticle forces.

Anderson was the discoverer of the positron (page 514).

The muon is sometimes called the μ meson.

Powell received the 1950 Nobel physics prize for the discovery of the pion.

23.3 CONSERVATION LAWS AND SYMMETRY

The decays and other interactions of particles are governed not only by certain intrinsic properties but by conservation laws. Some of these laws, such as conservation of *mass-energy, linear momentum, angular momentum,* and *electric charge,* are familiar because of their applications to macroscopic systems, whereas others were specifically formulated in order to correlate observations on elementary particles. As far as is known at

TABLE 23.2 PARTICLE INTERACTIONS

TYPE	FUNCTION	EXCHANGE PARTICLE	RELATIVE STRENGTH OF FORCE	LIFE OF TYPICAL PROCESS*
Strong	Holds nucleons together to form nuclei	Pion	1	10^{-23} sec
Electromagnetic	Binds electrons to atoms and joins atoms to form molecules and crystals	Photon	10^{-2}	10^{-8} sec
Weak	Makes leptons interact with each other and with heavier particles	Not yet identified	10^{-14}	10^{-9} sec
Gravitational	Causes universal attraction between macroscopic bodies	Graviton (not yet observed)	10^{-40}	Very long

* Nominal values; ranges in each category may extend over several powers of 10.

present, all elementary processes conform to the foregoing conservation principles.

Other conservation laws, applicable only to some groups of particles, have been added to the above list. For example, a law of *conservation of baryon number* states that in any process the total number of baryons remains constant. In applying this rule, each baryon must be given the number $+1$, and each antibaryon the number -1. A special case of this law is already familiar in the conservation of mass number in nuclear reactions (Sec. 22.9, page 588). There is also a law of *conservation of leptons,* analogous to that for baryons. However, there is no corresponding conservation principle for mesons.

The conservation laws are found to be closely connected with certain symmetries in the description of events, although it is not always clear whether the connection is a rigorous one. For example, the assumption that the laws of physics do not change their form with the passage of time is shown to lead logically to the conservation of mass-energy. Also, if a description of nature is not dependent on location or direction in space, the laws of conservation of linear and of rotational momentum follow. Conservation of electric charge is not as simply related to a symmetry principle, but it is known to have some connection with certain symmetry properties of the wave function. Symmetries that may be related to conservation of baryons and conservation of leptons have not been identified.

Certain additional conservation laws, applicable only to particular families of particles, will not be discussed in detail here except for one, described in the next paragraph, which deserves some attention because it led to an unexpected turn in recent theoretical developments.

The prevalence of symmetries in classical physics led naturally to the expectation that such regularities might be applicable to microphysical processes as well. For some time, this seemed to be valid, and physicists formulated a rule called **conservation of parity** to describe the situation.

to evolve. Progress in microphysics required the abandonment of visualizable models and their replacement by mathematical ones. Belief in visualization has been shown not to be a necessity but only a habit carried over from experience with the macroscopic world. According to Dirac (page 546n),

. . . the main object of physical science is not the provision of pictures but the formulation of laws governing the phenomena and the application of these laws to the discovery of new phenomena. If a picture exists, so much the better; but whether a picture exists or not is a matter of only secondary importance. In the case of atomic phenomena no picture can be expected to exist in the usual sense of the word 'picture,' by which is meant a model functioning essentially on classical lines. One may extend the meaning of the word 'picture' to include any way of looking at the fundamental laws which makes their self-consistency obvious. With this extension, one may acquire a picture of atomic phenomena by becoming familiar with the laws of quantum theory.

In a famous pronouncement, Einstein said: "God does not play dice!"

Although most physicists are reconciled to the increasing mathematization of their subject, there are many who have reservations. Einstein himself was reluctant to accept some of the implications of this trend, particularly the renunciation of strict determinism, and it is by no means impossible for future developments to swing back toward a more deterministic view.

Dirac. (*The Bettmann Archive, Inc., New York.*)

In recent years, several theorists, notably the American physicist David Bohm, have made attempts in this direction. From our present vantage point, however, it seems extremely unlikely that science will ever revert to the concept of a world whose processes and participating entities are visualizable in detail.

Programmed review

Instructions: See page 18.

1. Of what do primary cosmic rays consist?

    High-energy charged particles, mainly protons, along with some heavier nuclei.
    [23.1]

2. Name some secondary cosmic-ray components produced in the atmosphere.

    Photons, electrons, positrons, neutrinos, and various kinds of elementary particles.
    [23.1]

3. Classify elementary particles into families, grouped in order of increasing mass.

    Photons, gravitons, leptons, mesons, nucleons, and hyperons. The last two are collectively called baryons.
    [23.1]

4. What is the current view of the nature of the particle resonances?

    They appear to be short-lived excited states of known elementary particles.
    [23.1]

5. Describe the basic premise of quantum field theory.

    Each kind of interparticle force arises through the continual exchange of a characteristic field quantum between the interacting particles.
    [23.2]

6. Name the field quantum associated with each of the known types of force: (*a*) strong (nuclear), (*b*) electromagnetic, (*c*) weak, (*d*) gravitational.

    (*a*) Pion, (*b*) photon, (*c*) and (*d*) not yet observed.
    [23.2]

7. What is meant by the statement that parity is not conserved in weak interactions?

    There is an absence of complete mirrorlike symmetry in such processes, as first observed in the asymmetric ejection of electrons in beta decay.
    [23.3]

For further reading

23.1.   *Korff, S. A.*   The Origin and Implications of the Cosmic Radiation, in A. Beiser (ed.), "The World of Physics," pp. 229–246. By one of the foremost experimenters in this field.

23.2. *Burbidge, G.* The Origin of Cosmic Rays, *Sci. Am.,* August, 1966, p. 32.

23.3. *Frisch, O. R.* "Experimentation with Elementary Particles," Penguin Science Survey A, 1964, p. 9. The originator of the liquid-drop model of the nucleus discusses methods, instruments, and aims.

23.4. *Ford, K. W.* "The World of Elementary Particles." A readable, informative book.

23.5. *Park, D.* "Contemporary Physics." This brief book includes the background of classical physics needed to understand recent developments. Read especially chaps. 6, 7, and 10.

23.6. *Morrison, P.* The Overthrow of Parity, *Sci. Am.,* April, 1957, p. 45.

23.7. *March, A.,* and *Ira M. Freeman* "The New World of Physics." Read especially pp. 3–21, 139–147, and 174–189 in connection with this chapter.

23.8. *Dirac, P. A. M.* The Evolution of the Physicist's Picture of Nature, *Sci. Am.,* May, 1963, p. 45. An attempt by a master to predict the future course of physical theory.

Questions and problems

23.1. What is the maximum number of electron-positron pairs that can be produced in a cascade shower initiated by a photon of wavelength 5×10^{-5} Å?

23.2. A primary cosmic-ray proton with an energy of 5 GeV moves in a magnetic field of $B = 1 \times 10^{-6}$ Wb/m^2, which is the approximate value of the field high in the earth's atmosphere. Show that, if the proton approaches roughly in the plane of the magnetic equator, it has a good chance of circling the earth in the manner of a satellite. HINT: Find the radius of the path, using Eq. 18.4 (page 439), and compare it with the earth's radius. Since the particle is traveling at very nearly the speed c, the relativistic expression $E = pc$ must be used, where E is the energy of the particle and p is its momentum.

23.3. Neutral pions are produced when high-energy protons impinge on target protons, which may be assumed to be at rest, according to the equation

$$p + p \longrightarrow p + p + \pi^0$$

If the rest mass of the neutral pion is $264m_e$, where m_e is the rest mass of the electron, what minimum KE must the incoming proton supply? A relativistic calculation shows, however, that conservation of momentum can be satisfied only if the KE of the incident proton is about twice this amount.

23.4. Calculate the energy released in the decay of a positive muon (see Prob. 23.6 for the reaction equation).

23.5. One mode of decay of the positive kaon is

$$K^+ \longrightarrow \pi^+ + 2\pi_0$$

Compute the reaction energy Q. Consult Table 23.1 for mass values.

23.6. Positive and negative muons decay, respectively, according to the following schemes:

$$\mu^+ \longrightarrow e^+ + \nu + \bar{\nu}$$
$$\mu^- \longrightarrow e^- + \nu + \bar{\nu}$$

Remembering that each lepton has a lepton number $+1$ and each antilepton -1, assign such numbers to the particles in these equations with the help of Table 23.1, and show that the results are consistent with the conservation of leptons.

23.7. In harmony with the nonconservation of parity in weak interactions, the neutrino and the antineutrino must be taken to have their spins in opposite senses. In the equations for muon decay given in Prob. 23.6, both the neutrino and its antiparticle appear on the right. What conservation principle is thereby satisfied?

23.8. Verify the conservation of rotational momentum for the decay of Σ^+, which can break down into either a neutron and a positive pion or a proton and a neutral pion.

23.9. Examine the equations for the beta decay of the proton and of the neutron,

$$p \longrightarrow n + e^+ + \nu$$
$$n \longrightarrow p + e^- + \bar{\nu}$$

from the point of view of (a) conservation of baryons and (b) conservation of leptons, and show that both principles are satisfied.

23.10. In the first successful laboratory experiment in which antineutrinos were observed, these particles were detected by their reaction with protons:

$$\bar{\nu} + p \longrightarrow n + e^+$$

If the target protons are assumed to be free and at rest, what minimum KE must the antineutrinos have?

23.11. A stationary π^0 (neutral pion) decays into two photons. Explain why these photons have the same frequency and move off in exactly opposite directions. If the mass of a π^0 is $264m_e$, find the frequency of each of the photons.

23.12. What is the wavelength of the photon produced in the decay of a Σ^0 at rest into a Λ^0? Take the energy equivalents of the exact masses of these particles to be 1,193.2 and 1,115.4 MeV, respectively, and assume that all the energy released is given to the photon.

APPENDIX

PROPORTION AND VARIATION

In the study of physics and in the expression of its laws there frequently arises the need for stating the exact relation between two or more quantities, and most physical laws find their only adequate expression in this form. Such statements often involve the concept of *proportion*. In ordinary affairs we hear such expressions as "the greater the effort, the greater the gain" or "Less risk is involved if the investments are more diversified." In physics an effort is made to express quantitatively the nature of the variation of one quantity with one or more other quantities.

Direct proportion

In one of the simplest instances of a relation between two variables, one of them is directly proportional to the other. This means that if one quantity is doubled, the related quantity also becomes doubled; if one increases threefold, the other increases in value three times, etc. It may be said that *y is proportional to x*, where x and y are the things concerned. For example, x might represent the cost of one article of a given kind, and y might be the cost per dozen, per gross, or per hundred; or x could represent the length of a wire and y its electrical resistance.

The statement "y is proportional to x" may be put in symbols in a number of ways. Using $\propto$, the sign of proportion, the direct shorthand expression is

$$y \propto x \hspace{4cm} \text{A.1}$$

Another way of expressing the same fact, and one that lends itself to computation, is

$$y = kx \hspace{4cm} \text{A.2}$$

where k is a constant. This means that any y value is obtained by multiplying the corresponding x value by a number k which has a constant numerical value throughout the problem. Thus if y represents the distance covered by a helicopter traveling at constant speed and x is the elapsed time, we might have the following set of values:

x, hr	1	2	3	4	5	etc.
y, mi	250	500	750	1,000	1,250	

Any y divided by the corresponding x is equal to 250 mi/hr, and this is the value of k (Eq. A.2) appropriate to this example:

$$y = 250x \qquad \text{A.3}$$

If the value of any pair of entries in the above table is known, the statement of direct proportion may be expressed in still another way:

$$\frac{x_1}{y_1} = \frac{x_2}{y_2} \qquad \text{A.4}$$

to be read, "x_1 is to y_1 as x_2 is to y_2." Here x_1, y_1 and x_2, y_2 represent any two pairs of values. If any three of these quantities are known, the fourth may be found. For example, how far will the helicopter in the preceding problem go in 3 hr 24 min (3.4 hr)? Using any known pair of values of time and distance, say the second pair in the table, Eq. A.4 can be written in the form

$$\frac{2}{500} = \frac{3.4}{y}$$

Cross-multiplying (clearing of fractions) yields $2y = (3.4)(500)$, or $y = 850$ mi.

Note that the relation expressed by Eq. A.4 can also be written in any of the following ways:

$$x_1 y_2 = x_2 y_1 \qquad \frac{x_1}{x_2} = \frac{y_1}{y_2} \qquad \frac{y_1}{x_1} = \frac{y_2}{x_2} \qquad \text{A.5}$$

The numerical example could have been solved also by using Eq. A.3. Substitution gives $y = (3.4)(250) = 850$ mi, as found above.

If two quantities x and y are in direct proportion, a graph of y against x will be a *straight line* passing through the origin of the coordinate system. Plotting experimental data and obtaining points that lie, within experimental error, on a straight line is a convenient way of recognizing the existence of such proportionality between the two variables. Figure 17.4a (page 409) illustrates the *linear* relation (direct proportion) between electric current and applied potential for a metallic conductor.

Sometimes the straight lines do not pass through the origin, as illustrated by Fig. 5.11 (page 99). Then the relation between the variables must be written $y = a + kx$, where a and k are constant. Shifting the origin by an amount a will again give the simple proportionality relation $y = kx$.

The fact that the circumference of a circle is proportional to its radius is expressed by the formula

$$C = 2\pi r \qquad \text{A.6}$$

where C is the circumference and r is the radius. This equation expresses this proportionality in the form of Eq. A.2, with k equal to 2π ($= 2 \times 3.1416 \ldots$).

Quadratic and higher proportion

If the question is how the *area* of a circle is related to its radius, the answer is expressed by the geometric relation

$$A = \pi r^2 \qquad\qquad \text{A.7}$$

This formula is slightly more complex than the previous one. Here there is one quantity, A, proportional to the *square* of another quantity, r. This means that, if the radius is doubled, the area of the circle becomes 2^2, or 4, times as great as before; 3 times the radius means 9 times the area, etc. Corresponding to Eqs. A.1 and A.4, this fact could be written

$$A \propto r^2 \quad \text{or} \quad \frac{A_1}{A_2} = \frac{r_1^2}{r_2^2} \qquad\qquad \text{A.8}$$

Example

How much more paint does it take to cover a disk 10 ft in diameter than to cover one that is only 2 ft in diameter? Since areas are proportional to squares of diameters (or to squares of radii),

$$\frac{A_1}{A_2} = \frac{10^2}{2^2} \quad \text{or} \quad \frac{A_1}{A_2} = 25$$

The ratio of diameters is 5, the ratio of the areas is $5^2 = 25$, so that 25 times as much paint will be needed.

The volume of a sphere is proportional to the *cube* of the radius (or diameter):

$$V = \tfrac{4}{3}\pi r^3 \qquad\qquad \text{A.9}$$

Double the radius, and the volume increases $2^3 = 2 \times 2 \times 2 = 8$ times; treble the radius, and it increases 27 times. Thus, since the diameter of the sun is about 100 times that of the earth, the volume of the sun is about 1 million times that of the earth.

The surface areas of geometrically similar bodies are proportional to the squares of corresponding dimensions, and their volumes are proportional to the cubes of corresponding dimensions. For instance, if an airplane model is built to a linear scale of 1 to 20, the wing area of the actual plane is 400 ($= 20^2$) times that of the model. The cabin space (volume) is $20^3 = 8,000$ times that of the model.

Inverse proportion

Two magnitudes x and y so related that

$$\frac{x_1}{x_2} = \frac{y_2}{y_1} \qquad\qquad \text{A.10}$$

are said to be *inversely proportional* to each other. Alternative ways of expressing this are

$$x \propto \frac{1}{y} \qquad x = \frac{k}{y} \qquad xy = k \qquad\qquad \text{A.11}$$

where k is a constant.

In this kind of relation, in contrast with previous examples, one quantity *decreases* as the other increases. For instance, when the temperature of a gas is held fixed, the volume occupied is inversely proportional to the applied pressure (Boyle's law, page 223). This means that if the pressure is doubled, the new volume becomes half of the old; if the initial pressure is trebled, the volume changes to one-third its initial value, etc.

Inverse-square proportion; other forms

Another important form of relation between two quantities x and y is expressed by

$$\frac{x_1}{x_2} = \frac{y_2{}^2}{y_1{}^2} \qquad\qquad \text{A.12}$$

The quantity x is said to be *inversely proportional to the square* of y. Using the other forms of writing proportions, this is expressed

$$x \propto \frac{1}{y^2} \qquad x = \frac{k}{y^2} \qquad \text{or} \qquad xy^2 = k \qquad\qquad \text{A.13}$$

where k is a constant. The force of gravitational attraction between two particles of matter is inversely proportional to the square of the distance between them. Doubling the distance makes the force one-fourth as great; reducing the original distance to one-third increases the force nine times, etc.

In addition to the above forms which are among those most frequently used in elementary physics, many other types of variation may be encountered. In particular, several forms may be combined, as in the algebraic expression $a = kbc^2 \sqrt{d}/e^3$.

A good reference on the common sense of the subject of proportion is W. W. Sawyer, "Mathematician's Delight," chap. VIII.

SIGNIFICANT DIGITS

Every physical measurement inevitably has some uncertainty associated with it, and the number expressing the result should indicate not only the value of the measured quantity but the precision of the measurement as well. The precision of a series of determinations of a given quantity is a measure of how well the measurements agree among themselves.

Suppose it is required to compute the volume of a rectangular block of metal from measurements of its three dimensions: length, 20.02 cm; width, 7.11 cm; height, 3.67 cm. Since the smallest division on the rule that was used is 1 mm (0.1 cm), the last digit in each of these readings had to be *estimated* by dividing a millimeter interval by eye. Multiplying the three numbers, using direct, longhand arithmetic, gives a product 522.395874 cm³ as the volume of the block.

What does this result mean? How much trust can be placed in it? There are nine digits in this number, so that if all but the last is to be trusted, the volume has been found to within 0.000001 cm³. Since the volume of the block is a little over 500 cm³, this would mean that the volume has been determined to within 1 part in 500 million. There are very few scientific measurements that can be made to this exactness, and we suspect at once that relatively crude work with a measuring stick could not possibly be one of them. Just how precise is the result for the volume of the block? Should it be written 522.3958, 522.4, or something else?

There is nothing wrong with the rules of multiplication, and the nine-digit number obtained as a product is arithmetically correct. The difficulty arises from the circumstance that *each of the factors in this product represents a physical measurement whose last digit is uncertain.* This introduces uncertainty into the answer and, as a result, the nine-digit number obtained by arithmetic gives false information about the volume of the block. It gives more than was put into the problem in the form of the original measurements.

In any number representing a physical measurement, only a certain number of digits are what may be called significant. A significant digit is one that carries legitimate information. For example, there are three significant digits in 3.67 cm, the measured height of the block. To be sure, the "7" is only an estimate, but it adds something to our knowledge of this dimension. It is significant. By contrast, any digit that might be put after the "7" in the above number would be meaningless. The method of measurement used could not give such information. *In counting the number of significant digits, only the first uncertain digit is included.* It may be said that the number of significant digits is determined by the number of *decisions* made in arriving at the number.

The only practical way to increase the number of significant digits in measuring the dimensions of the block would be to replace the steel rule by a more exact instrument, such as the micrometer caliper. Then the reading for the height of the block may be 3.6723 cm, or perhaps 3.6681, the final digit having been estimated between the smallest divisions on the micrometer scale. Each of the numbers quoted has five significant digits, as compared with only three for the measurement made with the rule. Each one, when rounded off to three digits, becomes 3.67, which agrees with the steel-rule value.

With the steel rule, any block whose actual height was as small as 3.665 or as large as 3.675 could have this measurement recorded as 3.67. Similar statements are true for each of the other two dimensions.

The following are some principles to be observed in handling questions involving significant digits. There is an element of arbitrariness connected with such working rules, and a rigorous treatment must be based on the theory of probability. Nevertheless, consistent results that honestly represent the data can be obtained by following these rules.

1. *In dropping doubtful figures* ("rounding off"), if the first digit to be dropped is:
   a. less than 5, the last digit kept should be left as it is.
   b. greater than 5, increase the last retained digit by 1.
   c. 5 itself, leave the preceding digit unchanged if it is even, but increase it by 1 if odd. Doing this helps to avoid biasing the result when this number is used in a calculation (see item 4, below).

Examples

   3.5401 rounded off to two digits becomes 3.5.
   9.536 rounded off to three digits is 9.54.
   3.65 rounded off to two digits becomes 3.6, but 3.95 becomes 4.0.

2. *In multiplication and division,* keep the same number of significant digits in the result as there are in the *least precise* factor.

Examples

$(11.5)(0.063) = 0.7245$. This should be rounded off to two significant digits—the same number as in the second factor—and written 0.72. $457.6/8.3 = 55.1$ when the division is carried out to three digits (not decimal places). The result is then rounded off to two digits and is written 55.

3. *In determining powers or roots of a number,* retain the same number of significant digits in the result as there are in the given number. Since these processes are intrinsically multiplication, this rule follows from the preceding one.

Examples

$(13.1)^2 = 172$; $\sqrt[3]{106} = 4.73$.

4. *In addition or subtraction,* record the numbers in columnar form. Round off, retaining no column that is to the right of a column containing a doubtful digit, then perform the indicated operation.

Example

To add 6904.5, 0.055, 82.674, and 525.0, round off to give 6904.5, 0.1, 82.7, and 525.0. The sum is 7512.3.

Zeros are significant if they occur between other digits, as in 203.5, 40.09, 7.021×10^{-6}. Each of these numbers has four significant digits. If zeros follow other digits that are to the right of the decimal point, they are significant. For instance, writing 92.700 sec indicates that a time interval was measured to the nearest 0.001 sec, which is 100 times as precise as if it were stated as 92.7. In this case the second and third places after the decimal point happen to be zeros. They could just as well have been other integers, and so both are significant, and the original number has five significant digits.

Zeros are sometimes significant when they follow other digits to the left of the decimal point. If they were actually read from an instrument, they are significant, and that fact should be shown. One way to do this is to underscore such zeros, as in 120,000 or 86,000. On the other hand, zeros used merely to show the position of the decimal point are not significant. Examples are the three zeros following the underscored one in 120,000 or the final one in 86,000.

Zeros are never significant if they come between the decimal point and the only digits in the number. Examples are 0.004789 (four significant) and 0.0500 (three significant). As in the two examples in the preceding paragraph, these zeros serve only to show where to put the decimal point. They carry no numerical information. To see this another way, consider the measurement 0.31 mm, which has two significant digits. This length can just as well be expressed in meters by shifting the decimal point three places to the left, and it then becomes 0.00031 m. Merely changing units cannot change the exactness of a measurement, and so the new number must still have only two significant digits.

The best way to avoid confusion is to use the denary form (page 31), writing 3.1×10^{-1} mm, 3.1×10^{-4} m, etc. In this form, it is evident that both have the same number of significant figures. Remember that denary numbers can be added or subtracted directly only if both are written with the *same* power of 10. For example, $7.35 \times 10^{-3} + 4.2 \times 10^{-4} = (7.4 + 0.42) \times 10^{-3} = 7.8 \times 10^{-3}$.

There are certain numbers for which it is meaningless to specify the number of significant digits. These are *exact* numbers, such as the number of people in a room, or the "2" in the formula $C = 2\pi r$. Exact numbers, unlike numbers that represent measurements, do not limit the number of significant digits in a result.

It is often difficult to be consistent about significant digits, and in dealing with them in a practical problem, some judgment must be used, based on what is known of the situation. For instance, suppose a 10-lb weight is mentioned in a laboratory procedure. There can be no doubt that the "0" is significant, even if not so marked, for it certainly is possible to manufacture a metal weight of this size and be sure of its value to a small fraction of a pound. Most likely, in this case, the weight is known to within 0.01 lb, so that the expression "a ten-pound weight" really means one of magnitude 10.00 lb (four significant digits).

THE LAWS OF EXPONENTS

Many of the algebraic expressions used in physics involve quantities that are raised to certain powers. A simple integral *exponent* applied to a quantity is merely a shorthand notation indicating that the quantity is to be raised to the indicated power—that is, to be multiplied by itself the indicated number of times. For instance A^4 is a compact way of indicating $A \cdot A \cdot A \cdot A$. The power (or exponent) 4 used here indicates that A is to be used as a factor four times over.

The following rules then result for exponents that are integers:

1. If two or more quantities, each a power of the same base, are *multiplied,* the product is the base raised to a power equal to the *sum* of the separate powers. Thus $2^2 \times 2^3 = 2^5 = 32$.
2. If one power of a given quantity is *divided* by a smaller power of the same quantity, the quotient is given by the base raised to a power equal to the exponent of the dividend *minus* the exponent of the divisor. For example, $3^5/3^2 = 3^3 = 27$.
3. If a number given as a power of a certain base is raised to another power, the result is equal to the base with an exponent equal to the *product* of the two exponents. Thus

$$(2^4)^3 = 2^{12} = 4{,}096$$

The meaning of these rules for operations with exponents can readily be extended to include negative and fractional powers. Then the general rules become

(1) $A^m A^n = A^{m+n}$

(2) $\dfrac{A^m}{A^n} = A^{m-n}$ A.14

(3) $(A^m)^n = A^{mn}$

These rules hold when either m or n or both are any rational numbers. The exponent *zero* also has a meaning. Inasmuch as A^m/A^m always equals unity, it follows from rule 2 of Eqs. A.14 that $A^m/A^m = A^{m-m} = A^0 = 1$; that is, any quantity with the exponent 0 is equal to 1.

SOME PROPERTIES OF TRIANGLES

Similar triangles   Corresponding sides of any two geometrically *similar* figures are *proportional.* This fact is particularly useful in connection with triangles. For instance, in certain problems in Chap. 4 dealing with vectors there occur two triangles having their sides respectively parallel or two triangles with corresponding sides perpendicular to each other. In either case the triangles are *similar,* and their *sides are proportional.* Calling the two triangles ABC and $A'B'C'$, we may thus write

$$\frac{AB}{A'B'} = \frac{AC}{A'C'} = \frac{BC}{B'C'}$$

Right triangles   The sides of a right triangle are related very simply by the theorem of Pythagoras, which says that

$$c^2 = a^2 + b^2$$ A.15

where c is the hypotenuse and a and b are the sides adjoining the right angle. If the lengths of any pair of sides are known, the third side may be computed from this relation.

There are particular sets of *integer* values for the sides a, b, and c that satisfy Eq. A.15. Two of the most common combinations are 3, 4, 5 and 5, 12, 13. Equal multiples of these numbers will also satisfy Eq. A.15; for example, a right triangle can be constructed with sides of length 6, 8, and 10.

BASIC TRIGONOMETRIC FUNCTIONS

Trigonometry deals more formally with the relations between the sides and angles of triangles. The size of any acute angle may be specified in terms of certain ratios between the sides of a right triangle of which the given angle is a part. These ratios are called *trigonometric functions* of the angle. We shall restrict attention to the three most frequently used ones, called the *sine, cosine,* and *tangent,* respectively.

In Fig. A.1, θ is the angle to be described, and ACB is a right triangle constructed on it. The sine of theta, written $\sin \theta$, is defined as the ratio of the opposite side a to the hypotenuse c. The cosine of theta, written $\cos \theta$, is the ratio of the adjoining side b to the hypotenuse c. The tangent of theta, written $\tan \theta$, is the ratio of the opposite side a to the adjoining side b. Thus

$$\sin \theta = \frac{a}{c} \qquad \cos \theta = \frac{b}{c} \qquad \tan \theta = \frac{a}{b}$$ A.16

FIGURE A.1   The right triangle.

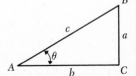

Values of the functions have been computed for all values of θ and are available in convenient tables for reference. Table A.5 is an abbreviated table (intervals of 1 deg) of the sine, cosine, and tangent values. From Fig. A.1, it is evident that for acute angles the range of numerical values of these functions is given by

	$\theta = 0°$	$\theta = 90°$
$\sin \theta$	0	1
$\cos \theta$	1	0
$\tan \theta$	0	∞

For certain angles, the values of the functions can be found geometrically, without recourse to tables. Thus sin 45 deg = cos 45 deg = $1/\sqrt{2}$, sin 30 deg = cos 60 deg = $\frac{1}{2}$, tan 45 deg = 1, etc.

From the way they were defined, the sine and cosine of any given angle are simply related. From Fig. A.1, $a^2 + b^2 = c^2$. Dividing through by c^2, we get $(a/c)^2 + (b/c)^2 = 1$, or

$$\sin^2 \theta + \cos^2 \theta = 1 \qquad\qquad \text{A.17}$$

(This is read: "Sine-squared theta plus cosine-squared theta. . . .") Thus, if the value of either the sine or the cosine of an angle is known, the other is easily found.

CONVERSION FACTORS

Length
1 meter (m) = 39.37 in. = 3.281 ft = 6.214×10^{-4} mi
1 in. = 0.02540000 m; 1 ft = 0.3048 m; 1 mi = 1,609 m
1 angstrom (Å) = 10^{-10} m

Area
1 m² = 10.76 ft² = 1,550 in.²
1 ft² = 929 cm²; 1 in.² = 6.452 cm²

Volume
1 m³ = 35.31 ft³ = 6.102×10^4 in.³
1 ft³ = 0.02832 m³

Time
1 mean solar day = 86,400 sec
1 sidereal day (period of earth's rotation with respect to the stars) = 86,164 sec
1 yr = 365.2422 days = 8.766×10^3 hr = 5.259×10^5 min = 3.156×10^7 sec

Speed
1 m/sec = 3.281 ft/sec = 3.6 km/hr = 2.237 mi/hr = 1.944 knot
1 km/hr = 0.2778 m/sec = 0.9113 ft/sec = 0.6214 mi/hr
1 mi/hr = 1.467 ft/sec = 1.609 km/hr = 0.8689 knot

Mass
1 kg = 2.205 lb-mass
1 lb-mass = 0.4536 kg

Density

1 kg/m³ = 0.001 g/cm³ = 0.06243 lb-mass/ft³
1 lb-mass/ft³ = 16.02 kg/m³

Force

1 newton = 10^5 dynes = 0.1020 kg-wt = 0.2248 lb
1 lb (force) = 4.448 newtons = 0.4536 kg-wt

Work, Energy

1 joule = 0.2389 cal = 9.481×10^{-4} Btu = 0.7376 ft-lb = 6.242×10^{12} MeV
1 kcal = 4,186 joules = 3.968 Btu = 3,087 ft-lb
1 kwhr = 3.6×10^6 joules = 3413 Btu = 860.1 kcal = 1.341 hp-hr

Power

1 hp = 2545 Btu/hr = 550 ft-lb/sec = 745.7 watts = 0.1782 kcal/sec
1 watt = 2.389×10^{-4} kcal/sec = 1.341×10^{-3} hp = 0.7376 ft-lb/sec

Electric Charge

1 electron charge e = 1.602×10^{-19} coulomb
1 coulomb = 6.242×10^{18} electron charges
1 faraday = 96,487 coulombs

TABLES

TABLE A.1 MATHEMATICAL SIGNS AND SYMBOLS

= equals
≅ approximately equal to
≠ is not equal to
> is greater than
≫ is much greater than; is large compared with
< is less than
≪ is much less than; is small compared with
≥ is more than or equal to; is not less than
≤ is less than or equal to; is not greater than
± plus or minus (Example: $\sqrt{9} = \pm 3$)
∝ is proportional to
$\bar{v}$ the average value of v
∞ infinity; large beyond all bounds

TABLE A.2 PREFIXES—INTERNATIONAL SYSTEM OF UNITS

FACTOR BY WHICH UNIT IS MULTIPLIED	PREFIX	SYMBOL
10^{12}	tera	T
10^9	giga	G
10^6	mega	M
10^3	kilo	k
10^2	hecto	h
10	deka	da
10^{-1}	deci	d
10^{-2}	centi	c
10^{-3}	milli	m
10^{-6}	micro	μ
10^{-9}	nano	n
10^{-12}	pico	p
10^{-15}	femto	f
10^{-18}	atto	a

TABLE A.3 THE GREEK ALPHABET

A	α	alpha	N	ν	nu
B	β	beta	Ξ	ξ	xi
Γ	γ	gamma	O	o	omicron
Δ	δ	de'ta	Π	π	pi
E	ϵ	epsilon	P	ρ	rho
Z	ζ	zeta	Σ	σ	sigma
H	η	eta	T	τ	tau
Θ	θ	theta	Υ	υ	upsilon
I	ι	iota	Φ	φ	phi
K	κ	kappa	X	χ	chi
Λ	λ	lambda	Ψ	ψ	psi
M	μ	mu	Ω	ω	omega

TABLE A.4   ASTRONOMICAL DATA

EARTH

mass	5.983×10^{24} kg
mean radius	6,371.22 km
mean distance from sun	1.49×10^8 km
period of rotation relative to:	
a. the sun (mean solar day)	86,400 sec = 24 hr
b. the stars (sidereal day)	86,164 sec = 23.94 hr
period of revolution in orbit	
(sidereal year)	3.156×10^7 sec = 365.25 days

MOON

mass	7.343×10^{22} kg
mean radius	1.74×10^3 km
mean distance from earth	3.84×10^5 km
period of revolution relative to:	
a. the earth	2.551×10^6 sec = 29.53 days
b. the stars	2.360×10^6 sec = 27.32 days

SUN

mass	1.987×10^{30} kg
mean radius	6.96×10^5 km
total radiation rate (luminosity)	3.90×10^{23} kW

TABLE A.5 VALUES OF THE TRIGONOMETRIC FUNCTIONS

ANGLE, deg	SINE	COSINE	TANGENT	ANGLE, deg	SINE	COSINE	TANGENT
1	0.0175	0.9998	0.0175	46	0.7193	0.6947	1.0355
2	0.0349	0.9994	0.0349	47	0.7314	0.6820	1.0724
3	0.0523	0.9986	0.0524	48	0.7431	0.6691	1.1106
4	0.0698	0.9976	0.0699	49	0.7547	0.6561	1.1504
5	0.0872	0.9962	0.0875	50	0.7660	0.6428	1.1918
6	0.1045	0.9945	0.1051	51	0.7771	0.6293	1.2349
7	0.1219	0.9925	0.1228	52	0.7880	0.6157	1.2799
8	0.1392	0.9903	0.1405	53	0.7986	0.6018	1.3270
9	0.1564	0.9877	0.1584	54	0.8090	0.5878	1.3764
10	0.1736	0.9848	0.1763	55	0.8192	0.5736	1.4281
11	0.1908	0.9816	0.1944	56	0.8290	0.5592	1.4826
12	0.2079	0.9781	0.2126	57	0.8387	0.5446	1.5399
13	0.2250	0.9744	0.2309	58	0.8480	0.5299	1.6003
14	0.2419	0.9703	0.2493	59	0.8572	0.5150	1.6643
15	0.2588	0.9659	0.2679	60	0.8660	0.5000	1.7321
16	0.2756	0.9613	0.2867	61	0.8746	0.4848	1.8040
17	0.2924	0.9563	0.3057	62	0.8829	0.4695	1.8807
18	0.3090	0.9511	0.3249	63	0.8910	0.4540	1.9626
19	0.3256	0.9455	0.3443	64	0.8988	0.4384	2.0503
20	0.3420	0.9397	0.3640	65	0.9063	0.4226	2.1445
21	0.3584	0.9336	0.3839	66	0.9135	0.4067	2.2460
22	0.3746	0.9272	0.4040	67	0.9205	0.3907	2.3559
23	0.3907	0.9205	0.4245	68	0.9272	0.3746	2.4751
24	0.4067	0.9135	0.4452	69	0.9336	0.3584	2.6051
25	0.4226	0.9063	0.4663	70	0.9397	0.3420	2.7475
26	0.4384	0.8988	0.4877	71	0.9455	0.3256	2.9042
27	0.4540	0.8910	0.5095	72	0.9511	0.3090	3.0777
28	0.4695	0.8829	0.5317	73	0.9563	0.2924	3.2709
29	0.4848	0.8746	0.5543	74	0.9613	0.2756	3.4874
30	0.5000	0.8660	0.5774	75	0.9659	0.2588	3.7321
31	0.5150	0.8572	0.6009	76	0.9703	0.2419	4.0108
32	0.5299	0.8480	0.6249	77	0.9744	0.2250	4.3315
33	0.5446	0.8387	0.6494	78	0.9781	0.2079	4.7046
34	0.5592	0.8290	0.6745	79	0.9816	0.1908	5.1446
35	0.5736	0.8192	0.7002	80	0.9848	0.1736	5.6713
36	0.5878	0.8090	0.7265	81	0.9877	0.1564	6.3138
37	0.6018	0.7986	0.7536	82	0.9903	0.1392	7.1154
38	0.6157	0.7880	0.7813	83	0.9925	0.1219	8.1443
39	0.6293	0.7771	0.8098	84	0.9945	0.1045	9.5144
40	0.6428	0.7660	0.8391	85	0.9962	0.0872	11.4301
41	0.6561	0.7547	0.8693	86	0.9976	0.0698	14.3007
42	0.6691	0.7431	0.9004	87	0.9986	0.0523	19.0811
43	0.6820	0.7314	0.9325	88	0.9994	0.0349	28.6363
44	0.6947	0.7193	0.9657	89	0.9998	0.0175	57.2900
45	0.7071	0.7071	1.0000	90	1.0000	0.0000	

TABLE A.6 NAMES AND SYMBOLS OF THE ELEMENTS

ELEMENT	SYMBOL	ELEMENT	SYMBOL	ELEMENT	SYMBOL	ELEMENT	SYMBOL
actinium	Ac	erbium	Er	mercury	Hg	samarium	Sm
aluminum	Al	europium	Eu	molybdenum	Mo	scandium	Sc
americium	Am	fermium	Fm	neodymium	Nd	selenium	Se
antimony	Sb	fluorine	F	neon	Ne	silicon	Si
argon	Ar	francium	Fr	neptunium	Np	silver	Ag
arsenic	As	gadolinium	Gd	nickel	Ni	sodium	Na
astatine	At	gallium	Ga	niobium (columbium)	Nb	strontium	Sr
barium	Ba	germanium	Ge	nitrogen	N	sulfur	S
berkelium	Bk	gold	Au	nobelium	No	tantalum	Ta
beryllium	Be	hafnium	Hf	osmium	Os	technetium	Tc
bismuth	Bi	helium	He	oxygen	O	tellurium	Te
boron	B	holmium	Ho	palladium	Pd	terbium	Tb
bromine	Br	hydrogen	H	phosphorus	P	thallium	Tl
cadmium	Cd	indium	In	platinum	Pt	thorium	Th
calcium	Ca	iodine	I	plutonium	Pu	thulium	Tm
californium	Cf	iridium	Ir	polonium	Po	tin	Sn
carbon	C	iron	Fe	potassium	K	titanium	Ti
cerium	Ce	krypton	Kr	praseodymium	Pr	tungsten (wolfram)	W
cesium	Cs	lanthanum	La	promethium	Pm		
chlorine	Cl	lawrencium	Lw	protactinium	Pa	uranium	U
chromium	Cr	lead	Pb	radium	Ra	vanadium	V
cobalt	Co	lithium	Li	radon	Rn	xenon	Xe
copper	Cu	lutetium	Lu	rhenium	Re	ytterbium	Yb
curium	Cm	magnesium	Mg	rhodium	Rh	yttrium	Y
dysprosium	Dy	manganese	Mn	rubidium	Rb	zinc	Zn
einsteinium	Es	mendelevium	Md	ruthenium	Ru	zirconium	Zr

TABLE A.7 PERIODIC TABLE OF THE ELEMENTS

n:	I_A	II_A	III_B	IV_B	V_B	VI_B	VII_B	VIII			I_B	II_B	III_A	IV_A	V_A	VI_A	VII_A	0
1	1 H 1.00797																	2 He 4.0026
2	3 Li 6.939	4 Be 9.0122											5 B 10.811	6 C 12.011	7 N 14.0067	8 O 15.9994	9 F 18.998	10 Ne 20.183
3	11 Na 22.990	12 Mg 24.312											13 Al 26.982	14 Si 28.086	15 P 30.974	16 S 32.064	17 Cl 35.453	18 Ar 39.948
4	19 K 39.102	20 Ca 40.08	21 Sc 44.956	22 Ti 47.90	23 V 50.942	24 Cr 51.996	25 Mn 54.938	26 Fe 55.847	27 Co 58.933	28 Ni 58.71	29 Cu 63.54	30 Zn 65.37	31 Ga 69.72	32 Ge 72.59	33 As 74.922	34 Se 78.96	35 Br 79.909	36 Kr 83.80
5	37 Rb 85.47	38 Sr 87.62	39 Y 88.905	40 Zr 91.22	41 Nb 92.906	42 Mo 95.94	43 Tc (99)	44 Ru 101.07	45 Rh 102.91	46 Pd 106.4	47 Ag 107.870	48 Cd 112.40	49 In 114.82	50 Sn 118.69	51 Sb 121.75	52 Te 127.60	53 I 126.90	54 Xe 131.30
6	55 Cs 132.91	56 Ba 137.34	57 La 138.91	72 Hf 178.49	73 Ta 180.95	74 W 183.85	75 Re 186.2	76 Os 190.2	77 Ir 192.2	78 Pt 195.09	79 Au 196.97	80 Hg 200.59	81 Tl 204.37	82 Pb 207.19	83 Bi 208.98	84 Po (210)	85 At (210)	86 Rn (222)
7	87 Fr (223)	88 Ra (226)	89 Ac (227)															

Lanthanide series

57 La 138.91	58 Ce 140.12	59 Pr 140.91	60 Nd 144.24	61 Pm (147)	62 Sm 150.35	63 Eu 151.96	64 Gd 157.25	65 Tb 158.92	66 Dy 162.50	67 Ho 164.93	68 Er 167.26	69 Tm 168.93	70 Yb 173.04	71 Lu 174.97

Actinide series

89 Ac (227)	90 Th 232.04	91 Pa (231)	92 U 238.03	93 Np (237)	94 Pu (244)	95 Am (243)	96 Cm (247)	97 Bk (247)	98 Cf (251)	99 Es (254)	100 Fm (253)	101 Md (255)	102 No (254)	103 Lw (257)

Notes:

1. The atomic weights are based on $C^{12} = 12$ exactly.
2. An atomic weight appearing in parentheses is that of the most stable known isotope.
3. For names of the elements, see Table A.6.

BIBLIOGRAPHY

All books listed in the section "For further reading" at the end of each chapter are arranged here alphabetically according to principal author or editor.

Andrade, E. N. "Rutherford and the Nature of the Atom," Anchor Science Study Series, Doubleday & Company, Inc., Garden City, N.Y., 1964.

Andrade, E. N. "Sir Isaac Newton," Anchor Science Study Series, Doubleday & Company, Inc., Garden City, N.Y., 1958.

Armitage, A. "The World of Copernicus," Mentor Books, New American Library of World Literature, Inc., New York, 1947.

Backus, J. "The Acoustical Foundations of Music," W. W. Norton & Company, Inc., New York, 1969.

Beiser, A. (ed.) "The World of Physics," McGraw-Hill Book Company, New York, 1960.

Benade, A. H. "Horns, Strings and Harmony," Anchor Science Study Series, Doubleday & Company, Inc., Garden City, N.Y., 1960.

Berlan, T. "The Fight for Quiet," Prentice-Hall, Inc., Englewood Cliffs, N.J., 1970.

Bleich, A. R. "The Story of X Rays," Dover Publications, Inc., New York, 1960.

Bohm, D. "Causality and Chance in Modern Physics," Routledge & Kegan Paul, Ltd., London, 1957.

Born, M. "The Restless Universe," Dover Publications, Inc., New York, 1951. Transl. by W. M. Deans.

Boys, C. V. "Soap Bubbles and the Forces Which Mould Them," Anchor Science Study Series, Doubleday & Company, Inc., Garden City, N.Y., 1959.

Bragg, W. "The Universe of Light," Dover Publications, Inc., New York, 1940.

Brown, S. "Count Rumford: Physicist Extraordinary," Anchor Science Study Series, Doubleday & Company, Inc., Garden City, N.Y., 1962.

Butterfield, H. "The Origins of Modern Science," rev. ed., Collier Books, The Macmillan Company, New York, 1962.

Campbell, N.   "What Is Science?" Dover Publications, Inc., New York, 1952.

Chalmers, T. W.   "Historic Researches," Charles Scribner's Sons, New York, 1952.

Clark, R. W.   "Einstein, the Life and Times," The World Publishing Company, Cleveland, 1971.

Cline, B. L.   "The Questioners: Physicists and the Quantum Theory," Thomas Y. Crowell Company, New York, 1965. Also paperback: "Men Who Made a New Physics: Physicists and the Quantum Theory," Signet Books, New American Library, Inc., New York, 1969.

Cohen, I. B. (ed.)   "Benjamin Franklin's Experiments," Harvard University Press, Cambridge, Mass., 1941.

Crew, H.   "The Rise of Modern Physics," The Williams & Wilkins Company, Baltimore, 1935.

De Santillana, G.   "The Crime of Galileo," The University of Chicago Press, Chicago, 1955.

Einstein, A.   "The World as I See It," Philosophical Library, Inc., New York, 1949. Transl. by A. Harris.

"Encyclopedia of Physics," Reinhold Publishing Corporation, New York, 1966.

Ford, K. W.   "The World of Elementary Particles," Blaisdell Publishing Co., Inc., New York, 1963.

Frank, P.   "The Philosophy of Science," Prentice-Hall, Inc., Englewood Cliffs, N.J., 1957.

Frisch, O. R.   "Atomic Physics Today," Basic Books, Inc., Publishers, New York, 1961.

Froome, K., and L. Essen   "The Velocity of Light and Radio Waves," Academic Press, Inc., New York, 1969.

Galilei, G.   "Two New Sciences," Dover Publications, Inc., New York, 1952. Transl. by Crew and de Salvio.

Gamow, G.   "Gravity," Anchor Science Study Series, Doubleday & Company, Inc., Garden City, N.Y., 1962.

Gamow, G.   "Mr. Tompkins in Paperback," Cambridge University Press, London, 1965.

Gillispie, C. C.   "The Edge of Objectivity," Princeton University Press, Princeton, N.J., 1960.

Gouiran, R.   "Particles and Accelerators," McGraw-Hill Book Company, New York, 1967.

Harsanyi, Z.   "The Star Gazer," G. P. Putnam's Sons, New York, 1939. Transl. by Paul Tabor.

Hoffman, B.   "The Strange Story of the Quantum," 2d ed., Dover Publications, Inc., New York, 1959.

Holden, A.   "Conductors and Semiconductors," Bell Telephone Laboratories, Inc., Murray Hill, N.J., 1964.

Holden, A., and P. Singer   "Crystals and Crystal Growing," Anchor Science Study Series, Doubleday & Company, Inc., Garden City, N.Y., 1960.

Holton, G., and D. H. D. Roller   "Foundations of Modern Physical Science," Addison-Wesley Publishing Company, Inc., Reading, Mass., 1958.

Jaffe, B.   "Crucibles: The Story of Chemistry," Premier Books, Fawcett Publications, Inc., Greenwich, Conn., 1960.

Jaffe, B. "Michelson and the Speed of Light," Anchor Science Study Series, Doubleday & Company, Inc., Garden City, N.Y., 1960.

Jaffe, B. "Men of Science in America," Simon & Schuster, Inc., New York, 1944.

Jeans, J. H. "Science and Music," Cambridge University Press, London, 1961.

Jones, G., J. Rotblat, and *G. Whitrow* "Atoms and the Universe," 2d ed., Charles Scribner's Sons, New York, 1963.

Kernan, W. J. "Accelerators," U.S. Atomic Energy Commission, Oak Ridge, Tenn., 1964.

Kitiagorodskiy, A. "Order and Disorder in the World of Atoms," Springer-Verlag New York Inc., New York, 1967.

Lee, O. J. "Measuring Our Universe," The Ronald Press Company, New York, 1950.

MacDonald, D. K. C. "Faraday, Maxwell and Kelvin," Anchor Science Study Series, Doubleday & Company, Inc., Garden City, N.Y., 1964.

Magie, F. W. "A Source Book in Physics," McGraw-Hill Book Company, New York, 1935.

March, A., and *Ira M. Freeman* "The New World of Physics," Random House, Inc., New York, 1962.

Massey, H. "The New Age in Physics," 2d ed., Basic Books, Inc., Publishers, New York, 1967.

Menzel, D. H. "Our Sun," rev. ed., Harvard University Press, Cambridge, Mass., 1959.

Millikan, R. A., D. Roller, and *E. C. Watson* "Mechanics, Molecular Physics, Heat and Sound," The M.I.T. Press, Cambridge, Mass., 1965.

Mott-Smith, M. "Heat and Its Workings," Dover Publications, Inc., New York, 1962.

Nash, L. K. "The Atomic-Molecular Theory," Harvard University Press, Cambridge, Mass., 1950.

Newman, J. R. "Science and Sensibility," 2 vols., Simon & Schuster, Inc., New York, 1961.

Newman, J. R. (ed.) "The World of Mathematics," 4 vols., Simon & Schuster, Inc., New York, 1956.

Newton, I. "Opticks," Dover Publications, Inc., New York, 1952.

Newton, I. "Principia Mathematica," University of California Press, Berkeley, 1934. Transl. by A. Motte; rev. by F. Cajori.

Park, D. "Contemporary Physics," Harcourt, Brace & World, Inc., New York, 1964.

Pierce, J. R. "Electrons and Waves," Anchor Science Study Series, Doubleday & Company, Inc., Garden City, N.Y., 1964.

Pierce, J. R., and *E. E. David, Jr.* "Man's World of Sound," Anchor Science Study Series, Doubleday & Company, Inc., Garden City, N.Y., 1958.

Price, D. J. "Little Science, Big Science," Columbia University Press, New York, 1963.

Ruhemann, M. "Power," Sigma Books, London, 1946.

Russell, B. "The ABC of Atoms," E. P. Dutton & Co., Inc., New York, 1923.

Sandfort, J. F. "Heat Engines: Thermodynamics in Theory and Practice," Anchor Science Study Series, Doubleday & Company, Inc., Garden City, N.Y., 1962.

Sawyer, W. W. "Mathematician's Delight," Pelican Books, Penguin Books, Inc., Baltimore, 1946.

Segré, E. "Enrico Fermi—Physicist," The University of Chicago Press, Chicago, 1970.

Semat, H. "Introduction to Atomic and Nuclear Physics," 3d ed., Holt, Rinehart and Winston, Inc., New York, 1960.

Shamos, M. H. (ed.) "Great Experiments in Physics," Holt, Rinehart and Winston, Inc., New York, 1959.

Shapiro, A. H. "Shape and Flow," Anchor Science Study Series, Doubleday & Company, Inc., Garden City, N.Y., 1961.

Shurcliff, W., and S. Ballard "Polarized Light," Momentum Books, D. Van Nostrand Company, Inc., Princeton, N.J., 1964.

Smyth, H. D. "Atomic Energy for Military Purposes," Princeton University Press, Princeton, N.J., 1945.

Snow, C. P. "The Two Cultures; and a Second Look," 2d ed., Cambridge University Press, London, 1969. Also available in paperback.

Swann, W. F. G., and I. M. Freeman "Physics," John Wiley & Sons, Inc., New York, 1941.

Taylor, L. W. "Physics, The Pioneer Science," 2 vols., Dover Publications, Inc., New York, 1959.

Terrell, J. Fission, "Encyclopedia of Physics," Reinhold Publishing Corporation, New York, 1966.

Tolansky, S. "Revolution in Optics," Penguin Books, Ltd., Harmondsworth, England, 1968.

Van Bergeijk, W., J. R. Pierce, and E. E. David, Jr. "Waves and the Ear," Anchor Science Study Series, Doubleday & Company, Inc., Garden City, N.Y., 1960.

Walker, M. "The Nature of Scientific Thought," Prentice-Hall, Inc., Englewood Cliffs, N.J., 1963.

Whitrow, G. J. "The Natural Philosophy of Time," Harper & Row, Publishers, Incorporated, New York, 1967.

Williams, L. P. "Michael Faraday," Basic Books, Inc., Publishers, New York, 1965.

Wood, A. "The Physics of Music," rev. by J. M. Bowsher, Dover Publications, Inc., New York, 1962.

Wood, A., and F. Oldham "Thomas Young, Natural Philosopher," Cambridge University Press, London, 1954.

Zemansky, M. "Temperatures, Very Low and Very High," D. Van Nostrand Company, Inc., Princeton, N.J., 1964.

Zworykin, V., and E. Ramberg "Photoelectricity and Its Applications," John Wiley & Sons, Inc., New York, 1949.

ANSWERS
TO SELECTED QUESTIONS AND PROBLEMS

CHAPTER 2

2.1. 13 cm.
2.6. 1,083 ft/sec; 738 mi/hr.
2.8. $19.70.
2.11. 985.
2.14. 7.86 ft^3.
2.17. 89.6 cm^3.
2.20. 4.4×10^{-2}.

CHAPTER 3

3.2. 2,300 lb/ft.
3.4. 0.11 ft.
3.11. 98.4 g; 51.0 m^2.
3.14. About 2,500 atoms thick.
3.16. 1.2×10^{-8} m; about 10^{-25} m^3.

CHAPTER 4

4.1. (a) 1 ft in a direction perpendicular to the original direction; (b) 2.5 ft in a direction 135° from the initial direction.
4.4. 1.732 units at an angle of 210° to the first vector.
4.7. 16° west of south.
4.9. $\phi = 23°$.
4.14. (a) 40 km; (b) 69 km.
4.16. 28°.
4.18. No.
4.20. 23.4 lb.
4.22 38.7 kg force.
4.24. The configuration changes until CA and AB have equal slopes. The two tensions then become equal. Give reasons for these statements.

4.26. 2 ft.
4.28. 0.14 kg force.
4.29. 4 ft.

CHAPTER 5

5.3.   30 mi/hr.
5.4.   12.5 mi.
5.8.   34.6 ft/sec.
5.9.   (a) Yes; (b) no. In (a) the trip takes 3.0 hr.
5.11. 1.1×10^8 m/sec in the backward direction.
5.13. 15,600 km/hr²; 60° south of east.
5.15. (a) -7.0 ft/sec²; (b) 224 ft; (c) 168 ft.
5.18. 56.8 sec.
5.19. (a) 11.4 sec; (b) 248.
5.21. (a) 6.1 m; (b) 9.8 m/sec.
5.28. 14.5 ft.

CHAPTER 6

6.9. $2f/m$, due east.
6.11. 1.4×10^{12}.
6.13. (a) 2.5 m/sec²; (b) 500 newtons; (c) 250 newtons.
6.14. (a) 6.67×10^4 newtons; (b) 1,390 m.
6.15. 2.3×10^{-14} newton.
6.18. 3,240 newtons; twice this amount.
6.20. (a) 0.50 newton; (b) 12 kg.
6.22. 1.5.
6.24. $\frac{1}{4}g$.
6.26. (a) 1.0 m/sec²; (b) 24 newtons.
6.28. 3.6×10^4 newtons.

CHAPTER 7

7.2. mg. No: There is no resistance at that point, since the body is instantaneously at rest.
7.5.   (a) 585 newtons; (b) 550 newtons; (c) 550 newtons.
7.10. 8,000 g-cm/sec; northward.
7.12. 820 m/sec.
7.14. 0.36 kg-m/sec.
7.16. (a) 20 cm from the heavier object; (b) 40 cm.
7.18. Boat starts backward at 4 cm/sec.
7.21. 108 m/sec.
7.23. 3.34×10^{-27} kg.

CHAPTER 8

8.2.   17.
8.6.   0.8 of the weight.
8.8.   (a) At the lowest point on the circle; the highest point. (b) 5.0 kg.

8.10. 1,400 newtons.
8.13. 8.6×10^8 mi^2/sec.
8.15. 5.3×10^9 km.
8.17. Orbit radius, 4.23×10^4 km; height above ground, 22,300 mi.
8.19. About six times as far.
8.21. 6.0×10^{24} kg.
8.23. 2.0×10^{30} kg.
8.25. 274 m/sec^2, or about 28 times g on earth.

CHAPTER 9

9.3. 90 ft-lb.
9.5. (a) Momentum change numerically twice original momentum; (b) no change in KE.
9.8. 17.7 m/sec.
9.11. Mechanical energy decreases by 20.25 joules.
9.15. 0.064.
9.17. (a) 54.5; (b) 1; (c) 54.5.
9.20. 100 newtons.
9.22. 12,800 newtons.
9.24. 3,200 newtons.
9.27. 80 m/sec.
9.29. 3,250 joules.
9.30. (a) 2.25×10^{10} joules; (b) 2.7×10^9 joules.
9.33. 60 hp.
9.35. 11 hp.

CHAPTER 10

10.2. 1.03 ft.
10.4. 1.13.
10.6. 15 percent increase.
10.9. 21.4 K.
10.11. 34,700 N/m^2.
10.14. A time of the order of 0.1 sec.
10.18. 719°C.
10.20. $v_{\mathrm{rms}} = 3,520$ m/sec.
10.24. 5×10^{16} km.
10.25. 27 million.

CHAPTER 11

11.2. (a) 100 percent; (b) 59.7°C.
11.5. 11.6 g.
11.7. 533 kcal/kg.
11.12. 0.085.
11.15. 76 watts.
11.16. 0.43.
11.18. 102 kcal.
11.19. 0.36 kcal.

11.21. 4.27×10^4 m.
11.26.   10.2 percent.
11.28.   (a) 30 percent; (b) 1.26×10^4 joules; (c) 7 kcal.

CHAPTER 12

12.3.   (a) Zero; (b) 197 cm/sec^2.
12.7.   8.8 sec.
12.10.   No; speed decreases toward the top.
12.13.   68 ft/sec.

CHAPTER 13

13.2.   32 yr.
13.3.   8,000 yr.
13.6.   2.5 cycles/sec.
13.7.   185 newtons.
13.9.   1:2.
13.10.   6.7 in.
13.13.   Top edge of mirror midway between level of eye and top of head; bottom edge midway between eye level and floor.
13.15.   85°.
13.17.   50°.

CHAPTER 14

14.5.   −20 in.; diverging.
14.7. $q = -13.3$ cm.
14.9.   0.25 cm.
14.11.   (a) 10 cm from object; 90 cm from object. (b) The two magnifications are reciprocals of each other.
14.15.   (a) 93 cm; (b) must be moved back.

CHAPTER 15

15.6.   6000 Å.
15.9.   5000 Å.
15.11.   6400 Å.
15.13.   0.50 cm.
15.16.   4500 Å.
15.18.   (a) 45°; (b) 60°.

CHAPTER 16

16.2.   0.036 newton; repulsion.
16.3. 2.7×10^{-9} coulomb.
16.5.   12,800 coulombs; 10 newtons.
16.9.   2 cm from smaller charge.
16.11.   19.4°.

16.15.   (a) 8.0×10^{-15} newton; (b) upward; (c) reduce potential difference to 80 volts.
16.19.   0.20 coulomb.
16.21. -4.5×10^5 coulombs.
16.23.   Constant negative acceleration; 0.25 cm from negative plate.
16.25.   0.035 eV.
16.27.   193 μf.
16.29.   2,860.
16.30.   0.21 μf.

CHAPTER 17

17.2. 1.9×10^{-7} ampere.
17.3.   0.052 ohm.
17.4.   17.
17.8.   (a) 2.25 ohms; (b) 2.92 ohms; (c) 2.67 ohms.
17.9.   1.75 ohms.
17.11.   59.5 watts.
17.14.   22 watts.
17.17.   8.0 volts.
17.19.   0.23 ohm.
17.20.   (a) 0.283 ampere; (b) 1.70 volts, 2.26 volts; (c) 3.96 volts.
17.22. $2\frac{1}{4}$ days.
17.24.   0.66 ampere, toward the right.

CHAPTER 18

18.3. 2.6×10^{-7} Wb.
18.6.   Very nearly equal.
18.7. 1.0×10^{-26} kg.
18.9.   50.
18.12.   0.6 newton.
18.13. 4×10^{-4} newton.
18.14.   2.04 g.
18.16. 6×10^{-5} Wb/m^2.
18.17. 8×10^{-4} newton, to the left.
18.20.   7.5 Wb/m^2.

CHAPTER 19

19.2. 3.3×10^{-4} volt.
19.3.   0.036 ampere.
19.5.   (a) Counterclockwise; (b) downward.
19.7.   0.50 volt; 0.010 ampere.
19.9.   (a) 0.5 volt; (b) clockwise.
19.13.   (a) 350; (b) 14 amperes.
19.14.   (a) 3.3×10^{-13} Wb/m^2; (b) 1.3×10^{-11} watt/m^2; (c) 4.3×10^{-20} N/m^2.
19.17.   (a) 2,300 earth diameters; (b) 1/6000 as much.

CHAPTER 20

20.2.   1500 K.
20.4. 2.9×10^{-31} joule.
20.7. 5×10^{21}.
20.9.   0.012 Å.
20.12. 7.5 eV.
20.14. 1.24×10^{-6} m-volt.
20.16. 120°.
20.18. (a) Only for the first; (b) 1.3×10^8 m/sec.
20.24. 1.8×10^{-34} m.
20.26. (a) 55 volts; (b) 0.015.
20.29. (a) Equal; (b) $K_e = 1{,}840 \, K_P$.

CHAPTER 21

21.1.   (a) 2,300; (b) about 4×10^{-8}.
21.4.   12.6 Wb/m².
21.5. 6.6×10^{15} cycles/sec; 1/8 as much.
21.6.   Nearly 10^7.
21.8.   To three significant digits, 1210 Å and 1020 Å.
21.10. −3.4 eV; −1.5 eV.
21.12. 1.06×10^{-10} m (1.06 Å).
21.16. About 4×10^{-8} eV.
21.18. About 7×10^{67} units, directed northward along axis.
21.19. (a) 3; (b) 9; (c) 18.

CHAPTER 22

22.1.   15.9905261 u.
22.3.   9.9.
22.7.   54.4 sec; Rn ($Z = 86$).
22.8.   About 30 g.
22.11. (a) 1.2 MeV; (b) zero.
22.14. −7.3 MeV.
22.16. 0.0017 Å.
22.18. 5.03 MeV; 22.4 MeV.
22.20. 2.6 eV.
22.22. (a) 8.61 MeV; (b) 17.3 MeV; (c) −4.96 MeV.
22.24. Residual nucleus is an isotope of plutonium.
22.25. $_{48}\text{Cd}^{108}$.
22.28. 11.5 cm (slightly larger than a standard softball).
22.32. About 6×10^{11} kg.

CHAPTER 23

23.2.   Path radius, 1.7×10^4 km.
23.4.   105 MeV.
23.5.   84.3 MeV.
23.10. 1.8 MeV.
23.11. 1.63×10^{22} cycles/sec.
23.12. 1.6×10^{-4} Å.

INDEX

Page numbers followed by ''N'' indicate that the reference occurs in a marginal note; followed by ''P'' indicate a problem or question.

TABLE A.8 PHYSICAL CONSTANTS

Gravitational constant	$G = 6.673 \times 10^{-11}$ N-m^2/kg^2
Year	1 yr = 365 days 5 hr 48 min 46 sec = 3.156×10^7 sec
Avogadro's number	$N = 6.0222 \times 10^{26}$ particles/kmol
Universal gas constant	$R = 8{,}314$ J/kmol-deg
Boltzmann constant	$k = R/N = 1.380 \times 10^{-23}$ J/deg
Ice point	$T_0 = 273.15$ K
Joule equivalent	$J = 4{,}186$ J/kcal
Speed of electromagnetic waves in free space	$c = 2.998 \times 10^8$ m/sec
Coulomb-force constant	$k_0 = 8.987 \times 10^9$ N-m^2/C^2
Electron charge	$e = 1.602 \times 10^{-19}$ coulomb
Planck's constant	$h = 6.626 \times 10^{-34}$ J-sec
Electron volt	1 eV = 1.60×10^{-19} joule
	1 MeV = 1.60×10^{-13} joule
Unified mass unit	1 u = 1.660×10^{-27} kg
Mass-energy conversion	1 u = 931.5 MeV
	1 MeV = 1.78×10^{-30} kg

TABLE A.9 SOME REST MASSES AND THEIR ENERGY EQUIVALENTS

	kg	u	MeV	joules
electron, m_e	9.109×10^{-31}	0.00055	0.5110	8.187×10^{-14}
proton, m_p	1.672×10^{-27}	1.00728	938.8	1.504×10^{-10}
neutron, m_n	1.675×10^{-27}	1.00866	939.6	1.505×10^{-10}